LINE LOSS ANALYSIS AND CALCULATION OF ELECTRIC POWER SYSTEMS

LINE LOSS ANALYSIS AND CALCULATION OF ELECTRIC POWER SYSTEMS

Anguan Wu

North China Electric Power University, China

Baoshan Ni

Zhejiang University, China

WILEY

中国电力出版社
CHINA ELECTRIC POWER PRESS

This edition first published 2016

Published by John Wiley & Sons Singapore Pte Ltd, 1 Fusionopolis Walk, #07-01 Solaris South Tower, Singapore 138628, under exclusive license granted by China Electric Power Press for all media and languages excluding Chinese and throughout the world excluding Mainland China, and with non-exclusive license for electronic versions in Mainland China.

Registered Office
John Wiley & Sons Singapore Pte Ltd, 1 Fusionopolis Walk, #07-01 Solaris South Tower, Singapore 138628

For details of our global editorial offices, for customer services and for information about how to apply for permission to reuse the copyright material in this book please see our website at www.wiley.com.

Library of Congress Cataloging-in-Publication data applied for

A catalogue record for this book is available from the British Library.

Set in 9/11pt Times by SPi Global, Pondicherry, India
Printed and bound in Singapore by Markono Print Media Pte Ltd

1 2016

Contents

Foreword

Energy saving and consumption reduction in the electric power system are important tasks to which generations of people have always paid attention and been committed. In recent years, I heard that several aspiring experts have been summarizing and writing (revising) works or papers based on their working practices in line loss. A few days ago, Wu Anguan told me that he had finished the revision of the second edition of *Analysis and Calculation of Line Loss in Electric Power Systems* (hereinafter referred to as *Line Loss*) and asked me to write a foreword for this second edition. I felt happy that new results were created against new challenges, and also afraid that my words could not express my thoughts. However, as I wanted to live up to an old friend's sincerity, I pleasantly promised to have a try, taking this opportunity to focus on loss reduction in power grids and to express my respect for line loss workers. I was filled with emotion when looking back on my participation in the line loss business, recalling the benefits I gained from the revision and review of the line loss management guidelines (provisions) and the loss reduction plans of the State Grid Corporation of China, and reading Wu's second edition of *Line Loss* and his papers published in recent years.

Over the past 30 years since the reform and opening-up policy, China has implemented both energy development and savings, promoting the rapid development and increasing the benefits of the electric power system. As of 2011, the installed power-generating capacity has ranked second in the world for consecutive years, and the size of power grids and the total electricity consumption nationwide have ranked first in the world. The economic efficiency of the electric power industry was significantly improved: the average electric supply coal consumption rate of coal-fired units reached 330 g/(kW·h), and some large units even realized a coal consumption rate of 300 g/(kW·h); the electric supply line loss rate in power grids was 6.39% nationwide, and even maintained 5.5% or below in some provinces and cities; and overall, the line loss rate was at an internationally advanced level.

Looking back to the progress of power grid loss reduction in China after the reform and opening-up, the first 20 years saw a remarkable reduction of line loss rate from 9.64% in 1978 to 7.81% in 2000, but the curve of the annual line loss rate revealed several saddle-shaped fluctuations, showing that loss reduction was still difficult. During the first 12 years of the new century, with progress in science, technology, and equipment, especially strengthened line loss management, the line loss rate kept a steady decline without any big fluctuation, and was reduced by 1.42% in 2011 over 2000. Based on the size of the electric supply in 2010, if the line loss rate was reduced by 1%, equal to an annual electric loss reduction of about 42×10^{12} kW·h, then over 13.8×10^{6} t of coal was saved each year and over 27.6×10^{6} t of carbon emissions were reduced, marking a great achievement in energy conservation and emission reduction. Due to my personal experiences, I pay close attention to and feel delighted with each year's new achievements in the loss reduction in power grids. I deeply respect the technicians and managers engaged in the loss reduction for their steadfast and diligent professional ethics.

These remarkable achievements in the reduction of loss in power grids are mainly attributable to improvements in the theoretical analysis and calculation of line loss, the implementation of loss reduction technologies and measures, and effective policy management, as well as the assiduous efforts of relevant personnel. Since the 1980s, many professors, scientific and technical personnel, and business personnel working in the frontline of power grid enterprises, who were engaged in theoretical research into power grid line loss and in practical analysis and calculation, have worked hard and closely with each other. They have made outstanding contributions to the promotion of the theoretical research, analysis, and calculation of line loss in China, the implementation of loss reduction measures and technical management, and the realization of a continuous and steady decline in the line loss rate. In the 1980s, Zhejiang University and Nanjing Administration of Power Supply cooperated in preparing *Theoretical Calculation of Line Loss and Technical Measures of Loss Reduction* (written by Professor Ni Baoshan); Professor Yang Xiutai of Chongqing University compiled *Theoretical Calculation and Analysis of Line Loss in Power Grids*; Wu Anguan (Senior Engineer of Taiyuan Administration of Power Supply) and Professor Ni Baoshan coauthored *Line Loss in Electric Power System* (1996); Yu Zhongian (Senior Engineer of Jiangsu Electric Power Company) and Professor Chen Xingying of Hohai University wrote *Electric Energy Loss in Power Grids* (2000). They studied, learnt, and introduced the theoretical results of Europe, America, and the former Soviet Union in the calculation of line loss in the electric power system. Based on the actual situations of power grids in China, they discussed and proposed many methods for the theoretical calculation of line loss, and promoted the application of microcomputers in the calculation of line loss. In addition, they participated in many seminars, exchange meetings, and information sessions, making good technical preparations and doing real work for the formulation of line loss calculation guidelines, line loss management provisions, and other industrial standards or corporate regulations, both for training in line loss management and reviewing loss reduction plans. Their efforts have had an enduring influence on the theoretical calculation of line loss by power grid enterprises, the preparation of loss reduction plans, and improvements in the technical management of line loss in power grid enterprises. Meanwhile, the national competent authority and the State Grid Corporation of China, by following the scientific development perspective and adhering to the energy development and conservation policy, worked together to promote an industry–university–research cooperation, to formulate, propagate, and implement regulations and guidelines, to take effective input and technical measures, and to play an organizational leadership and regulatory role. This is an important reason why great achievements have been made in loss reduction for so many years, and an experience that should be learnt.

Since the twenty-first century, power grids have developed fast in China, with a continuously expanded size, more complicated network, and personnel increase or turnover. In addition, China has boosted the development and utilization of renewable energy sources, and the electricity market is increasingly growing. All of these developments and changes have posed new challenges to the objective of line loss management and loss reduction. Therefore, even after his retirement, the author of this book, with diligent study and tireless exploration, was still dedicated to the work of line loss and spent over eight years in making this new achievement.

In the second revision of *Line Loss*, the author retained or improved the basic knowledge and methods of theoretical calculation of line loss from the first edition, such as the line loss calculation considering power factor, the concept of load loss coefficient, the change rule of line loss, the methods of calculating line loss in loss calculation units and multi-branch lines, the methods of prediction of electricity line loss and line loss rate, and the principles of optimal distribution of an electric supply. New additions in this revision include the calculation of loss in complicated power grids, the link electricity price, the balance between electricity flow and capital flow within power grid enterprises, the analysis of coal-electricity price linkage, and the utilization of mass information in line loss. In addition, new results are theoretically explored and presented in terms of the three section division of the load curve, the loss calculation and loss allocation in complicated power grids, and the composition and application of a multi-section electricity price. Real analysis and calculation cases or parameters are also selected and provided. These studies and results are instructive when focusing on energy conservation and loss reduction against the back drop of a market-oriented economy, and when seeking ways to improve benefits by power grid enterprises.

Wu Anguan and I are schoolmates, and we exchanged and cooperated much with each other on the line loss in power grids after the 1990s. Wu studies hard, assiduously and ambitiously, and has always been keen on the professional theories of line loss and the practical application of research. Despite being over 60, Wu tirelessly writes papers and makes friends with others for sharing and improvement, which I respect very much.

I sincerely recommend the second edition of *Line Loss*, and believe that it will be conducive to professionals and college students involved in line loss management practices. I hope that more young scholars can constantly dig into the ancient but young subject of "line loss", to help resolve many complicated technical and economic problems arising from electricity marketization and the accessability of renewable energy sources to the network. Witnessing the development of line loss management in China, I would like to take this opportunity to share my thoughts and experiences with all young scholars.

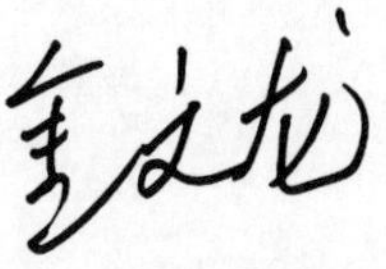

Wenlong Jin
Senior Engineer; Senior Academic Adviser
Chinese Society for Electrical Engineering

Preface

It has been over a decade since the Chinese publication of *Line Loss in Electric Power Systems*. To keep pace with technological developments, I started a revision as early as 2002, following the main principles that the theoretical framework and characteristics of the first edition should be retained, with new contents added according to new problems after the reform of electric power systems and the new requirements for line loss management practices and in combination with practical experience.

The theoretical framework of the first edition used the loss factor method as the mainline and the quadratic trinomial $\Delta A = B + CA^2$ as the subline. The random test method was used to verify the wide applicability and validity of the load factor formula put forward by Liu Yingkuan, a Chinese scholar. The equivalent load curve method was extended and expanded. The line loss calculation curve was given for various designs of voltage class lines. The conditions at which the minimum line loss rate was achieved were set out. The relationship between the load loss coefficient and six variables was demonstrated by using the concept of the loss factor of the equivalent load curve. The inherent law between the electricity line loss and the electric supply change was analyzed, and then the electricity line loss prediction formula was put forward.

The concept of marginal line loss rate was first introduced, and based on this, the optimal distribution of an electric supply was proposed. The first edition provided and compared several methods of theoretical calculation of line loss in a multi-branch distribution network. As the construction of smart distribution networks is being vigorously promoted today, these methods provide multiple choices for the intelligent calculation and control of line loss in the distribution network. The first edition comprehensively explained the technical measures of loss reduction. In addition, this book gives many examples to help readers understand the content and refer to this book during their work.

Maintaining the theoretical system of the first edition, this revision makes small adjustments to the sequence of chapters and supplements these with new content. The second edition contains 15 chapters as compared to 10 chapters in the first edition, and 36 examples as compared to 27 in the first edition. The title of the revision is changed to *Analysis and Calculation of Line Loss in Electric Power Systems* which is more consistent with the contents of this book.

This revision mainly includes the following supplementary contents:

1. The maximum, normal, and minimum modes are classified from the perspective of probability. The calculation formulas of the three section power point of division and the electric supply are given, to provide a new way to analyze load prediction, loss prediction, or time of use electricity pricing by considering changes of the three modes (Section 3.4 in Chapter 3).

2. A chapter is specially provided to explain the calculation of loss in high-voltage power grids. Various calculation methods of loss in high-voltage power grids are compared, and a new calculation method based on the three mode section division is put forward (Chapter 10).
3. The essence of loss allocation is elaborated by a simple power supply model. A new method of inter-provincial loss allocation within the regional network is created with a marginal loss electricity price (the product of marginal line loss rate and transmission price) as the economic signal. The Shapley value method and the generation quantity multiplier method used in California are introduced regarding the calculation of reasonable loss allocation between high-voltage customers, both direct supply and under a complex trading mode. The example of calculating the loss allocation in a five node power network is given (Chapter 11).
4. The prediction of electricity line loss and line loss rate in power grids and loss reduction plans are introduced. On the basis of the prediction formula of electricity line loss set out in the first edition, the prediction formula of line loss rate is added, and the requirements and preparation methods for loss reduction plans are introduced (Chapter 13).
5. The cost–volume–profit analysis model for power grid enterprises is established. The balance between electricity flow and capital flow among main production links within power grid enterprises is analyzed. The model of link cost and link electricity price within power grid enterprises is put forward, which satisfies the lean management requirements of these enterprises (Sections 14.1 and 14.2).
6. To control both the direct supply electricity price for high-voltage customers and the sales electricity price for low-voltage customers, the multi-section electricity price model is established by taking the line loss effect into account. The underlying cause of implementing the coal–electricity price linkage in China is explained. The double component ratio coefficient is used to build the coal–electricity price linkage model, to provide a quantitative analysis method to control the on-grid price level (Sections 14.3 and 14.4).
7. The method of analysis of price markup at the power sales end is studied, to provide conditions for completely assessing the production benefits of a transmission and transformation project during the post-project assessment (Section 14.5).
8. The current application situations of integrating loss mass information with other information are introduced. The integration of line loss information with other information and the use of this integrated information for control of voltage quality are explained. To realize the utilization of mass information on line loss, the design concepts of relevant data warehouse and data mining are introduced (Chapter 15).

Many of the above contents have been published in national professional meetings or magazines, and some contents are published first in this revision. I hope that years of my thoughts and accumulated experience can benefit readers who focus on such issues.

When revising this book, I was counselled and encouraged by Professor Liu Qingguo of North China Electric Power University and supported and helped by upperclassman Jin Wenlong. Some contents of this revision were reviewed by Zhang Zuping, Senior Engineer of China Electric Power Research Institute, who proposed many valuable suggestions. Zhang Youmin of Shanxi Electric Power Exploration and Design Institute provided full support to the proof calculation of the three mode section division. I, hereby, would like to express my heartfelt thanks to all my teachers, upperclassmen, and peers for their sincere help.

As entrusted by my teacher Ni Baoshan and thanks to the help of everybody, this revision is finally submitted for publication. Although Teacher Ni, who guided and coauthored this revision, has passed away and was unable to witness this result by himself, I can console my teacher as I have made all efforts to academically pass on the knowledge to a new generation.

Due to the limitation of my level of knowledge, this book may have some inaccuracies and areas for improvement. Readers are invited to comment and correct.

Wu Anguan

Introduction

This book is a revision of the authors' earlier work, *Line Loss in Electric Power Systems*. Basic contents are retained in the revision, including the basic knowledge and methods of theoretical calculation of line loss in electric power systems; the line loss calculation considering power factor, the concept of load loss coefficient, the change rule of line loss, and the methods of calculation of line loss in loss calculation units and multi-branch lines; the methods of prediction of electricity line loss and line loss rate, and the principles of optimal distribution of electric supply; as well as some technical measures to reduce line loss.

New additions in the revision include the calculation of loss in complicated power grids, the calculation of loss allocation, the link electricity price and the balance between electricity flow and capital flow within power grid enterprises, the composition and application of multi-section electricity price, the coal-electricity price linkage, and the utilization of mass information on line loss.

This book is intended as a reference for professionals working on line loss management with power grid enterprises, power grid designers and operators, and for graduate students majoring in the study of electric power systems.

1

Overview

1.1 Active Power Loss and Electric Energy Loss

In an electric supply area, electric energy is supplied to customers through transmission, substation, and power grid distribution. During the transmission and distribution of electric energy, a certain quantity of active power loss and electric energy loss will be generated in all units of the power grids.

1.1.1 Main Types of Active Power Loss

According to the analysis based on electromagnetic field theory, the energy of an electromagnetic field is transmitted from the power source to the loads through the dielectric space of the electromagnetic field, and wires lead the energy of the electromagnetic field. The electric energy loss that goes into the wires and is then converted into heat energy is also supplied by the electromagnetic field.

According to the results of the analysis of a single core coaxial cable by using the Poynting vector of energy flow density in the case of AC transmission, while power is needed to transmit loads in the dielectric space, four types of active power loss are produced in the cable:

1. *Resistance heat loss* ΔP_1 *(W)*

   This is in direct proportion to the square of current, that is

$$\Delta P_1 = I^2 R \tag{1.1}$$

   Wherein: I – current passing the cable core (A);

   R – the sum of resistance of both the cable core and tegmen (Ω).

2. *Leakage loss* ΔP_2 *(W)*

   This is in direct proportion to the square of voltage, that is

$$\Delta P_2 = U^2 G \tag{1.2}$$

Line Loss Analysis and Calculation of Electric Power Systems, First Edition. Anguan Wu and Baoshan Ni.

$$G=\frac{2\pi lr}{\ln\frac{r_2}{r_1}} \tag{1.3}$$

Wherein: U – voltage between the cable core and tegmen (V);
G – leakage conductance of dielectric (1/Ω);
r – conductivity [1/(Ω·m)];
l – length of the cable (m);
r_1 – radius of the cable core (cm);
r_2 – inside radius of the cable tegmen (cm).

3. *Dielectric magnetizing loss* ΔP_3 *(W)*
This is in direct proportion to the square of current and the frequency, that is

$$\Delta P_3=I^2\omega L\tan\delta \tag{1.4}$$

$$L=\frac{l\mu}{2\pi}\ln\frac{r_2}{r_1} \tag{1.5}$$

Wherein: ω – AC angular frequency (1/s);
L – inductance of the cable (Wb/A);
μ – magnetic conductivity of the cable dielectric (Ω·s/m);
$\tan\delta$ – repeated magnetizing loss tangent of the cable dielectric.

4. *Dielectric polarization loss* ΔP_4 *(W)*
This is in direct proportion to the square of voltage and the frequency, that is

$$\Delta P_4=U^2\omega C\tan\delta \tag{1.6}$$

$$C=\varepsilon\frac{2\pi l}{\ln\frac{r_2}{r_1}} \tag{1.7}$$

Wherein: C – capacitance of the cable (F);
ε – dielectric constant of the cable dielectric (F/m);
$\tan\delta$ – repeated magnetizing loss tangent of the cable dielectric.

The above four types of active power loss represent the basic types of active power loss in the electric power system. In addition, corona loss may occur in high-voltage lines and high-voltage motors. This is a special type of active power loss caused by ionization of dielectric particles outside a conductor when the electric field intensity is too high in the surface of the conductor. It is related to the surface field intensity of the conductor and the air density. See Chapter 8, Section 8.2 for details.

1.1.2 *Calculation of Electric Energy Loss*

Electric energy loss ΔA (kW·h) is the integral of active power loss to time within a period, that is

$$\Delta A=\int_0^T \Delta P(t)\mathrm{d}t\times 10^{-3} \tag{1.8}$$

For resistance heat loss, Formula (1.8) can be rewritten to

$$\Delta A = \int_0^T I^2(t)R(t)\mathrm{d}t \times 10^{-3} \tag{1.9}$$

Within the period T, the load current and conductor resistance may vary, so it is more complicated to calculate the electric energy loss than the active power loss. When the period for calculation is long, it is difficult to use the method of point by point square accumulation to calculate the electric energy loss. If relevant parameters of the current load curve $I(t)$ or the active load curve $P(t)$ are used to calculate the electric energy loss, it is difficult to obtain satisfactorily accurate calculation results. This is an issue that should be focused on when studying the theory and calculation method of electric energy loss.

1.1.3 Electricity Line Loss and Line Loss Rate

The total quantity of electricity loss (including the allocated electricity loss in power grids, the electricity consumed by electric reactors and reactive compensation equipment, and unknown electricity loss) in transmission, substation, and distribution within a given period (day, month, quarter, or year) in an electric supply area or power grids is called electricity line loss or line loss. Although part of the electricity line loss can be determined by theoretical calculation or measured by tailor-made line loss meters, the total electricity line loss cannot be accurately determined. Therefore, the electricity line loss is usually calculated by subtracting the total "power sales quantity" from the total "electric supply" measured by the electric energy meter. In other words, the line loss is a margin, and its accuracy relies both on the accuracy of the electric energy metering system used to measure the electric supply and the power sales quantity, and on the scientific and reasonable system for recording the statistics of power sales quantity to customers.

The electric supply and the power sales quantity are a pair of interrelated concepts and are closely related to the scope of line loss. In the 1990s, before power plants and power grids were not separated, China's power supply enterprises and power generation enterprises were managed by a competent authority and administered by an electricity bureau at the provincial level. The line loss statistics were conducted at national, provincial, and prefectural levels. For prefectural power generation enterprises, their electric energy meters at the generation side and the supply side were managed by power supply agencies entrusted to the provincial electricity bureau. As a result, the electric supply in a given area refers to the electricity supplied by power plants, power supply areas, or power grids to customers, including the electricity line loss in transmission and distribution of electric energy. The formula(1) for calculating the electric supply is as follows:

$$A_{e.s} = A_{e.p} - A_{e.c} - A_{out} + A_{in} \tag{1.10}$$

Wherein: $A_{e.s}$ – electric supply of an electric supply area or power grids;
$A_{e.p}$ – electric energy production of power plants in a local area or local power grids;
$A_{e.c}$ – electricity consumption of power plants;
A_{out} – electricity output to other power grids;
A_{in} – electricity input from other power grids (including purchased electricity).

Power sales quantity refers to the electricity sold by electric power enterprises to customers and the electricity supplied by electric power enterprises for internal non-generation use (such as capital construction departments). Non-electricity generation departments of electric power enterprises should be treated as customers. Therefore, the power sales quantity in an electric supply area or power grids is the total electricity measured by the electric energy meters of customers.

The percentage of the electricity line loss in the electric supply is called the line loss rate, and its calculation formula is as follows:

$$\text{Line loss rate }(\%) = \frac{\text{electric supply} - \text{power sales quantity}}{\text{electric supply}} \times 100\% \tag{1.11}$$

During the operation management of power grids, the electricity line loss obtained by subtracting the total power sales quantity from the total electric supply is called the statistical electricity line loss, and the corresponding line loss rate is called the statistical line loss rate.

Part of the statistical electricity line loss cannot be avoided during the transmission and distribution of electric energy and is determined by the load conditions of power grids and the parameters of power supply equipment. Such electricity loss is called technical electricity loss and can be obtained by theoretical calculation. Therefore, it is also called the theoretical electricity line loss, and the corresponding line loss rate is called the theoretical line loss rate.

The electricity used by substations is also included in the statistical electricity line loss. This part of electricity is similar to the electricity used by power plants and is necessary for production. It does not really belong to the electricity line loss, but due to historical reasons, it is not managed by power grid enterprises as a production cost. Instead, it is included in the line loss for management and control. Given the lean management requirements, the electricity used by substations can be excluded from the electricity line loss and included in the production cost for management. This may play a positive role in standardizing cost management and promoting the assessment and theoretical calculation of line loss.

Part of the statistical electricity line loss is an unknown loss, also known as management loss, which can and should be avoided or reduced by means of necessary measures.

In the 1990s, after the reform of the separation between power plants and power grids, the power plants and power grid enterprises became independent operators. Power grid enterprises used the on-grid electricity of power plants as the main electric supply. China's two major power grid enterprises, namely the State Grid Corporation of China and China Southern Power Grid, manage their internal line loss by several levels of large regional power grid enterprises covering multiple provinces, provincial power grid enterprises, and prefectural power grid enterprises. The electric supply and power sales quantity in different ranges differ from Formula (1.10).

1. National, large regional, and provincial power grid enterprises do not sell electricity directly, and their power sales quantity is the sum of electricity sold by prefectural power grid enterprises. The exchange electricity between provincial power grid enterprises within a large region and between prefectural power grid enterprises within a province can, in a broad sense, be considered as the electric supply or power sales quantity. The electricity loss in power grid loss calculation units directly administered by large regional and provincial power grid enterprises plus the electricity line loss within the subordinate administration range is the total electricity loss at this level. The line loss rate at this level is calculated by comparing the electric supply at this level with the total electricity loss at this level.
2. For a prefectural power grid enterprise, in addition to calculating the on-grid electricity of power plants in the local area, the electricity input from other areas in the local province or other provinces should be considered as the electric supply, and the electricity output to other areas in the local province or other provinces should be considered as the power sales quantity. Accordingly, the electricity line loss in the local area is calculated.

   Given the current electric power management system and the statistical criteria, the calculation formula of statistical line loss rate is as follows:

$$\begin{aligned}\text{Statistical line loss rate }(\%) &= (\text{statistical electricity line loss}) / (\text{electric supply}) \times 100\% \\ &= (\text{on-grid electricity of power plants} + \text{electricity input} \\ &\quad - \text{electricity output} - \text{power sales quantity}) / \\ &\quad (\text{on-grid electricity of power plants} + \text{electricity input}) \times 100\%\end{aligned} \tag{1.12}$$

The calculation formula of theoretical line loss rate is as follows:

$$\text{Theoretical line loss rate (\%)} = \text{theoretically calculated electricity line loss} / \text{theoretically calculated electric supply} \times 100\% \quad (1.13)$$

$$\text{Theoretically calculated electric supply} = \text{on-grid electricity of power plants} + \text{electricity input}$$

3. When a large regional power grid enterprise assesses the planned loss rate of one of its provincial power grid enterprises, the influence of the difference between the actual electricity of mutual supply and the planned electricity supply on the loss rate should be analyzed. This is a complicated loss allocation problem, which is addressed in Chapter 11, Section 11.3.

In line loss management, each level of a power grid enterprise should summarize the line loss information of their subordinate level of power grid enterprises and then include such information in the information of power grid loss calculation units at this level, thereby calculating the statistical line loss rate or the theoretical line loss rate of power grids at this level. The structure of the electricity line loss management information system can be designed to satisfy the requirements of such line loss management.

1.1.4 Calculation and Analysis of Line Loss

The planning of power grids, the comparison of power grid connection programs, and the design of substations require the theoretical calculation of line loss. The accuracy required for the line loss calculation during such planning and design is not high, but the calculation methods need to be simple and practical. Therefore, tabular methods and calculated curve methods are preferred. Local theoretical calculation of line loss can be used to predict the benefits of some technical measures of loss reduction, and the comparison in technology and economy is conducted to select an economical and reasonable loss reduction program. Relatively comprehensive and detailed theoretical calculation of line loss can determine the quantity and composition of the electricity line loss, and can also reveal the relationship between the technical electricity line loss and factors such as operating voltage level, load rate, and average power factor so as to establish technical measures of loss reduction more scientifically. The results of comprehensive theoretical calculation of line loss can also be compared with the statistical electricity line loss, so as to estimate the quantity of management electricity loss and provide a basis for reducing the management electricity loss.

The above three types of theoretical calculations of line loss are necessary for power supply agencies and industrial enterprises with independent power supply systems. Therefore, a comprehensive discussion of the theoretical calculation of line loss is very necessary.

The analysis of line loss with the results of a theoretical calculation of line loss is important for line loss management. According to Reference [88], three types of line loss analysis are required, namely statistical analysis, indicator analysis, and economic analysis.

1.1.4.1 Statistical Analysis

The electric energy loss during the transmission and substation in the main system is called power grid loss, and the electric energy loss during the transmission, substation and distribution in regional power grids is called regional line loss.

a. *Analysis of the composition of regional line loss.* The line loss in transmission and substation should be analyzed by voltage and line; the line loss in distribution should be analyzed by region, substation, line or distribution area (divided by the supply range of a distribution transformer). In addition, the no-load loss and load loss in regional power grids should be analyzed separately to calculate the no-load line loss rate and the load line loss rate.

b. *Analysis of the structure of power grids.* The line loss rate should be analyzed by voltage, and the electric supply and line loss rate for different electric supply structures should be analyzed, especially for the various step-down and coupling modes of two- and three-winding transformers, so as to find ways to improve the electric supply structure and reduce line loss.
c. *Analysis of the composition of power sales quantity.* The electric supply output to adjacent regions will increase the line loss in the local region, and the influence of such transit electric supply should be further analyzed. The power sales quantity of dedicated lines wherein the line loss is borne by customers and the bulk power sales quantity not considering loss are collectively called power sales quantity without loss. Obviously, the percentage of the power sales quantity without loss in the total power sales quantity directly affects the value of statistical line loss rate and should also be analyzed.

The results of the above three types of statistical analysis should be compared with the results of theoretical calculation of line loss, in order to identify where the line loss is the biggest in transmission, substation and distribution systems and determine main measures of loss reduction.

1.1.4.2 Indicator Analysis

The indicator analysis basically includes the comparison of line loss rate indicators in the current period with those in last period, and the comparison of the difference between the statistical value of line loss rate in the current period and the planned value in last period. The indicator analysis can follow the five considerations below:

a. The increase/decrease in the power sales quantity, and changes in the electricity utilization category and the voltage composition.
b. Changes in the operating mode of the electric power system, the load flow distribution, and the structure of the power grids.
c. The influence of loss reduction measures and project production.
d. The influence of new large customers.
e. The influence of replacement of main system units.

1.1.4.3 Economic Analysis

The economic analysis mainly includes two types, namely the analysis of loss reduction benefits achieved in reactive compensation equipment intensively installed in substations and reactive compensation equipment dispersedly installed in distribution lines, and the analysis of benefits achieved in the assessment of peak power factors of large customers or the assessment of peak/valley power factors.

The Provisions issued in 2004 by the State Grid Corporation of China [75] set out the following requirements on the annual report summary and analysis of line loss:

1. The performance of line loss indicators.
2. The analysis of the composition of line loss by comprehensive line loss rate, loss rate and regional line loss rate; the analysis of line loss rate by voltage class; the analysis of line loss deducting the power sales quantity without loss and the bulk power sales quantity.
3. Existing problems and measures; the quantitative analysis of reasons for the increase/decrease in the line loss rate and the degree of influence of the increase/decrease in the line loss rate.
4. Solutions to the problems and key measures of subsequent work.

1.2 Calculation of AC Resistance

Overhead power lines generally use bare conductors which have larger AC resistance than DC resistance as a result of the skin effect of alternating current. Steel-cored aluminum wires have even larger AC resistance due to iron loss caused by magnetization of their steel cores. The increased resistance due to the skin effect can be theoretically calculated, while the increased resistance caused by the magnetization of steel cores must be determined through actual measurement. The calculation formula of AC resistance is as follows:

$$R = K_1 K_2 R_{dc}^{(2)} \tag{1.14}$$

$$K_1 = 0.99609 + 0.018578X_1 - 0.030263X_1^2 + 0.020735X_1^3 \tag{1.15}$$

$$K_2 = 0.99947 + 0.028895X_2 - 0.005934X_2^2 + 0.00042259X_2^3 \tag{1.16}$$

$$X_1 = \frac{D+2d}{D+d} \times 0.01\sqrt{\frac{8\pi f(D-d)}{R_{dc}(D+d)}} \tag{1.17}$$

$$X_2 = \frac{I}{S} \tag{1.18}$$

Wherein: R_{dc} – DC resistance of conductors at the calculation temperature (Ω/km);
K_1 – skin effect coefficient of conductors;
X_1 – parameter used to calculate the skin effect coefficient of conductors;
D, d – outer diameter and inner diameter of conductors (cm);
f – frequency of alternating current (Hz);
K_2 – iron loss coefficient of conductors;
X_2 – parameter used to calculate the iron loss coefficient of conductors;
I – current passing conductors (A);
S – sectional area of conductors (mm^2).

According to the calculation, the AC resistance is only ~0.02 to ~5.0% higher than the DC resistance of all aluminum conductors whose sectional areas range from 50 to 240 mm^2; the AC resistance is 1.3–4.6% higher than the DC resistance of steel-cored aluminum wires whose sectional areas range 25–240 mm^2. The lower limits above are calculated when the current-carrying capacity is 20% of the allowable value, while the upper limits above are calculated when the current-carrying capacity is the allowable value. This shows that, when the current of overhead line conductors is close to or exceeds the allowable value, factors causing higher AC resistance must be taken into account. In other circumstances, the DC resistance can be directly used to calculate the line loss, without leading to any significant error.

1.3 Influence of Temperature and Voltage Changes on Line Loss in the Measuring Period

1.3.1 Influence of Temperature Change on Line Loss in the Measuring Period

According to Formula (1.9), not only loads change with time, but also the resistance of conductors changes with temperature within a measuring period. Apparently, it is extremely complicated to take into account the two change factors at the same time for integral operation. To easily calculate the line loss, the influence of temperature change on the variable resistance can be considered first.

It is generally known that the resistance of conductors with temperature change can be calculated as per the following formula:

$$R_T = R_0(1+\alpha T) \tag{1.19}$$

Wherein: R_0 – resistance of conductors at 20 °C (Ω);
α – temperature coefficient of conductor resistance: generally $\alpha = 0.004$ for copper, aluminum and steel-cored aluminum wires;
T – air temperature (°C).

The record data of load current and temperature within one day (24 h) are substituted into Formulas (1.19) and (1.9), resulting in:

$$\begin{aligned} \Delta A &= \int_0^{24} I^2(t)R(t)\mathrm{d}t \times 10^{-3} \\ &= \left[I_1^2R_0(1+\alpha T_1) + I_2^2R_0(1+\alpha T_2) + \cdots + I_{24}^2R_0(1+\alpha T_{24})\right] \times 10^{-3} \\ &= \left[\left(I_1^2 + I_2^2 + \cdots + I_{24}^2\right) + \alpha\left(I_1^2T_1 + I_2^2T_2 + \cdots + I_{24}^2T_{24}\right)\right]R_0 \times 10^{-3} \end{aligned} \tag{1.20}$$

The weighted average temperature is defined as follows:

$$T_{we} = \frac{I_1^2T_1 + I_2^2T_2 + \cdots + I_{24}^2T_{24}}{I_1^2 + I_2^2 + \cdots + I_{24}^2} \tag{1.21}$$

Then, Formula (1.20) can be rewritten to:

$$\begin{aligned} \Delta A &= \left(I_1^2 + I_2^2 + \cdots + I_{24}^2\right)(1+\alpha T_{we})R_0 \times 10^{-3} \\ &= \left(I_1^2 + I_2^2 + \cdots + I_{24}^2\right)R_{we} \times 10^{-3} \end{aligned} \tag{1.22}$$

$$R_{we} = R_0(1+\alpha T_{we}) \tag{1.23}$$

Wherein: R_{we} – conductor resistance at the weighted average temperature.

As mentioned above, if the electric energy loss is calculated based on the weighted average temperature and Formula (1.22), then the influence of temperature change is completely considered.

As per Formula (1.21), if the load is constant, then $T_{\mathrm{we}} = T_{\mathrm{av}}$ (average temperature). As the daily temperature changes in a unimodal manner and the daily load normally changes with two different peaks, T_{av} is very close to T_{we} within a period of one day and night or more than one day and night, and the replacement of T_{we} with T_{av} will not produce any larger negative error.

According to the analysis of relative error of resistance as per Formula (1.23), as the temperature coefficient α of conductor resistance is very small, even if the replacement of T_{we} with T_{av} produces a certain relative error, relative errors of resistance and electric energy loss are still very small. If the measuring period is one month or one year, Formula (1.9) used to calculate the electric energy loss of a three-phase symmetrical unit can be rewritten to the following by using T_{av} to calculate the resistance:

$$\Delta A = 3R\int_0^T I^2(t)\mathrm{d}t \times 10^{-3} \tag{1.24}$$

1.3.2 Influence of Voltage Change on Line Loss in the Measuring Period

When the measured load data represents active power and reactive power instead of the current, the voltage change should be considered for the calculation of line loss. If the measuring period is one day and night, then Formula (1.24) can be rewritten to:

$$\Delta A = R\int_0^{24} \frac{[P_2(t)+Q^2(t)]}{U^2(t)}\mathrm{d}t \times 10^{-3} \tag{1.25}$$

Wherein: R – resistance calculated per Formula (1.23) by considering the temperature change (Ω);
$P(t)$, $Q(t)$ – active power (kW) and reactive power (kvar) at the same measuring point;
$U(t)$ – voltage at active and reactive power measuring point (kV).

The active power weighted average voltage and the reactive power weighted average voltage can be defined with the squared values of active power and reactive power of one day and night as the weight, that is:

$$\left.\begin{aligned} \frac{1}{U^2_{we.P}} &= \left(\frac{P_1^2}{U_1^2}+\frac{P_2^2}{U_2^2}+\cdots+\frac{P_{24}^2}{U_{24}^2}\right)\Big/(P_1^2+P_2^2+\cdots+P_{24}^2) \\ \frac{1}{U^2_{we.Q}} &= \left(\frac{Q_1^2}{U_1^2}+\frac{Q_2^2}{U_2^2}+\cdots+\frac{Q_{24}^2}{U_{24}^2}\right)\Big/Q_1^2+Q_2^2+\cdots+Q_{24}^2 \end{aligned}\right\} \tag{1.26}$$

Then, Formula (1.25) can be rewritten to:

$$\Delta A = R\left[\int_0^{24} P^2(t)\mathrm{d}t/U^2_{we.P} + \int_0^{24} Q^2(t)\mathrm{d}t/U^2_{we.Q}\right] \times 10^{-3} \tag{1.27}$$

According to the calculations with the measured data of 220, 110, and 35 kV systems with different voltage and load variations, if the average voltage U_{av} is used to replace weighted average voltages $U_{we.P}$ and $U_{we.Q}$, the error generally does not exceed minus 1%, so Formula (1.27) can be further rewritten to:

$$\Delta A = \frac{R}{U^2_{av}}\left[\int_0^{24} P^2(t)\mathrm{d}t + \int_0^{24} Q^2(t)\mathrm{d}t\right] \times 10^{-3} \tag{1.28}$$

Under the condition of normal operation, as the long-term voltage variation is not large, the replacement of $U_{\text{we.}P}$ and $U_{\text{we.}Q}$ with U_{av} is still feasible, that is

$$\Delta A = \frac{R}{U_{\text{av}}^2}\left[\int_0^T P^2(t)\text{d}t + \int_0^T Q^2(t)\text{d}t\right] \times 10^{-3} \tag{1.29}$$

A former Soviet Union scholar once studied the relationship between voltage deviation and electric energy loss in the distribution network [3]. The results show that, regardless of the voltage deviation, if the average voltage is used to calculate the electric energy loss the error is related to the correlation coefficient between the voltage and current changes, the signs, and the voltage deviation, that is

$$\left.\begin{aligned} \delta(\Delta A)\% &= |r_{U/I}|\sigma_U\% \\ \delta(\Delta A)\% &= \left(\frac{\Delta A_2 - \Delta A_1}{\Delta A_1}\right) \times 100\% \end{aligned}\right\} \tag{1.30}$$

Wherein: ΔA_1 – electric energy loss considering the voltage deviation (kW·h);
ΔA_2 – electric energy loss calculated with the average voltage (kW·h);
$r_{U/I}$ – correlation coefficient between voltage and current changes, $-1 \le r_{U/I} \le 1$;
$\sigma_U\%$ – mean square error of voltage change, which is calculated by percentage.

The distribution of voltage deviations in the distribution network is close to a "normal distribution", so

$$\sigma U\% = \frac{1}{6}\left(\frac{U_{\max} - U_{\min}}{U_{\text{av}}}\right) \times 100\% \tag{1.31}$$

Wherein $U_{\max}$ and $U_{\min}$ – maximum and minimum voltages.

According to Formula (1.31), when the voltage change in the distribution network reaches up to 20%, the error resulted from the calculation of line loss with the average voltage will not exceed 3.3% which is allowable in engineering calculation.

1.4 Influence of Load Curve Shape on Line Loss

1.4.1 Load Curve and Load Duration Curve

Load changes are recorded with time sequence, which can be used to produce a normal load curve. Within a period T, a derivative curve arranged by load value and its duration rather than time sequence is called load duration curve.

Figure 1.1a shows the load curve and Figure 1.1b shows the load duration curve, as follows:

$$\begin{cases} \overline{cd} = \overline{c'd'}, \overline{ef} + \overline{gh} = \overline{e'h'} \\ I_0 = I'_0 \; I_1 = I'_1, \; I_2 = I'_2 \end{cases}$$

Obviously, the load square curve has its corresponding load duration square curve, so

$$I_0^2 = (I'_0)^2, I_1^2 = (I'_1)^2, I_2^2 = (I'_2)^2$$

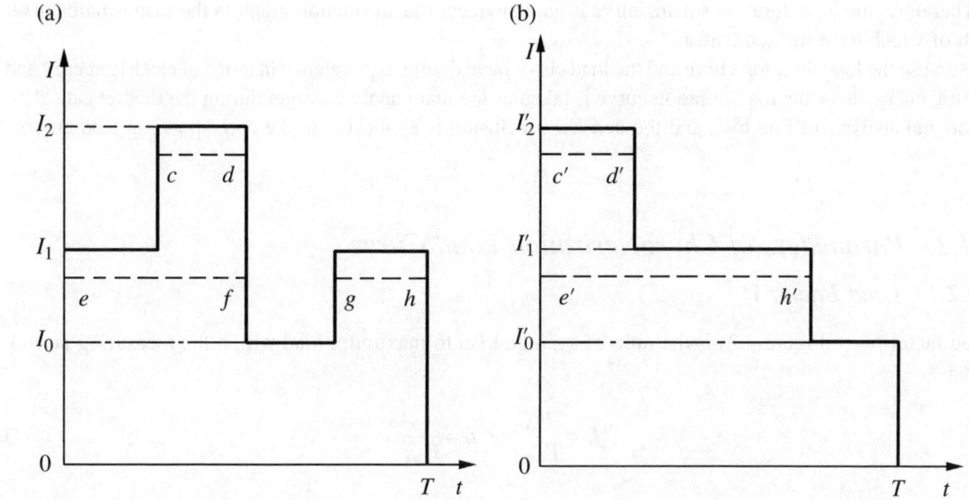

Figure 1.1 Load curve and load duration curve. (a) Load curve. (b) Load duration curve.

Assume that the power factor of a three-phase symmetrical unit is 1.0, and that the voltage is constant within the measuring period, with the load curve as shown in Figure 1.1a. Then, the electric energy passing through this unit is $A=\sqrt{3}U\int_0^T I(t)\mathrm{d}t$ within the measuring period. As per Figure 1.1,

$$\begin{aligned}A&=\sqrt{3}U\int_0^T I(t)\mathrm{d}t\\&=\sqrt{3}U\left[I_0T+(I_1-I_0)\left(\overline{ef}+\overline{gh}\right)+(I_2-I_1)\overline{cd}\right]\\&=\sqrt{3}U\left[I'_0T+(I'_1-I'_0)\overline{e'h'}+(I'_2-I'_1)\overline{c'd'}\right.\\&=\sqrt{3}U\int_0^T I'(t)\mathrm{d}t=A'\end{aligned}$$

Therefore, the load duration curve is an equivalent transformation graph to the load curve, both of which have the same area.

Assume that the resistance of the above unit is R per phase. Then, the electric energy loss of this unit is $\Delta A=3R\int_0^T I^2(t)\mathrm{d}t\times10^{-3}$ within the measuring period. According to the definition of the load duration square curve, the following relationship can be derived as per Figure 1.1:

$$\begin{aligned}\Delta A&=3R\int_0^T I^2(t)\mathrm{d}t\times10^{-3}\\&=3R\left[I_0^2T+(I_1-I_0)^2\left(\overline{ef}+\overline{gh}\right)+(I_2-I_1)^2\overline{cd}\right]\\&=3R\left[(I'_0)^2T+(I'_1-I'_0)^2\overline{e'h'}+(I'_2-I'_1)^2\overline{c'd'}\right.\\&=3R\int_0^T[I'(t)]^2\mathrm{d}t=\Delta A'\end{aligned}$$

Therefore, the load duration square curve is an equivalent transformation graph to the load square curve, both of which have the same area.

Because the load duration curve and the load curve have double equivalence in terms of electric energy and electric energy loss, the load duration curve is taken as the main analysis target during the theoretical calculation and analysis of line loss, and the derived conclusion is applicable to the corresponding load curve.

1.4.2 Parameters of Characterization Load Curve

1.4.2.1 *Load Factor* f

Load factor f (see Figure 1.2) is the ratio of average load to maximum load within the measuring period, that is

$$f_P = \frac{P_{av}}{P_{max}} \text{ or } f_I = \frac{I_{av}}{I_{max}} \tag{1.32}$$

Wherein: P_{av} and P_{max} – average active power and maximum active power of load (kW);
I_{av} and I_{max} – average current and maximum current of load (A).

The load rate reflects the average utilization of power system equipment and serves as an important indicator to assess the operation of the power system.

1.4.2.2 *Minimum Load Rate* β

Minimum load rate β (see Figure 1.2) is the ratio of minimum load to maximum load within the measuring period, that is

$$\beta = \frac{I_{min}}{I_{max}} \tag{1.33}$$

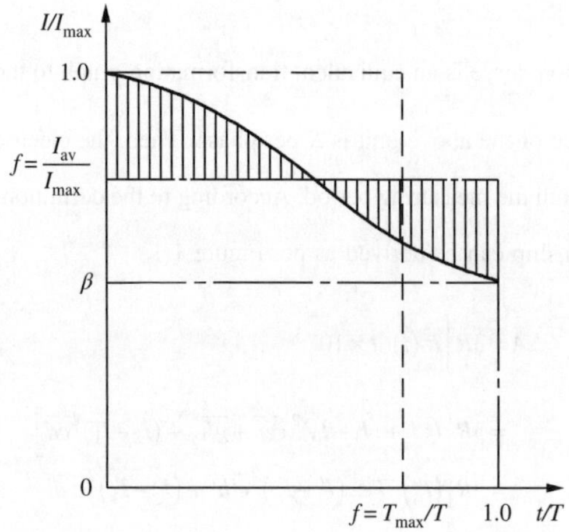

Figure 1.2 Meaning of load factor f and minimum load rate β.

1.4.2.3 *Maximum Load Utilization Time* T_{max} *and Maximum Loss Time* τ_{max}

Assume that the voltage and power factor remain the same within the measuring period, and that the electric energy passing through a unit under varying current within the period T equals the electric energy passing through the unit under maximum current within $T_{\max}$.

Then, $T_{\max}$ is called the maximum load utilization time (see Figure 1.3), that is

$$\sqrt{3}U\cos\varphi\int_0^T I(t)\mathrm{d}t = \sqrt{3}UI_{\max}\cos\varphi T_{\max}$$

$$T_{\max} = \frac{\int_0^T I(t)\mathrm{d}t}{I_{\max}} \tag{1.34}$$

$\tau_{\max}$, the maximum loss time, is defined as: when the voltage and power factor remain unchanged within the measuring period, if the electric energy loss of current passing through a unit under varying current equals the electric energy loss of current passing through the unit under maximum current within the time of $\tau_{\max}$ (see Figure 1.3), that is

$$\tau_{\max}I_{\max}^2 3R = 3R\int_0^T I^2(t)\mathrm{d}t$$

$$\tau_{\max} = \frac{\int_0^T I^2(t)\mathrm{d}t}{I_{\max}^2} \tag{1.35}$$

As $I_{\mathrm{av}} = \int_0^T I(t)\mathrm{d}t/T$, $I_{\max} = \int_0^T I(t)\mathrm{d}t/T_{\max}$ is derived from Formula (1.34) and substituted into Formula (1.32), resulting in:

$$f = \frac{I_{\mathrm{av}}}{I_{\max}} = \frac{T_{\max}}{T} \tag{1.36}$$

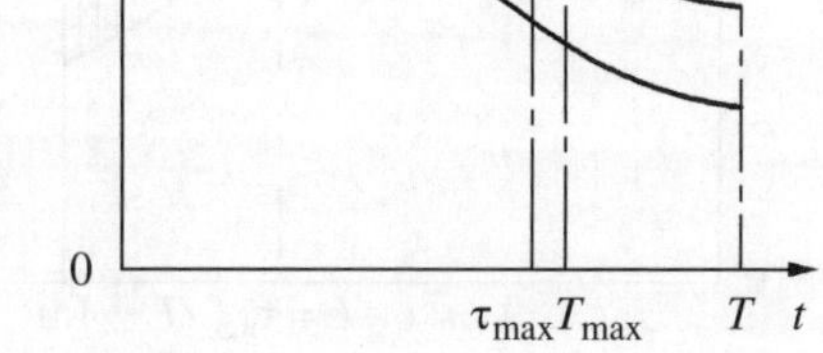

Figure 1.3 Definition of $T_{\max}$ and $\tau_{\max}$.

According to Formula (1.36), the load rate is equal to the per unit value at the maximum load utilization time.

1.4.2.4 *Loss Factor* F

Loss factor F is the ratio of maximum loss time to measuring period T, that is

$$F = \frac{\tau \max}{T} = \frac{\int_0^T I^2(t)\mathrm{d}t / I_{\max}^2}{T} = \frac{I_{\mathrm{rms}}^2}{I_{\max}^2} \tag{1.37}$$

$$I_{\mathrm{rms}} = \sqrt{\int_0^T I^2(t)\mathrm{d}t / T} \tag{1.38}$$

Wherein I_{rms} – rms current.

According to Formula (1.37), the loss factor is also equal to the ratio of rms current square to maximum current square. The meaning of loss factor F is shown in Figure 1.4.

The maximum load of power supply equipment is the focus of operation monitoring. The parameter of loss factor together with the maximum load can be used to calculate the line loss, so Formula (1.24) can be rewritten to:

$$\Delta A = 3I_{\max}^2 FRT \times 10^{-3} \tag{1.39}$$

1.4.2.5 *Form Coefficient* K

$\sqrt{F} = I_{\mathrm{rms}} / I_{\max}$ can be derived from Formula (1.37) and it is called loss equivalent load factor in Japan. As I_{rms} is the "equivalent loss" current, the ratio of it to $I_{\max}$ can be regarded as another type of load factor.

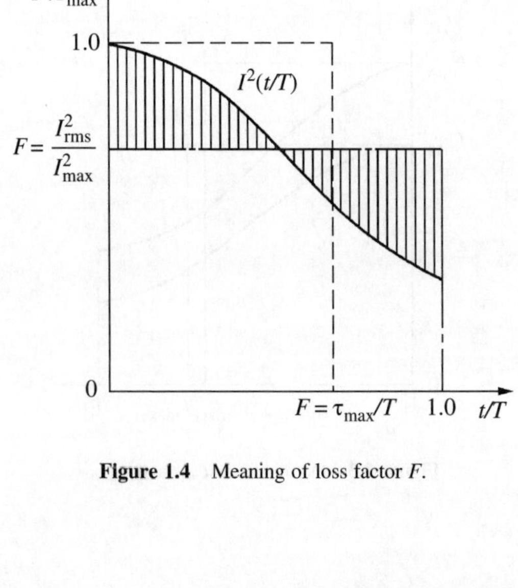

Figure 1.4 Meaning of loss factor F.

Form coefficient is defined as the ratio of rms current to average current, that is

$$K = I_{rmx}/I_{av} \tag{1.40}$$

$I_{rms} = \sqrt{F} I_{max}$, so

$$K = \sqrt{F}/(I_{av}/I_{max}) = \sqrt{F}/f \tag{1.41}$$

In summary, the loss equivalent load rate is the ratio between rms current of equivalent loss and the maximum current, and the load rate is the ratio of average value to the maximum value in the current load curve. The form coefficient is the ratio of such two load rates and also the ratio of the loss equivalent current to the electric energy equivalent current, comprehensively reflecting the features of load loss (square) curve and load curve.

The above five parameters are defined based on the current variable. Similar coefficients can be derived for three load curves expressed by active power, reactive power and apparent power. For the sake of distinction, different subscripts may be used.

1.4.3 Relationship Between Loss Factor and Load Factor

Figure 1.5a shows two special load curves.

For load curve 1, the maximum load duration is $t_{max} \approx 0$, and the load is constant as the minimum load during most of the time. For load curve 2, the maximum load duration is t_{max}, and the load is close to zero during the rest of the time. The y-coordinate of the load curve (see Figure 1.5a) refers to the per unit value of load current based on the maximum current, and the x-coordinate refers to the per unit value of time based on the measuring period T.

According to load curve 1 in Figure 1.5,

$$f = \frac{\beta \times 1.0}{1.0} = \beta, F = \frac{(\beta \times 1.0)^2}{1.0^2} = \beta^2$$

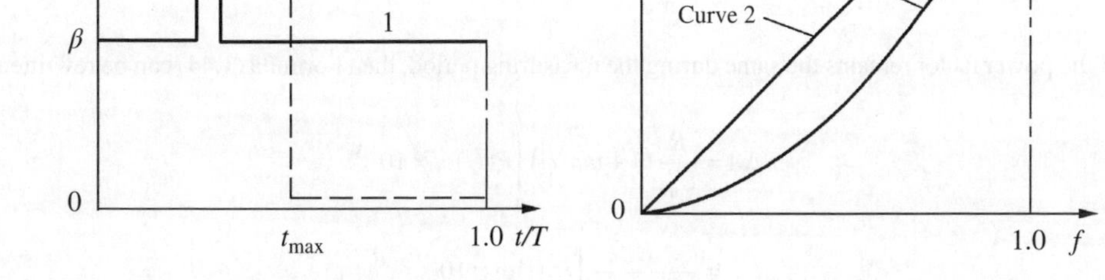

Figure 1.5 Relationship between loss factor and load rate. (a) Two special load curves. (b) Value range of F and f.

So if any value is taken for β, then

$$F = f^2 \tag{1.42}$$

According to load curve 2 in Figure 1.5,

$$f = \frac{1.0 \times t_{max}}{1.0} = t_{max},\ F = \frac{1.0^2 \times t_{max}}{1.0^2} = t_{max}$$

So if any value is taken for t_{max}, then

$$F = f \tag{1.43}$$

The $F(f)$ curve expressed by Formula (1.42) and Formula (1.43) is shown in Figure 1.5b. When the load curve is mutant (curve 1), $F = f^2$; when the load curve is stable (curve 2), $F = f$. Any actual load curve falls between such two extreme cases, so the relationship between F and f for any load curve must fall in the range between $F = f$ and $F = f^2$.

Figure 1.5b also indicates that, in addition to such two end values as 0 and 1, the same load factor f can correspond to different F values, *which shows that a single load factor parameter cannot accurately reflect the loss conditions of a unit.* In other words, even with the same load factor, the loss factor varies with the shape of the load curve. When $f = 0.5$, the loss factor uncertainly leads to the largest absolute error and relative error, each of which is as follows:

$$\Delta F_{max} = F_2 - F_1 = 0.5 - 0.5^2 = 0.25$$

$$(\delta F)_{max} = \frac{\Delta F_{max}}{F_1} \times 100\% = \frac{0.25}{0.25} \times 100\% = 100\%$$

This indicates that, as the load curve differs in thousands of ways, a significant error must be produced if a single load factor parameter is used to calculate the loss factor. To avoid this, a loss factor formula with multiple parameters is required, which will be further analyzed in Chapter 2.

1.5 Influence of Load Power Factor and Load Distribution on Line Loss

1.5.1 Influence of Load Power Factor

If the average voltage during the measuring period T is U_{av} and the conductor resistance is R, then the electric energy loss of a three-phase line can be calculated by:

$$\begin{aligned}\Delta A &= \frac{R}{U_{av}^2}\left\{\int_0^T \left[P^2(t) + Q^2(t)\right]\mathrm{d}t\right\} \times 10^{-3} \\ &= \frac{R}{U_{av}^2}\int_0^T P^2(t)\left[1 + \tan^2\varphi(t)\right]\mathrm{d}t \times 10^{-3}\end{aligned} \tag{1.44}$$

1. If the power factor remains the same during the measuring period, then Formula (1.44) can be rewritten to:

$$\begin{aligned}\Delta A &= \frac{R}{U_{av}^2}\left(1 + \tan^2\varphi\right)\int_0^T P^2(t)\mathrm{d}t \times 10^{-3} \\ &= \frac{R}{U_{av}^2\cos^2\varphi}\int_0^T P^2(t)\mathrm{d}t \times 10^{-3}\end{aligned} \tag{1.45}$$

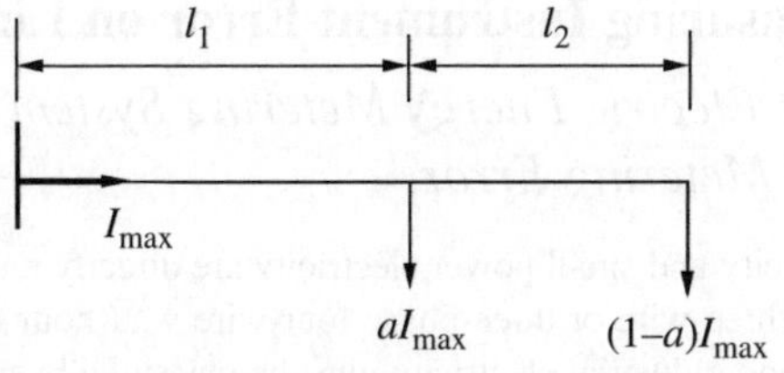

Figure 1.6 Influence of load distribution.

2. If the power factor is varying during the measuring period, then Formula (1.44) can be rewritten to:

$$\Delta A = \frac{R}{U_{av}^2}\left[\int_0^T P^2(t)\mathrm{d}t + \int_0^T P^2(t)\tan^2\varphi(t)\mathrm{d}t\right] \times 10^{-3} \quad (1.46)$$

According to the above formulas, the load power factor has a complicated influence on the calculation of line loss, which will be analyzed in detail in Chapter 4.

1.5.2 *Influence of Load Distribution of Multi-Branch Line*

Assume that there are two customers in a line, both of which have the same shape of load curve. The whole line uses conductors with the same sectional area. The distance from the branch point to the start end and the tail end of the line is l_1 and l_2, respectively. The maximum current at the start end and at the two customers is I_{max}, aI_{max}, and $(1–a)I_{max}$, respectively, as shown in Figure 1.6.

Because the load curves of the two customers are exactly the same, $F_1 = F_2 = F$. The electric energy loss of the whole line during the measuring period T can be calculated as per the following formula:

$$\begin{aligned}\Delta A &= \Delta A_1 + \Delta A_2 \\ &= 3\left[I_{max}^2 F l_1 r_0 + (1-a)^2 I_{max}^2 F l_2 r_0\right] \times T \times 10^{-3} \\ &= 3I_{max}^2 F r_0\left[l_1 + (1-a)^2 l_2\right] T \times 10^{-3}\end{aligned} \quad (1.47)$$

Wherein r_0 – resistance per unit length of conductor (Ω/km).

According to Formula (1.47), the smaller a is, the bigger the electric energy loss ΔA is. When a is positivecertain, the bigger l_2 is, and the bigger ΔA is. Therefore, two factors influence the electric energy loss of a multi-branch line: one is the shunt condition, that is the distribution of load values; the other is the distance from each load point to the start end of the line, that is the special distribution of load. Given the differences in the shape of load curve and power factor of multi-branch loads and in the voltage at multi-branch points, it is very complex to accurately calculate the electric energy loss of a multi-branch line. To do the line loss calculation, a series of approximate assumptions must be made for simplified calculation and analyzed in terms of their influence on the calculation results. These contents will be explained in Chapter 5.

1.6 Influence of Measuring Instrument Error on Line Loss

1.6.1 Composition of Electric Energy Metering System and Constitution of Metering Error

Low-voltage residential electricity and small power electricity are directly measured by a single-phase watt-hour meter and a three-phase three-wire or three-phase four-wire watt-hour meter respectively, while large power commercial electricity and industrial electricity must be measured by a current transformer. Industrial and agricultural irrigation electricity with voltage over 0.4 kV may be measured by a current transformer together with a voltage transformer. In this case, a voltage transformer with a rated secondary voltage of 100 V, a current transformer with a rated secondary current of 5 A, a supporting watt-hour meter, and wires connecting the foregoing three constitute a complete electric energy metering system.

The metering error of a complete electric energy metering system consists of the error of the watt-hour meter, the ratio error, the phasor angle error of transformers, and the error caused by the voltage drop of connecting wires. For star three-phase loads of star connection, if the electric energy metering system uses the three-phase, four-wire watt-hour meter as the main measuring element, then its composition error consists of the error of the transformers, the error caused by the secondary voltage drop of wires, and the error of the watt-hour meter, as shown below [34]:

$$\varepsilon = 0.003(f_1 + f_2 + f_3) + 0.0097(\delta_1 + \delta_2 + \delta_3)\tan\varphi + (\Delta f + \tan\delta') + \varepsilon_0 \tag{1.48}$$

$$f_i = f_{Ui} + f_{Ii}$$

$$\delta_i = \alpha_i - \beta_i$$

Wherein: f_i – sum of voltage transformer ratio error and current transformer ratio error of various phases;
δ_i – difference between current transformer angle error and voltage transformer angle error of various phases;
Δf, $\tan\delta'$ – additional ratio error and angle error caused by secondary voltage drop of wires;
ε_0 – error of the watt-hour meter.

1.6.2 Composition of Electronic Watt-Hour Meter Error

1.6.2.1 Structural Features of Electronic Watt-Hour Meter

The multiplier used in the hardware system of the increasingly widely used electronic watt-hour meter includes time division multiplier and AC/DC direct sampling multiplier. The time division multiplier uses field effect transistors (FETs) to control amplitude modulation circuits, which easily suffers interference from the spike effect of FETs. The production of a high-precision time division multiplier imposes quite high requirements on components and entails complicated auxiliary circuits and higher costs. Complicated auxiliary circuits are generally not set for the electronic watt-hour meter which is largely used for civil purpose with low precision, so its anti-interference capability is rather weak.

The traditional AC/DC direct sampling multiplier generally adopts a broken line approximate expression function generator composed of high-frequency diodes to obtain the product of voltage signal and current signal. Temperature compensation is required for the diodes. Its circuits are complicated and its precision is not high. In recent years, the A/D converter has been used to digitize voltage, current, and their differences; and microcomputer software has been used to realize square operation, forming a digital square multiplier; as complicated analog circuits are not used to realize square operation and the accuracy of the converter is highly improved, the high precision pulse (electronic) watt-hour meter based on the principle of digital square multiplier has been widely used for gateway electric energy metering in the electric power system.

1.6.2.2 Metering Error of Electronic Watt-Hour Meter

Both of the above multipliers have input circuits and generally use transformer input, so each electronic watt-hour meter has ratio error and angle error in voltage, current input voltage change coefficients K_u and K_i, and the square multiplier has amplitude error and phase error in the transfer coefficient. As a result, an electronic watt-hour meter itself has many factors that may cause errors. Even if instrument initialization and calibration of standard constants are well conducted at delivery, errors may also occur due to external interference in the operating environment, a change in the performance of internal transformers, and other random factors. Therefore, like a traditional electromechanical watt-hour meter, an electronic watt-hour meter may have random errors.

1.6.3 Influence of Metering System Error on the Calculation of Line Loss Rate

The existing electric energy measuring procedures set accuracy classes for various watt-hour meters, current transformers, and voltage transformers of electric energy metering systems by their quantity of electric energy within a measuring period. Since the reform of the separation between power plants and power grids in the electric power management system, power grid enterprises have had higher requirements for the accuracy of power purchase quantity metering systems than for power sales quantity metering systems. Therefore, in recent years, provincial, local, and municipal power grid enterprises have installed centralized meter reading systems featuring real-time monitoring. With respect to power purchase quantity and power sales quantity metering facilities with larger electric energy quantity, the satisfactory accuracy of electric energy metering can be achieved by selecting appropriate transformer accuracy, secondary wire design, and electronic watt-hour meter. Because a complete electric energy metering system has random errors caused by various factors, there are metering errors in the power purchase quantity and power sales quantity of a relatively independent municipal power grid enterprise, affecting the statistical accuracy of the line loss rate within a certain measuring period.

According to the results of long-term testing of an electromechanical watt-hour meter, positive and negative random errors of a heavily operated watt-hour meter are similar, and the long-term statistics show that the random errors of a watt-hour meter comply with a normal distribution in probability theory. In other words, the accuracy of aggregate metering values of n watt-hour meters with the same accuracy class is $1/\sqrt{n}$ of the accuracy of a single watt-hour meter. Given the lack of in-depth understanding of the error statistical law of an electronic watt-hour meter, assuming there are both electromechanical and electronic watt-hour meters and that metering errors in the power purchase quantity and power sales quantity still comply with the normal distribution statistical law, the influence of electric energy metering errors on the line loss rate may be estimated within a certain statistical period.

For example, if the total number of power purchase meters of a power grid enterprise is n_1 with a weighted average accuracy of Δ_1, the total number of power sales meters is n_2 with a weighted average accuracy of Δ_2, and the actual line loss rate is $\Delta A\%$, then the relative error can be calculated by the following formula as the meter accuracy affects the accuracy of the statistical line loss rate:

$$\delta(\Delta A\%) = \left(\frac{1-\Delta A\%}{\Delta A\%}\right)\left(\frac{\Delta_1}{\sqrt{n_1}} + \frac{\Delta_2}{\sqrt{n_2}}\right) \tag{1.49}$$

According to Formula (1.49), there is a metering error between the statistical line loss rate and the actual line loss rate, so a certain fluctuation range is allowed for the line loss rate. For example, considering a municipal power supply enterprise, the number of its power purchase gateway meters is $n_1 = 50$, $\Delta_1 = 0.50\%$; the total number of 10 kV distribution transformer electric energy meters and 10 kV large customers is $n_2 = 1300$, $\Delta_2 = 2.5\%$; and the planned line loss rate of up to 10 kV level is $\Delta A\% = 4.3\%$. Substituting these into Formula (1.49), $\delta(\Delta A\%) = 3.12\%$ is obtained. The allowable range of statistical line loss rate is 4.434 – 4.166%, and the allowable fluctuation range is ±3.12% of the planned value.

2

Calculation of Line Loss by Current Load Curve

For the purpose of operation monitoring, the load current of lines and transformers should be recorded over time at the substations of power supply enterprises and industrial customers. This chapter introduces two basic methods to calculate the line loss by current load curve, namely, the rms current method and the loss factor method, and explains the relationship between other methods and these two methods.

2.1 RMS Current Method and Loss Factor Method

2.1.1 RMS Current Method

According to Formula (1.38), $I_{\text{rms}} = \sqrt{\int_0^T I^2(t)\mathrm{d}t/T}$, the electric energy loss ΔA (kW·h) of a three-phase unit is:

$$\Delta A = 3R\int_0^T I^2(t)\mathrm{d}t \times 10^{-3} = 3I_{\text{rms}}^2 RT \times 10^{-3} \tag{2.1}$$

Formula (2.1) is the basic calculation formula of rms current method.

If the rms current of a representative day is used to calculate the electric energy loss of a whole month, then the calculation result should be corrected by the ratio of the average daily electric supply of a whole month to the electric supply of the representative day, that is

$$\Delta A_{\text{m}} = \Delta A_{\text{d}}\left(\frac{A_{\text{m}}/D}{A_{\text{d}}}\right)^2 D \tag{2.2}$$

$$\Delta A_{\text{d}} = 3I_{\text{rms. d}}^2 R \times 24 \times 10^{-3} \tag{2.3}$$

Line Loss Analysis and Calculation of Electric Power Systems, First Edition. Anguan Wu and Baoshan Ni.

$$I_{\mathrm{rms.\,d}} = \sqrt{\sum_{i=1}^{24} I_i^2 / 24}$$

Wherein ΔA_{m} – electricity loss of a month (kW·h);
D – calendar days of a whole month;
A_{m}, A_{d} – electric supply of a whole month and of the representative day (kW·h);
ΔA_{d} – electricity line loss of the representative day (kW·h);
$I_{\mathrm{rms.d}}$ – rms current of the representative day, which can be calculated by the 24 h current of the representative day.

2.1.2 Loss Factor Method

As previously mentioned, the loss factor being $F = \int_0^T I^2(t)\mathrm{d}t / \left(I_{\max}^2 T\right)$, the electric energy loss ΔA of a three-phase unit is:

$$\Delta A = 3 I_{\max}^2 FRT \times 10^{-3}$$

As the above formula uses the maximum current, the loss factor method is also called maximum current method in some literature.

To prevent damage to power supply and consumption equipment due to overload, power operation enterprises pay more attention to monitoring current carefully, and even install special instruments such as maximum demand meters for recording. Therefore, the key to the calculation of line loss by the loss factor method does not lie in obtaining the maximum current, but in how to acquire the loss factor.

2.1.3 Other Calculation Methods

2.1.3.1 Average Current Method

Records of active and reactive watt-hour meters within the measuring period can be used to calculate the average current I_{av} during the measuring period; then, the form coefficient K and the average current I_{av} can be used to calculate the rms current I_{rms}; and finally, the foregoing values can be used to calculate the line loss. Relevant formulas are shown as follows:

$$I_{\mathrm{av}} = \frac{\sqrt{A_P^2 + A_Q^2}}{\sqrt{3} U_{\mathrm{av}} T}$$

$$I_{\mathrm{rms}} = K I_{\mathrm{av}}$$

$$\Delta A = 3 I_{\mathrm{av}}^2 K^2 RT \times 10^{-3}$$

Wherein A_P, A_Q – active electricity (kW·h) and reactive electricity (kvar·h) obtained from readings of watt-hour meters;
U_{av} – average voltage (kV);
T – duration of the measuring period (h).

Formula (1.41) shows $K^2=F/f^2$, so

$$\Delta A = 3I_{av}^2 \frac{F}{f^2} RT \times 10^{-3} \tag{2.4}$$

Also, due to $I_{av} = I_{max} f$, Formula (2.4) is equivalent to Formula (1.39). Obviously, the average current method and the loss factor method both require that K or F should be calculated first in a simple and accurate way.

2.1.3.2 Variance Current Method

Based on relevant formulas of the probability theory, the calculation formula of variance current can be obtained as follows:

$$D(I) = I_{rms}^2 - I_{av}^2 \tag{2.5}$$

If the variance current is known, the average current can be calculated according to the records of a watt-hour meter and the average voltage, and then the rms current can be calculated by Formula (2.5). Finally, the electric energy loss can be derived.

If the load duration curve can be expressed by a mathematical analytical formula, the formula of $D(I)$ can be directly derived according to the definition of variance. See Chapter 3 for details.

2.2 Derivation of Functional Relationship $F(f)$ by Ideal Load Curve

2.2.1 Derivation of F(f) *Formula by Ideal Load Curve with Two Variables*

In 1960, Putein[4] from the former Soviet Union put forward four typical linear load duration curves according to the characteristics of outgoing loads of local thermal power plants and local hydroelectric power plants, as shown in Figure 2.1. The x-coordinate refers to the per unit value of time based on the measuring period T, and the y-coordinate refers to the per unit value of current based on the maximum current.

For the four linear load duration curves shown in Figure 2.1, the following loss factor formulas can be derived, respectively:

For Figure 2.1a,

$$F = \frac{2}{3} f \tag{2.6}$$

For Figure 2.1b,

$$F = \frac{1}{3}\left(1 - 2f + 4f^2\right) \tag{2.7}$$

For Figure 2.1c,

$$F = \frac{2}{3} f + \frac{1}{3} t_{max} \tag{2.8}$$

For Figure 2.1d,

$$F = 1 - 2(1-f)\left(1 - \frac{2}{3}\frac{1-f}{1-t_{max}}\right) \tag{2.9}$$

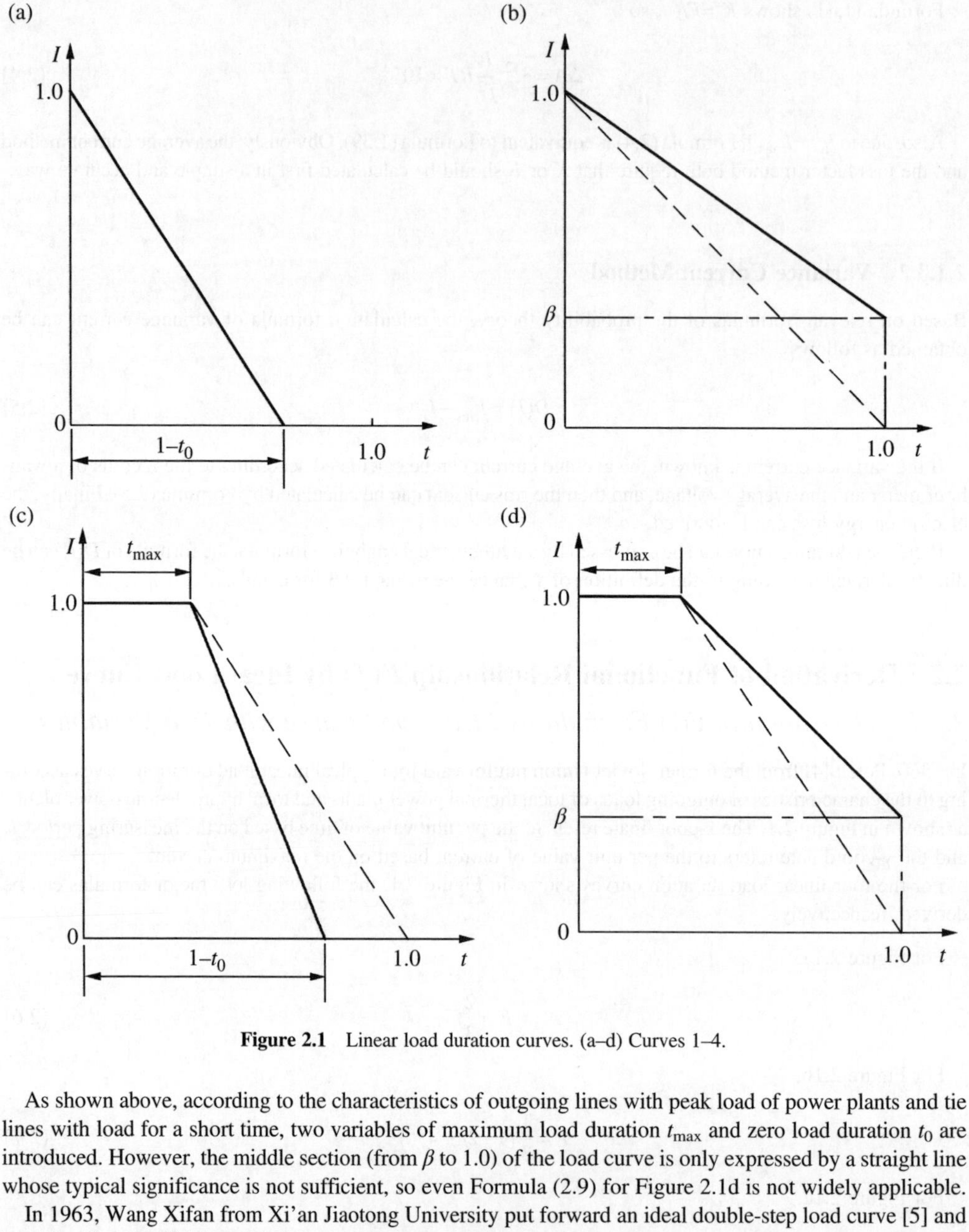

Figure 2.1 Linear load duration curves. (a–d) Curves 1–4.

As shown above, according to the characteristics of outgoing lines with peak load of power plants and tie lines with load for a short time, two variables of maximum load duration t_{max} and zero load duration t_0 are introduced. However, the middle section (from β to 1.0) of the load curve is only expressed by a straight line whose typical significance is not sufficient, so even Formula (2.9) for Figure 2.1d is not widely applicable.

In 1963, Wang Xifan from Xi'an Jiaotong University put forward an ideal double-step load curve [5] and introduced two variables of β and t_{max}, as shown in Figure 2.2.

According to Figure 2.2,

$$\begin{cases} f = t_{max} + \beta(1 - t_{max}) \\ F = t_{max} + \beta^2(1 - t_{max}) \end{cases}$$

The loss factor formula of the double-step load curve can be derived by eliminating β in the above two formulas, that is

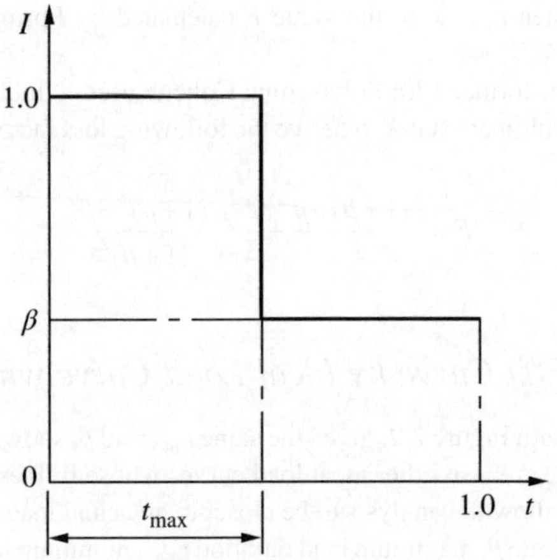

Figure 2.2 Double-step load curve.

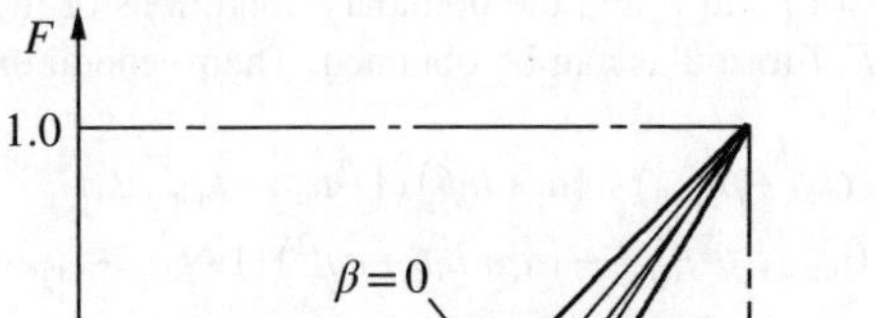

Figure 2.3 Bundle of straight lines for $F = \varphi(\beta, f)$.

$$F = \frac{t_{max} + 2t_{max} f + f^2}{1 - t_{max}} \tag{2.10}$$

If t_{max} is eliminated in the above two formulas, then

$$F = f - \beta(1 - f) \tag{2.11}$$

It is thus clear that different loss factor formulas can be derived for a certain ideal load curve. Take the different β values to Formula (2.11), and a bundle of straight lines can be derived for $F = \varphi(\beta, f)$, as shown in Figure 2.3. It is obvious that, because the loss factor formula contains the parameter β that reflects the characteristics of the load curve, the value range (half willow area) of (f, F) is divided regularly by the bundle of straight lines, thus significantly reducing possible errors resulted from the calculation of F by only the load rate f. It must be noted that the fullness of the middle section of an actual load curve is better than that of a double-step load curve, so when $t_{max} \neq 0$, the value F calculated by Formula (2.11) is smaller than that calculated by Formula (2.9).

In 1982, Xiao Yuankai from former Jilin Polytechnic College used a double-step ideal load curve and a trapezoidal ideal load curve as ultimate states to derive the following loss factor formula through analysis [6]:

$$F=\frac{f(1+\beta)-\beta}{2}+\frac{2}{3}f^2\frac{(1+\beta)^2-\beta}{(1+\beta)^2} \tag{2.12}$$

2.2.2 *Derivation of* F(f) *Curve by Ideal Load Curve with Four Variables*

A comparison of Figure 2.1d with Figure 2.2, given the same t_{max} and β, shows the linear load curve is "more full" than the step curve. There are also other ideal load curves whose fullness is either worse or better than that of a linear load curve. To allow the analysis to be closer to an actual load curve, four variables are introduced, namely, minimum load rate β, maximum load duration t_{max}, minimum load duration t_{min}, and zero load duration t_0. The following nine ideal load curves (see Figure 2.4) can be selected, and the types of their middle sections are: (1) step curve; (2) concave exponential curve; (3) concave quadratic parabola; (4) sine curve; (5) linear; (6) cosine curve; (7) convex quadratic parabola; (8) convex exponential curve; (9) rectangle.

According to the definitions of f and F and the boundary conditions of the middle sections of ideal load curves, nine combinations of F, f formulas can be obtained. Their general expressions are shown below:

$$f=(t_{max}+\beta t_{min})+(a_j+b_j\beta)(1-t_{max}-t_{min}-t_0)$$
$$F=(t_{max}+\beta^2 t_{min})+(a_i+b_i\beta+c_i\beta^2)(1-t_{max}-t_{min}-t_0)$$

Coefficients in the above F, f formulas are listed in Table 2.1.

According to conditions of actual load curves, take such values as 0, 0.1, 0.2, and 0.3 for t_{max}, t_{min} and t_0, and take values ranging from 0 to 1.0 for β. Nine f, F values are calculated for various combinations of variables by the nine F, f formulas, and connected into a $F(f)$ curve by the smooth curve method. If values are taken regularly for the four variables, then a family of $F(f)$ curves with such variables being layers can be derived. Figure 2.5 shows some of the curves.

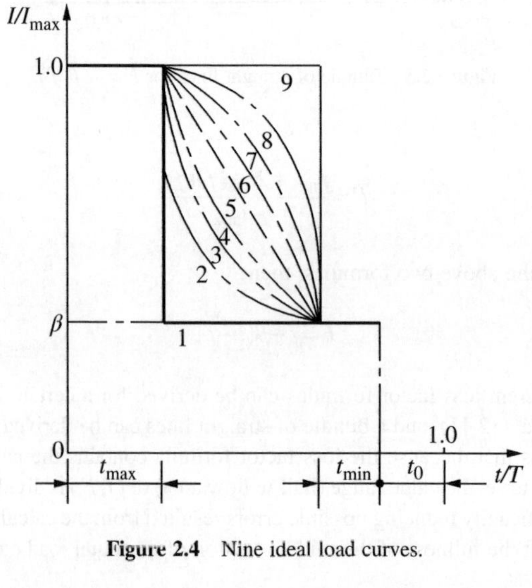

Figure 2.4 Nine ideal load curves.

Table 2.1 Coefficients in F, f formulas for nine ideal load curves with four variables.

Type of middle section of ideal load curve	1	2	3	4	5	6	7	8	9
a_j	0.000 0	0.222 2	0.333 3	0.363 4	0.500 0	0.636 6	0.666 7	0.777 8	1.000 0
b_j	1.000 0	0.777 8	0.666 7	0.636 6	0.500 0	0.363 4	0.333 3	0.222 2	0.000 0
a_i	0.000 0	0.111 1	0.200 0	0.226 8	0.333 3	0.500 0	0.533 3	0.666 7	1.000 0
b_i	0.000 0	0.222 2	0.266 7	0.273 2	0.333 3	0.273 2	0.266 7	0.222 2	0.000 0
c_i	1.000 0	0.666 7	0.533 3	0.500 0	0.333 3	0.226 8	0.200 0	0.111 1	0.000 0

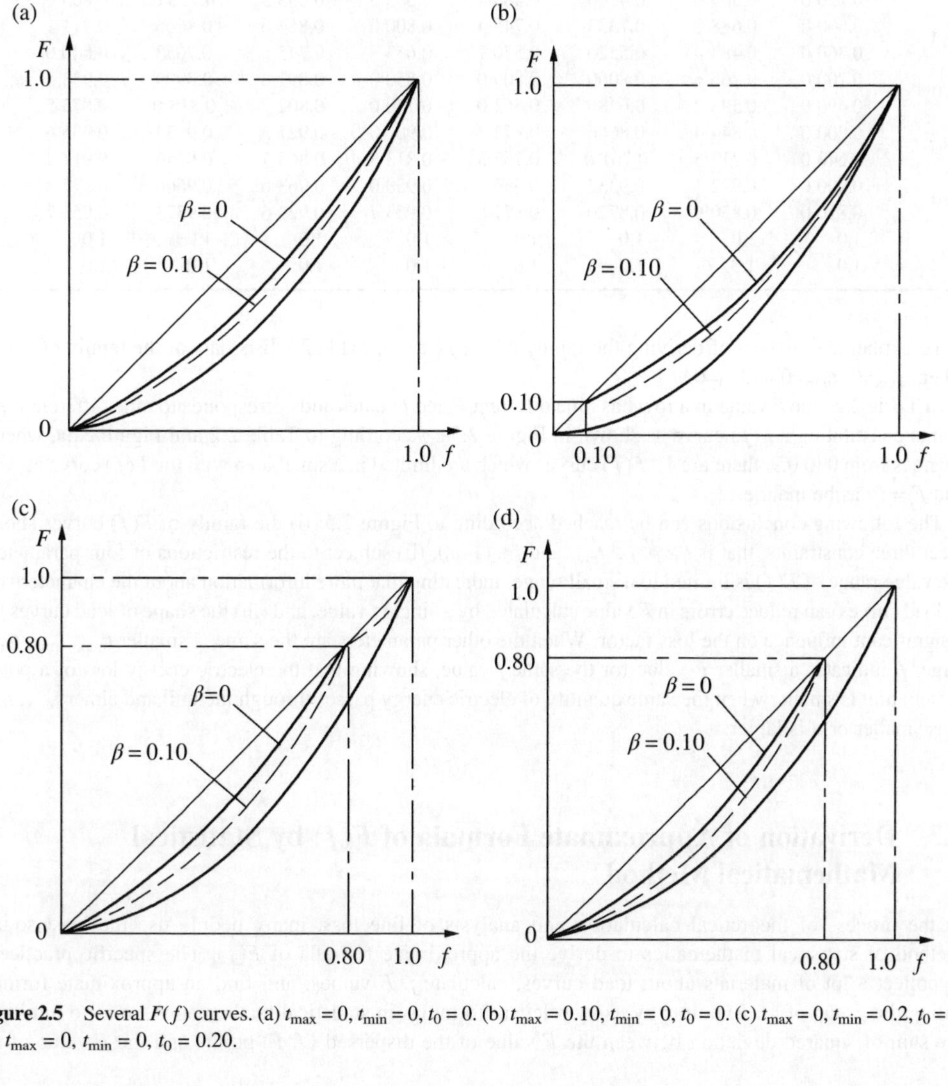

Figure 2.5 Several $F(f)$ curves. (a) $t_{max} = 0$, $t_{min} = 0$, $t_0 = 0$. (b) $t_{max} = 0.10$, $t_{min} = 0$, $t_0 = 0$. (c) $t_{max} = 0$, $t_{min} = 0.2$, $t_0 = 0$. (d) $t_{max} = 0$, $t_{min} = 0$, $t_0 = 0.20$.

Table 2.2 Table of *f*/*F* values for nine ideal load curves with four variables (t_{max} = 0, t_{min} = 0, t_0 = 0).

Type / β value	1	2	3	4	5	6	7	8	9
0	0.000 0	0.222 2	0.3333	0.363 4	0.500 0	0.636 6	0.6667	0.777 8	1.0
	0.000 0	0.111 1	0.2000	0.226 8	0.333 3	0.500 0	0.5333	0.666 7	1.0
0.10	0.100 0	0.300 0	0.4000	0.427 1	0.550 0	0.672 9	0.7000	0.800 0	1.0
	0.010 0	0.140 0	0.2320	0.259 1	0.370 0	0.529 6	0.5620	0.690 0	1.0
0.20	0.200 0	0.377 8	0.4667	0.490 8	0.600 0	0.709 3	0.7332	0.822 2	1.0
	0.040 0	0.182 2	0.2747	0.301 4	0.413 3	0.563 7	0.5946	0.715 6	1.0
0.30	0.300 0	0.455 5	0.5333	0.554 4	0.650 0	0.745 6	0.7666	0.844 5	1.0
	0.090 0	0.237 8	0.3280	0.353 8	0.463 3	0.602 4	0.631 3	0.743 4	1.0
0.40	0.400 0	0.533 3	0.6000	0.618 0	0.700 0	0.781 9	0.8000	0.866 7	1.0
	0.160 0	0.306 6	0.3920	0.415 1	0.519 9	0.645 6	0.671 0	0.773 4	1.0
0.50	0.500 0	0.611 1	0.6666	0.681 7	0.750 0	0.818 3	0.8333	0.888 9	1.0
	0.250 0	0.389 0	0.4666	0.488 4	0.583 3	0.693 3	0.716 6	0.805 6	1.0
0.60	0.600 0	0.688 8	0.7333	0.745 4	0.800 0	0.854 6	0.8666	0.911 1	1.0
	0.360 0	0.484 4	0.5520	0.570 7	0.653 3	0.745 6	0.7653	0.840 0	1.0
0.70	0.700 0	0.766 6	0.8000	0.809 0	0.850 0	0.891 0	0.8999	0.933 3	1.0
	0.490 0	0.593 2	0.6480	0.663 0	0.730 0	0.802 7	0.818 0	0.875 5	1.0
0.80	0.800 0	0.844 4	0.8666	0.872 7	0.900 0	0.927 3	0.9333	0.955 6	1.0
	0.640 0	0.715 5	0.761 0	0.765 3	0.813 3	0.863 7	0.8746	0.914 4	1.0
0.90	0.900 0	0.922 1	0.9333	0.936 3	0.950 0	0.963 6	0.9666	0.977 8	1.0
	0.810 0	0.850 9	0.8720	0.877 6	0.903 7	0.929 6	0.9373	0.956 7	1.0
1.0	1.0	1.0	1.0	1.0	1.0	1.0	1.0	1.0	1.0
	1.0	1.0	1.0	1.0	1.0	1.0	1.0	1.0	1.0

To explain the process of deriving the family of $F(f)$ curves, Table 2.2 lists data of the family of curves when t_{max}=0, t_{min}=0 and t_0=0.

In Table 2.2, one β value in a row has nine different *F* and *f* values and corresponds to nine different (f, F) points constituting a $F(f)$ curve as shown in Figure 2.5a. According to Table 2.2 and Figure 2.5a, when β changes from 0 to 0.9, there are 10 $F(f)$ curves which are limited in a small area with the $F(f)$ curve at $\beta = 0$ and $F = f^2$ as boundaries.

The following conclusions can be reached according to Figure 2.5: (i) the family of $F(f)$ curves should meet three constraints, that is $f \geq \beta$, $f \geq t_{max}$ and $f \leq (1-t_0)$; (ii) subject to the restrictions of four parameters, the value range of $F(f)$ is limited to a small range, indicating that more information about the characteristics of load curves can reduce errors in *F* value calculated by a single *f* value; and (iii) the shape of load curves has a significant influence on the loss factor. When the other parameters are the same, a smaller t_{max}, t_{min}, t_0, or larger β indicates a smaller *F* value for the same *f* value, showing that the electric energy loss of a power system unit is smaller when the same quantity of electric energy passes through the unit and either t_{max}, t_{min}, t_0 is smaller or β is larger.

2.3 Derivation of Approximate Formula of $F(f)$ by Statistical Mathematical Method

In the process of theoretical calculation and analysis of line loss, many people use the least square method of statistical mathematics to derive the approximate formula of $F(f)$. The specific practice is to collect a lot of materials about load curves, calculate *f*, *F* values, and find an approximate formula fitting the group of (f, F) points by using relevant formulas in statistical mathematics, so as to minimize the sum of squared deviation between the *F* value of the dispersed (f, F) points and the *F* value of the

(f, F) points in the $F(f)$ curve of the approximate formula. The key to this method is to select an appropriate mathematical model and to use data statistics to determine the coefficient in the approximate formula. Nevertheless, the approximate formula derived by this method is only applicable when the characteristics of the load curve to be calculated are close to those of the original load curve serving as the basis for statistics. Otherwise, errors in the line loss calculated by such an approximate formula may be too large to be accepted.

2.3.1 *Binomial Approximate Formula of* F(f)

In the early stages of studying the theoretical calculation of line loss, the following mathematical model was proposed by compromising between two extreme conditions of $F = f$ and $F = f^2$:

$$F = kf + (1-k)f^2 \tag{2.13}$$

Formula (2.13) is called the Hoebel Formula [35] in American literature.

In 1926, Jansen from Germany calculated many (f, F) points according to the load curves of the outgoing lines of many power plants and obtained $k = 0.5$ by means of a statistical mathematical method [7], so

$$F = \frac{1}{2}(f + f^2) \tag{2.14}$$

In 1928, Buller from the United States, according to the load curves of a distribution system, derived 18 (f, F) point sets and obtained $k = 0.3$ by statistical calculation [8], that is

$$F = 0.3f + 0.7f^2 \tag{2.15}$$

Around the 1950s, Formula (2.15) was used by electric power companies for line loss calculation in Shanghai, Tianjin, and other metropolises.

In the 1970s, Chinese line loss workers also derived similar formulas according to statistical information on loads and line losses in distribution networks. For example, Shenyang once put forward

$$F = 0.2f + 0.8f^2 \tag{2.16}$$

Shanghai proposed

$$F = 0.175f + 0.875f^2 \tag{2.17}$$

In the late 1980s, the United States Electric Power Research Institute (EPRI) brought up the following formulas in its Technical Assessment Guide (TAG):

$$F = f^{1.912} \tag{2.18}$$

And

$$F = 0.08f + 0.92f^2 \tag{2.19}$$

In North America, if the f range of 0.438–0.845 is used for calculation, the average error in the loss factor is less than 0.5%; while when the old Formula (2.15) with $k = 0.30$ is used, the error is up to 28%. The comparison indicates that more a dramatic change in the load leads to a smaller k in Formula (2.13).

2.3.2 *Trinomial Approximate Formula of* F(f)

Some other loss factor formulas usually include constant terms or higher-order terms of load factor f.

In 1920, Tröger from Germany put forward the following approximate formula [9]:

$$F = 0.12 - 0.24f + 1.12f^2 \tag{2.20}$$

According to the calculation of substituting $t_{max} = 0.107$ in Formula (2.10), Formula (2.20) can be derived. The approximate formula proposed by Wolf from Germany in 1931 and confirmed again in 1958 is [10]:

$$F = 0.083f + 1.036f^2 - 0.119f^3 \tag{2.21}$$

Wolf conducted statistical analysis of nearly 100 daily load curves, 25 monthly load curves and 25 annual load curves to derive Formula (2.21) which is generally accepted due to the huge number of statistical "samples". This formula has been so far widely used in many countries of Western Europe and Eastern Europe. This formula is also used in power transmission and distribution works in Taiwan, China. It should be noted that the f cubic term in this formula only allows the $F(f)$ curve of the approximate formula to approach the group of (f, F) points and has no definite physical significance.

In 1948, Kezevich from the former Soviet Union put forward the following formula [11]:

$$F = (0.124 + 0.876f)^2 \tag{2.22}$$

Both Formula (2.22) and Formula (2.20) have constant terms, which certainly does not meet the end value conditions of $f = 0$, $F = 0$. This indicates that the characteristic of load curves based on which Kezevich derived the formula is $t_{max} \neq 0$, that is $f \neq 0$, which limits the range of application of Formulas (2.21) and (2.22). Kezevich once clearly pointed out that Formula (2.22) was only applicable to the line loss calculation when the maximum load duration lasted for some time ($t_{max} \neq 0$).

2.3.3 *Approximate Formula of Family of* F(f) *Curves with Four Variables*

Some approximate formulas can be derived when the family of $F(f)$ curves with four variables mentioned in the last section are subject to mathematical treatment. To use the least square method to derive the approximate formula of dispersed point sets, the "moderation principle" in statistical mathematics is first applied to test whether the distribution of point sets is a normal distribution. Taking data listed in Table 2.2 for example, f, F values are calculated by β in the table with an interval of 0.02, so that 80 (f, F) points in the table will be increased to 400 (f, F) points. Forty F_i values at $f = 0.80$ in the table are listed and substituted into the following formula to calculate the standard deviation σ of normal distribution and the standard deviation σ' of the actual distribution.

$$\sigma = \sqrt{\sum_{i=1}^{n}(F_i - F_{av})^2/(n-1)} = 1.904 \times 10^{-2}$$

$$\sigma' = \sqrt{\frac{\pi}{2}}\,\frac{\sum_{i=1}^{n}|F_i - F_{av}|}{\sqrt{n(n-1)}} = 2.08 \times 10^{-2}$$

Because σ and σ' are very close, the distribution of (f, F) points within a small range limited by four variables complies with a normal distribution. Therefore, we can use the least square method to derive the approximate formula of $F(f)$.

Given the variable t_{max} in the formula, two end value conditions of $F(t_{max}) = t_{max}$ and $F(1.0) = 1.0$ should be satisfied (see Figure 2.5b), so the mathematical model of the approximate formula of the loss factor is:

$$F = k(f - t_{max}) + \frac{1-k}{1-t_{max}}(f - t_{max})^2 + t_{max}\,(f \geq t_{max}) \tag{2.23}$$

The following results can be obtained based on data similar to the (f, F) calculation table with four variables as listed in Table 2.2 and by using relevant formulas of the least square method: when $t_{min} = 0$ and $t_0 = 0$, k is calculated to be 0.2, 0.2, and 0.225, respectively if t_{max} is 0, 0.1, and 0.2. These values are respectively substituted into Formula (2.23), obtaining:

$$F = 0.2f + 0.8f^2,\ f \geq 0$$

$$F = 0.089 + 0.022f + 0.889f^2,\ f \geq 0.10 \tag{2.24}$$

$$F = 0.194 + 0.163f + 0.969f^2,\ f \geq 0.20 \tag{2.25}$$

As mentioned earlier, Formula (2.16) was recommended for line loss calculation of distribution networks in some Chinese metropolises, which could be attributed to the situation where the maximum load duration was very short ($t_{max} \approx 0$) for distribution lines with comprehensive loads in such cities. This shows that Formula (2.16) is applicable to the calculation of a loss factor for an instantaneous peak ($t_{max} = 0$) load curve in a statistical sense.

According to Formulas (2.24) and (2.25), constant terms in $F(f)$ formulas are as a result of $t_{max} \neq 0$ for the load curves on which statistics is based and are calculated by Formula (2.23) after coefficient k is determined. The use condition of the above two formulas is $f \geq t_{max}$. Such two aspects theoretically explain that $F(f)$ approximate formulas with constant terms are only applicable to a line loss calculation under certain conditions.

2.4 Derivation of $F(f)$ Formula by Mathematical Analysis Method

2.4.1 Direct Integration Method

Raymond from the United States put forward the following analytical formula of load duration curves in 1980 [12]:

$$I_* = \beta + 1.128\,3(f - \beta)\sqrt{\ln\frac{1}{t}} \tag{2.26}$$

According to Formula (2.26)[1], $F = \int_0^1 I_*^2(t)\mathrm{d}t/T$ is used for integral calculation, and the Raymond loss factor formula is obtained as follows:

$$F = f^2 + 0.273(f - \beta)^2 \tag{2.27}$$

Raymond pointed out that Formula (2.27) is applicable when $f \leq 0.80$. When $f > 0.80$, $F = f^2$ is used to calculate the loss factor.

[1] The coefficient of the second term in the original text is $\frac{4}{\pi}$, and according to the basic requirements of $\int_0^1 I_*\,\mathrm{d}t = f$, the coefficient is revised to 1.128 3.

As Formula (2.27) is simple in its form and includes two parameters of f and β, it is expected to obtain accurate calculation results. Therefore, it was soon cited by some works [13] upon its release.

2.4.2 *Subsection Integration Method*

In 1981, Liu Yingkuan of the former Xining Administration of Power Supply used four sections of broken lines formed by double moving points to express a family of load curves, and applied the subsection integration method to derive the loss factor formula [14], that is

$$\begin{aligned} F &= \frac{23}{36}f^2 + \frac{13}{36}(f + f\beta - \beta) \\ &= 0.639f^2 + 0.361(f + f\beta - \beta) \end{aligned} \tag{2.28}$$

See Appendix C for a detailed derivation.

3

Probability Theory Analysis of Current Load Curve

This chapter uses probability theory as a mathematical tool to analyze a current load curve and its parameters, derive a loss factor formula, and compare various loss factor formulas, thereby dividing an active load duration curve into three mode sections.

3.1 Probability Meanings of Load Curve and Its Parameters

3.1.1 Probability Meaning of Load Duration Curve

The load of a public line at any time is a random variable which cannot be predetermined. Within a measuring period, a sequential set of such random variables constitutes a load curve. When the load curve is changed to a load duration curve, a different load current can be expressed by probabilities.

In Figure 3.1, by means of uneven segmentation, the sum of rectangle areas is used to approximate the area under the load duration curve, and each different current $I_1, I_2, \ldots, I_n$ matches a different time $t_1, t_2, \ldots, t_n$. If the duration Δt_k is taken as the probability value, then the probability value of I_1 is t_1, so

$$P(I=I_1)=(t_1-0)=\Delta t_1,$$

The probabilities of other current values can be obtained in the same way:

$$P(I=I_2)=(t_2-t_1)=\Delta t_2,$$
$$P(I=I_3)=(t_3-t_2)=\Delta t_3\ldots$$

Obviously, $\sum_{k=1}^{n}\Delta t_k=\Delta t_1+\Delta t_2+\cdots+\Delta t_n=1$. In other words, within the current range of I_1~I_n, the sum of probabilities of various current values is 1, meeting the constraint conditions of random variable probability. This also indicates that the load duration curve expressed by per unit value reflects the probability law of current random variables.

Line Loss Analysis and Calculation of Electric Power Systems, First Edition. Anguan Wu and Baoshan Ni.

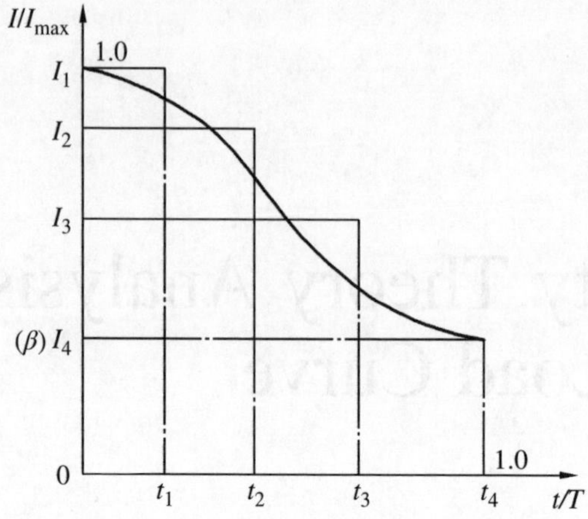

Figure 3.1 Approximation of load duration curve.

3.1.2 Probability Meanings of Minimum Load Rate and Load Rate

3.1.2.1 Probability Meaning of Minimum Load Rate

As shown in Figure 3.1, I_4 is the minimum current, and if expressed by per unit value, is the minimum load rate β. The probability that current is greater than or equal to I_4 is 1, representing a certain event. This is the probability meaning of the minimum load rate.

3.1.2.2 Explanation of Load Rate by Probability Theory

According to the area approximation shown in Figure 3.1,

$$I_1 t_1 + I_2(t_2 - t_1) + I_3(t_3 - t_2) + \cdots + I_n(t_n - t_{n-1}) = \int_0^1 I(t)\mathrm{d}t$$

As

$$\int_0^1 I(t)\mathrm{d}t = I_{\mathrm{av}} = f$$

then

$$\sum_{k=1}^{n} I_k \Delta t_k = I_{\mathrm{av}} = f$$

Because of $\sum_{k=1}^{n} \Delta t_k = 1$, allowing $\Delta t_k / \sum_{k=1}^{n} \Delta t_k = p_k$(called time probability), the mathematical expectation $E(I)$ is obtained as:

$$E(I) = I_{\mathrm{av}} = f = \sum_{k=1}^{n} I_k p_k \tag{3.1}$$

Formula (3.1) shows that the mathematical expectation of current random variables is the average current within the measuring period, that is the load rate when the current is expressed per unit value. According to the probability theory, the mathematical expectation is a weighted average with probability as the weight, so the load factor of the current load curve is a weighted average of current random variables with occurrence time probability as the weight.

3.1.3 Barth Formula of Loss Factor

The probability theory proves that variance is equal to the difference between the mathematical expectation of random variable square and the square of mathematical expectation of random variables, that is $D(I) = E(I^2) - [E(I)]^2$, so the formula of mathematical expectation of current square is obtained as follows:

$$E(I^2) = D(I) - [E(I)]^2 \tag{3.2}$$

$$D(I) = E\left\{[I - E(I)]^2\right\} \tag{3.3}$$

Wherein $D(I)$ – variance of current random variables.

In Formula (3.3), $I - E(I)$ is called deviation, so the variance is defined as the mathematical expectation of the square of random variable deviation. The square root of variance is called the mean square error, expressed as $\sigma(I)$.

According to the definition of mathematical expectation and the definition of rms current,

$$E(I^2) = \sum_{k=1}^{n} I_k^2 \Delta t_k \sum_{k=1}^{n} \Delta t_k = I_{\text{rms}}^2$$

Due to $E(I) = I_{\text{av}}$ in Formula (3.1), so Formula (3.2) can be rewritten to:

$$I_{\text{rms}}^2 = D(I) + I_{\text{av}}^2 \tag{3.4}$$

Because loss factors are $F = I^2_{\text{rms}}/I^2_{\max}$, $f = I_{\text{av}}/I_{\max}$, $\sigma^2(I) = D(I)$, divide $I^2_{\max}$ at both sides of Formula (3.4), and allow $D_*(I) = D(I)/I^2_{\max}$, $\sigma^2{}_*(I) = \sigma^2(I)/I^2_{\max}$, and then

$$F = D_*(I) + f^2 = \sigma_*^2(I) + f^2$$

Or

$$\left(\sqrt{F}\right)^2 = \sigma_*^2(I) + f^2 \tag{3.5}$$

According to Formula (3.5), the square root of loss factor, the mean square error of current, and the load factor comply with Pythagoras' Theorem. This was firstly put forward by the Norwegian scholar Barth in 1965 [15], so Formula (3.5) is called the Barth Formula.

3.2 Analysis of Rossander Formula as Distribution Function

3.2.1 Rossander Formula of Load Duration Curve

The German scholar Rossander put forward an exponential approximate formula in the early twentieth century, and this formula has been widely used by countries around the world, known as the general formula, that is

$$I_{*1}(t) = 1-(1-\beta)\ (t/T)^{\lambda} \tag{3.6}$$

$$I_{*2}(t) = \beta+(1-\beta)\left(1-\frac{t}{T}\right)^{\frac{1}{\lambda}} \tag{3.7}$$

$$\frac{1}{\lambda} = \frac{1-f}{f-\beta} = \frac{I_{\max}-I_{\text{av}}}{I_{\text{av}}-I_{\min}} = \frac{\text{peak height}}{\text{valley depth}} \tag{3.8}$$

Wherein $I_*(t)$ – per unit value of current based on $I_{\max}$;
T – measuring period; t/T is the per unit value of time;
$\frac{1}{\lambda}$ – load change index.

If Formula (3.6) is integrated with the measuring period as boundary conditions, then

$$\int_0^1 \left[1-(1-\beta)\ (t/T)^{\lambda}\right] \mathrm{d}t = 1-\frac{1-\beta}{\lambda+1} = \frac{\lambda+\beta}{\lambda+1} = \frac{f(1-\beta)}{(1-\beta)} = f$$

The above calculation shows that when the maximum current and the minimum current are definite, the Rossander Formula is an equivalent formula in which the area under the load duration curve remains the same. Thanks to this nature, the Rossander Formula has been applied in many problems such as the theoretical calculation of line loss and the calculation of reactive compensation benefits.

3.2.2 *Exponential Distribution Function*

Formulas (3.6) and (3.7) are both decreasing functions and do not meet the requirement that a distribution function should be a non-decreasing function, so both of them should be changed to the following form of increasing function:

$$I_{*1}(t) = 1-(1-\beta)\left(1-\frac{t}{T}\right)^{\lambda} \tag{3.9}$$

$$I_{*2}(t) = \beta+(1-\beta)\left(\frac{t}{T}\right)^{\frac{1}{\lambda}} \tag{3.10}$$

According to Formulas (3.9) and (3.10), $I(0) = \beta$ and $I(1) = 1$, which does not meet the requirement that the value range of a distribution function should be [0, 1], so the inverse function of the current per unit value function can be defined as the distribution function, that is

$$T_{*1} = 1-\left(\frac{1-i}{1-\beta}\right)^{\frac{1}{\lambda}} = F_1(i) \tag{3.11}$$

$$T_{*2} = \left(\frac{i-\beta}{1-\beta}\right)^{\lambda} = F_2(i) \tag{3.12}$$

Obviously, in Formulas (3.11) and (3.12), $T(\beta) = 0$ and $T(1) = 1$, which meets the requirement of a distribution function. Then, it can be concluded that the inverse function (i.e. time function) of a function expressed by the Rossander Formula of an increasing load duration curve can be defined as the distribution

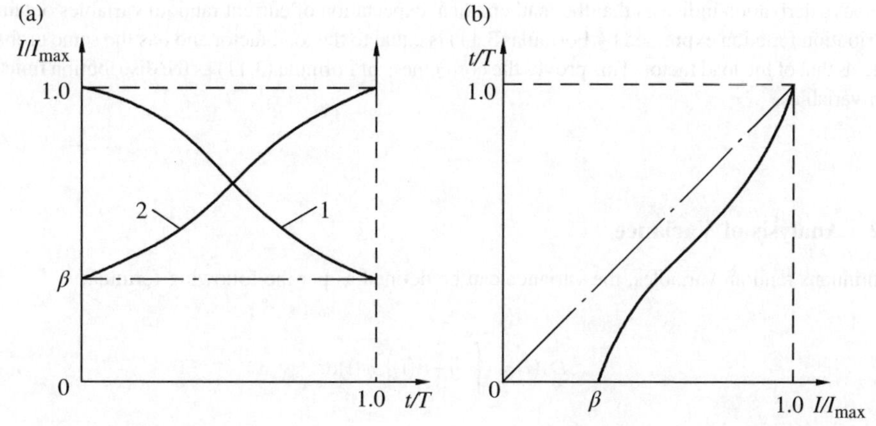

Figure 3.2 Current function and distribution function: (a) current function (1 – decreasing; 2 – increasing); (b) distribution function.

function in the probability theory. The decreasing and increasing current functions and the time function are shown in Figure 3.2.

3.2.3 Derivation of Loss Factor Formula

According to the Barth Formula (3.5), the formula of variance $D(I)$ is derived, and then the loss factor formula can be derived.

3.2.3.1 Analysis of Mathematical Expectation

The derived function of a distribution function of continuous random variables is called distribution density function, expressed by $\varphi(i)$. The mathematical expectation of continuous random variables is defined as an improper integral of the product of random variable i and distribution density function within an infinite interval (–∞, +∞), that is

$$E(i)=\int_{-\infty}^{+\infty} i\varphi(i)\mathrm{d}i=\int_{\beta}^{1} i\mathrm{d}F(i)$$

Wherein $F(i)$ – distribution function of continuous random variable i.

If the distribution function expressed by Formula (3.11) is substituted into it, then

$$E(i)=\int_{\beta}^{1} i\mathrm{d}\left[1-\left(\frac{1-i}{1-\beta}\right)^{\frac{1}{\lambda}}\right]=i\left[1-\left(\frac{1-i}{1-\beta}\right)^{\frac{1}{\lambda}}\right]\Bigg|_{\beta}^{1}-\int_{\beta}^{1}\left[1-\left(\frac{1-i}{1-\beta}\right)^{\frac{1}{\lambda}}\right]\mathrm{d}i$$

$$=1-\int_{\beta}^{1}\mathrm{d}i+\int_{\beta}^{1}\left(\frac{1-i}{1-\beta}\right)^{\frac{1}{\lambda}}\mathrm{d}i=1-(1-\beta)+\frac{\lambda(1-\beta)}{1+\lambda}$$

$$=f$$

The above derivation indicates that the mathematical expectation of current random variables obtained by the distribution function expressed by Formula (3.11) is equal to the load factor and has the same probability meaning as that of the load factor. This proves the correctness of Formula (3.11) as the distribution function of random variables.

3.2.3.2 Analysis of Variance

For continuous random variables, the variance can be defined as per the following formula:

$$D_*(i) = \int_{-\infty}^{+\infty} [i - E(i)]^2 \varphi(i) \mathrm{d}i$$

Substitute the mathematical expectation $E(i) = f$ into the above formula to obtain

$$D_*(i) = \int_{-\infty}^{+\infty} (i-f)^2 \varphi(i) \mathrm{d}i = \int_{\beta}^{1} (i-f)^2 \mathrm{d}F(i)$$

Now, substitute the distribution function expressed by Formula (3.11) into the above formula to obtain

$$\begin{aligned}
D_*(i) &= \int_{\beta}^{1} (i-f)^2 \mathrm{d}\left[1 - \left(\frac{1-i}{1-\beta}\right)^{\frac{1}{\lambda}}\right] \\
&= (i-f)^2 \left[1 - \left(\frac{1-i}{1-\beta}\right)^{\frac{1}{\lambda}}\right] \Bigg|_{\beta}^{1} - \int_{\beta}^{1} \left[1 - \left(\frac{1-i}{1-\beta}\right)^{\frac{1}{\lambda}}\right] \mathrm{d}(i-f)^2 \\
&= (1-f)^2 - \int_{\beta}^{1} \mathrm{d}(i-f)^2 + \int_{\beta}^{1} \left(\frac{1-i}{1-\beta}\right)^{\frac{1}{\lambda}} \times 2(i-f)\mathrm{d}i \\
&= (1-f)^2 - \left[(1-f)^2 - (\beta-f)^2\right] - 2(f-\beta)^2 - \frac{2(f-\beta)(1-\beta)}{\frac{1}{\lambda}+2} \left(\frac{1-i}{1-\beta}\right)^{\frac{1}{\lambda}+2} \Bigg|_{\beta}^{1} \\
&= -(f-\beta)^2 + \frac{2(f-\beta)^2(1-\beta)}{1+f-2\beta} \\
&= \frac{(f-\beta)^2(1-f)}{1+f-2\beta}
\end{aligned}$$

The above formula indicates that the variance of current has something to do with the load factor f and the minimum load rate β. If β is closer to f and f is closer to 1, then the variance is smaller. This coincides with the situation where the variance reflects the degree of deviation of random variables from the average.

3.2.3.3 Rossander Loss Factor Formula

According to the Barth Formula (3.5), with respect to the Type 1 Rossander Formula of the load duration curve, the following loss factor formula can be obtained:

$$F_1 = D_*(I) + f^2 = \frac{(f-\beta)^2(1-f)}{1+f-2\beta} + f^2$$
$$= f - \frac{(1-f)\left(f-\beta^2\right)}{1+f-2\beta}$$

In the same way, with respect to the Type 2 Rossander Formula of the load duration curve, the distribution function, variance, and loss factor formula can be obtained, as shown in Table 3.1.

Formulas listed in the fifth column of Table 3.1 are called Rossander loss factor formulas and should be selected based on the load change index $\frac{1}{\lambda}$.

$$F_1 = f - \frac{\left(f-\beta^2\right)(1-f)}{1+f-2\beta} \quad (3.13)$$

$$F_2 = f - \frac{(1-f)(f+\beta-2f\beta)}{2-f-\beta} \quad (3.14)$$

When $\frac{1}{\lambda} \leq 1$, that is $f \geq \frac{1+\beta}{2}$, where the peak height is smaller than the valley depth and the load curve changes slowly, then F_1 in Formula (3.13) is used to calculate the loss factor. When $\frac{1}{\lambda} \geq 1$, that is $f \leq \frac{1+\beta}{2}$, where the peak height is greater than the valley depth and the load curve changes dramatically, then F_2 in Formula (3.14) is used to calculate the loss factor. When $\lambda = 1$, that is $f = \frac{1+\beta}{2}$, where the peak height is equal to the valley depth, then

$$F_2 = F_1 = \frac{1}{3}\left(1-2f+4f^2\right)$$

The above formula is Formula (2.7) in Section 2.2 of Chapter 2, and its load duration curve is an oblique line shown in Figure 2.1b.

To further understand the relationship between $F(f, \beta)$ curves expressed by F_1 and F_2 formulas, take $\beta = 0$ and $\beta = 0.1$ to respectively derive $F_1(f)$ and $F_2(f)$ curves, as shown in Figure 3.3. This shows that $F_1(f)$ is located where the f value is larger; in contrast, $F_2(f)$ is located where the f value is smaller.

Table 3.1 Derivation process of two types of Rossander loss factor formula.

Type	Increasing load duration curve formula $I_*(t)$	Distribution function $T(i)$	Variance $D_*(I)$	Loss factor formula F
Type 1	$1-(1-\beta)\left(1-\frac{t}{T}\right)^{\lambda}$	$1-\left(\frac{1-i}{1-\beta}\right)^{\frac{1}{\lambda}}$	$\frac{(f-\beta)^2(1-f)}{1+f-2\beta}$	$f-\frac{\left(f-\beta^2\right)(1-f)}{1+f-2\beta}$
Type 2	$\beta+(1-\beta)\left(\frac{t}{T}\right)^{\frac{1}{\lambda}}$	$\left(\frac{i-\beta}{1-\beta}\right)^{\lambda}$	$\frac{(f-\beta)(1-f)^2}{2-f-\beta}$	$f-\frac{(1-f)(f+\phi-2f\beta)}{2-f-\beta}$

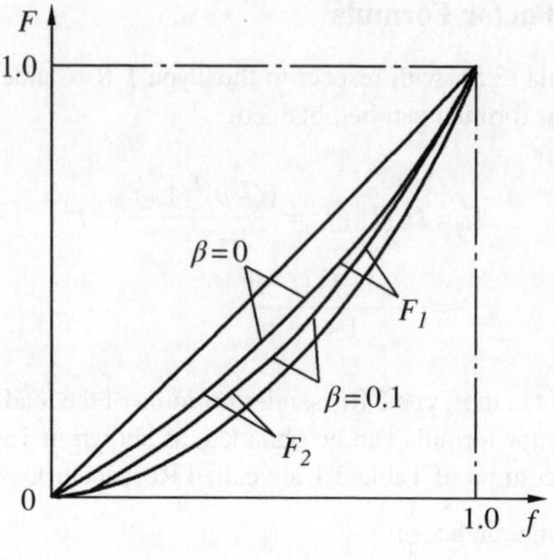

Figure 3.3 Rossander loss factor curves with two parameters.

3.2.4 Comparison of Direct Integration Method and Distribution Function Analysis Method

The Rossander load duration curve formula expressed by Formula (3.6) can be also used to derive the following loss factor formula by means of direct integration of function square:

$$
\begin{aligned}
F_1 &= \int_0^T I_*^2(t)\mathrm{d}t/T = \int_0^T \left[1-(1-\beta)\left(\frac{t}{T}\right)^{\lambda}\right]^2 \mathrm{d}t/T \\
&= 1-\frac{2(1-\beta)}{(\lambda+1)}+\frac{(1-\beta)^2}{2\lambda+1} \\
&= 2f-1+\frac{(1-\beta)^2(1-f)}{1+f-2\beta} \\
&= f-\frac{(f-\beta^2)\,(1-f)}{1+f-2\beta}
\end{aligned}
$$

The above formula is exactly the same as the loss factor formula derived by variance calculation of distribution function.

3.3 Comparison of Various Loss Factor Formulas

According to studies of probability theory, various moments (mathematical expectation is first moment, and variance is second-order central moment) of general random variables cannot uniquely determine the distribution function. Therefore, neither load factor nor loss factor alone can uniquely determine the load duration curve, or correctly calculate the electric energy loss of a certain load duration curve. Hence it can be concluded that any loss factor formula used to individually calculate the line loss within a certain period cannot be absolutely correct. Any loss factor formula should be assessed for not only rational derivation, but also the value and symbol of any error resulting from its application over a wide range, so it is necessary to use simulated or actual load curves for calculating over a wide range and comparing various loss factor formulas.

3.3.1 Loss Factor Formula Comparison Procedures Prepared by Monte Carlo Method

The Monte Carlo method, also known as random test method or statistical test method, is a very important method in modern numerical calculation and uses random numbers to address some mathematical problems regarding certainty.

Use a microcomputer to generate random numbers within an internal of (0, 1) to form a large number of random daily load curves and corresponding square curves. Calculate the true values of their load factors and loss factors as the basis for comparing the calculation results of various loss factor formulas.

Calculate the minimum load rate β according to the maximum value of the random daily load curves, and calculate the load factor f according to the cumulative results within 24 h. Substitute the two parameters f and β of random daily load curves into each loss factor formula to obtain various loss factor values. Compare these loss factor values with the true values, and take the sum of squares of errors as the assessment criteria for comparing various loss factor formulas.

To do the comparison within a wide range, take the maximum load duration t_{max} as 0, 2, 3, and 4 h; take the minimum load fixed part e as 0, 0.2, 0.4, 0.6, and 0.8. For each case, take $n = 100$, that is 100 random daily load curves, for the above calculation and comparison. For the formation of random daily load curves, as well as the meanings and values of the two variables t_{max} and e, see Figure 3.4. Computer programs can be written based on the above ideas.

3.3.2 Comparison Results of Various Loss Factor Formulas

Six loss factor formulas listed in Table 3.2 are selected according to their origins and their degree of wide application and calculated for comparison by computer programs. There are 20 groups and a total of 2000 random daily load curves. See Table 3.3 for the calculation and comparison results.

According to the variation range of average error and the total average error, the fifth and sixth formulas are both ideal loss factor formulas.

According to the analyses collected by the authors regarding load curves of multiple voltage public lines and customer dedicated lines, the frequency of occurrence of 20 groups of random daily load curves is determined and its weight is calculated. Then, the average error and weight of each group are used to calculate the total weighted average error. The results show that the total weighted average errors of the second, fifth, and sixth formulas are all below 5%, and the fourth formula has the maximum total weighted average error,

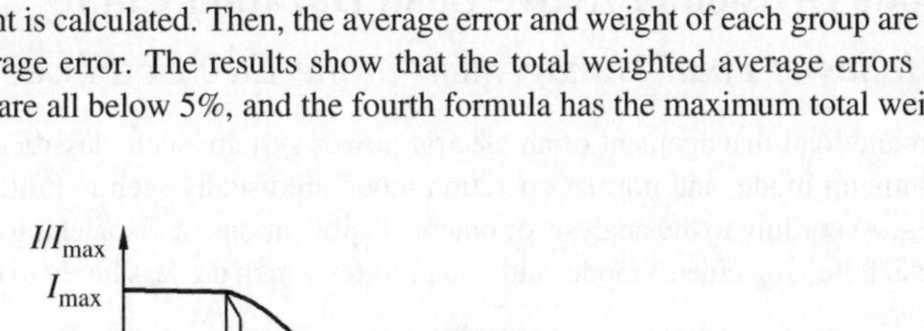

Figure 3.4 Random daily load curve (t_{max} = 0, 2, 3, 4h; e = 0, 0.2, 0.4, 0.6, 0.8). $RND(t)$ = random number in [t_{max}, 24].

Table 3.2 Compared loss factor formulas.

No.	Loss factor formula	Source
1	$F=\frac{1}{2}[f(1+\beta)-\beta]-\frac{2}{3}f^2\left[(1+\beta)^2-\beta\right]/(1+\beta)^2$	Formula (2.12)
2	$F=0.2f+0.8f^2$	Formula (2.16)
3	$F=0.083f+1.036f^2-0.119f^3$	Formula (2.21)
4	$F=f^2+0.273\ (f-\beta)^2$	Formula (2.27)
5	$F=0.639f^2+0.361\ (f+f\beta-\beta)$	Formula (2.28)
6	$F=f-\left(f-\beta^2\right)\ (1-f)/(1+f-2\beta)\ \ f\geq\frac{1}{2}(1+\beta)$	Formula (3.13)
	$F=f-(1-f)\ (f+\beta-2f\beta)/(2-f-\beta)\ \ f\leq\frac{1}{2}(1+\beta)$	Formula (3.14)

Table 3.3 Comparison of average error of loss factors (%).

Formula No.	1	2	3
Range of average error of groups	1.58–19.37	0.73–12.59	0.62–14.98
Total average error of 20 groups	6.72	4.18	5.21
Formula No.	4	5	6
Range of average error of groups	3.84–8.24	2.57–6.17	2.91–5.80
Total average error of 20 groups	5.61	4.72	4.08

that is 5.8%. According to the weighted average errors, the six loss factor formulas are all suitable for engineering calculations.

The fifth formula, that is Formula (2.28), is recommended due to its rational derivation and demonstration, simple expression form, easy use, and smaller average error resulting from its application in a wide range.

3.4 Three-Mode Division of Active Load Duration Curve

3.4.1 Three Modes of Load Management in the Electric Power System

During the operation and load management of an electric power system, such classification expressions as maximum mode, minimum mode, and normal operation mode are usually seen in protection setting calculation and dispatching. According to the analysis of practical applications of dispatching and operation, there are maintenance mode, failure (accident) mode, and other modes which are less likely to occur, in addition to the above three modes.

From the perspective of load management, the maximum mode means that, when the maximum load occurs within a dispatching area, the system offers the maximum feasible capacity, and maximum load flow occurs in the system. Reactive compensation equipment is put into operation to meet the voltage level demands. At this time, the short-circuit current level of buses in substations is high, and the power (energy) losses are high in the whole network. The minimum mode generally means that the total loads reach their lowest within a dispatching area, or the system connection is not tight. Meanwhile, the system impedance is large, and the short-circuit current level of buses in substations is low. The transmission power at transmission sections is low, and the power (energy) losses are low in the whole network. The voltage level at local areas may exceed the upper limit. The normal operation mode means an arrangement of load level and generator capacity, which is most likely to occur in the system.

To allow a wide range of calculation results of losses in high-voltage power grids to predict values under different operation modes (maximum, minimum, normal, maintenance, etc.) rather than just being general check and statistics results, it is necessary to study the relations between the three modes and to find a way of division based on the probability interpretation, so as to discuss a calculation method of losses in high-voltage power grids based on the power loss rate under the three modes.

3.4.2 *Differences and Relations of the Three Operation Modes*

The time of occurrence of the maximum and minimum operation modes in the system is occasional, but the rough period of occurrence can be estimated. Within a longer period (a year or a quarter), the period during which the maximum load occurs every day may change slowly.

For an annual active load duration curve, loads at the peak section are sorted by the load values at the maximum operation mode during different calendar periods of the year; loads at the valley section are sorted by load values at the minimum operation mode during different calendar periods of the year; and between these two sections are loads which are sorted by the load values at normal operation mode which last the longest and are most likely to occur.

Because the maximum and minimum loads change slowly day by day at the maximum and minimum operation modes, and peak loads in winter only represent normal load levels in summer in regions with high air conditioning loads in summer, the power differences between the minimum mode and the normal mode and between the normal mode and the maximum mode are vague. However, with respect to the continuity of actual load changes, within a long-term period (a year or a quarter), there should be a dividing point in the probability sense between the three modes. Such a division may have a certain reference value for the time related to load values at different modes, the estimate of voltage quality at three modes, the prediction of line losses in the whole network, and even the price ratio and time division of peak–valley price. Therefore, to find the probability division of the three operation modes from the perspective of studying system load changes has a certain significance.

3.4.3 *Probability Division of Three Operation Modes*

3.4.3.1 Introduction to PERT

Program Evaluation Review Technique (PERT) is a scientific management method first used by the United States Navy in the development of Polaris weapon systems in the 1960s, and which now has become an important management technique adopted in large and complicated engineering projects. PERT has the following key points:

1. All single tasks required to complete a program must be visualized in a very clear form and included in a network chart composed of "events" and "activities".
2. The sequence of events and activities in the network chart must conform to a set of strict logic rules which are followed to find the crucial critical path and the near-critical path.
3. The three-point estimation technique is used to estimate the time required by each activity in the network chart. In other words, a person who is very familiar with such activity estimates the optimistic time a, most likely time m, and pessimistic time b, and the three values are then simplified to the expected time t_e and statistical variance.
4. The time and time difference of the critical path are calculated.

3.4.3.2 *Expected Time* t_e *and Probability Interpretation*

The expected time t_e and the standard deviation are calculated by the following two formulas, respectively:

$$t_e = \frac{1}{6}(a + 4m + b) \tag{3.15}$$

$$\sigma = \frac{1}{6}(b-a) \tag{3.16}$$

According to the explanation of Hua Luogeng, a famous deceased Chinese mathematician, in respect of the calculation of the expected time [36], *a* is the optimistic time estimate and *m* is the most likely time estimate. Assuming the probability of *m* is two times that of *a*, the average of *a* and *m* is $(a + 2m)/3$; likewise, the average of *m* and *b* is $(2m + b)/3$. The distribution of two such averages that respectively occur with a probability of $\frac{1}{2}$ can approximate the actual distribution, so the mean value of two such averages is the expected time t_e, which can be expressed by the following formula:

$$t_e = \frac{1}{2}\left(\frac{a+2m}{3} + \frac{2m+b}{3}\right) = \frac{1}{6}(a+4m+b)$$

Formulas (3.15) and (3.16) can be used to obtain $t_e - \sigma = \frac{1}{3}(a+2m)$ and $t_e + \sigma = \frac{1}{3}(2m+b)$, indicating that the variation range of the expected time is the average of the extreme estimated time (*a* or *b*) and the most likely time *m* of multiplying probability.

3.4.3.3 Three-mode Division Under Multiplying Probability Assumption

According to the calculation of the expected time t_e by means of PERT, if two extreme conditions and the most likely condition are known for a probability distribution, then the expected value of such a distribution can be obtained under the conditions of a multiplying probability assumption. A load duration curve based on the maximum load over a long period has two parameters: minimum load rate β and load factor *f*. In a curve composed using these two parameters, to identify three sections of minimum mode, normal mode, and maximum mode, it is necessary to find out the power point of division $P_{x.zh}$ between the minimum mode and the normal mode as well as the power point of division $P_{zh.d}$ between the normal mode and the maximum mode (subscripts x, zh, and d refer to minimum, normal, and maximum modes, respectively). Relevant analysis shows that two power points of division $P_{x.zh}$ and $P_{zh.d}$ cannot be determined by the load duration curve alone and the multiplying probability assumption.

Through calculation and thinking, the authors have put forward a method of combined analysis of the load duration curve and the load duration square curve of relevant electric energy losses. As shown in Formula (1.36), the area under the load duration curve based on the maximum load is the load factor *f*; as shown in Formula (1.37), the area under the load duration square curve based on the maximum load square is the loss factor *F*. Within the same per unit value coordinate system, these two curves have a mapping relation. If the assumption of multiplying probability can be used in the load duration square curve reflecting losses to realize the division of three sections, then the power points of division $P_{x.zh}$ and $P_{zh.d}$ for the three mode sections can be found in the load duration curve.

To realize the division of three mode sections, the electric energy loss curve is taken as the carrier of the multiplying probability assumption. As shown in Figure 3.5, in the increasing loss curve, ΔA_1, ΔA_2 and ΔA_3 refer to electric energy losses of sections for the minimum, normal, and maximum modes, respectively, and the time points of division are taken as $t_{x.zh}$ and $t_{zh.d}$. When the long time is *T*, then the per unit values are $t_{x.zh}/T$ and $t_{zh.d}/T$.

Take the area under the $P^2(t)$ curve for the $(0 - t_{x.zh})$ section as b_P, the area under the $P^2(t)$ curve for the $(t_{x.zh} - t_{zh.d})$ section $4m_P$, and the area under the $P^2(t)$ curve for the $(t_{zh.d} - 1.0)$ section a_P. The following three formulas can be obtained with the electric energy loss ratio as the probability:

$$b_P/F = \int_0^{t_{x.zh}} P^2(t)\mathrm{d}t \Big/ \int_0^{1.0} P^2(t)\mathrm{d}t$$

$$a_P/F = \int_{t_{zh.d}}^{1.0} P^2(t)\mathrm{d}t \Big/ \int_0^{1.0} P^2(t)\mathrm{d}t$$

$$4m_P/F = \int_{t_{x.zh}}^{t_{zh.d}} P^2(t)\mathrm{d}t \Big/ \int_0^{1.0} P^2(t)\mathrm{d}t$$

The above three formulas are electric energy loss probabilities under the minimum, maximum, and normal modes, respectively. The following formulas are obtained if the principle of multiplying probability is satisfied:

$$b_P = a_P = m_P = \int_{t_{x.zh}}^{t_{zh.d}} P^2(t)\mathrm{d}t/4 \tag{3.17}$$

Therefore, the time points $t_{x.zh}$ and $t_{zh.d}$ for the three-mode division can be calculated by the following two formulas:

$$\int_0^{t_{x.zh}} P^2(t)\mathrm{d}t = (b_P + 4m_P + a_P)/6 = F/6 \tag{3.18}$$

$$\int_0^{t_{zh.d}} P^2(t)\mathrm{d}t = \int_0^{1.0} P^2(t)\mathrm{d}t - \int_{t_{zh.d}}^{1.0} P^2(t)\mathrm{d}t = F - \int_0^{t_{x.zh}} P^2(t)\mathrm{d}t = 5/6F \tag{3.19}$$

If the parameters of the load curve are given, then use the Liu Yingkuan Formula (2.28) to calculate F. Solve the above two integral equations to obtain two time points $t_{x.zh}$ and $t_{zh.d}$ which divide the three modes. Substitute them into the load duration curve, Rossander Formula (3.6) or Formula (3.7), to find the power

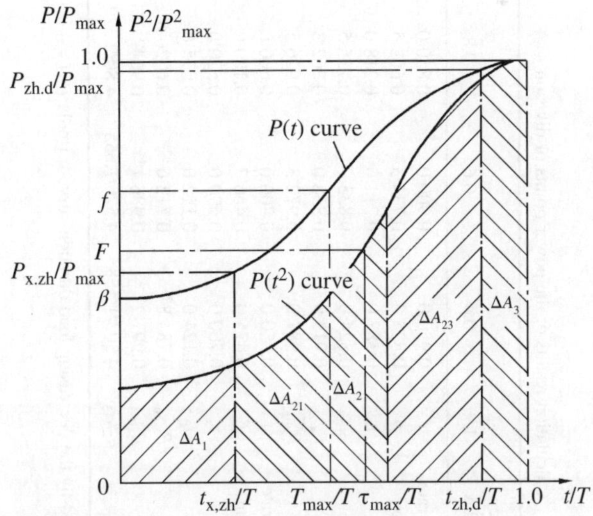

Figure 3.5 Schematic diagram of three-mode division. T_{max} – Maximum load utilization time; τ_{max} – maximum loss time. $\Delta A_{21} = \Delta A_{23} = \frac{1}{2}\Delta A_2$; $\Delta A_{21} = 2\Delta A_1$; $\Delta A_{23} = 2\Delta A_3$

Table 3.4 Calculation results of the power points of division.

Load curve change index	$1/\lambda$	0.4	0.6	0.8	1.0	2.0	3.0	4.0	4.5	5.0	6.0
High load rate combination	f	0.950 0	0.900 0	0.850 0	0.850 0	0.750 0	0.700 0	0.650 0	0.600 0	0.550 0	0.500 0
	β	0.825 0	0.733 3	0.662 5	0.700 0	0.625 0	0.600 0	0.562 5	0.511 1	0.460 0	0.416 7
	t_x	0.187 0	0.211 0	0.218 0	0.227 0	0.240 0	0.231 0	0.228 0	0.239 0	0.250 0	0.254 0
	t_d	0.824 0	0.833 0	0.735 5	0.873 0	0.895 0	0.907 0	0.930 0	0.946 0	0.958 0	0.970 0
	$P_{x.zh}$	0.913 9	0.838 0	0.762 3	0.768 1	0.644 6	0.604 9	0.563 7	0.511 8	0.460 5	0.416 9
	$P_{zh.d}$	0.987 0	0.972 3	0.926 5	0.961 9	0.925 4	0.898 5	0.889 8	0.899 7	0.895 7	0.902 6
Low load rate combination	f	0.850 0	0.800 0	0.600 0	0.700 0	0.650 0	0.600 0	0.550 0	0.500 0	0.450 0	0.400 0
	β	0.475 0	0.466 7	0.100 0	0.400 0	0.475 0	0.466 7	0.437 5	0.388 9	0.340 0	0.300 0
	t_x	0.700 0	0.690 0	0.520 0	0.341 0	0.308 0	0.287 0	0.279 0	0.297 0	0.322 0	0.335 0
	t_d	0.124 0	0.112 0	0.075 0	0.908 0	0.920 0	0.929 0	0.939 0	0.948 0	0.950 0	0.960 0
	$P_{x.zh}$	0.784 8[a]	0.712 6[a]	0.6026 6[a]	0.604 6	0.524 8	0.479 3	0.440 9	0.391 6	0.342 3	0.301 0
	$P_{zh.d}$	0.997 2[a]	0.986 1[a]	0.9647 7[a]	0.944 8	0.919 4	0.894 3	0.874 8	0.869 4	0.850 7	0.847 9
	α_2/α_3	4.303/1.558	4.526/1.655	4.80/2.329	2.56/0.522	2.79/0.507	2.95/0.498	3.00/0.467	2.90/0.419	2.68/0.420	2.61/0.360

[a] Calculated by the decreasing load duration curve formula $P(t) = 1 - (1 - \beta)(t/T)^{\lambda}$.

points of division of three modes $P_{x.zh}$ and $P_{zh.d}$. The calculation of the time points and power points of division of the three mode sections is shown in Figure 3.5.

To sum up, with the ratio of the electricity losses of the three modes to the total losses as the probability, and with the probabilities of electric energy losses in the normal mode section being two times those in the minimum mode section and the maximum mode section, respectively (with a hidden assumption that the probabilities of occurrence of the normal mode are two times those of the minimum mode and the maximum mode, respectively), the division of three mode sections in the load duration curve keeps all characteristic parameters of the load duration curve and the load duration square curve (loss curve) unchanged. Therefore, the result of the division of three sections can be applied in analyses of technical and economic problems of load, loss, electricity price, and so on.

3.4.3.4 Calculation of Power Points of Division of Three Modes

To find the rule of calculating power points for the division of three modes, taking the load curve change index $1/\lambda$ shown in Formula (3.8) as the variable, and based on two types by the relationship between f and β, use the Rossander load duration curve formula and the loss curve formula shown in Formulas (3.20) and (3.21), as well as Formulas (3.18) and (3.19) to calculate the time points $t_{x.zh}$ and $t_{zh.d}$ and power points $P_{x.zh}$ and $P_{zh.d}$ of division of three mode sections by means of successive approximation.

$$P(t)=\beta+(1-\beta)\ (t/T)^{\frac{1}{\lambda}} \tag{3.20}$$

$$P^2(t)=\beta^2+2\beta(1-\beta)\ (t/T)^{\frac{1}{\lambda}}+(1-\beta)^2(t/T)^2 \tag{3.21}$$

Table 3.4 shows the calculation results.

The calculation results listed in Table 3.4 can be used to obtain the approximate formula of power point of division of three modes by means of curve fitting.

1. For the parameter combination (such as monthly or quarterly load curve) with a high load factor f_P in the measuring period

$$P_{x.\,zh}=0.008(1/\lambda)^2-0.126\ 6(1/\lambda)+0.904\ 3, R=0.980\ 8 \tag{3.22}$$

$$P_{zh.\,d}=0.005\ 2(1/\lambda)^2-0.045\ 6(1/\lambda)+0.991\ 8, R=0.934\ 3 \tag{3.23}$$

2. For the parameter combination (such as yearly load curve) with a low load factor f_P in the measuring period

$$P_{x.\,zh}=0.007\ 7(1/\lambda)^2-0.119\ 4(1/\lambda)+0.7585, R=0.966\ 9 \tag{3.24}$$

$$P_{zh.\,d}=0.004(1/\lambda)^2-0.050\ 4(1/\lambda)+1.007\ 1, R=0.990\ 1 \tag{3.25}$$

In the above four formulas, R is the correlation coefficient, and larger R indicates better total fitting effect.

After the time points $t_{x.zh}$ and $t_{zh.d}$ of division of three modes are obtained, a single integral calculation of the load duration curve formula is made to obtain the electric quantity ratio of the three mode sections $\alpha_1{:}\alpha_2{:}\alpha_3$. When α_1 takes the cardinal number 1.0, $1{:}\alpha_2{:}\alpha_3$ in the case of a low load factor parameter combination is listed in the last row of Table 3.4, which can be chosen for relevant calculations [54].

4

Calculation of Line Loss by Power Load Curve

4.1 Line Loss Calculation Considering Power Factor

If the power factor of apparent power at the time of maximum load is $\cos\varphi_{\max}$, then the calculation of electric energy losses considering the power factor can be analyzed by the following two cases.

4.1.1 *The Maximum Apparent Power is Caused by the Maximum Active Power*

As

$$\Delta A = \frac{R}{U_{av}^2}\int_0^T \left[P^2(t)+Q^2(t)\right]\mathrm{d}t \times 10^{-3}$$

So

$$\begin{aligned}\Delta A &= \frac{R}{U_{av}^2}S_{\max}^2\left(\frac{P_{\max}^2}{S_{\max}^2}\int_0^T \frac{P^2(t)}{P_{\max}^2}\mathrm{d}t + \frac{Q_{\max}^2}{S_{\max}^2}\int_0^T \frac{Q^2(t)}{Q_{\max}^2}\mathrm{d}t\right)\times 10^{-3}\\ &= \frac{R}{U_{av}^2}S_{\max}^2\left(\cos^2\varphi_{\max}F_P + \sin^2\varphi' F_Q\right)T\times 10^{-3}\end{aligned} \quad (4.1)$$

$$\cos\varphi_{\max} = P_{\max}/S_{\max}$$

$$\sin\varphi' = Q_{\max}/S_{\max}$$

$$F_P = \frac{1}{T}\int_0^T P^2(t)\mathrm{d}t/P_{\max}^2$$

$$F_Q = \frac{1}{T}\int_0^T Q^2(t)\mathrm{d}t/Q_{\max}^2$$

Line Loss Analysis and Calculation of Electric Power Systems, First Edition. Anguan Wu and Baoshan Ni.

Wherein: F_P – active power loss factor;
F_Q – reactive power loss factor;
$\cos\varphi_{max}$ – power factor at the time of maximum apparent power;
$\sin\varphi'$ – ratio of maximum reactive power and maximum apparent power.

Because the maximum reactive power and the maximum active power do not always occur at the same time, the power factor angles φ_{max} and φ' are not always the same.

If the loss factor of the apparent power is F_S, then

$$\Delta A = \frac{R}{U_{av}^2} S_{max}^2 F_S T \times 10^{-3} \tag{4.2}$$

According to Formula (4.1),

$$F_S = F_P \cos^2\varphi_{max} + F_Q \sin^2\varphi' \tag{4.3}$$

When S_{max}, P_{max}, and Q_{max} occur at the same time, allowing $F_S = F_{max}$, then Formula (4.3) is changed into

$$F_{max} = F_P \cos^2\varphi_{max} + F_Q \sin^2\varphi_{max} \tag{4.4}$$

4.1.2 The Maximum Apparent Power is Caused by the Maximum Reactive Power

In the phasing operation of a hydroelectric power plant, the plant mainly transmits reactive power outward, and the maximum apparent power of power grid units (main transformers, lines) is caused by the maximum reactive power. At this time, the following formula conversion can be conducted:

$$\begin{aligned}\Delta A &= \frac{R}{U_{av}^2} S_{max}^2 \left(\frac{P_{max}^2}{S_{max}^2} \int_0^T P^2(t)\mathrm{d}t \Big/ (P_{max}^2 T) + \frac{Q_{max}^2}{S_{max}^2} \int_0^T Q^2(t)\mathrm{d}t \Big/ Q_{max}^2 T) \right) \times T \times 10^{-3} \\ &= \frac{R}{U_{av}^2} S_{max}^2 \left[\cos^2\varphi' F_P + \sin^2\varphi_{max} F_Q\right] T \times 10^{-3}\end{aligned} \tag{4.5}$$

$$\sin\varphi_{max} = Q_{max}/S_{max} = \sqrt{1-\cos^2\varphi_{max}}$$

$$\cos\varphi' = P_{max}/S_{max}$$

Wherein: $\cos\varphi'$ – the ratio of maximum active power and maximum apparent power; it is not the power factor at the time of maximum apparent power;
$\sin\varphi_{max}$ – sine value of power factor angle at the time of maximum apparent power.

Formula (4.5) can be rewritten to:

$$\Delta A = \frac{R}{U_{av}^2} S_{max}^2 F_S T \times 10^{-3} \tag{4.6}$$

$$F_S = F_P \cos^2\varphi' + F_Q \sin^2\varphi_{max}$$

According to Formulas (4.3) and (4.6), the loss factor of apparent power depends on not only the loss factors of active power and reactive power, but also the power factor and power ratios ($\sin\varphi'$, $\cos\varphi'$) at the time of maximum apparent power, indicating that the line loss calculation considering the power factor is a complicated issue.

4.2 Maximum Load Power Factor Method of Tröger

Tröger from Germany first put forward the method of calculating line losses by the power factor at the time of maximum load in 1920 [9]. In his opinion, reactive power changes of power lines fall between two situations where the power factor remains unchanged and where the reactive power remains unchanged. After analyzing the method of calculating line losses under such two situations, he conceived that a coefficient C could be used to express actual changes in reactive power. Taking $C = 1/2$, he derived the formula of line loss calculation considering reactive power changes, that is

$$F_S = \frac{1}{4}\left[(1+2f_P) - \cos^2\varphi_{\max}(1+2f_P-3F_P)\right] \tag{4.7}$$

Tröger substituted the loss factor of Formula (2.18), $F_P = 0.12 - 0.24f_P + 1.12f_P^2$, as derived by means of statistical mathematics, into Formula (4.7), obtaining the calculation formula of apparent power loss factor, that is

$$F_S = 0.28 + 0.44f_P + 0.28f_P^2 - \cos^2\varphi_{\max}(0.16 + 0.68f_P - 0.84f_P) \tag{4.8}$$

The table of calculation of $F_S(f_p, \cos\varphi_{\max})$ can be made based on the above formula, as shown in Table 4.1.

According to the analysis, the assumption of a constant value for C in fact artificially defines the relationship between changed active loads and reactive loads. Although this facilitates calculation, the physical significance of the C coefficient is not definite.

To be sure, Formula (4.8) calculates the loss factor by the power factor at the time of maximum load, so that the calculation result of load flow can be directly used to calculate electric energy losses. Therefore, this method was highly praised by designers.

4.3 Annual Average Power Factor Method of Glazynov

In 1947, Glazynov [16] from the former Soviet Union put forward line loss calculation curves for calculating the annual maximum loss time $\tau_{\max}$ by the annual average power factor $\cos\varphi_{\text{av}}$ and the maximum load utilization time $T_{\max}$. These curves were first cited in teaching materials of Chinese higher colleges and universities, and then were applied for a long time in the Chinese design and operation departments. Even now these curves are still used.

Table 4.1 Loss factor calculation table, $F_S(f_p, \cos\varphi_{\max})$.

$\cos\varphi_{\max}$ / f_p	1.0	0.90	0.80	0.70	0.60	0.50	0.40	0.30	0.20	0.10	0
0.90	0.811	0.829	0.844	0.858	0.870	0.880	0.888	0.895	0.899	0.902	0.903
0.80	0.645	0.677	0.705	0.730	0.751	0.770	0.785	0.796	0.804	0.809	0.811
0.70	0.501	0.543	0.581	0.615	0.644	0.669	0.689	0.705	0.716	0.723	0.725
0.60	0.379	0.430	0.475	0.515	0.549	0.578	0.602	0.621	0.634	0.642	0.645
0.50	0.280	0.335	0.385	0.428	0.466	0.498	0.524	0.544	0.558	0.567	0.570
0.40	0.203	0.259	0.310	0.355	0.394	0.426	0.453	0.474	0.489	0.498	0.501
0.30	0.149	0.204	0.253	0.296	0.333	0.365	0.391	0.411	0.426	0.434	0.437

Table 4.2 Original data of T_{max}–τ_{max} curves.

T_{max}(h)	P_{max}	P_{min}	P_{ill}(illumination)
2000	1.00	0.10	1.0
3000	1.00	0.20	0.70
4000	1.00	0.30	0.50
5000	1.00	0.40	0.35
6000	1.00	0.50	0.20
7000	1.00	0.60	0.10

Table 4.3 Comparison of τ_{max1} and τ_{max2} (h).

$\cos\varphi_{av}$	T_{max}	3000	4000	5000	6000	7000
0.60	τ_{max1}	2620	3380	4110	5010	5920
	τ_{max2}	2560	3415	4450	5440	6310
0.70	τ_{max1}	2350	3100	3890	4890	5870
	τ_{max2}	2093	2990	3880	5150	6190
0.80	τ_{max1}	2000	2780	3690	4680	5790
	τ_{max2}	1715	2610	3680	4855	6040
1.0	τ_{max1}	1300	1970	2940	4290	5680
	τ_{max2}	1150	1960	2985	4250	5700

According to the original data in Table 4.2, Glazynov used the Rossander Formula (3.6), and Formula (3.7) established an analytical formula of active load duration curves. He removed the illumination load electric quantity at $T = 2000$ h (the per unit value is 0.228), and derived reactive power load duration curves from the remaining load duration curves as per the annual average power factor ($\cos\varphi_{av}$).

Then, he summed square curves of active power and reactive power load duration curves and obtained apparent power square curves. He calculated the τ_{max} value, and calculated several combinations with T_{max} range 3000–7000 h and $\cos\varphi_{av}$ range 0.6–1.0. Table 4.3 lists $\tau_{max}(T_{max}, \cos\varphi_{av})$ values, and Figure 4.1 shows T_{max}–τ_{max} curves.

The τ_{max1} values in Table 4.3 are found as per the T_{max}–τ_{max} curves of Glazynov (see Figure 4.1), and the τ_{max2} values are calculated by the authors. According to the comparison of τ_{max1} and τ_{max2}, smaller T_{max} leads to a larger difference between the two. It is estimated that Glazynov corrected some values according to calculation results of actual operation information.

For 6–35 kV regional power transmission and distribution lines with comprehensive loads, the maximum loads are generally caused by their illumination loads. Excluding the illumination loads, the remaining active load duration curves and reactive load duration curves are similar. The original data used by Glazynov in Table 4.2 are derived from the investigation results of the former Soviet Union and are not in line with the characteristics of comprehensive loads in different Chinese regions, so these T_{max}–τ_{max} curves cannot be copied. In addition, after reactive compensation of power grids, the reactive load curves no longer have similar graphs to active load curves according to the relationship of average power factor (see Section 5.1 in Chapter 5 for details), so the T_{max}–τ_{max} curves of Glazynov are not suitable for calculating line losses of power grids subject to reactive compensation. The average power factor of loads of the outgoing lines of a hydroelectric power plant during its phasing operation is very low, and the calculation of electric energy losses in such lines cannot adopt the T_{max}–τ_{max} curves of Glazynov. Due to these problems, since the 1960s, people in some Chinese higher colleges and universities and design departments and even in the former Soviet Union pointed out that these curves proposed by Glazynov should be abandoned and that theoretical

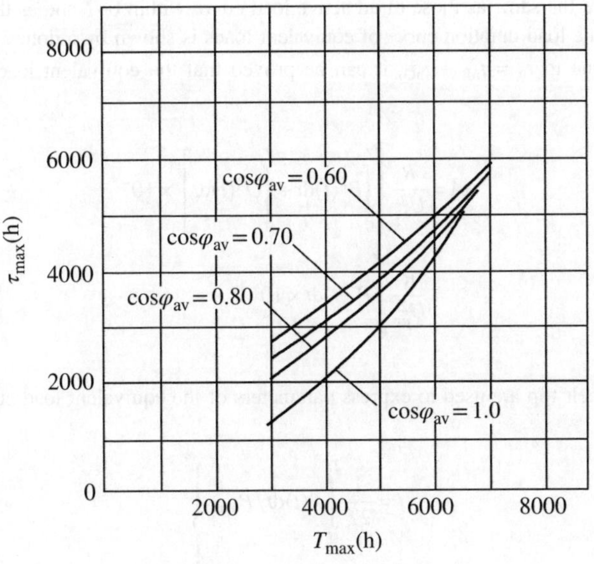

Figure 4.1 T_{max}–τ_{max} curves.

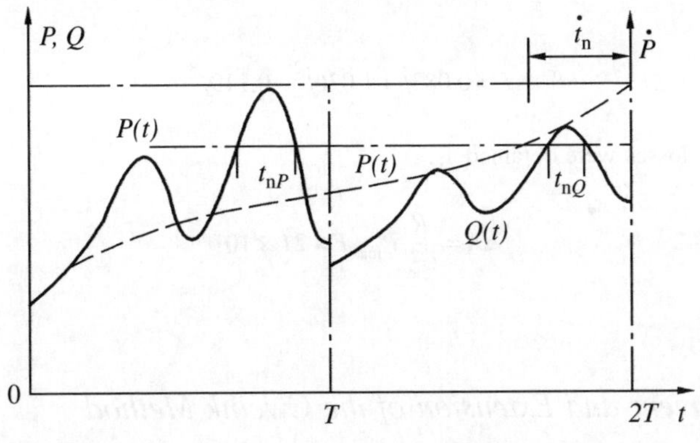

Figure 4.2 Load duration curve for equivalent load curve.

restudies should be conducted to find new line loss calculation methods and line loss calculation curves considering the power factor.

4.4 Equivalent Load Curve Method

4.4.1 Equivalent Load Curve Method of Cweink

Cweink from Poland put forward a new method of line loss calculation considering the power factor in 1962 [17]. According to this method, electric energy losses caused by active power and reactive power within a calculation period T are deemed as those caused by an equivalent load within $2T$, as shown in Figure 4.2.

Equivalent loads are the same as those of an active load curve within 0~T and as those of a reactive load curve within T~$2T$. The load duration curve of equivalent loads is shown by a dotted line in Figure 4.2 and expressed by $\dot{P}(t)$. Due to $\dot{t}_n = t_{nP} + t_{nQ}$, it can be proved that the equivalent loads meet the following conditions:

$$
\begin{aligned}
\Delta A &= \frac{R}{U_{\text{av}}^2}\left[\int_0^T P^2(t)\mathrm{d}t + \int_0^T Q^2(t)\mathrm{d}t\right]\times 10^{-3} \\
&= \frac{R}{U_{av}^2}\int_0^{2T} \dot{P}^2(t)\mathrm{d}t \times 10^{-3}
\end{aligned}
\tag{4.9}
$$

Letters with "··" on their top are used to express parameters of the equivalent load curve, so

$$
\left.
\begin{aligned}
\dot{f} &= \frac{1}{2T}\int_0^{2T} \dot{P}(t)\mathrm{d}t / \dot{P}_{\max} \\
\dot{\beta} &= \dot{P}_{\min}/\dot{P}_{\max}
\end{aligned}
\right\}
\tag{4.10}
$$

Cweink directly used the Wolf Formula (2.21) and calculated the loss factor $\dot{F}$ by the load rate $\dot{f}$ of equivalent load curve, that is

$$\dot{F} = 0.083\dot{f} + 1.036\dot{f}^2 - 0.119\dot{f}^3 \tag{4.11}$$

Then, electric energy losses were obtained as:

$$\Delta A = \frac{R}{U_{\text{av}}^2}\dot{P}_{\max}^2 \dot{F} \times 2T \times 10^{-3} \tag{4.12}$$

4.4.2 Improvement and Extension of the Cweink Method

According to Figure 4.2, when the minimum value of the equivalent load curve is determined by the reactive power ($\dot{P}_{\min} = Q_{\min}$), the two parameters $\dot{f}$ and $\dot{\beta}$ should represent two characteristics of active and reactive load curves. However, the parameter $\dot{\beta}$ is not included in the Wolf Formula (4.11), so the equivalence of the line loss calculation cannot be guaranteed. An ideal loss factor formula including both parameters $\dot{f}$ and $\dot{\beta}$ can improve the equivalence of the equivalent load curve method.

Given the actual needs of the theoretical line loss calculation considering the power factor, line loss calculation curves can be divided into two categories, namely, design curves with the power factor at the time of maximum load as a parameter, and operation curves with the average power factor as a parameter. According to actual practices of line loss calculation, with respect to electric loads passing electric equipment, $\dot{P}_{\max} = Q_{\max}$ and $\dot{P}_{\min} = P_{\min}$ may occur; or a complicated situation may occur, where active power is only transmitted and reactive power is not transmitted at some time (such as tie lines), or where reactive power is only transmitted and active power is not transmitted (such as outgoing lines at the time of phasing operation of a hydroelectric power plant). Therefore, the equivalent load curve method proposed by

Table 4.4 Calculation formulas of parameters $\dot{f}$ and $\dot{\beta}$ of the equal time equivalent load curve.

Parameter	Type	$\dot{P}_{max}=P_{max}(\cos\varphi_{max}>0.707)$		$\dot{P}_{max}=Q_{max}(\cos\varphi_{max}<0.707)$	
$\dot{f}$	For operation	$\dot{f}=\frac{1}{2}f_P(1+\tan\varphi_{av})$		$\dot{f}=\frac{1}{2}f_Q(1+1/\tan\varphi_{av})$	
	For design	$\dot{f}=\frac{1}{2}(f_P+f_Q\tan\varphi_{max})$		$\dot{f}=\frac{1}{2}(f_Q+f_P/\tan\varphi_{max})$	
Parameter	Type	$\dot{P}_{min}=P_{min}$	$\dot{P}_{min}=Q_{min}$	$\dot{P}_{min}=Q_{min}$	$\dot{P}_{min}=P_{min}$
$\dot{\beta}$	For operation	$\dot{\beta}=\beta_P$	$\dot{\beta}=\beta_Q\tan\varphi_{av}/k_f$	$\dot{\beta}=\beta_Q$	$\dot{\beta}=\beta_P k_f/\tan\varphi_{av}$
	For design	$\dot{\beta}=\beta_P$	$\dot{\beta}=\beta_Q\tan\varphi_{max}$	$\dot{\beta}=\beta_Q$	$\dot{\beta}=\beta_P/\tan\varphi_{max}$

Cweink should be extended to accommodate different situations during the active load duration or the reactive load duration.

4.4.3 *Equal Time Equivalent Load Curve Method*

Table 4.4 lists the calculation formulas of equivalent load curve parameters $\dot{f}$ and $\dot{\beta}$, which are divided into two categories by $\dot{P}_{max} = P_{max}$ and $\dot{P}_{max} = Q_{max}$. These parameters are respectively calculated by the average power factor and the calculation power factor at the time of maximum load, so that design and operation departments can use them conveniently.

In Table 4.4, k_f refers to the ratio of reactive load factor and active load factor; $k_f = f_Q/f_P \cdot \cos\varphi_{max} = \cos[\tan^{-1}(Q_{max}/P_{max})]$ is called the calculation power factor of maximum load. When the maximum values of active power, reactive power and apparent power occur at the same time, $\cos\varphi_{max}$ is the power factor at the time of maximum load. If the above conditions are not satisfied, the power factor is not an actual power factor at a certain time, but the calculation power factor of a hypothetical apparent power ($S_{max \cdot jx} = \sqrt{P_{max}^2 + Q_{max}^2}$). $\dot{f}$ and $\dot{\beta}$ are calculated by formulas in Table 4.4 and substituted into Formula (2.28) to obtain the loss factor $\dot{F}$ of the equivalent load curve:

$$\dot{F}=0.639\dot{f}^2+0.361\,(\dot{f}+\dot{f}\dot{\beta}-\dot{\beta}) \tag{4.13}$$

Then, $\dot{F}$ and $\dot{P}_{max}$ are substituted into Formula (4.12) to calculate the electric energy losses ΔA.

According to the formulas listed in Table 4.4 and Formulas (4.12) and (4.13), the theoretical line loss calculation considering the power factor can be completed without any assumption condition. This is a difficult issue which was studied by Chinese scientific and technical personnel from the 1960s and completely solved in the late 1990s.

Example 4.1 A certain 110 kV transmission line uses LGJ240 conductors and has a length of 10 km and resistance of 1.32 Ω at 20 °C. The quantity of active electricity quantity passing through the line within T = 720 h in a month is 36 000 MW·h, and the quantity of reactive electricity 18 000 Mvar·h. During this period, the maximum active power is 62.5 MW, the minimum active power 31.255 MW, the maximum reactive power 50 Mvar, and the minimum reactive power 20 Mvar. The maximum active power and the maximum reactive power occur at the same time. *Try out different methods to calculate the electric energy losses of the line within the measuring period, and compare the calculation results.*

Solutions

1. *Parameter calculation.*

The calculation of various parameters is as follows:

$$f_P = P_{av}/P_{max} = 36\,000/(720\times 62.5) = 0.80$$

$$\beta_P = 31.25/62.5 = 0.5$$

$$f_Q = Q_{av}/Q_{max} = 18\,000/(720\times 50) = 0.50$$

$$\beta_Q = 20/50 = 0.4$$

2. *Accurate calculation.*

As $f_p > (1+\beta_P)/2$, that is $0.8 > (1+0.5)/2$, use the Type 1 Rossander Formula (3.6) of the load duration curve for the calculation. As $\lambda_P = \frac{f_P - \beta_P}{1-f_P} = \frac{0.3}{0.2} = 1.5$, the expression of the active load curve is $P_*(t) = 1 - 0.5t^{1.5}$. Integrate the square of active power to obtain

$$\int_0^1 P^2(t)\mathrm{d}t = \int_0^1 P_{max}^2 \left(1-0.5t^{1.5}\right)^2 \mathrm{d}t = P_{max}^2 \times 0.662\,5$$

As $f_Q < (1+\beta_Q)/2$, that is $0.5 < (1+0.4)/2$, use the Type 2 Rossander Formula (3.7) of the load duration curve for the calculation. As $\frac{1}{\lambda_Q} = \frac{1-f_Q}{f_Q-\beta_Q} = \frac{0.5}{0.1} = 5$, $Q_*(t) = 0.4 + 0.6(1-t)^5$. Integrate the square of reactive power to obtain

$$\int_0^1 Q^2(t)\mathrm{d}t = \int_0^1 Q_{max}^2 \left(0.4+0.6(1-t)^5\right)^2 \mathrm{d}t = Q_{max}^2 \times 0.272\,7$$

Due to $\Delta A = \frac{R}{U_{av}^2}\left(\int_0^T P^2(t)\mathrm{d}t + \int_0^T Q^2(t)\mathrm{d}t\right) \times 10^{-3} = \frac{R}{U_{av}^2}\left(\int_0^1 P^2(t)\mathrm{d}t + \int_0^1 Q^2(t)\mathrm{d}t\right) T\times 10^{-3}$, substitute $\int_0^1 P^2(t)\mathrm{d}t$, $\int_0^1 Q^2(t)\mathrm{d}t$ and $T = 720$ h into the formula above to obtain

$$\begin{aligned}
\Delta A &= \frac{R}{U_{av}^2}\left(P_{max}^2 \times 0.662\,5 + Q_{max}^2 \times 0.272\,7\right) T \times 10^{-3} \\
&= \frac{R}{U_{av}^2} P_{max}^2 \left(0.662\,5 + \frac{Q_{max}^2}{P_{max}^2} \times 0.272\,7\right) T \times 10^{-3} \\
&= \frac{1.32}{1.21\times 10^{10}} \times \left(62.5\times 10^6\right)^2 \times 0.837\,0 \times 720 \times 10^{-3} \\
&= 2.568 \times 10^5 (\mathrm{kW\cdot h})
\end{aligned}$$

3. *Calculation by the loss factor of apparent power.*

According to the above calculation, $F_P = 0.662\,5$, $F_Q = 0.272\,7$. Calculate the loss factor of the apparent power as per Formula (4.4), obtaining

$$F_S = F_P\cos^2\varphi_{max} + F_A\sin^2\varphi_{max} = 0.662\,5 \times \frac{62.5^2}{62.5^2+50^2} + 0.272\,7 \times \frac{50^2}{62.5^2+50^2} = 0.510\,4$$

Substitute F_S into Formula (4.2) to obtain

$$\Delta A = \frac{R}{U_{\text{av}}^2} S_{\max}^2 F_S T \times 10^{-3} = \frac{1.32}{1.21 \times 10^{10}} \times (62.5^2 + 50^2) \times 10^{12}$$

$$\times 0.510\,4 \times 720 \times 10^{-3} = 2.568 \times 10^5 (\text{kW} \cdot \text{h})$$

The result is the same as that of the accurate calculation.

The above calculations show that both of the calculation methods use all "information" of the load curve, so naturally they derive the same results.

4. *Calculation by equivalent load curve method.*

Use formulas for operation in Table 4.4 to calculate the followings:

$$\dot{f} = \frac{1}{2} f_P (1 + \tan\varphi_{\text{av}}) = \frac{1}{2} \times 0.8 \times \left(1 + \frac{18\,000}{36\,000}\right) = 0.60$$

$$\dot{\beta} = \beta_Q \frac{\tan\varphi_{\text{av}}}{k_f} = 0.40 \times \frac{18\,000/36\,000}{0.5/0.8} = 0.32$$

When formulas for design in Table 4.4 are used, the same $\dot{f}$ and $\dot{\beta}$ values can be obtained.

Substitute the above $\dot{f}$ and $\dot{\beta}$ values into Formula (4.13) to obtain

$$\dot{F} = 0.639\dot{f}^2 + 0.361\,(\dot{f} + \dot{f}\dot{\beta} - \dot{\beta}) = 0.639 \times 0.60^2 + 0.361\,(0.60 + 0.60 \times 0.32 - 0.32) = 0.400\,4$$

Substitute the above values into Formula (4.12) to obtain

$$\Delta A = \frac{R}{U_{\text{av}}^2} \dot{P}_{\max}^2 \dot{F} \times 2T \times 10^{-3} = \frac{1.32}{1.21 \times 10^{10}} \times (62.5 \times 10^6)^2 \times 0.400\,4 \times 2 \times 720 \times 10^{-3}$$

$$= 2.457 \times 10^5 (\text{kW} \cdot \text{h})$$

Compared with the result of accurate calculation, the error of the calculation result of the equivalent load curve method is

$$\delta(\Delta A)\% = \frac{2.457 - 2.568}{2.568} \times 100\% = -4.32\%$$

The error is below 5% which is within the allowable error range of engineering calculation.

4.4.4 *Unequal Time Equivalent Load Curve Method*

Tie lines of the electric power system transmit only active power and do not transmit reactive power at any time (such as tie lines), and outgoing lines transmit only reactive power and do not transmit active power during the phasing operation of some hydroelectric power plants. As a result, it is necessary to study the calculation of electric energy losses at different periods in the active power and reactive power load duration curves.

Assume that both active power and reactive power are transmitted during the period t_1, while only active power (or reactive power) is transmitted during the period $(1 - t_1)$, as shown in Figure 4.3.

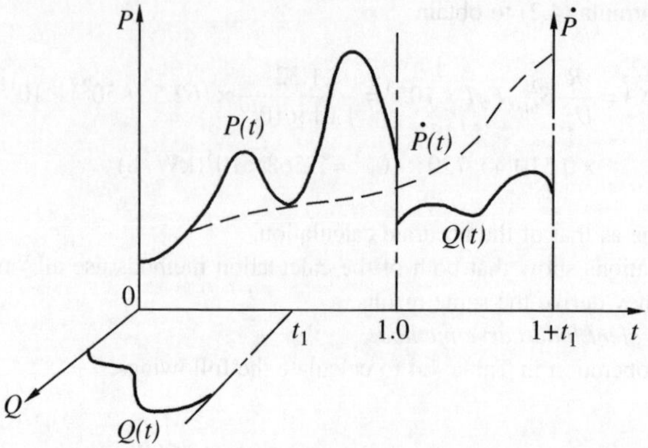

Figure 4.3 Unequal time equivalent load curve method.

In this case, the measuring period of the equivalent load curve is $(1 + t_1)$. Two cases will be considered for analysis as follows.

1. Only active power is transmitted during the period $(1 - t_1)$, with $\dot{P}_{max} = P_{max}$. For example, tie lines among provincial regions only transmit active power instead of reactive power. According to the definition of load rate,

$$\dot{f} = \int_0^{1+t} \dot{P}(t)\mathrm{d}t \Big/ \dot{P}_{max} = \frac{1}{1+t_1}\left[\int_0^1 P(t)\mathrm{d}t + \int_0^{t_1} Q(t)\mathrm{d}t\right] \Big/ \dot{P}_{max}$$

Substitute $\int_0^1 P(t)\mathrm{d}t/\dot{P}_{max} = f_P, \int_0^{t_1} Q(t)\mathrm{d}t/\dot{P}_{max} = (t_1 Q_{av}/P_{av})(P_{av}/P_{max}) = t_1 \tan\varphi_{av} f_P$ into the above formula to obtain

$$\left.\begin{aligned} \dot{f} &= \frac{1}{1+t_1}(f_P + t_1 f_P \tan\varphi_{av}) \\ \text{Or} \quad \dot{f} &= \frac{1}{1+t_1}(f_P + t_1 f_Q \tan\varphi_{max}) \end{aligned}\right\} \tag{4.14}$$

2. Only reactive power is transmitted during the period $(1 - t_1)$. For example, if a hydraulic power plant is mainly undergoing phasing operation during a period, $\dot{P}_{max} = Q_{max}$ at this time. Similar to the previous analysis, the calculation formula of load factor of the equivalent load curve can be obtained:

$$\left.\begin{aligned} \dot{f} &= \frac{1}{1+t_1}(f_Q + t_1 f_Q \tan\varphi_{av}) \\ \text{Or} \quad \dot{f} &= \frac{1}{1+t_1}(f_Q + t_1 f_P \tan\varphi_{max}) \end{aligned}\right\} \tag{4.15}$$

However, a hydraulic power plant or a pumped storage power plant is mainly undergoing peak regulation and occasional phasing operation most of the time, with $\dot{P}_{max} = P_{max}$ at this time.

Similar to the previous analysis, the calculation formula of load factor of the equivalent load curve is obtained:

$$\dot{f} = \frac{1}{1+t_1}(f_Q \tan\varphi_{\max} + t_1 f_P)$$

Or (4.16)

$$\dot{f} = \frac{1}{1+t_1}(f_P \tan\varphi_{\text{av}} + t_1 f_P)$$

Under the above two cases, $\dot{\beta} = Q_{\min}/\dot{P}_{\max}$ (or $P_{\min}/P_{\max}$), and is not necessarily equal to 0. $\dot{F}$ can be calculated by Formula (4.13), and the electric energy loss is calculated by the following formula:

$$\Delta A = \frac{R}{U_{\text{av}}^2}\dot{P}_{\max}^2 \dot{F}(1+t_1)T \times 10^{-3} \quad (t_1 \leq 1.0) \tag{4.17}$$

If $t_1 = 1$ in Formula (4.17), then the above formula is the same as Formula (4.12). Therefore, the unequal time equivalent load curve method is an extension of the equal time equivalent load curve method and is applicable to the calculation of electric energy losses under various complicated situations.

Example 4.2 A 110 kV line has the resistance of 1 Ω per phase, and transmits 50 MW active loads and 20 Mvar reactive loads in 5 h morning and evening peaks. During the rest of the time, it transmits 30 MW active loads and 0 reactive load. *Try to use the equal time and unequal time load curve methods to calculate electricity line losses within 24 h and compare them with accurate values.*

Solutions

1. *Accurate value calculation method.*

As shown in Figure 4.4, the maximum apparent power is $S_{\max} = \sqrt{50^2 + 20^2} = 53.85(\text{MVA})$, and the apparent power load duration curve is expressed by dotted lines in Figure 4.4.

Two periods are considered for accumulation to obtain the integral of S^2, and then the accurate daily electricity loss can be calculated as

$$\Delta A = \frac{R}{U_{\text{av}}^2}\int_0^{24} S^2(t)\text{d}t \times 10^{-3} = \frac{1}{1.21 \times 10^{10}} \times \left[30^2 \times 19 \times + \left(50^2 + 20^2\right) \times 5\right]$$

$$\times 10^{12} \times 10^{-3} = 2.612 \times 10^3 (\text{kW} \cdot \text{h})$$

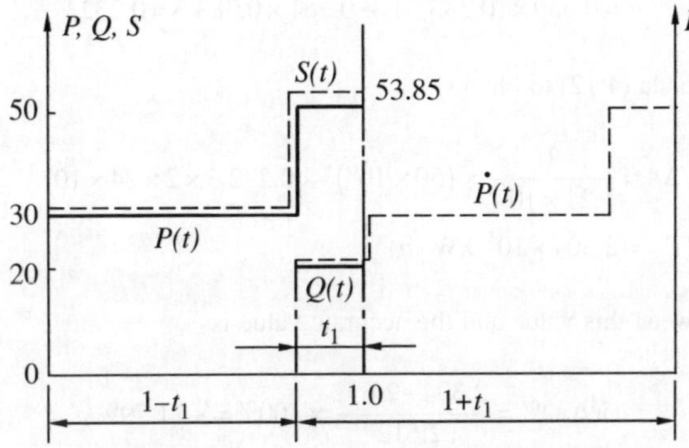

Figure 4.4 $P(t)$, $Q(t)$, $\dot{P}(t)$ load duration curves.

2. *Unequal time equivalent load curve method.*
$t_1 = 5/24 = 0.208\ 3$, $P_{av} = (19 \times 30 + 5 \times 50)/24 = 34.167$ (MW), $f_P = (19 \times 30 + 5 \times 50)/(24 \times 50) = 0.683\ 3$, $\tan\varphi_{av} = Q_{av}/P_{av} = 20/34.167 = 0.585\ 4$, $\dot{\beta} = Q_{min}/\dot{P}_{max} = 20/50 = 0.40$. Substitute the above values into Formula (4.14) to obtain

$$\dot{f} = \frac{1}{1+0.208\ 3} \times (0.683\ 3 + 0.208\ 3 \times 0.683\ 3 \times 0.585\ 4) = 0.634\ 5$$

Substitute the above $\dot{f}$ and $\dot{\beta}$ into Formula (4.13) to obtain

$$\begin{aligned}\dot{F} &= 0.639 \times 0.634\ 5^2 + 0.361 \times (0.634\ 5 + 0.634\ 5 \times 0.4 - 0.4)\\ &= 0.433\ 5\end{aligned}$$

Substitute $\dot{F}$ into Formula (4.17) to obtain

$$\begin{aligned}\Delta A &= \frac{1}{1.21 \times 10^{10}} \times \left(50 \times 10^6\right)^2 \times 0.433\ 5 \times (1 + 0.208\ 3) \times 24 \times 10^{-3}\\ &= 2.597 \times 10^3 (\text{kW} \cdot \text{h})\end{aligned}$$

The relative error between this value and the accurate value is

$$\delta(\Delta A)\% = \frac{2.597 - 2.612}{2.612} \times 100\% = -0.57\%$$

3. *Equal time equivalent load curve method.*
As the zero load time is not considered for reactive power, $\tan\varphi_{av}$ is calculated according to the ratio of reactive electricity and active electricity, that is $\tan\varphi_{av} = (20 \times 5)/(5.50 + 19.30) = 0.122\ 0$. As $Q_{min} = 0$ during the whole period, $\dot{\beta} = 0$, and calculate $\dot{f}$ as per the formula in Table 4.4 to obtain

$$\dot{f} = \frac{1}{2} f_P (1 + \tan\varphi_{av}) = \frac{1}{2} \times 0.683\ 3 \times (1 + 0.122\ 0) = 0.383\ 3$$

Calculate $\dot{F}$ as per Formula (4.13) as

$$\dot{F} = 0.639 \times (0.383\ 3)^2 + 0.361 \times 0.383\ 3 = 0.2323$$

Substitute $\dot{F}$ into Formula (4.12) to obtain

$$\begin{aligned}\Delta A &= \frac{1}{1.21 \times 10^{10}} \times \left(50 \times 10^6\right)^2 \times 0.232\ 3 \times 2 \times 24 \times 10^{-3}\\ &= 2.304 \times 10^3 (\text{kW} \cdot \text{h})\end{aligned}$$

The relative error between this value and the accurate value is

$$\delta(\Delta A)\% = \frac{2.304 - 2.612}{2.612} \times 100\% = -11.79\%$$

Based on Example 4.2, take eight values from 0.10~0.833 for t_1 for similar calculations. With respect to the error between the electricity line loss calculated by the equal time formula and the accurate value, the relationship between the error percentage $\delta(\Delta A)\%$ and t_1 can be expressed by a linear formula $\delta(\Delta A)\% = (-21.19 + 40.68t_1)\%$, with a correlation coefficient of $\gamma = 0.9895$. According to the above calculations, when $t_1 = 0.40$–0.64, that is the per unit value of zero load duration of a component (P or Q) range of 0.36–0.60, the error in the direct calculation of electricity line loss by the equal time method is not more than 5%; a smaller t_1 indicates a smaller negative error, while a larger t_1 indicates a larger positive error.

Example 4.3 A hydroelectric power plant is equipped with four 12 MW generator units. In the month of November, the plant was undergoing full phasing operation, with a maximum daily reactive load of 30 Mvar, and reactive load rates of $f_Q = 0.80$, $\beta_Q = 0.30$. It transmitted active load for 4 h at evening peak of each day. The monthly average power factor is derived from records of the watt-hour meter of the generators, that is $\cos\varphi'_{av} = 0.15$. The 110 kV outgoing line has a resistance of 1 Ω per phase and a monthly average operating voltage of 115 kV. *Try to calculate the electricity losses of the line in the whole month.*

Solution

As the period during which the outgoing line of the hydroelectric power plant transmitted both reactive power and active power is 4 h, $t_1 = 4/24 = 0.166\,7$. $\cos\varphi'_{av} = 0.15$, so $\tan\varphi'_{av} = 6.591\,2$. $\tan\varphi'_{av} = f_Q/(t_1 f_P)$, so $\tan\varphi_{av} = \dfrac{Q_{av}}{P_{av}} = t_1 \tan\varphi'_{av} = 0.166\,7 \times 6.591\,2 = 1.099$. Then, the load rate of the equivalent load curve is obtained by Formula (4.15), that is

$$\dot{f} = \frac{1}{1+0.166\,7} \times (0.80 + 0.166\,7 \times 0.80/1.099) = 0.7897$$

As $\dot{\beta} = \beta_Q = 0.30$, substitute $\dot{f}$ and $\dot{\beta}$ into Formula (4.13) to obtain $\dot{F} = 0.660\,8$. Calculate the monthly electricity losses as per Formula (4.17), that is

$$\Delta A = \frac{1.0}{(1.15 \times 10^5)^2} \times (30 \times 10^6)^2 \times 0.660\,8 \times (1 + 0.166\,7) \times 720 \times 10^{-3}$$
$$= 3.778 \times 10^4 (\text{kW} \cdot \text{h})$$

Example 4.4 A pumped storage power plant is equipped with four 300 MW pumped storage units and generates electricity for 6 h every day, with the maximum generation load of 1200 MW and the generated electricity of 526.7×10^4 kW·h. It operates pumped storage for 8.5 h/day, with a maximum consumption load of 1250 MW and an electricity consumption of 721.51×10^4 kW·h. Except during generation and pumping time, one unit is undergoing a phasing operation, with a maximum capacity of 300 Mvar and minimum capacity of 200 Mvar. The plant is connected to the system through a double-circuit 500 kV line with a length of 100 km. The resistance of quad bundled conductors is $\gamma_0 = 0.024$ Ω/km per unit length, subject to double-circuit parallel operation. The rated power factors of generation and consumption are both 0.85, and the average voltage of the 500 kV bus is $1.025U_N$ and U_N at generation and pumping, respectively. Try to calculate the daily electricity loss and the line loss rate of the line.

Solutions

According to the above description, Figure 4.5 shows the positions of the operating modes of the pumped storage power plant during different periods, as well as the shape of the active and reactive load duration curves during different periods.

Figure 4.6 shows the active and reactive load duration curves and equivalent load curve.

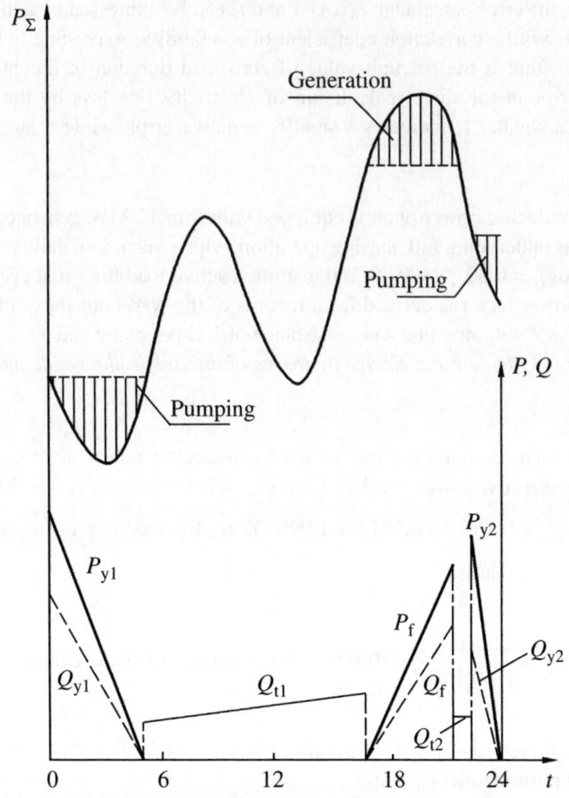

Figure 4.5 Three operating conditions of the pumped storage power plant. Upper: positions of three operating conditions in the load curve of the system. Lower: active and reactive load duration curves under three operating conditions.

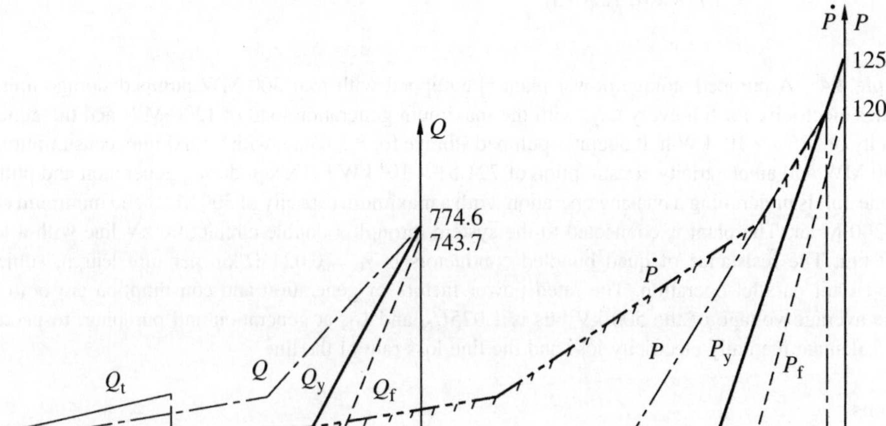

Figure 4.6 Active and reactive load duration curves and equivalent load curve.

1. *Accurate value calculation.* Three periods of consumption, generation, and phasing are considered for the calculation. The consumption period is t_y = 8.5 h, and the electricity loss is calculated as per Formula (4.2), that is $\Delta A_1 = \frac{R}{U_{av}^2} S_{max}^2 F_{S1} t_1 \times 10^{-3}$. During the period based on t_1, the load duration curve is an oblique line. Substitute f_y = 0.50 into Formula (2.7) to obtain $F_{S1} = \frac{1}{3}$. During the double-circuit parallel operation, $R_S = R_{1/2}$ = (0.024 × 100)/2 = 1.2 (Ω), S_{max1} = 1250/0.85 = 1470.58 (MVA). During consumption, U_{av} = 500 kV, and substitute it into the above formula to obtain $\Delta A_1 = \frac{1.2}{(500\times 10^3)^2} \times (1470.58 \times 10^6)^2 \times 0.333 \times 8.5 \times 10^{-3} = 29.36 \times 10^3$ (kW · h).
   During the generation period, t_f = 6h. Similar to the above, F_{S2} = 0.333, S_{max2} = 1200/0.85 = 1411.76 (MVA). At generation, U_{av} is 2.5% higher than the rated voltage, so U_{av} = 512.5 kV, obtaining

$$\Delta A_2 = \frac{1.2}{(512.5 \times 10^3)^2} \times (1411.76 \times 10^6)^2 \times 0.333 \times 6 \times 10^{-3} = 18.19 \times 10^3 \text{ (kW} \cdot \text{h)}$$

   During the phasing period, t_3 = 9.5 h, and the reactive load duration curve is of trapezoid shape. Use Formula (2.7) to calculate the loss factor F_3. As $f_Q = (1 + \beta_Q)/2$ = [1 + (200/300)]/2 = 0.833 3, substitute f_Q into $F_3 = \frac{1}{3}\left(1 - 2f_Q + 4f_Q^2\right)$ to obtain F_3 = 0.703 6, thus $\Delta A_3 = \frac{1.2}{(512.5\times 10^3)^2} \times (300 \times 10^6)^2 \times 0.703\,6 \times 9.5 \times 10^{-3} = 2.75 \times 10^3$ (kW · h).
   The electric energy loss of the line during the whole day is ΔA_2 = (29.36 + 18.19 + 2.75) × 10^3 = 50.30 × 10^3 (kW·h).
   The total electric supply for generation and consumption lines is A = (721.51 + 526.70) × 10^4 = 1248.21 × 10^4 (kW·h), so the daily line loss rate of the plant is $\Delta A\%$ = 50.30 × 10^3/1248.21 × 10^4 = 0.40%.
2. *Calculation by unequal time method.*
   As shown in Figure 4.6, the active load duration curve is composed of P_y and P_f during consumption and generation periods, and the reactive load duration curve is composed of Q_y, Q_f and Q_t during consumption, generation, and phasing periods. The active load curve includes the zero load period, so the equivalent load curve $\dot{P}(t)$ composed of active and reactive load duration curves also has zero load time.

   i. Calculate the knee point time t_{Px} of the active load duration curve. The formula of load duration curve for consumption at pumping is $P_y(t)$ = 1250(1 − t_{Px}/8.5). Take $P_y(t)$ = 1200, and calculate the knee point time t_{Px} = 0.34*h*.
   ii. Calculate the load factor of the active load duration curve.

$$\begin{aligned} f_P &= \int_0^{t_{Px}} P(t)\mathrm{d}t/(P_{max} t_{Px}) \\ &= \left[1200 \times 0.34 + \frac{1}{2} \times 0.34 \times 50 + \frac{1}{2} \times 1200 \times (14.5 - 0.34)\right] \Big/ (1250 \times 14.5) \\ &= 8912.5/(1250 \times 14.5) = 0.4917 \end{aligned}$$

   iii. Calculate the load factor of the active load duration curve $f_Q = \int_0^1 Q(t)\mathrm{d}t/(Q_{max} \cdot 24)$. Similar to the above, calculate the knee point time t_{Qx} = 0.34h. As cosφ_y = cosφ_f = 0.85, tanφ_y = tanφ_f = 0.619 7, obtaining Q_{maxy} = 1250 × 0.619 7 = 774.6 (Mvar), Q_{maxf} = 1200 × 0.619 7 = 743.64 (Mvar). According to Figure 4.6,

$$\begin{aligned} \int_0^{1.0} Q(t)\mathrm{d}t &= \frac{1}{2}(774.7 - 743.6) \times 0.34 + 743.6 \times 0.34 + \frac{1}{2} \times 743.6 \\ &\times (14.5 - 0.34) + \frac{1}{2}(30 + 20) \times (24 - 14.5) \\ &= 5760.3\,(\text{Mvar} \cdot \text{h}) \end{aligned}$$

iv. Calculate the load factor $\dot{f}$ of the equivalent load curve. As $t_1 = 14.5/24 = 0.604\ 2$, $\tan\varphi_{max} = \tan\varphi_f = \tan\varphi_y = 0.619\ 7$, the load rate is obtained as per Formula (4.16) as follows:

$$\dot{f} = \frac{1}{1.604\ 2} \times (0.309\ 8 \times 0.619\ 7 + 0.604\ 2 \times 0.491\ 7) = 0.304\ 9$$

As $\dot{\beta} = 0$, the loss coefficient of the equivalent load curve is

$$\dot{F} = 0.639\dot{f}^2 + 0.361\dot{f} = 0.639 \times (0.304\ 9)^2 + 0.361 \times 0.304\ 9 = 0.169\ 5$$

Substitute it into Formula (4.17) to obtain

$$\Delta A = \frac{1.2}{(506.25 \times 10^3)^2} \times (1250 \times 10^6)^2 \times 0.169\ 5 \times 1.604\ 2 \times 24 \times 10^{-3}$$
$$= 47.74 \times 10^3 (\text{kW} \cdot \text{h})$$

In the above formula, $U_{av} = 506.25$ kV and is the average voltage under three operating conditions of generation, consumption, and phasing. The line loss calculated by the unequal time method is 5.09% smaller than the accurate line loss value, which is in the allowable range of engineering calculation.

4.5 Analysis of Errors of Various Line Loss Calculation Methods

This section analyzes different relative errors in the results of theoretical line loss calculation which can be based on records of ampere meters or watt-hour meters and can adopt different methods.

4.5.1 Analysis of Relative Error in Line Loss Calculated by rms Current Method

4.5.1.1 Analysis of Line Loss Calculation Based on Records of Ampere Meter

According to Formula (2.1), $\Delta A = 3I^2_{rms}RT \times 10^{-3}$ (kW·h). Take the total derivative of this formula to obtain

$$d(\Delta A) = (3I^2_{rms}T \times 10^{-3})dR + (3RT \times 10^{-3}) \cdot 2I_{rms}dI_{rms}$$

Divide by Formula (2.1) on both sides to obtain

$$\delta(\Delta A) = d(\Delta A)/\Delta A = dR/R + 2dI_{rms}/I_{rms} \approx \delta R + 2\delta I_{rms}$$

As

$$I_{rms} = \sqrt{\sum_{i=1}^{n} I_i^2/n}$$

then

$$dI_{rms} = \frac{1}{2}\left(1 \middle/ \sqrt{\sum_{i=1}^{n} I_i^2/n}\right)(2I_i/n)dI_i$$

obtaining $dI_{rms}/I_{rms} = (I_i/n)/(I_i^2/n)dI_i = dI_i/I_i$. Substitute the above formula into this to obtain

$$\delta(\Delta A) \approx \delta R + 2\delta I_i \tag{4.18}$$

According to Formula (4.18), the relative error in the line loss calculation by rms current method is the relative error in resistance calculation plus 2× the relative error in current meter measurements.

The resistance calculation can be corrected by taking into account the average temperature in the measuring period, so δR can be taken as 0. As to the current measurement of 10 kV and above loss calculation units, the accuracy classes of current transformer and current meter are Class 0.5 and Class 2.5, with errors of ±0.5% and ±2.5%, respectively. According to Formula (4.18), the maximum relative error in the line loss calculation by rms current method can be up to ±6% if the error in current meter measurements is considered.

4.5.1.2 Analysis of Line Loss Calculation Based on Records of Watt-hour Meter

First of all, use the records of a watt-hour meter to calculate the average current of a loss calculation unit within the measuring period, and then use the formula below to calculate the line loss.

$$\Delta A = 3I_{av}^2 K^2 RT \times 10^{-3}$$

Use the above method to obtain

$$\delta(\Delta A) \approx \delta R + 2\delta I_{av} + 2\delta K \tag{4.19}$$

If the resistance value is corrected based on the average temperature within the measuring period, and the form coefficient K is accurately calculated according to the records of the watt-hour meter, then δR and δK can be both taken as 0, so δI_{av} needs to be analyzed.

As

$$I_{av} = S_{av}/\left(\sqrt{3}U_{av}\right), dI_{av} = dS_{av}/\left(\sqrt{3}U_{av}\right) - S_{av}dU_{av}/\left(\sqrt{3}U_{av}^2\right)$$

Divide one formula by the other above to obtain

$$\delta I_{av} = dS_{av}/S_{av} - dU_{av}/U_{av} \approx \delta S_{av} - \delta U_{av} \tag{4.20}$$

Take total derivative of $S^2{}_{av} = P^2{}_{av} + Q^2{}_{av}$ to obtain $2S_{av}dS_{av} = 2P_{av}dP_{av} + 2Q_{av}dQ_{av}$. Divide one formula by the other above to obtain

$$\delta S_{av} = \frac{P_{av}dP_{av}}{P_{av}^2 + Q_{av}^2} + \frac{Q_{av}dQ_{av}}{P_{av}^2 + Q_{av}^2} = \cos^2\varphi_{av}\frac{dP_{av}}{P_{av}} + \sin\varphi_{av}\frac{dQ_{av}}{Q_{av}} \tag{4.21}$$

Municipal power grid enterprises have the largest number of watt-hour meters with monthly electricity of 1×10^6 kW·h and below. The accuracy class of such Type II active watt-hour meters is Class 1, and reactive watt-hour meters Class 2. The accuracy class of current and voltage transformers is Class 0.5. When $\cos\varphi_{av} = 0.85$, $\sin\varphi_{av} = 0.527$, and $\delta S_{av} = \pm 2.27\%$ calculated by Formula (4.21).

When accuracy class of voltage transformer is Class 2.5 and current transformer Class 0.5, $\delta U_{av} = \pm 3.0\%$, and $\delta I_{av} \approx (2.27 - 3.0)\% = \pm 0.73\%$ calculated by Formula (4.20), then $(\Delta A) \approx 2\delta I_{av} = \pm 1.5\%$. Therefore, by the rms current method, the relative error in the line loss calculation with the readings of a watt-hour meter is only one-quarter of that in the line loss calculation with the readings of a current meter.

4.5.2 Analysis of Relative Error in Line Loss Calculated by Loss Factor Method

4.5.2.1 Analysis of Line Loss Calculation Based on Ampere Meter Records

As mentioned in Chapter 2, the loss factor method uses the maximum current and the loss factor for line loss calculation, as shown in the formula below:

$$\Delta A = 3I_{\max}^2 F_I RT \times 10^{-3}$$

Take the total derivative of the above formula and then divide by the above formula to obtain

$$\delta(\Delta A) = \mathrm{d}(\Delta A)/\Delta A \approx \delta R + 2\delta I_{\max} + \delta F_I \tag{4.22}$$

The quantity of electricity passing a loss calculation unit during the measuring period can be calculated as per the following formula:

$$A = \sqrt{3}U_{\mathrm{av}}(I_{\max} f_I)\cos\varphi_{\mathrm{av}} T \times 10^{-3}$$

Then

$$f_I = A\Big/\left(\sqrt{3}U_{\mathrm{av}} I_{\max}\cos\varphi_{\mathrm{av}} T \times 10^{-3}\right)$$

Take the rate of change of the current load factor f_I to the maximum current.

$$\mathrm{d}f_I/\mathrm{d}I_{\max} = A\Big/\left(\sqrt{3}U_{\mathrm{av}}\cos\varphi_{\mathrm{av}} T \times 10^{-3}\right)\left(-\frac{1}{I_{\max}^2}\right)$$

Divide by the above formula to obtain

$$\mathrm{d}f/f_I = -(\mathrm{d}I_{\max}/I_{\max}) = -\delta I_{\max}$$

Calculate the differential of the simple loss factor formula $F = 0.2f_I + 0.8f^2{}_I$ to obtain

$$\mathrm{d}F = (0.2 + 1.6f_I)\mathrm{d}f$$

Then

$$\delta F_I \approx \mathrm{d}F/F_I = (0.2 + 0.16f_I)\mathrm{d}f_I/\left(0.2f_I + 0.8f_I^2\right) = k(\mathrm{d}f/f_I)$$

In the formula, $k = (0.2 + 1.6f_I)/(0.2 + 0.8f_I) > 1.0$, and substitute the formula into Formula (4.22) to obtain

$$\delta(\Delta A) \approx \delta R + (2-k)\delta I_{\max} \tag{4.23}$$

If $f_I = 0.50 \sim 0.80$, then $k = 1.67 \sim 0.76$. Assume $\delta R = 0$ and the maximum relative error in current meter measurements is $\delta I_{\max} = \pm 3\%$, and use Formula (4.23) to calculate

$$\delta(\Delta A) = (0.33 \sim 0.24) \times \pm 3\% = \pm 0.99\% \sim \pm 0.72\%$$

As a result, with current meter measurements, the relative error in the line loss calculation by the loss factor method is one-sixth to one-eighth of that by the rms current method, because there is an inverse relation between the maximum value and the load factor, and relative errors in these two values can offset each other.

4.5.2.2 Analysis of Line Loss Calculation Based on Watt-hour Meter Records

The maximum apparent power S_{max} calculated with accurate records of the watt-hour meter and the calculated apparent power loss factor F_S can be used for the line loss calculation within the measuring period T.

$$\Delta A = \left(R/U_{av}^2\right) S_{max}^2 F_S T \times 10^{-3} \tag{4.24}$$

Similar to the above, take the total derivative of Formula (4.24) and then divide by Formula (4.24) to obtain the formula of relative error in the line loss calculation:

$$\delta(\Delta A) \approx \delta R + 2\delta S_{max} + \delta F_S - 2\delta U_{av} \tag{4.25}$$

The quantity of electricity passing a loss calculation unit can be calculated as per the following formula:

$$A = P_{max} f_P T \times 10^{-3} = (S_{max} \cos\varphi_S) f_P T \times 10^{-3}$$

So $f_P = A/(S_{max}\cos\varphi_S T \times 10^{-3})$, $df_P/dS_{max} = -A/(S^2_{max}\cos\varphi_S T \times 10^{-3})$, obtaining

$$df_P/f_P = -dS_{max}/S_{max}$$

According to the loss factor calculation table $F_S(f_P, \cos\varphi_{max})$ in Table 4.1, when f_P is reduced from 0.90 to 0.8 and $\cos\varphi_{max} = 0.80$, the corresponding F_S is reduced from 0.844 to 0.705, resulting in $dF_S/F_S = 1.48 df_P/f_P$. Substitute $F_P = 0.2f_P + 0.8f^2_P$ into Formula (4.7) for a similar calculation. When $\cos\varphi_{max} = 0.80$, $dF_S/F_S = 1.59 df_P/f_P$. When analyzing and calculating the error, take $dF_S/F_S \approx 1.50(df_P/f_P) = -1.50(dS_{max}/S_{max})$, and substitute it into Formula (4.25) to obtain the following formula:

$$\delta(\Delta A) \approx \delta R + (2 - 1.5)\delta S_{max} - 2\delta U_{av} \tag{4.26}$$

Similar to Formula (4.21), the formula of relative error in the maximum apparent power can be obtained.

$$\delta S_{max} = \cos\varphi_S (dP_{max}/P_{max}) + \sin^2\varphi_S (dQ_{max}/Q_{max}) \tag{4.27}$$

Under normal circumstances, the power factor at maximum apparent power time is smaller than the average power factor during the measuring period. For the consistency of comparison, when $\cos\varphi_{av} = 0.85$, $\cos\varphi_S = 0.80$. With the same accuracy of the watt-hour meter and transformer, use Formula (4.27) to obtain $\delta S_{max} = \pm 2.36\%$. Substitute $\delta U_{av} = \pm 3.0\%$ into Formula (4.26) to obtain $\delta(\Delta A) \approx \pm 4.82\%$.

According to the above analysis of relative error:

1. Based on the current meter measurements, the relative error in the line loss calculation by the loss factor method is smaller than that by the rms current method. If only the current meter measurements are available, the loss factor method is recommended for the theoretical line loss calculation.
2. By means of the rms current method, the relative error in the line loss calculation based on watt-hour meter measurements is much smaller than that based on current meter measurements.
3. Based on the watt-hour meter measurements for in the line loss calculation, the relative errors obtained by the above two methods are generally below 5%, and both methods can be selected depending on specific data conditions and problems to be analyzed.

5

Line Loss Calculation after Reactive Compensation

5.1 Calculation of Load Curve Parameters after Reactive Compensation

The ratio of the actual active compensation capacity to the minimum reactive load is called the reactive compensation degree p; $p = Q_{com}/Q_{min}$. When $p < 1$, it is called under-compensation; when $p = 1$, it is called full compensation; when $p > 1$, it is called over-compensation. Subscripts 1 and 2 can be used to express parameters before and after the compensation.

5.1.1 Calculation of Reactive Load Curve Parameters at Under-Compensation

As shown in Figure 5.1, when $p\beta_Q < \beta_Q$,

$$f_{Q2} = \frac{f_{Q1}Q_{\max 1} - Q_{com} \times 1}{(Q_{\max 1} - Q_{com}) \times 1} = \frac{f_{Q1} - p\beta_{Q1}}{1 - p\beta_{Q1}} \tag{5.1}$$

$$\beta_{Q2} = \frac{(\beta_{Q1} - p\beta_{Q1})Q_{\max 1}}{Q_{\max 2}} = \frac{\beta_{Q1}(1-p)}{1 - p\beta_{Q1}} \tag{5.2}$$

$$\tan\varphi_{av2} = \frac{f_{Q2}Q_{\max 2}}{f_P P_{\max}} = \left(1 - p\frac{\beta_{Q1}}{f_{Q1}}\right)\tan\varphi_{av1} \tag{5.3}$$

According to the above formulas, reactive compensation changes the shape and characteristic parameters of the reactive load curve, so $f_{Q2} < f_{Q1}$, $\beta_{Q2} < \beta_{Q1}$, and $\tan\varphi_{av2} < \tan\varphi_{av1}$ after under-compensation, thus $\cos\varphi_{av2} > \cos\varphi_{av1}$. This indicates a smaller reactive load factor and smaller minimum load rate but a larger average power factor after compensation.

Line Loss Analysis and Calculation of Electric Power Systems, First Edition. Anguan Wu and Baoshan Ni.

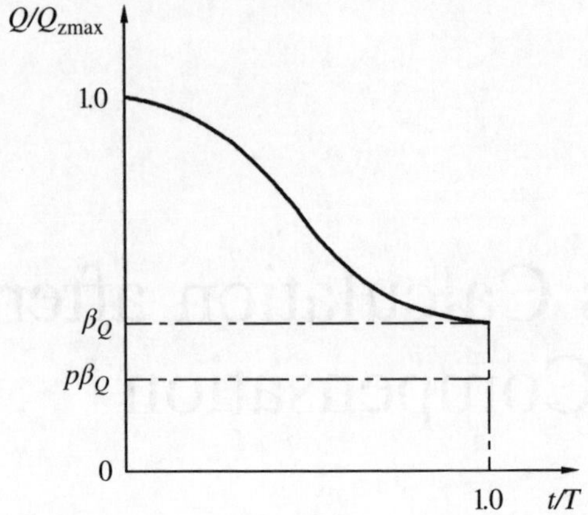

Figure 5.1 Under-compensation, $p < 1$.

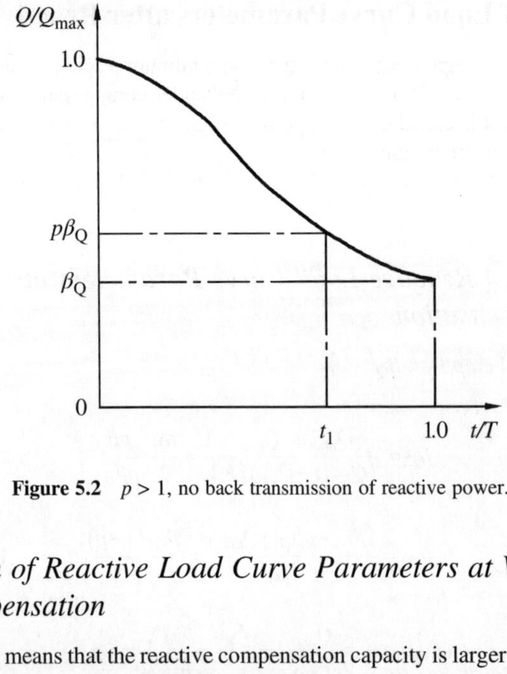

Figure 5.2 $p > 1$, no back transmission of reactive power.

5.1.2 Calculation of Reactive Load Curve Parameters at Weak Over-Compensation

Weak over-compensation means that the reactive compensation capacity is larger than the minimum reactive load but smaller than the average reactive load and includes the following two possible situations.

1. If the compensation capacity can be automatically regulated, there is no back transmission of reactive power in the line. As shown in Figure 5.2, $Q(t_1) = p\beta_{Q1}$, t_1 should be first obtained to calculate the load factor f_{Q2} of the full period.

   The analysis of reactive load curves of many lines indicates that, when $f_{Q1} < 0.70$, the condition $f_{Q1} < \frac{1}{2}(1+\beta_{Q1})$ can be satisfied, and the Type 2 Rossander Formula (3.7) of the load duration curve can be used to determine t_1, that is

$$p\beta_{Q1} = \beta_{Q1} + \left(1-\beta_{Q1}\right)\left(1-t_1\right)^{\frac{1}{\lambda}}$$

So

$$1 - t_1 = [\beta_{Q1}(p-1)/(1-\beta_{Q1})]^{\lambda}$$

that is

$$t_1 = 1 - \left[\frac{\beta_{Q1}(p-1)}{1-\beta_{Q1}}\right]^{\lambda} \tag{5.4}$$

After integral operation and simplification,

$$\begin{aligned} f_{Q2} &= \left[\int_0^{t_1} Q(t)\mathrm{d}t - p\beta_{Q1}t_1\right] \Big/ Q_{\max 2} \\ &= \frac{f_{Q1}-\beta_{Q1}}{1-\beta_{Q1}} - t_1 \frac{\beta_{Q1}(p-1)(1-f_{Q1})}{(1-\beta_{Q1})(1-p\beta_{Q1})} \end{aligned} \tag{5.5}$$

According to Figure 5.2, $\beta_{Q2} = 0$ at this time.

2. If the compensation capacity cannot be regulated, there is a brief back transmission of reactive power in the line.

   In Figure 5.3, the dashed area is the reactive electricity back-transmitted; and this can be calculated as per $\int_{t_1}^{1} [Q_{\text{com}} - Q(t)]\mathrm{d}t$, so f_{Q2} can be calculated as per the following formula:

$$f_{Q2} = \left\{ \int_0^{t_1} [Q(t) - Q_{\text{com}}]\,\mathrm{d}t + \int_0^{1} [Q_{\text{com}} - Q(t)]\,\mathrm{d}t \right\} \Big/ Q_{\max 2}$$

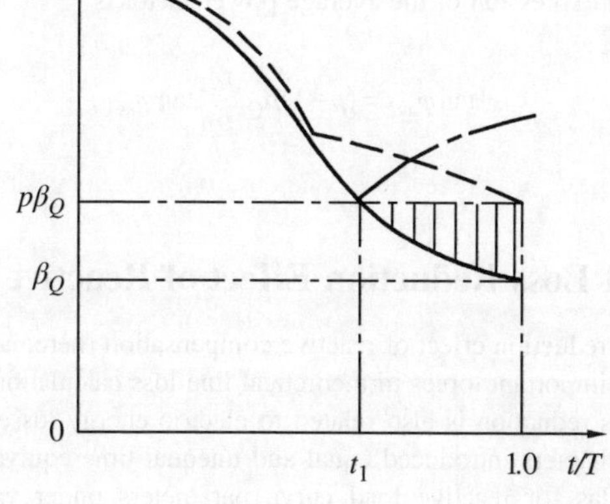

Figure 5.3 $p > 1$, back transmission of reactive power (dashed line is the load duration curve considering back transmission of reactive power).

After integral operation and simplification,

$$f_{Q2} = \frac{f_{Q1} - p\beta_{Q1}}{1 - p\beta_{Q1}} + 2(1 - t_1)\frac{\beta_{Q1}(p-1)(1 - f_{Q1})}{(1 - \beta_{Q1})(1 - p\beta_{Q1})} \tag{5.6}$$

$\beta_{Q2} = 0$ at this time.

Under the above two situations, $Q_{max2} = (1 - p\beta_{Q1})Q_{max1}$, so

$$\tan\varphi_{av2} = (1 - p\beta_{Q1})\frac{f_{Q2}}{f_{Q1}}\tan\varphi_{av1} \tag{5.7}$$

5.1.3 *Calculation of Reactive Load Curve Parameters at Strong Over-Compensation*

Strong over-compensation means that the reactive compensation capacity is larger than the average reactive load but smaller than the maximum reactive load and it includes the following two possible situations.

1. If the compensation capacity can be automatically regulated, there is no back transmission of reactive power in the line. f_{Q2} can be calculated as per Formula (5.5), and the average power factor can still be calculated as per Formula (5.7).
2. If the compensation capacity cannot be regulated, there is lengthy back transmission of reactive power in the line. After integral operation and simplification,

$$f_{Q2} = 2(1 - t_1)\frac{(1 - f_{Q1})}{(1 - \beta_{Q1})} - \frac{(p\beta_{Q1} - f_{Q1})}{(1 - p\beta_{Q1})} \tag{5.8}$$

The maximum reactive power back-transmitted is larger than the maximum reactive power forward-transmitted, so

$$Q_{max\,2} = (Q_{com} - Q_{min\,1}) = (p\beta_{Q1} - \beta_{Q1})Q_{max\,1}$$

Therefore, the relational expression of the average power factor is

$$\tan\varphi_{av2} = (p-1)\beta_{Q1}\frac{f_{Q2}}{f_{Q1}}\tan\varphi_{av1} \tag{5.9}$$

5.2 Calculation of Loss Reduction Effect of Reactive Compensation

The calculation of the loss reduction effect of reactive compensation (hereinafter referred to as the reactive compensation effect) is an important topics in theoretical line loss calculation, and it is also a complicated issue because the total loss reduction is also related to electric energy losses caused by the transmission of active loads. Previous chapters introduced equal and unequal time equivalent load curve methods and derived calculation formulas for reactive load curve parameters under various compensations, which provided conditions for the calculation of the reactive compensation effect.

5.2.1 Calculation of Compensation Effect at High Natural Power Factor

The power factor before compensation is generally called the natural power factor. Urban public lines and some dedicated lines have a high natural power factor and generally meet the condition of $\cos\varphi_{av} > 0.7$. Therefore, the line loss calculation before and after the compensation should use relevant formulas under $\dot{P}_{max} = P_{max}$ listed in the left column of Table 4.4. The calculation of reactive compensation effect is introduced below by three situations.

1. In the case of under-compensation, first calculate the compensation degree p based on the average power factor after compensation, that is

$$p = \frac{f_{Q1}}{\beta_{Q1}}\left(1 - \frac{\tan\varphi_{av2}}{\tan\varphi_{av1}}\right) \tag{5.10}$$

Then, use p to calculate the parameters f_{Q2} and β_{Q2} of the reactive load curve after compensation.

Note that $\beta_{Q2} \neq \dot{\beta}$, and because β_{Q2} is based on Q_{max2}, and $\dot{\beta}$ is based on $\dot{P}_{max} = P_{max}$, so $\dot{\beta}$ should be calculated as per the following formula:

$$\dot{\beta} = \beta_{Q2}\tan\varphi_{av2}/k_{f2} \tag{5.11}$$

$$k_{f2} = f_{Q2}/f_P$$

Wherein k_{f2} – ratio of reactive load factor to active load factor after compensation.

2. In the case of over-compensation, the compensation capacity can be switched by group without back transmission of reactive power. First of all, determine the compensation degree p. When $1 < p < f_{Q1}/\beta_{Q1}$, it is weak over-compensation; when $f_{Q1}/\beta_{Q1} < p < \frac{1}{\beta_{Q1}}$, it is strong over-compensation. Calculate the period t_1 during which both active power and reactive power are transmitted in the line, and then calculate the parameter f_{Q2} of the reactive load curve after compensation and $\tan\varphi_{av2}$ of the average power factor after compensation. Use t_1, $\tan\varphi_{av2}$, and f_P to calculate the parameter $\dot{f}_2$ of the equivalent load curve after compensation.

   Obviously, $\beta_{Q2} = 0$ after over-compensation, so $\dot{\beta} = 0$, and the calculation formula of loss factor $\dot{F}_2$ of the equivalent load curve can be simplified to that including only one parameter, $\dot{f}_2$. The electric energy loss ΔA can be calculated as per Formula (4.17).

3. In the case of over-compensation, the compensation capacity cannot be regulated and back transmission of reactive power occurs. The calculation process is the same as the above. Note that $\tan\varphi_{av2}$ after strong over-compensation can be calculated as per Formula (5.9), and f_{Q2} values at weak over-compensation and strong over-compensation can be calculated as per Formulas (5.6) and (5.8), respectively.

   For ease of use, the calculation process and the formulas applied under the above three situations are shown in Table 5.1.

Example 5.1 For a 10 kV dedicated line in a certain city, the parameters of its quarterly load curve are as follows: $f_P = 0.90$, $\beta_P = 0.712$, $f_{Q1} = 0.887$, $\beta_{Q1} = 0.676$, $\cos\varphi_{av1} = 0.70$, $P_{max} = 4000$ kW, and the resistance of the line is 1 Ω per phase (the temperature factor is taken into account).

1. *Try to calculate reactive compensation capacities needed when the average power factor is increased to 0.80, 0.85 and 0.90.*
2. *Based on a quarter as the calculation period, try to compare loss reduction effects of various compensation capacities.*

Table 5.1 Summary of calculation formulas for electric energy losses after reactive compensation.

Reactive load curve parameters	Equivalent load curve parameters	Loss factor	Electric energy loss
$f_{Q2}=(f_{Q1}-\beta_{Q1})/(1-p\beta_{Q1})$ $\beta_{Q2}=(\beta_{Q1}-p\beta_{Q1})/(1-p\beta_{Q1})$	$\dot{f}_2=\frac{1}{2}f_P(1+\tan\varphi_{av2})$ $\dot{\beta}_2=\beta_{Q2}\tan\varphi_{av2}/k_{f2}$	$\dot{F}_2=0.639\dot{f}_2^2+0.361(\dot{f}_2+\dot{f}_2\dot{\beta}_2-\dot{\beta}_2)$	$\Delta A=\frac{R}{U_{av}^2}\dot{P}_{max}^2\dot{F}_2\times 2T\times 10^{-3}$
$t_1=1-[\beta_{Q1}(p-1)/(1-\beta_{Q1})]^{\lambda}$; $\lambda=\frac{f_{Q1}-\beta_{Q1}}{1-f_{Q1}}$	$\dot{f}_2=\frac{1}{1+t_1}(f_P+t_1f_P\tan\varphi_{av2})$	$\dot{F}_2=0.639\dot{f}_2^2+0.361\dot{f}_2$	$\Delta A=\frac{R}{U_{av}^2}\dot{P}_{max}^2\times\dot{F}_2(1+t_1)T\times 10^{-3}$
$f_{Q2}=\frac{f_{Q1}-\beta_{Q1}}{1-\beta_{Q1}}-t_1\frac{\beta_{Q1}(p-1)(1-f_{Q1})}{(1-\beta_{Q1})(1-p\beta_{Q1})}$ $f_{Q2}=\frac{f_{Q1}-\beta_{Q1}}{1-\beta_{Q1}}+2(1-t_1)\frac{\beta_{Q1}(p-1)(1-f_{Q1})}{(1-\beta_{Q1})(1-p\beta_{Q1})}$	$\tan\varphi_{av2}=(1-p\beta_{Q1})\frac{f_{Q2}}{f_{Q1}}\tan\varphi_{av1}$		
$f_{Q2}=2(1-t_1)\frac{(1-f_{Q1})}{(1-\beta_Q)}-\frac{(p\beta_{Q1}-f_{Q1})}{(1-p\beta_{Q1})}$	$\tan\varphi_{av2}=(p-1)\beta_{Q1}\frac{f_{Q2}}{f_{Q1}}\tan\varphi_{av1}$		
$\beta_{Q2}=0$	$\dot{\beta}_2=0$		

Solutions

1. *Calculation of electric energy loss before compensation.* $\tan\varphi_{av1}$ = 1.020 2 is given, so

$$\dot{f}=\frac{1}{2}f_P(1+\tan\varphi_{av1})=\frac{1}{2}\times 0.90\times(1+1.020\ 2)=0.909\ 1$$

$$\dot{\beta}=\beta_{Q1}\tan\varphi_{av1}/k_{f1}=\frac{0.676\times 1.020\ 2}{0.887/0.90}=0.699\ 8$$

$$\begin{aligned}\dot{F}&=0639\dot{f}^2+0.361(\dot{f}+\dot{f}\dot{\beta}-\dot{\beta})\\&=0.639\times 0.909\ 1^2+0.361\times(0.909\ 1+0.909\ 1\times 0.699\ 8-0.699\ 8)\\&=0.833\ 3\end{aligned}$$

Use Formula (4.16) to calculate the electric energy loss before compensation as

$$\begin{aligned}\Delta A_1&=\frac{R}{U_{av}^2}\dot{P}_{max}^2\dot{F}\times 2T\times 10^{-3}\\&=\frac{1}{(10^4)^2}\times(4\times 10^6)^2\times 0.833\ 3\times 2\times(8760/4)\times 10^{-3}\\&=58.398\times 10^4(\text{kW}\cdot\text{h})\end{aligned}$$

2. *Calculation of compensation degree and compensation capacity.* Use Formula (5.10) to calculate the compensation degree p. $Q_{max1}=P_{max}\tan\varphi_{av1}/k_{f1}$, so the calculation formula of the compensation capacity can be obtained as

$$Q_{com}=p\beta_{max1}Q_{max1}=p\beta_{Q1}P_{max}\tan\varphi_{av1}/k_{f1}$$

The calculation results of the compensation degree p and the compensation capacity are shown in Tables 5.2 and 5.3.

In Table 5.2, the loss reduction rate of reactive compensation is

$$\delta(\Delta A)\% = \frac{\Delta A_1 - \Delta A_2}{\Delta A_1} \times 100\% = \frac{\dot{F}_2 - \dot{F}_1}{\dot{F}_1} \times 100\%$$

$$\delta(\Delta P)\% = \left(1 - \frac{\cos^2\varphi_{av1}}{\cos^2\varphi_{av2}}\right) \times 100\%$$

According to Table 5.2, when f_P is closer to 1.0, $\delta(\Delta P)\%$ is closer to the loss reduction rate $\delta(\Delta A)\%$.

$\delta(\Delta A)$ and Q_{com} values listed in Tables 5.2 and 5.3 can be used to calculate the average electric energy loss C_A reduced by installation of reactive compensation equipment per unit capacity. If the calculation period is one year (8760 h), then the following results can be obtained:

$\cos\varphi_{av}$ 0.70 → 0.80

$$C_{A1} = 4 \times \frac{131\ 629}{972.7} = 541.3(\text{kW·h/kvar})$$

$\cos\varphi_{av}$ 0.80 → 0.85

$$C_{A2} = 4 \times \frac{184\ 946 - 131\ 629}{1441.8 - 972.7} = 454.63(\text{kW·h/kvar})$$

$\cos\varphi_{av}$ 0.85 → 0.90

$$C_{A3} = 4 \times \frac{233\ 534 - 184\ 946}{1929.1 - 1441.8} = 398.83(\text{kW·h/kvar})$$

According to the above results, $C_{A2} = 0.84C_{A1}$, $C_{A3} = 0.737C_{A1}$. This indicates that the effect of installation of the first kilovar reactive compensation equipment is better than that of installation of subsequent kilovar reactive compensation equipment. Therefore, if the amount of investment is certain, the distribution of reactive compensation capacities must be optimized, which should not only meet other constraint conditions such as voltage quality, but also gain the optimal loss reduction benefits.

5.2.2 *Calculation of Reactive Compensation Effect at Low Natural Power Factor*

If the natural power factor is $\cos\varphi_{av} < 0.70$, $\dot{P}_{max} = Q_{max1}$, the formulas in the right column of Table 4.4 should be used to calculate $\dot{f}$, $\dot{\beta}$ and the electric energy loss, that is

$$\Delta A_1 = \frac{R}{U_{av}^2}\dot{P}_{max}^2 \dot{F}_1 \times 2T \times 10^{-3} = \frac{R}{U_{av}^2} Q_{max1}^2 \dot{F}_1 \times 2T \times 10^{-3}$$

If the average power factor $\cos\varphi_{av2}$ is still smaller than 0.70 after compensation, at this time $Q_{max2} > P_{max}$, so $\dot{f} = \frac{1}{2} f_{Q2}\left(1 + \frac{1}{\tan\phi_{av2}}\right)$, $\dot{\beta} = \beta_{Q2}$. The electric energy loss is

$$\Delta A_2 = \frac{R}{U_{av}^2}\dot{P}_{max}^2 \dot{F}_2 \times 2T \times 10^{-3} = \frac{R}{U_{av}^2} Q_{max2}^2 \dot{F}_2 \times 2T \times 10^{-3}$$

Table 5.2 Calculation of loss reduction effect of reactive compensation (part 1).

$\cos\varphi_{av2}$	$\tan\varphi_{av2}$	p	Q_{com} (kvar)	$\dot{f}$	β_{Q2}	f_{Q2}	k_{f2}	$\dot{\beta}$	$\dot{F}_2$	$\delta(\Delta A)$ (kW·h)	$\delta(\Delta A)$%	$\delta(\Delta P)$%
0.80	0.75	0.3475	972.7	0.787 5	0.5765	0.852 3	0.9470	0.456 6	0.645 5	131 629	22.54	23.44
0.85	0.619 7	0.515 1	1441.8	0.728 9	0.5029	0.826 6	0.9184	0.339 3	0.569 4	184 946	31.67	32.18
0.90	0.484 3	0.6892	1929.1	0.667 9	0.3934	0.788 4	0.8760	0.217 5	0.500 1	233 534	39.99	39.51

Table 5.3 Calculation of loss reduction effect of reactive compensation (part 2).

$\cos\varphi_{av2}$	$\tan\varphi_{av2}$	p	Q_{com} (kvar)	$\dot{f}$	β_{Q2}	f_{Q2}	k_{f2}	$\dot{\beta}$	$\dot{F}_2$	$\dot{F}_{2inc}$	$\delta(\Delta A)$%	$\delta(\Delta P)$%
0.70	1.020 2	0.4486	781.9	0.554 6	0.211 1	0.560 1	1.120 2	0.211 1	0.362 8	0.264 3	25.80	26.53
0.75	0.881 9	0.6470	1127.5	0.470 5	0.1462	0.524 0	1.0480	0.123 0	0.287 8	0.2527*	29.06	36.0
0.80	0.75	0.8361	1457.5	0.437 5	0.0737	0.483 5	0.9670	0.057 2	0.268 6	0.2359*	33.77	43.75

Note: $\dot{F}_{2inc} = (k_{f1}/\tan\varphi_{av1})^2 \dot{F}_2 = 0.878\ 2\dot{F}_2$

Table 5.4 Calculation curves $F_y(f_P, \cos\varphi_{max}, \dot{\beta}_y)$ of annual electric energy losses of 35–110 kV transmission lines.

		f_P										
β_y	$\cos\varphi_{max}$	0.35	0.40	0.45	0.50	0.55	0.60	0.70	0.80	0.85	0.90	0.95
	0.75	0.155 5	0.197 9	0.243 1	0.291 2	0.341 9	0.395 5	0.511 1	0.638 0	0.705 4	0.775 9	0.848 9
	0.80	0.133 3	0.171 2	0.211 6	0.254 4	0.299 6	0.347 2	0.449 6	0.561 9	0.621 6	0.683 8	0.748 3
0.20	0.85	0.112 2	0.145 9	0.181 8	0.219 7	0.259 6	0.301 7	0.391 9	0.490 5	0.543 0	0.597 5	0.654 0
	0.90	0.091 1	0.120 8	0.152 1	0.185 2	0.220 0	0.256 7	0.335 1	0.420 5	0.465 8	0.512 9	0.561 7
	0.95	0.068 1	0.093 2	0.119 8	0.147 7	0.177 1	0.207 9	0.273 7	0.345 0	0.382 8	0.422 0	0.462 5
	0.75	0.131 9	0.176 0	0.222 9	0.272 7	0.325 1	0.380 4	0.499 4	0.629 6	0.698 8	0.770 9	0.845 6
	0.80	0.108 8	0.148 3	0.190 2	0.234 6	0.281 3	0.330 5	0.436 1	0.551 6	0.612 9	0.676 5	0.742 7
0.30	0.85	—	0.122 0	0.159 3	0.198 6	0.240 0	0.283 6	0.376 7	0.478 2	0.532 1	0.588 0	0.646 1
	0.90	—	0.095 7	0.128 4	0.162 9	0.199 0	0.237 0	0.318 1	0.406 2	0.452 8	0.501 2	0.551 3
	0.95	—	—	0.094 7	0.123 9	0.154 4	0.186 4	0.254 6	0.328 3	0.367 3	0.407 7	0.449 4
	0.75			0.202 8	0.254 2	0.308 3	0.365 3	0.487 6	0.621 2	0.692 1	0.765 9	0.842 3
	0.80			0.168 9	0.214 8	0.263 1	0.313 9	0.422 7	0.541 2	0.604 1	0.669 3	0.737 0
0.40	0.85			—	0.177 6	0.220 5	0.265 4	0.361 4	0.465 9	0.521 2	0.578 6	0.638 1
	0.90			—	—	0.178 0	0.217 3	0.301 1	0.391 8	0.439 7	0.489 5	0.540 9
	0.95			—	—	—	0.164 9	0.235 5	0.311 6	0.35 18	0.393 3	0.436 3
	0.75					0.291 5	0.350 2	0.475 9	0.612 9	0.685 4	0.760 9	0.839 0
	0.80					—	0.297 3	0.409 2	0.530 9	0.595 3	0.662 1	0.731 4
0.50	0.85					—	0.247 3	0.346 2	0.453 6	0.510 4	0.569 2	0.630 2
	0.90					—	—	0.284 0	0.377 5	0.426 7	0.477 8	0.530 6
	0.95					—	—	—	0.294 9	0.336 3	0.379 0	0.423 2
	0.75							0.464 1	0.604 5	0.678 7	0.755 9	0.835 7
	0.80							0.395 7	0.520 5	0.586 5	0.654 9	0.725 7
0.60	0.85							—	0.441 3	0.499 5	0.559 8	0.622 2
	0.90							—	0.363 1	0.413 7	0.466 2	0.520 2
	0.95							—	—	—	0.364 7	0.410 1
	0.75										0.750 9	0.832 4
	0.80										0.647 7	0.720 1
0.70	0.85										0.550 4	0.614 3
	0.90										—	0.509 9
	0.95										—	—

The loss reduction rate after compensation is

$$\delta(\Delta A)\% = \frac{\Delta A_1 - \Delta A_2}{\Delta A_1} \times 100\% = \left(1 - \frac{Q_{\max 2}^2 \dot{F}_2}{Q_{\max 1}^2 \dot{F}_1}\right) \times 100\%$$
$$= \left(1 - \frac{\dot{F}_{2\text{inc}}}{\dot{F}_1}\right) \times 100\% \tag{5.12}$$

Wherein $\dot{F}_{2\text{inc}}$ – loss factor included by the maximum value of the equivalent load curve after reactive compensation; the subscript "inc" refers to inclusive. $Q_{\max2} = Q_{\max1}(1 - p\beta_{Q1})$, so

$$\dot{F}_{2\text{inc}} = (1 - p\beta_{Q1})^2 \dot{F}_2 \tag{5.13}$$

After compensation, if the average power factor is $\cos\varphi_{\text{av2}} > 0.70$, at this time $Q_{\max2} < P_{\max}$, $\dot{P}_{\max 2} = P_{\max}$, the electric energy loss is

$$\Delta A_2 = \frac{R}{U_{\text{av}}^2} \dot{P}_{\max 2}^2 \dot{F}_2 \times 2T \times 10^{-3} = \frac{R}{U_{\text{av}}^2} P_{\max}^2 \dot{F}_2 \times 2T \times 10^{-3}$$

The loss reduction rate after compensation is

$$\delta(\Delta A)\% = \frac{\Delta A_1 - \Delta A_2}{\Delta A_1} \times 100\% = \left(1 - \frac{P_{\max}^2 \dot{F}_2}{Q_{\max 1}^2 \dot{F}_1}\right) \times 100\%$$
$$= \left(1 - \frac{\dot{F}_{2\text{inc}}}{F_1}\right) \times 100\%$$

$Q_{\max1} = P_{\max}\tan\varphi_{\text{av1}}/k_{\text{f1}}$, so

$$\dot{F}_{2\text{inc}} = \left(\frac{k_{\text{f1}}}{\tan\varphi_{\text{av1}}}\right)^2 \dot{F}_2 \tag{5.14}$$

Example 5.2 For a certain 35 kV agricultural electric line, the parameters of its quarterly load curve are as follows: $f_P = 0.50$, $\beta_P = 0.208\,2$, $f_Q = 0.624\,6$, $\beta_{Q1} = 0.326\,7$, $\cos\varphi_{\text{av1}} = 0.60$, $P_{\max} = 10$ MW, and the resistance of the line is $R = 1\ \Omega$ per phase (the temperature factor is taken into account).

1. *Try to calculate compensation capacities needed when the average power factor is increased to 0.70, 0.75 and 0.80.*
2. *If a quarter is taken as the loss calculation period, try to compare loss reduction effects of various compensations.*

Solutions

1. *Calculation of electric energy loss before compensation.* $\cos\varphi_{\text{av1}} < 0.70$, so $\dot{P}_{\max} = Q_{\max}$, obtaining $\dot{f} = \frac{1}{2} f_Q \left(1 + \frac{1}{\tan\varphi_{\text{av}}}\right) = \frac{1}{2} \times 0.624\,6 \times \left(1 + \frac{1}{1.333}\right) = 0.546\,5$. Since $\dot{P}_{\min} = P_{\min}$, $\dot{\beta} = \beta_P k_{\text{f1}} / \tan\varphi_{\text{av1}} = 0.208\,2 \times \frac{0.624\,6/0.50}{1.333} = 0.195\,1$. So the loss factor is

$$\begin{aligned}\dot{F} &= 0.639\dot{f}^2 + 0.361\left(\dot{f} + \dot{f}\dot{\beta} - \dot{\beta}\right)\\ &= 0.639 \times (0.546\ 5)^2 + 0.361 \times (0.546\ 5 + 0.546\ 5 \times 0.195\ 1 - 0.195\ 1)\\ &= 0.356\ 2\end{aligned}$$

As mentioned in Example 5.1, $Q_{\max 1} = P_{\max}\tan\varphi_{av1}/k_{f1} = (10 \times 10^6) \times \dfrac{1.333}{(0.624\ 6/0.50)} = 10.67 \times 10^6$ (Mvar), so the electric energy loss is

$$\begin{aligned}\Delta A_1 &= \frac{R}{U_{av}^2}\dot{P}_{\max}^2 F_1 \times 2T \times 10^{-3} = \frac{R}{U_{av}^2}Q_{\max 1}^2 \dot{F}_1 \times 2T \times 10^{-3}\\ &= \frac{1}{\left(3.5 \times 10^4\right)^2} \times \left(10.67 \times 10^6\right)^2 \times 0.356\ 2 \times 2\\ &\quad \times (8760/4) \times 10^{-3} = 14.50 \times 14^4 (\text{kW·h})\end{aligned}$$

2. *Calculation of compensation degree and compensation capacity.* See Table 5.3 for the calculation results. According to the comparison between $\delta(\Delta A)\%$ and $\delta(\Delta P)\%$ in Table 5.3, when the load factor is low, the replacement of $\delta(\Delta A)\%$ with $\delta(\Delta P)\%$ will lead to larger positive errors.

5.3 Calculation Curves of Annual Electric Energy Losses for Power Grid Planning and Design

During the planning and design stages of power grids, annual electric energy losses and annual operating costs of some units (main transformers, lines, etc.) need to be calculated, so as to compare the technology and economy of different programs and select parameters of electrical equipment or sectional area of line conductors. Because the calculation of annual electric energy losses is based on estimates of the annual load curve, the requirement on the accuracy of calculation results is not high, but the calculation method must be simple and reasonable.

5.3.1 Calculation Curves of Annual Electric Energy Losses of 35–110 kV Transmission Lines

The authors collected and analyzed data about loads of 35 and 110 kV transmission lines, finding that larger reactive loads pass medium-voltage transmission lines and change less significantly than active loads. Under general situations, the two load factors of such lines are $f_Q > f_P$. Statistical calculation can derive the following formula:

$$f_Q = 0.045\ 6 + 0.987\ 1 f_P (\text{correlation coefficient}\ r = 0.858) \quad (5.15)$$

Substitute Formula (5.15) into $\dot{f} = \frac{1}{2}(f_P + f_Q \tan\varphi_{\max})$, obtaining the calculation formula of the load rate of the annual equivalent load curve:

$$\dot{f}_y = f_P(0.5 + 0.493\ 6 \tan\phi_{\max}) + 0.022\ 8 \tan\varphi_{\max} \quad (5.16)$$

Substitute Formula (5.16) into Formula (4.13), obtaining the calculation formula of the loss factor of the annual equivalent load curve:

$$\dot{F}_y = 0.639\dot{f}_y + 0.361\left(\dot{f}_y + \dot{f}_y\dot{\beta}_y - \dot{\beta}_y\right) \quad (5.17)$$

Wherein $\dot{\beta}_y$ – minimum load rate of annual equivalent load curve; generally $\dot{\beta}_y = \beta_P$.

Use the annual active load factor f_P, the power factor at maximum load time $\cos\varphi_{max}$ (it is believed that the maximum values of active load and reactive load occur at the same time) and $\dot{\beta}_y$ as variables to calculate $\dot{F}_y$, as shown in Table 5.4. According to the data listed in Table 5.4, the calculation curves of annual electric energy losses of 35–110 kV transmission lines can be drawn as shown in Figure 5.4.

According to Table 5.4, when f_P and $\cos\varphi_{max}$ are certain values, a larger difference in $\dot{\beta}_y$ indicates a larger difference in $\dot{F}_y$. For example, when $f_P = 0.60$, $\cos\varphi_{max} = 0.85$, $\dot{\beta}_y$ at 0.2 and 0.5 matches $\dot{F}_y$ at 0.301 7 and

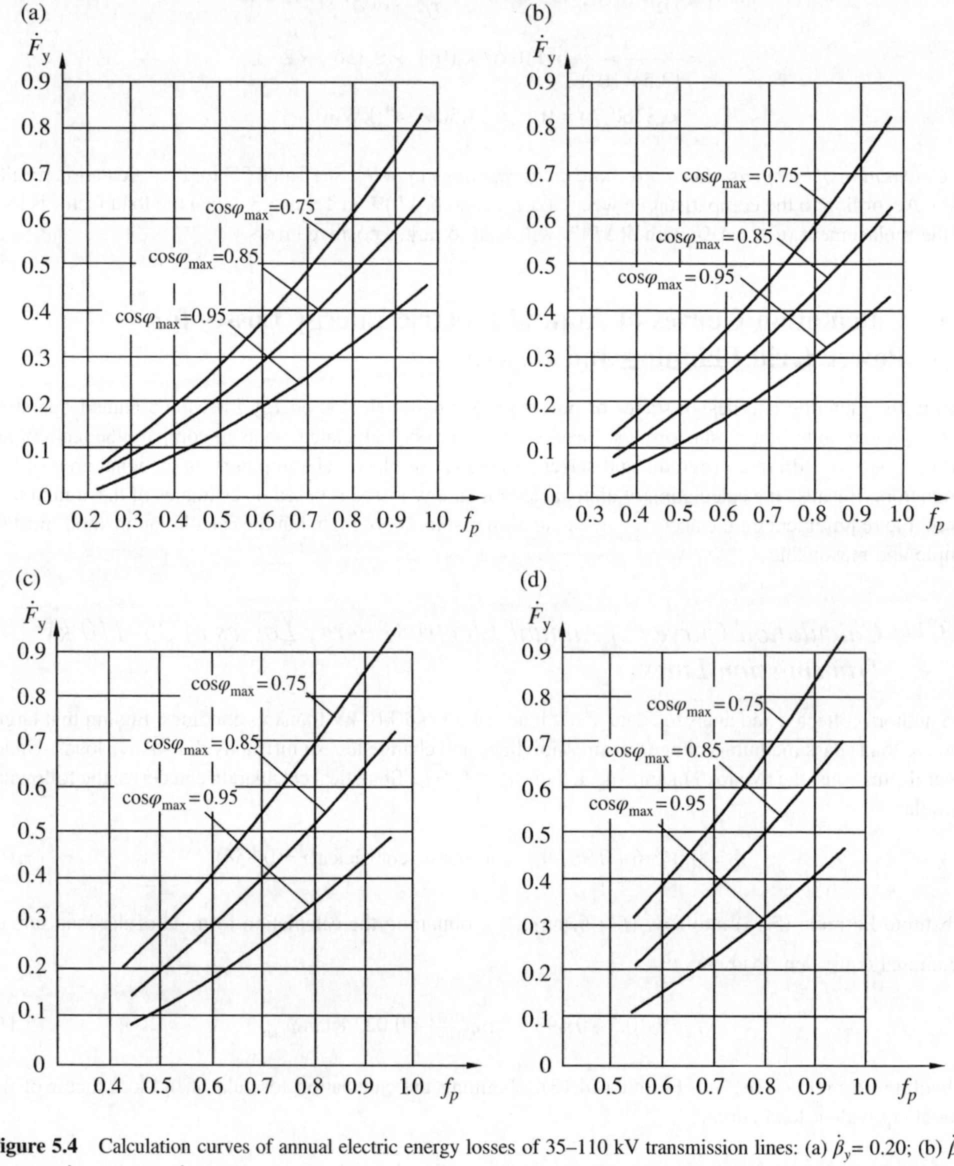

Figure 5.4 Calculation curves of annual electric energy losses of 35–110 kV transmission lines: (a) $\dot{\beta}_y$= 0.20; (b) $\dot{\beta}_y$= 0.30; (c) $\dot{\beta}_y$= 0.40; (d) $\dot{\beta}_y$= 0.50.

0.247 3, respectively, and the difference reaches 22%. This indicates that it is necessary to consider the minimum load rate $\dot{\beta}_y$ in the calculation of annual electric energy losses.

5.3.2 *Calculation Curves of Annual Electric Energy Losses of 220 kV Transmission Lines*

The authors also collected and analyzed daily load, quarterly load and annual load curves of some 220 kV transmission lines, indicating that centralized reactive compensation equipment is always installed in the receiving ends of such 220 kV lines to transmit more active power and reduce losses. Therefore, smaller reactive loads pass such 220 kV transmission lines and no transmission of reactive power often occurs for short time. These load curves have two characteristics, that is $f_Q < f_P$, $\beta_Q = 0$.

According to relevant analysis and calculation of the collected data,

$$f_Q = 0.787\,4 f_P - 0.0144 (\text{correlation coefficient}\, r = 0.759\,9) \tag{5.18}$$

Substitute Formula (5.18) into $\dot{f} = \frac{1}{2}(f_P + f_Q \tan\varphi_{\max})$, obtaining the calculation formula of the load rate $\dot{f}_y$ of the annual equivalent load curve:

$$\dot{f}_y = f_P(0.5 + 0.393\,7 \tan\varphi_{\max}) - 0.007\,2 \tan\varphi_{\max} \tag{5.19}$$

Given $\beta_Q = 0$, $\dot{\beta}_y = 0$, so the formula of the loss factor of the annual equivalent load curve is obtained:

$$\dot{F}_y = 0.639\dot{f}_y^2 + 0.361\dot{f}_y \tag{5.20}$$

Use the annual active load factor f_P and the power factor $\cos\varphi_{\max}$ at maximum load time as variables to calculate $\dot{F}_y$ as per Formulas (5.19) to (5.20), as shown in Table 5.5. According to the data listed in Table 5.5, the calculation curves of annual electric energy losses of 220 kV transmission lines can be drawn as shown in Figure 5.5.

Table 5.5 $\dot{F}_y$ calculation values of 220 kV transmission lines.

	f_P					
$\cos\varphi_{\max}$	0.40	0.45	0.50	0.55	0.60	0.65
0.85	0.160 7	0.189 0	0.219 0	0.250 8	0.284 3	0.319 6
0.90	0.146 0	0.171 3	0.198 0	0.226 4	0.256 2	0.287 6
0.95	0.129 8	0.151 8	0.175 1	0.199 6	0.225 5	0.252 5
	f_P					
$\cos\varphi_{\max}$	0.70	0.75	0.80	0.85	0.90	0.95
0.85	0.356 7	0.395 6	0.436 2	0.478 6	0.522 8	0.568 7
0.90	0.320 5	0.354 9	0.390 8	0.428 3	0.467 3	0.507 7
0.95	0.280 9	0.310 5	0.341 5	0.373 5	0.407 0	0.441 7

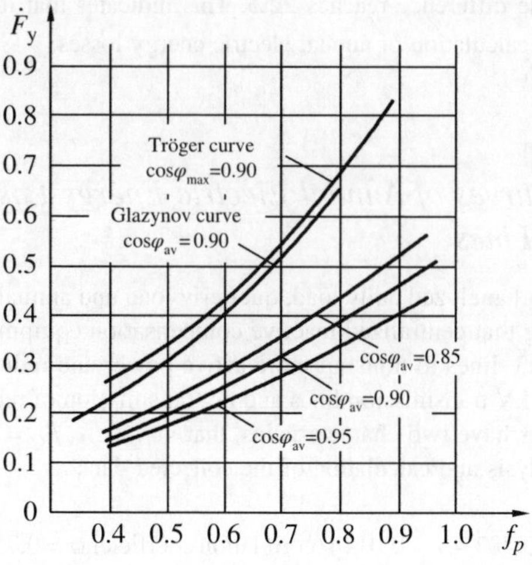

Figure 5.5 Calculation curves of annual electric energy losses of 220 kV transmission lines.

For comparison, Figure 5.5 shows a Tröger line loss calculation curve at $\cos\varphi_{max} = 0.90$ drawn using data in Table 4.1 and a Glazynov line loss calculation curve at $\cos\varphi_{av} = 0.90$ drawn using the interpolation method. According to the comparison, the latter line loss calculation curves have higher $\dot{F}_y$ values. Within the annual load rate range of 0.50–0.70, such value may be higher than 39–62%, so Glazynov curves should no longer be used for line loss calculations in the engineering design of high-voltage and urban power grids.

5.3.3 Calculation Curves of Annual Electric Energy Losses of Agricultural Electric Lines Consuming Electricity on a Quarterly Basis

The authors collected and analyzed data for annual load curves of agricultural electric lines consuming electricity on a quarterly basis in northern China, indicating that such lines have a lower annual active load factor, lower minimum active load rate, but higher annual reactive load factor due to the characteristics of quarterly consumption. Relevant analysis of the materials about load curves leads to

$$\left.\begin{aligned} f_Q &= 0.195\,0 + 0.829\,8 f_P \quad (\text{correlation coefficient } r = 0.9482) \\ \beta_P &= 0.527\,8 f_P - 0.081\,9 \end{aligned}\right\} \tag{5.21}$$

As per Formula (5.21), f_P serving as a variable that can constitute corresponding load curve parameters, as shown in Table 5.6.

When $\cos\varphi_{max} < 0.707$, use $\dot{f} = \frac{1}{2}(f_Q + f_P \tan\varphi_{max})$ and $\dot{\beta} = \beta_P / \tan\varphi_{max}$ to calculate $\dot{f}$ and $\dot{\beta}$, and use Formula (4.13) to calculate the loss factor $\dot{F}$ of the equivalent load curve. As $\cos\varphi_{max} < 0.707$, the maximum value of the equivalent load curve depends on the maximum reactive power, that is $\dot{P}_{max} = Q_{max}$, obtaining

$$\Delta A = \frac{R}{U_{av}^2} Q_{max}^2 \dot{F} \times 2T \times 10^{-3} \tag{5.22}$$

Table 5.6 Parameter calculations of annual electric energy losses of agricultural electric lines with quarterly loads.

f_P		0.10	0.15	0.20	0.25	0.30	0.35	0.40	0.45	0.50	0.55	0.60
β_P		0	0	0.023 7	0.050 0	0.076 4	0.102 8	0.129 2	0.155 6	0.182 0	0.208 4	0.234 8
f_Q		0.278 0	0.319 5	0.361 0	0.402 5	0.443 9	0.485 4	0.526 9	0.568 4	0.609 9	0.651 4	0.692 9
$\cos\varphi_{max} = 0.50$	$\dot{f}$	0.167 9	0.203 1	0.238 2	0.273 4	0.308 6	0.343 7	0.378 9	0.414 1	0.449 3	0.484 5	0.519 7
	$\dot{\beta}$	0	0	0.013 7	0.028 9	0.044 1	0.059 4	0.074 6	0.089 8	0.105 1	0.120 3	0.135 6
	$\dot{F}$	0.078 6	0.099 7	0.118 5	0.138 9	0.161 3	0.185 5	0.211 8	0.240 1	0.270 3	0.302 5	0.336 7
	$\dot{F}_{inc}$	0.235 8	0.299 1	0.355 5	0.416 7	0.483 9	0.556 5	0.635 4	0.720 3	0.810 9	0.907 5	1.010 0
$\cos\varphi_{max} = 0.60$	$\dot{f}$	0.176 5	0.216 0	0.255 5	0.295 0	0.334 5	0.374 0	0.413 5	0.453 0	0.492 5	0.532 0	0.571 5
	$\dot{\beta}$	0	0	0.017 8	0.037 5	0.057 3	0.077 1	0.096 9	0.116 7	0.136 5	0.156 3	0.176 1
	$\dot{F}$	0.083 6	0.107 8	0.129 2	0.152 6	0.178 5	0.207 0	0.238 0	0.271 6	0.307 8	0.346 5	0.387 8
	$\dot{F}_{inc}$	0.148 5	0.191 5	0.229 6	0.271 2	0.317 2	0.367 8	0.422 9	0.482 6	0.546 9	0.615 7	0.689 1
$\cos\varphi_{max} = 0.70$	$\dot{f}$	0.188 0	0.233 3	0.278 5	0.323 8	0.369 0	0.414 2	0.459 5	0.504 7	0.550 0	0.595 3	0.640 5
	$\dot{\beta}$	0	0	0.023 2	0.049 0	0.074 9	0.100 8	0.126 6	0.152 5	0.178 4	0.204 3	0.230 2
	$\dot{F}$	0.090 5	0.119 0	0.144 1	0.171 9	0.203 2	0.237 8	0.276 1	0.317 7	0.362 9	0.411 5	0.463 5
	$\dot{F}_{inc}$	0.094 2	0.123 9	0.150 0	0.178 9	0.211 5	0.247 5	0.287 4	0.330 7	0.377 7	0.428 3	0.482 4
$\cos\varphi_{max} = 0.80$	$\dot{f}$	0.154 3	0.194 8	0.235 4	0.275 9	0.316 5	0.357 0	0.397 6	0.438 2	0.478 7	0.519 3	0.559 8
	$\dot{\beta}$	0	0	0.023 7	0.050 0	0.076 4	0.102 8	0.129 2	0.155 6	0.182 0	0.208 4	0.234 8
	$\dot{F}$	0.070 9	0.094 6	0.113 8	0.135 2	0.159 4	0.186 5	0.216 5	0.249 3	0.285 0	0.323 6	0.365 0

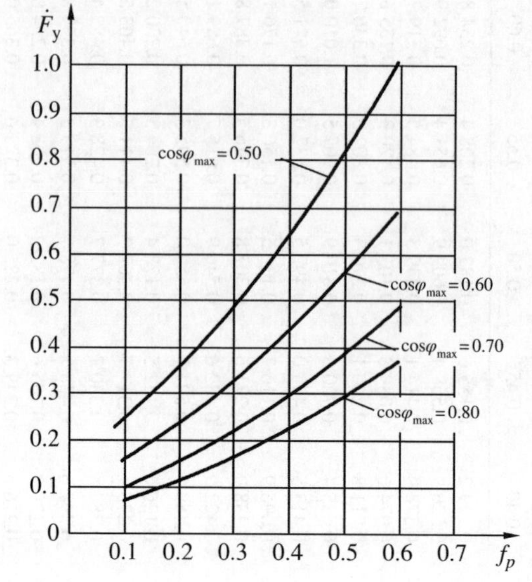

Figure 5.6 Calculation curves of annual electric energy losses of agricultural electric lines with quarterly loads.

When the maximum active power and the maximum reactive power occur at the same time, obtain $Q^2{}_{\max} = P^2{}_{\max}\tan^2\varphi_{\max}$ and substitute it into Formula (5.22) to obtain

$$\Delta A = \frac{R}{U_{\mathrm{av}}^2} P_{\max}^2 \tan^2\varphi_{\max} \dot{F} \times 2T \times 10^{-3} = \frac{R}{U_{\mathrm{av}}^2} P_{\max}^2 \dot{F}_{\mathrm{inc}} \times 2T \times 10^{-3} \tag{5.23}$$

$$\dot{F}_{\mathrm{inc}} = \tan^2\varphi_{\max} \dot{F} \tag{5.24}$$

Wherein $\dot{F}_{\mathrm{inc}}$ – loss factor of the equivalent load curve which is included to the maximum active load.

When $\cos\varphi_{\max} > 0.707$, use $\dot{f} = \frac{1}{2}(f_P + f_Q \tan\varphi_{\max})$ and $\dot{\beta} = \beta_P$ to calculate $\dot{f}$ and $\dot{\beta}$, and then to derive $\dot{F}$. Use Formula (4.12) to calculate ΔA.

Table 5.6 shows the calculation results of $\dot{F}$ and $\dot{F}_{\mathrm{inc}}$ when $\cos\varphi_{\max}$ is 0.50, 0.60, 0.70, 0.80. Figure 5.6 shows line loss calculation curves drawn by the data in Table 5.6.

Example 5.3 There is a 110 kV line with conductor model LGJ150, length of 50 km, and $r_0 = 0.21$ Ω/km. The parameters of its annual load curve are as follow: $f_P = 0.505\,4$, $\beta_P = 0.2$, $f_Q = 0.490\,1$, $\beta_Q = 0.263\,2$, $S_{\max} = 30.81$ MVA, $P_{\max} = 25$ MW, $Q_{\max} = 19$ Mvar. The maximum active power and the maximum reactive power do not occur at the same time. *Try to calculate its annual electric energy loss and compare the results calculated by different methods.*

Solutions

1. *Calculation of the electric energy loss by S_{max} and F_S.* Substitute f_P, β_P and f_Q, β_Q into Formula (2.28) respectively to obtain $F_P = 0.310\,0$, $F_Q = 0.282\,0$. The maximum active power and the maximum reactive power do not occur at the same time, so

$$\cos\varphi_{\max} = P_{\max}/S_{\max} = 25/30.81 = 0.811\,4$$
$$\sin\varphi' = Q_{\max}/S_{\max} = 19/30.81 = 0.616\,7$$

According to Formula (4.3),

$$
\begin{aligned}
F_S &= F_P \cos^2\varphi_{\max} + F_Q \sin^2\varphi' \\
&= 0.310\,0 \times (0.811\,4)^2 + 0.282\,0 \times (0.616\,7)^2 \\
&= 0.311\,3
\end{aligned}
$$

Since $R = 0.21 \times 50 = 10.5\ \Omega$, use Formula (4.2) to obtain the electric energy loss as

$$
\begin{aligned}
\Delta A &= \frac{10.5}{(1.1 \times 10^5)^2} \times (30.81 \times 10^6)^2 \times 0.311\,3 \times 8760 \times 10^{-3} \\
&= 224.63 \times 10^4 (\text{kW·h})
\end{aligned}
$$

2. *Calculation curve method of annual electric energy loss.* As $\beta_P = 0.2$, according to calculation curves shown in Figure 5.4a, $f_P = 0.505\,4$, $\cos\varphi_{\max} = 0.811\,4$. Use the interpolation method to calculate $\dot{F} = 0.234$. Use Formula (4.12) to calculate the electric energy loss as

$$
\begin{aligned}
\Delta A &= \frac{10.5}{(1.1 \times 10^5)^2} \times (25 \times 10^6)^2 \times 0.234 \times 2 \times 8760 \times 10^{-3} \\
&= 222.3 \times 10^4 (\text{kW·h})
\end{aligned}
$$

Compared with the result in Solution 1., the calculation error is

$$
\delta(\Delta A)\% = \left(\frac{222.3 - 224.63}{224.63}\right) \times 100\% = -1.02\%
$$

3. *Tröger curve method.* Substitute f_P and $\cos\varphi_{\max}$ into Formula (4.8) to obtain $F_S = 0.383\,5$. Compared with F_S obtained by Solution 1., the calculation error in the electric energy loss is

$$
\delta(\Delta A)\% = \left(\frac{0.383\,5 - 0.311\,3}{0.311\,3}\right) \times 100\% = 23.19\%
$$

4. *Glazynov curve method.* As $f_P = 0.505\,4$, the corresponding values are $T_{\max} = 4427$ h and $\cos\varphi_{\max} = 0.811\,4$. According to $T_{\max}$–$\tau_{\max}$ curves in Figure 5.1, use the interpolation method to obtain $\tau_{\max} = 2900$ h, and then $F_S = 2900/8760 = 0.331\,0$. Compared with F_S obtained by Solution 1., the calculation error in the electric energy loss is

$$
\delta(\Delta A)\% = \left(\frac{0.331\,0 - 0.311\,3}{0.311\,3}\right) \times 100\% = +6.33\%
$$

The above calculations in the example were prepared using load materials collected in China; they indicate that the calculation curve of annual electric energy loss proposed by the authors is simple and it derives a smaller error than that derived from other line loss calculation curves, so it is recommended in the planning and design stages of power grids.

6

Change Law for the Electric Energy Losses of Power Grids

There are many substations, lines, and other power grid units within any region governed by a power supply enterprise, so it is very complicated and difficult to accurately analyze the relationship between the total electric energy loss of regional power grids and the operating parameters such as average operating voltage, load factor, and average power factor. A deeper understanding of this issue is likely to predict and control the regional line loss rate with multiple variables, thus improving line loss management to a new level. This chapter will introduce the line loss binomial, analyze the relationship between no-load loss and operating voltage, and calculate the load loss coefficients for various voltage classes of power grids. This chapter will also analyze the reduction of electric energy losses by controlling the voltage level based on the above contents.

6.1 Basis of Analysis of Line Loss Changes

6.1.1 Line Loss Binomial

It is well-known that power and energy are scalar quantities. Power losses and energy losses in various power grid units within a region are cumulative, and the power loss and electric energy loss of any power grid unit are generally composed of no-load loss and load loss. Therefore, the total electric energy losses of power grids in a region are also composed of total no-load losses and total load losses. Based on this judgment, the following line loss binomial can be quoted:

$$\Delta A = B + CA^2 \tag{6.1}$$

Wherein ΔA – theoretical electricity line loss of power grids (kW·h);
B – theoretical no-load electricity loss (fixed loss; kW·h);
C – coefficient determined by electric supply and load loss (variable loss) within the theoretical calculation period [1/(kW·h)];
A – electric supply of power grids within the theoretical calculation period (kW·h).

Line Loss Analysis and Calculation of Electric Power Systems, First Edition. Anguan Wu and Baoshan Ni.

For the analysis of operating electricity line loss, B should also include unknown loss (management loss). The following agreement can be reached from the perspective of operational analysis:

B – no-load loss of power grids or units; unknown loss is included for operational statistics (kW·h);
C – load loss coefficient of power grids or units [1/(kW·h)];
A – electric supply of power grids or units (kW·h).

6.1.2 Condition of Minimum Line Loss Rate

Divide by electric supply on both sides of Formula (6.1) to obtain

$$\frac{\Delta A}{A} = \frac{B}{A} + CA$$

Then

$$\Delta A\% = \Delta A_0\% + \Delta A_L\% \tag{6.2}$$

Wherein $\Delta A\%$ – line loss rate;
$\Delta A_0\%$ – no-load line loss rate;
$\Delta A_L\%$ – load line loss rate.

Calculate the derivative of the variable A in Formula (6.2) to obtain

$$\frac{\mathrm{d}\,(\Delta A\%)}{\mathrm{d}A} = -\frac{B}{A^2} + C$$

According to the above formula, the condition of minimum line loss rate $\Delta A\%$ is

$$C_{\mathrm{op}} = \frac{B}{A^2} \tag{6.3}$$

Substitute Formula (6.3) into Formula (6.1) to obtain

$$\Delta A_0 = \Delta A_L = \frac{1}{2}\Delta A$$

When the no-load electricity loss is equal to the load electricity loss, and both of these are equal to one-half of the total electricity losses, the total line losses are minimal, and the line loss rate is at its least.

For ease of estimation, Formula (6.3) can be converted to

$$C_{\mathrm{op}} = \frac{B/A}{A} = \frac{\Delta A_0\%}{A} \tag{6.4}$$

According to Formula (6.4), when the load loss coefficient is equal to the no-load line loss rate per unit power supply, the total line loss rate is the lowest and equal to two times the no-load line loss rate.

Assume that $\theta = B/CA^2$ is the loss constituent ratio, and the condition of the minimum line loss rate is $\theta = 1.0$. To measure the "difference" between the actual line loss rate and the minimum line loss rate, the deviation coefficient $K_{non.op}$ can be defined to measure such difference. When $\theta = 1.0$, $K_{pj} = 0$, so the calculation formula of the deviation coefficient can be obtained

$$K_{non.op} = 0.50\ (1/\theta - 1)$$

If θ is changed from 0.60 to 0.80, $K_{non.op}$ is reduced from 0.333 to 0.125, indicating that a larger θ leads to a smaller difference between the actual line loss rate and the minimum line loss rate.

The condition of minimum line loss rate can be satisfied for some power grid units (such as distribution transformers). The condition of minimum line loss rate is also likely to be realized for regional power grids. This concept can guide line loss management personnel to optimize technical measures of loss reduction.

6.2 Calculation and Analysis of No-load Loss

In general, there are three typical mathematical expressions to define the relationship between the no-load loss and the operating voltage of power grid units, namely, higher-order expression, square expression, and quasi square expression.

6.2.1 *Higher-Order Expression*

The relationship between the no-load power loss and the operating voltage of a distribution transformer whose core is made of hot-rolled silicon-steel sheet was once measured [23]. The least square method can be used with the measurement to derive the following empirical formula (see Appendix D).

$$\Delta P_{0*} \approx 1.08 U^4{}_* \ (U_* \neq 1.0, \text{correlation coefficient } r = 0.9514) \tag{6.5}$$

Wherein ΔP_{0*} – per unit value of power loss based on the no-load power loss ΔP_0 under rated voltage; U_* – per unit value of operating voltage based on rated voltage.

If the voltage change within the measuring period is $u_*(t)$, then the no-load electric energy loss of the distribution transformer within the measuring period can be calculated as per the following formula:

$$\begin{aligned}\Delta A_0 &= \Delta P_0 \times 10^{-3} \int_0^T 1.08[u_*(t)]^4 \mathrm{d}t \\ &= 1.08 \Delta P_0 \times 10^{-3} \int_0^T \left[u_*^2(t)\right]^2 \mathrm{d}t \\ &= 1.08 \Delta P_0 \times 10^{-3} \int_0^T [U(t)]^2 \mathrm{d}t\end{aligned}$$

Wherein $U(t)$ – voltage square curve, that is $U(t) = u^2{}_*(t)$.

With such concepts as load factor, minimum load rate and loss factor in the loss factor method, the average f_U and the minimum value β_U of $U(t)$ within the measuring period can be calculated, that is

$$f_U = \frac{\int_0^T U(t)\mathrm{d}t}{U_{max}T} = \frac{\int_0^T u_*^2(t)\mathrm{d}t}{u_{*max}^2 T} = \frac{u_{*max}^2 F_u T}{u_{*max}^2 T} = F_u$$

$$\beta_U = \frac{U_{min}}{U_{max}} = \frac{u_{*min}^2}{u_{*max}^2} = \beta_u^2$$

The above analysis shows that the average of the voltage square curve is equal to the loss factor of the voltage curve; the minimum value of the voltage square curve is equal to the square of the minimum value of the voltage curve. Substitute $f_U = F_u$ and $\beta_U = \beta^2_u$ into Formula (2.28) to derive the loss factor of the $U(t)$ curve and finally obtain ΔA_0, that is

$$\Delta A_0 = 1.08\Delta P_0 \int_0^T [U(t)]^2 \mathrm{d}t \times 10^{-3} = 1.08\Delta P_0 U_{max}^2 F_U T \times 10^{-3} \tag{6.6}$$

When the deviation of operating voltage is controlled within $\pm 5\% U_N$, $u_{*max} = 1.05$, $u_{*min} = 0.95$, then $\beta_u = 0.95/1.05 = 0.905$. Because β_u is closer to 1, $f_u = (1 + \beta_u)/2 = 0.952$. Substitute β_u and f_u into Formula (2.28) to obtain F_u

$$f_U = F_u = 0.639 \times (0.952)^2 + 0.361 \times (0.952 + 0.952 \times 0.905 - 0.905) = 0.907\ 1$$

$$\beta_U = \beta_u^2 = 0.905^2 = 0.819\ 0$$

Substitute f_U and β_U into Formula (2.28) to obtain $F_U = 0.8258$. As $U^2_{max} = (u^2_{*max})^2 = 1.102\ 5^2 = 1.215\ 5$, substitute all of them into Formula (6.6) to obtain

$$\Delta A_0 = 1.08\Delta P_0 \times 1.215\ 5 \times 0.825\ 8 \times T \times 10^{-3} = 1.084\Delta P_0 T \times 10^{-3}$$

The above result indicates that as to the distribution transformer made of hot-rolled silicon-steel sheet, if its operating voltage is close to the rated value and changes within ±5% of the rated value, its no-load electric energy loss is about 8% higher than that when its voltage is constant and equal to the rated value.

When the operating voltage changes within (0.90–1.10)U_N, do the relevant calculation to obtain $F_u \approx 1.003\ 5f^2_u$, $F_U = 1.014(f_U)^2 = 1.021\ 1f^4_u$. Then, Formula (6.6) can be converted to the following approximate formula:

$$\Delta A_0 = 1.08 \times 1.021\ 1 u_{*max}^4 f_u^4 \Delta P_k T \times 10^{-3} = 1.103 U_{*av}^4 \Delta P_0 T \times 10^{-3} \tag{6.7}$$

According to Formula (6.7), when the voltage level is reduced to $0.95U_N$, $\Delta A_0 = 0.90\Delta P_0 T \times 10^{-3}$, that is the no-load electric energy loss of the distribution transformer is probably reduced by up to 10% less than that when its voltage is constant and equal to the rated value; when the voltage level is increased to $1.05U_N$, $\Delta A_0 = 1.341\Delta P_0 T \times 10^{-3}$, that is, the no-load electric energy loss is increased by about 34.1% more than that when its voltage is constant and equal to the rated value. This theoretically proves that reasonable reduction of the operating voltage can reduce the no-load electric energy loss of the distribution transformer.

6.2.2 *Square Expression*

The dielectric losses in the power cable and high-voltage bushing are in direct proportion to the voltages squared and can be calculated respectively as per the following two formulas:

$$\begin{aligned}\Delta A_{\mathrm{ca.0}} &= 3U_{\mathrm{av}}^2 \omega C_0 \tan\delta T \times 10^{-3} \\ \Delta A_{\mathrm{bush.0}} &= U^2 \omega C \tan\delta T \times 10^{-3}\end{aligned} \tag{6.8}$$

Wherein C_0 – capacitance per unit length of power cable (μF/km);
C – capacitance of high-voltage bushing (pF);
$\tan\delta$ – tangent of dielectric loss angle of power cable and high-voltage bushing.

Here concepts in the loss factor method can be used again to analyze the no-load dielectric loss of the high-voltage bushing under voltage change, so the formula below is derived:

$$\begin{aligned}\Delta A_{\mathrm{bush.0}} &= \omega C \tan\delta \int_0^T U^2(t)\mathrm{d}t \times 10^{-3} \\ &= \omega C \tan\delta U_{\max}^2 F_{\mathrm{u}} T \times 10^{-3}\end{aligned}$$

When the deviation of operating voltage is not more than $\pm 5\% U_{\mathrm{N}}$, as mentioned above $\beta_{\mathrm{u}} = 0.905$, $f_{\mathrm{u}} = 0.952$, while $F_{\mathrm{u}} = 0.907\,1$, $f^2_{\mathrm{u}} = (0.952)^2 = 0.906\,3$, so take $F_{\mathrm{u}} \approx f^2_{\mathrm{u}}$ approximately. Also, as $U^2_{\mathrm{av}} = U^2_{\max} f^2_{\mathrm{u}}$, substitute the above into this to obtain

$$\begin{aligned}\Delta A_{\mathrm{bush.0}} &= \omega C \tan\delta U_{\mathrm{av}}^2 T \times 10^{-3} = \Delta P_{\mathrm{bush.0}} U_{*\mathrm{av}}^2 T \times 10^{-3} \\ \Delta P_{\mathrm{bush.0}} &= \omega C \tan\delta U_{\mathrm{N}}^2\end{aligned} \tag{6.9}$$

Wherein $U_{*\mathrm{av}}$ – per unit value of the average operating voltage based on a rated voltage;
$\Delta P_{\mathrm{bush.0}}$ – dielectric power loss of the bushing when the operating voltage is a rated voltage (W).

According to Formula (6.9), when the voltage changes within $(0.95–1.05)U_{\mathrm{N}}$, the no-load electric energy losses of some power grid units can be calculated with the per unit value of average voltage.

6.2.3 *Quasi Square Expression*

The relationship between the no-load power loss and the voltage of a low loss series distribution transformer whose core is made of cold-rolled silicon-steel sheet was also measured [23]. The measurement data can be collected and used to derive the following empirical formula (see Appendix D).

$$\Delta P_{0*} = 1.012 U^{1.84}_{*} \quad (U_* \neq 1.0,\ \text{correlation coefficient } r = 0.932) \tag{6.10}$$

According to the above analysis, the calculation formula of no-load electric energy loss ΔA_0 (kW·h) considering the operating voltage change can be obtained

$$\Delta A_0 = 1.012 U_{*\mathrm{av}}^{1.84} \Delta P_{\mathrm{k}} T \times 10^{-3} \tag{6.11}$$

When the operating voltage is reduced or increased by 5% less/more than the rated voltage, the difference between the no-load electric energy loss under either of these voltages and that under the rated voltage is smaller than that for the distribution transformer whose core is made of hot-rolled silicon-steel sheet.

In a preliminary conclusion, other than high-loss series distribution transformers whose cores are made of hot-rolled silicon-steel sheets, when the deviation of operating voltage is not big, the no-load electric energy losses of general distribution transformers can be calculated by multiplying the no-load power losses $\Delta P_0(U = U_{av})$ under the average voltage by the number of hours during the measuring period.

6.3 Calculation and Analysis of Load Loss Coefficient *C*

6.3.1 Calculation of Load Loss Coefficient

The loss factor formula recommended in Chapter 2 and the load rate calculation formula of an equivalent load curve proposed by the equivalent load curve method in Chapter 4 can be used to derive the calculation formula of the load loss coefficient *C*.

If the operation meets the conditions set forth in the left column of Table 4.4, then the $\dot{f}$ calculation formula can be rewritten to

$$\begin{aligned}\dot{f} &= \frac{1}{2}(1+\tan\varphi_{av})f_P = af_P \\ a &= \frac{1}{2}(1+\tan\varphi_{av})\end{aligned} \tag{6.12}$$

Wherein a – calculation coefficient reflecting the average power factor.

Formulas (4.13) and (4.12) in Chapter 4 are now rewritten to

$$\dot{F} = 0.639\dot{f}^2 + 0.361(\dot{f} + \dot{f}\dot{\beta} - \dot{\beta})$$

$$\Delta A = \frac{R}{U_{av}^2}\dot{P}_{max}^2 \dot{F} \times 2T \times 10^{-3}$$

Substitute Formulas (6.12) and (4.13) into Formula (4.12), considering $\dot{\beta} = \beta_P$ and $\dot{P}_{max} = P_{max}$ in general situations and $P_{max} = \frac{A}{f_P T}$, to obtain the following formula after simplification:

$$\Delta A = \frac{R}{U_{av}^2 f_P T}\left[1.278a^2 f_P + 0.722a\left(1 + \beta_P - \frac{\beta_P}{af_P}\right)\right] \times 10^{-3} \times A^2 = CA^2 \tag{6.13}$$

$$C = \frac{R}{U_{av}^2 f_P T}\left[1.278a^2 f_P + 0.722a\left(1 + \beta_P - \frac{\beta_P}{af_P}\right)\right] \times 10^{-3} \tag{6.14}$$

Wherein U_{av} – average operating voltage (kV);
C – load loss coefficient [1/(kW·h)].

According to Formula (6.14), when the resistance R of the power grid unit, the average operating voltage U_{av} and the measuring period T are definite values, the load loss coefficient (hereinafter referred to as coefficient C) depends on three operating parameters: the load factor f_P, the average power factor within the calculation period, and the minimum load rate β_P.

For ease of application, the authors calculated C values under various voltage power grids, normal load curve parameters, and average power factors within a measuring period of one month (T = 720 h). See Tables 6.1 and 6.2.

Table 6.1 Table of load loss coefficients of 10 kV system (R = 1 Ω, T = 720 h) $10^{-8} \times 1/(\text{kW·h})$.

	β_P	0				0.10				0.20				0.30			
	f_P																
$\cos\varphi_{av}$	α	0.30	0.40	0.50	0.60	0.30	0.40	0.50	0.60	0.40	0.50	0.60	0.70	0.50	0.60	0.70	0.80
0.60	1.167 7	6.323	5.348	4.762	4.372	5.599	5.014	4.595	4.288	4.680	4.428	4.205	4.018	4.261	4.122	3.981	3.853
0.70	1.010 1	5.187	4.343	3.837	3.499	4.411	3.970	3.638	3.389	3.596	3.440	3.280	3.138	3.241	3.170	3.078	2.987
0.80	0.875 0	4.284	3.553	3.114	2.821	3.462	3.145	2.888	2.689	2.738	2.663	2.557	2.454	2.437	2.424	2.375	2.315
0.90	0.742 2	3.459	2.838	2.466	2.218	2.593	2.398	2.214	2.064	—	1.962	1.909	1.844	—	1.755	1.746	1.717

Table 6.2 Table of load loss coefficients of 35–110 kV system (R = 1 Ω, T = 720 h) $10^{-10} \times 1/(\text{kW·h})$.

	β_P	0.30				0.40				0.50				0.60			
U_{av}	f_P																
(kV)	$\cos\varphi_{av}$	0.50	0.60	0.70	0.80	0.60	0.70	0.80	0.90	0.60	0.70	0.80	0.90	0.70	0.80	0.90	0.95
35	0.60	34.79	33.65	32.50	3.45	32.97	3.19	31.37	30.58	3.28	31.89	31.28	30.64	31.58	31.20	30.69	30.41
	0.70	26.46	25.88	25.13	24.38	24.98	24.64	24.14	23.60	—	24.15	23.89	23.51	23.66	23.65	23.42	23.27
	0.80	19.89	19.79	19.38	18.90	—	18.74	18.51	18.19	—	—	18.13	17.98	—	—	17.76	17.72
	0.90	—	14.32	14.25	14.02	—	—	13.50	13.39	—	—	—	13.05	—	—	—	—
110	0.70	2.679	2.620	2.544	2.469	2.529	2.494	2.444	2.390	—	2.445	2.419	2.380	—	2.394	2.371	2.356
	0.80	—	2.004	1.962	1.913	—	1.897	1.874	1.842	—	—	1.835	1.820	—	—	1.798	1.793
	0.90	—	—	1.443	1.419	—	—	1.367	1.356	—	—	—	1.322	—	—	—	—
	0.95	—	—	1.163	1.154	—	—	—	1.095	—	—	—	—	—	—	—	—

Note: When T = 8760 h, C is equal to the table value divided by 12.17.

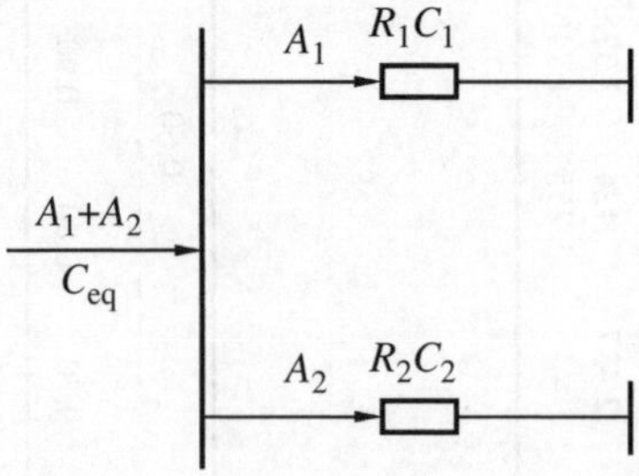

Figure 6.1 Inclusion of coefficient C of multiple units in the same bus.

According to Tables 6.1 and 6.2, C is monotonically decreased as f_P and $\cos\varphi_{av}$ are increased. Loss reduction measures can be optimized based on the change law of the C value.

6.3.2 *Inclusion of Load Loss Coefficient*

6.3.2.1 Calculation of Load Loss Coefficient of Equivalence C of Multiple Units in the Same Bus

As shown in Figure 6.1, the electric supplies and coefficients C of R_1 and R_2 are A_1, A_2 and C_1, C_2, respectively. Allow $\alpha = A_1/A_2$ to obtain the following formula:

$$C_{eq}(A_1 + A_2)^2 = C_1 A_1^2 + C_2 A_2^2$$

So

$$C_{eq} = \frac{C_1 A_1^2 + C_2 A_2^2}{(A_1 + A_2)^2} = \frac{C_1 \alpha^2 + C_2}{(\alpha + 1)^2} \tag{6.15}$$

Wherein C_{eq} – coefficient of equivalence.

According to Formula (6.15), when α is the larger, C_{eq} is closer to C_1, that is the total load loss coefficient of equivalence C_{eq} is close to the coefficient C_1 of the unit with higher electric supply.

When there are n units in the same point of supply, the coefficient of equivalence C_{eq} can be calculated as per the following formula:

$$C_{eq} = \sum_{i=1}^{n} C_i A_i^2 \bigg/ \left(\sum_{i=1}^{n} A_i\right)^2 \tag{6.16}$$

6.3.2.2 Calculation of Coefficient of Equivalence C_{eq} of Multiple Units in Adjacent Voltage Buses

As shown in Figure 6.2, there is a multi-unit system composed of adjacent voltage buses on both sides of the transformer.

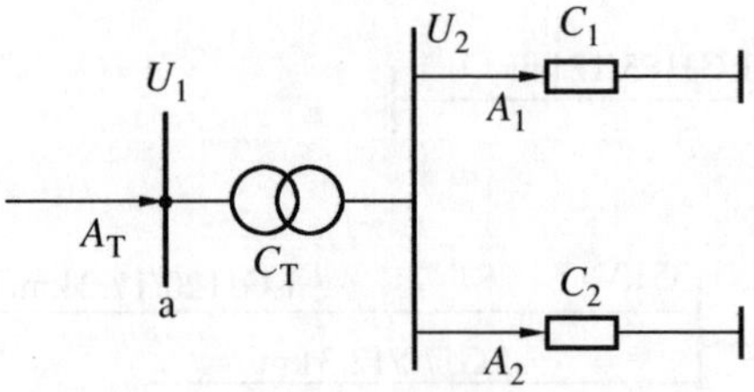

Figure 6.2 Inclusion of coefficient C of multiple units in adjacent voltage buses.

Allow the load loss and low-voltage outgoing load losses of the transformer to be ΔA_T, ΔA_1, and ΔA_2, so

$$\Delta A_T = C_T A_T^2,\ \Delta A_1 = C_1 A_1^2,\ \Delta A_2 = C_2 A_2^2$$

With respect to bus a, the coefficient of equivalence of multiple units on high and low voltage sides can be expressed by C_{eq}, and then the following relational expression can be derived:

$$C_{eq} A_T^2 = C_T A_T^2 + C_1 A_1^2 + C_2 A_2^2$$

So

$$C_{eq} = C_T + C_1 \left(\frac{A_1}{A_T}\right)^2 + C_2 \left(\frac{A_2}{A_T}\right)^2$$

When n units are connected in the bus on the other side of the transformer, the formula below can be derived:

$$C_{eq} = C_T + \sum_{i=1}^{n} C_i \left(\frac{A_i}{A_T}\right)^2 \tag{6.17}$$

Formulas (6.16) and (6.17) can be used to include load losses of power grid units with different connection relations so as to calculate the load loss coefficient of equivalence C_{eq}.

Example 6.1 Figure 6.3 shows the 35 kV and above power supply network, line parameters, and the capacity and number of main transformers for all substations of a county.

Table 6.3 lists the rated loss of each main transformer, the maximum load at each side of substations, and load parameters f_P, β_P, $\cos\varphi_{av}$.

The measuring period is 10 months (T = 7296 h). *Try to use the calculation method of coefficient C to calculate the loss rate of the whole region, and line loss rates of 35 kV and 110 kV lines, and to derive corresponding line loss calculation formula.*

Solutions

1. *Calculation of line losses of 35 kV system main transformers and lines.*

1.1 *Line loss calculation of No. 3 substation subsystem.*

As $\Delta A_{0.3b} = (24 + 29) \times 7296 = 38.67 \times 10^4$ (kW·h), assume that the form coefficient of apparent power is the same as that of the active power, so $K_S \approx K_P = \sqrt{F_P}/f_P$. Substitute f_P and β_P values in Table 6.3 into Formula (2.26) to obtain $F_P = 0.668\ 9$, so $K_S \approx \sqrt{0.668\ 9}/0.80 = 1.022$.

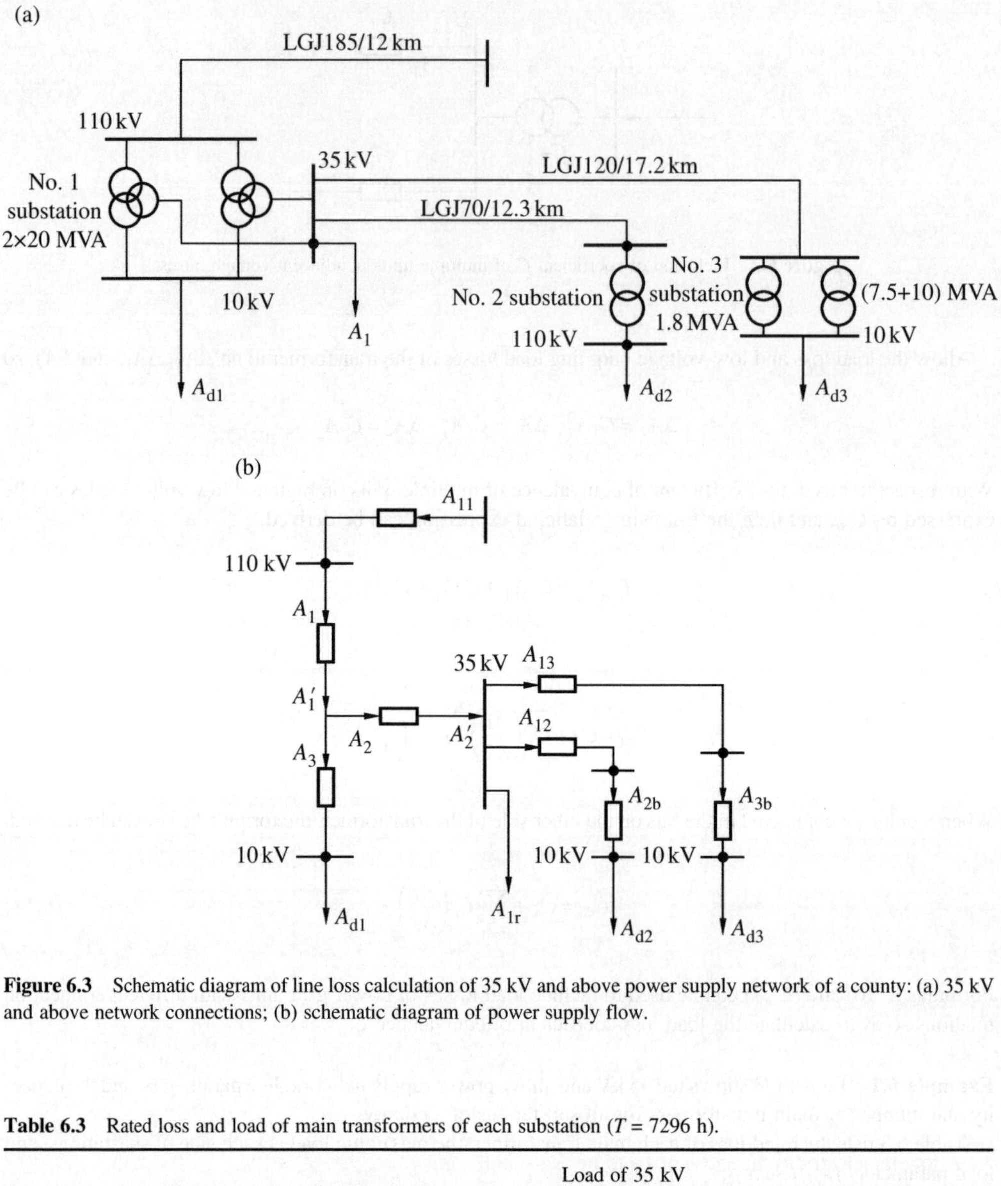

Figure 6.3 Schematic diagram of line loss calculation of 35 kV and above power supply network of a county: (a) 35 kV and above network connections; (b) schematic diagram of power supply flow.

Table 6.3 Rated loss and load of main transformers of each substation (T = 7296 h).

Substation number	Capacity of main trans-former (MVA)	Rated loss (kW) ΔP_0	ΔP_{LN} h.m	h.l	m.l	Load of 35 kV dedicated line P_{max} (kW)	β_P	f_P	$\cos\varphi_{av}$	Load of 10 kV public line β_P	$\cos\varphi_{av}$	P_{max} (kW)	f_P
1	2 × 20	32.6	138.1	142.2	101.28	8000	0.40	0.90		0.40	0.80	2000	0.75
2	1 × 1.8	8.3		24					0.85	0.30	0.75	1000	0.70
3	7.5 (a)	24		75						0.40	0.85	15 000	0.80
	10.0 (b)	29		92									

Loads undertaken by two transformers are approximately distributed by capacity to obtain the rms load as below:

$$S_{\mathrm{rms.a}} = K_S \frac{P_{\mathrm{av}}}{\cos\varphi_{\mathrm{av}}}\left(\frac{S_{\mathrm{a}}}{S_{\Sigma}}\right) = 1.022 \times \frac{15\,000 \times 0.80}{0.85} \times \frac{7.5}{17.5}$$
$$= 6183.5\ (\mathrm{kVA})$$

$$S_{\mathrm{rms.b}} = 1.022 \times \frac{15\,000 \times 0.80}{0.85} \times \frac{10}{17.5} = 8244.7\ (\mathrm{kVA})$$

The load loss of transformers is calculated as per $\Delta A_{\mathrm{L.b}} = \Delta P_{\mathrm{f.e}}\left(\frac{S_{\mathrm{rms}}}{S_{\mathrm{N}}}\right)^2 T$, that is

$$\Delta A_{\mathrm{L.3b}} = \left[75 \times \left(\frac{6183.5}{7500}\right)^2 + 92 \times \left(\frac{8244.7}{10\,000}\right)^2\right] \times 7296$$
$$= 113.52 \times 7296 = 82.82 \times 10^4\ (\mathrm{kW \cdot h})$$

Look for information about the LGJ120 conductor, obtaining $r_0 = 0.27\ \Omega/\mathrm{km}$, $R = Lr_0 = 17.2 \times 0.27 = 4.644\ (\Omega)$. Use f_P, β_P, $\cos\varphi_{\mathrm{av}}$ values in Table 6.3 to refer to Table 6.2, and use the interpolation method to obtain $C'_{13} = 16.01 \times 10^{-10}$ [1/(kW·h)]. Considering the conversion of the two factors of measuring period and resistance, the load loss coefficient of line L_3 is obtained

$$C_{13} = C'_{13} R \frac{T_0}{T} = 16.01 \times 10^{-10} \times 4.644 \times \frac{720}{7296}$$
$$= 7.34 \times 10^{-10} [1/(\mathrm{kW \cdot h})]$$

Allow the inlet electric supply of No. 3 substation to be A_{3b}, obtaining

$$A_{3b} = A_{d3} + \Delta A_{3b} = A_{d3} + (\Delta A_{0.3b} + \Delta A_{\mathrm{L.\,3b}})$$
$$= (15\,000 \times 0.80 \times 7296) + (38.67 + 82.82 \times 10^4$$
$$= 8755.2 \times 10^4 + 121.49 \times 10^4$$
$$= 8876.69 \times 10^4 (\mathrm{kW \cdot h})$$

Allow the electricity line loss of L_3 to be ΔA_{13}, obtaining

$$\Delta A_{13} = C_{13} A_{3b}^2 = 7.34 \times 10^{-10} \times \left(8876.69 \times 10^4\right)^2$$
$$= 578.36 \times 10^4 (\mathrm{kW \cdot h})$$

Allow the head end electric supply of L_3 to be A_{13}, obtaining

$$A_{13} = A_{3b} + \Delta A_{13} = (8876.69 + 578.36) \times 10^4$$
$$= 9455.05 \times 10^4 (\mathrm{kW \cdot h})$$

The load loss, total loss, and respective line loss rate of No. 3 substation subsystem are

$$\begin{aligned}\Delta A_{L.3bl} &= \Delta A_{L.3b} + \Delta A_{13}\\ &= (82.82 + 578.36) \times 10^4\\ &= 661.18 \times 10^4 (\text{kW}\cdot\text{h})\end{aligned}$$

$$\begin{aligned}\Delta A_{3bl} &= \Delta A_{0.3b} + \Delta A_{L.3bl}\\ &= (38.67 + 661.18) \times 10^4\\ &= 699.85 \times 10^4 (\text{kW}\cdot\text{h})\end{aligned}$$

$$\begin{aligned}\Delta A_{0.3bl}\% &= \frac{\Delta A_{0.3b}}{A_{13}} \times 100\%\\ &= \frac{38.67 \times 10^4}{9455.05 \times 10^4} \times 100\% = 0.41\%\end{aligned}$$

$$\begin{aligned}\Delta A_{L.3bl}\% &= \frac{\Delta A_{L.3bl}}{A_{13}} \times 100\%\\ &= \frac{661.18 \times 10^4}{9455.05 \times 10^4} \times 100\% = 6.99\%\end{aligned}$$

so $$\Delta A_{3bl}\% = \Delta A_{0.3bl}\% + \Delta A_{L.3bl}\% = 0.41\% + 6.99\% = 7.40\%$$

1.2 *Line loss calculation of No. 2 substation subsystem.*

As $\Delta A_{0.2b}$ = 8.3 × 7296 = 6.06 × 10^4 (kW·h), use f_P and β_P to obtain F_P = 0.533 3, so $K_S \approx \sqrt{0.533\,3}/0.70 = 1.043$, thus

$$S_{rms.2b} = 1.043 \times \frac{1000 \times 0.70}{0.75} = 973.5(\text{kVA})$$

$$\Delta A_{L.2b} = 24 \times \left(\frac{973.5}{1800}\right)^2 \times 7296 = 5.12 \times 10^4 (\text{kW}\cdot\text{h})$$

$$\begin{aligned}\Delta A_{2b} &= \Delta A_{0.2b} + \Delta A_{L.2b} = (6.06 + 5.12) \times 10^4\\ &= 11.18 \times 10^4 (\text{kW}\cdot\text{h})\end{aligned}$$

Look for information about LGJ70 conductor, obtaining $r_0 = 0.46$ Ω/km, $R = 12.3 \times 0.46 = 5.66$ Ω. Use f_P, β_P, $\cos\varphi_{av}$ to refer to Table 6.2, and use the interpolation method to obtain $C'_{12} = 22.26 \times 10^{-10}$ [1/(kW·h)]. After conversion,

$$\begin{aligned}C_{12} &= 22.26 \times 10^{-10} \times \left(\frac{720}{7296}\right) \times 5.66\\ &= 12.43 \times 10^{-10} [1/(\text{kW}\cdot\text{h})]\end{aligned}$$

Allow the inlet electric supply of No. 2 substation to be A_{2b}, obtaining

$$\begin{aligned}A_{2b} &= A_{ep2} + \Delta A_{2b}\\ &= (1000 \times 0.70 \times 7296) + 11.18 \times 10^4\\ &= 510.72 \times 10^4 + 11.18 \times 10^4\\ &= 521.90 \times 10^4 (\text{kW}\cdot\text{h})\end{aligned}$$

Allow the electric energy loss of L_2 to be ΔA_{12}, obtaining

$$\Delta A_{12} = C_{12} A_{2b}^2 = 12.43 \times 10^{-10} \times \left(521.90 \times 10^4\right)^2 = 3.39 \times 10^4 (\text{kW} \cdot \text{h})$$

Allow the head end electric supply of L_2 to be A_{12}, obtaining

$$A_{12} = A_{2b} + \Delta A_{12} = (521.90 + 3.39) \times 10^4 = 525.29 \times 10^4 (\text{kW} \cdot \text{h})$$

The load loss, total loss, and respective line loss rate of No. 2 substation subsystem are

$$\Delta A_{L.2bl} = \Delta A_{L.2b} + \Delta A_{12} = (5.12 + 3.39) \times 10^4 = 8.51 \times 10^4 (\text{kW} \cdot \text{h})$$

$$\Delta A_{2bl} = \Delta A_{0.2b} + \Delta A_{L.2bl} = (6.06 + 8.51) \times 10^4 = 14.57 \times 10^4 (\text{kW} \cdot \text{h})$$

$$\Delta A_{0.2bl}\% = \frac{6.06 \times 10^4}{525.29 \times 10^4} \times 100\% = 1.15\%$$

$$\Delta A_{L.2bl}\% = \frac{8.51 \times 10^4}{525.29 \times 10^4} \times 100\% = 1.62\%$$

$$\Delta A_{2bl}\% = \Delta A_{0.2bl}\% + \Delta A_{L.2b}\% = 1.15\% + 1.62\% = 2.77\%$$

1.3 *Line loss calculation of No. 1 substation subsystem.*

1.3.1 *Calculation of rated load power loss at tertiary side of main transformer.*

See below for the calculation:

$$\Delta P_{hN} = \frac{1}{2}(\Delta P_{h.m} + \Delta P_{h.l} - \Delta P_{m.l} = \frac{1}{2}(138.1 + 142.2 - 101.28) = 89.51 (\text{kW})$$

$$\Delta P_{mN} = \frac{1}{2}(138.1 + 101.28 - 142.2) = 48.59 (\text{kW})$$

$$\Delta P_{lN} = \frac{1}{2}(142.2 + 101.28 - 138.1) = 52.69 (\text{kW})$$

1.3.2 *Calculation of load loss at medium-voltage side.*

Allow the outlet electric supply of 35 kV line to be A'_2, obtaining

$$A'_2 = A_y + A_{12} + A_{13} = (8000 \times 0.90 \times 7296) + (525.29 + 9455.05) \times 10^4 = (5253.12 + 9980.34) \times 10^4 = 15\,233.46 \times 10^4 (\text{kW} \cdot \text{h})$$

If the difference between the active load rate and the average power factor at inlet and outlet ends of the line and transformers which is caused by the electric energy loss is ignored, then the approximate values of load factor and average power factor at the outlet end of the medium-voltage side can be calculated according to the following formulas:

$$f_{P.m}=\frac{P_{max.y}f_{P.y}+P_{max.d2}f_{P.d2}+P_{max.d3}f_{P.d3}}{P_{max.y}+P_{max.d2}+P_{max.d3}}$$

$$=\frac{8000\times0.9+1000\times0.7+15\ 000\times0.80}{8000+1000+15\ 000}=0.829$$

$$\cos\varphi_{av.m}=\frac{P_{av.y}+P_{av.d2}+P_{av/d3}}{P_{av.y}/\cos\varphi_{av.y}+P_{av.d2}/\cos\varphi_{av.d2}+P_{av.d3}/\cos\varphi_{av.d3}}$$

$$=\frac{7200+700+12\ 000}{7200/0.85+700/0.75+12\ 000/0.85}=0.846$$

Approximately take $\beta_{P.m}$ = 0.40, and use $f_{P.m}$ and $\beta_{P.m}$ to obtain $F_{P.m}$ = 0.713 7, so $K_S=K_P\frac{\sqrt{0.713\ 7}}{0.829}=1.019$. Then,

$$S_{rms.m}=K_SS_{av.m}=1.019\times\frac{15\ 233.46\times10^4/7296}{0.846}$$

$$=1.019\times24\ 679.9=25\ 148.8(\text{kVA})$$

So

$$\Delta A_{L.m}=2\times48.59\times\left(\frac{25\ 148.8/2}{20\ 000}\right)^2\times7296$$

$$=2\times14.01\times10^4=28.02\times10^4(\text{kW}\cdot\text{h})$$

Allow the head end electric supply of the medium-voltage winding to be A_2, obtaining

$$A_2=A_2'+\Delta A_{L.m}=(15\ 233.46+28.02)\times10^4$$

$$=15\ 261.48\times10^4(\text{kW}\cdot\text{h})$$

1.3.3 *Calculation of load loss at low-voltage side.*

Give $f_{P.dl}$ = 0.75, $\beta_{P.dl}$ = 0.40 to obtain $F_{P.dl}$ = 0.594 1, $K_S\approx K_P=\sqrt{0.594\ 1}/0.75=1.028$. Then, the load loss of low-voltage winding is obtained

$$S_{rms.l}=K_SS_{av.l}=1.028\times\left(\frac{2000\times0.75}{0.80}\right)=1927.5(\text{kVA})$$

$$\Delta A_{L.1}=2\times52.69\times\left(\frac{1927.5/2}{20\ 000}\right)^2\times7296=1785(\text{kW}\cdot\text{h})$$

Allow the head end electric supply of the low-voltage winding to be A_3, obtaining

$$A_3=A_{dl}+\Delta A_{L.1}=2000\times0.75\times7296+1785$$

$$=1094.4\times10^4+0.18\times10^4$$

$$=1094.58\times10^4(\text{kW}\cdot\text{h})$$

1.3.4 *Calculation of load loss at high-voltage side.*

The weighted average method can still be used to calculate the approximate values of $f_{P.h}$ and $\cos\varphi_{av.h}$

$$f_{P.h}=\frac{\cos\varphi_{av.zh}S_{av.zh}+P_{max.dl}f_{dl}}{\cos\varphi_{av.zh}S_{av.zh}/f_{p.zh}+P_{max.dl}}$$

$$=\frac{0.846\times 24\ 679.9+2000\times 0.75}{\dfrac{0.846\times 24\ 679.9}{0.829}+2000}=0.823$$

$$\cos\varphi_{av.h}=\frac{\cos\varphi_{av.m}S_{av.m}+P_{max.dl}f_{dl}}{S_{av.m}+P_{max.dl}f_{dl}/\cos\varphi_{av.dl}}$$

$$=\frac{0.846\times 24\ 679.9+2000\times 0.75}{24\ 679.9+\dfrac{2000\times 0.75}{0.80}}=0.843$$

Approximately take $\beta_{P.h}=0.40$ to obtain $F_{P.h}=0.704\ 3$, $K_S\approx K_P=\sqrt{0.704\ 3}/0.823=1.020$.

Allow the outlet electric supply of the high-voltage winding to be A'_1, obtaining

$$A'_1=A_2+A_3=(15\ 261.48+1094.58)\times 10^4$$
$$=16\ 356.06\times 10^4(\text{kW}\cdot\text{h})$$

Based on this, the load loss of the high-voltage winding can be obtained

$$S_{rms.h}=K_S S_{av.h}=1.020\times\frac{16\ 356.06\times 10^4/7296}{0.843}$$
$$=27\ 124.8(\text{kVA})$$

$$\Delta A_{L.h}=2\times 89.51\times\left(\frac{27\ 124.8/2}{20\ 000}\right)^2\times 7296$$
$$=60.06\times 10^4(\text{kW}\cdot\text{h})$$

1.3.5 *Calculation of inlet electric supply of 110 kV main transformers.*

See below for the calculation:

$$\Delta A_{0.1b}=2\times 32.6\times 7296$$
$$=47.57\times 10^4(\text{kW}\cdot\text{h})$$

$$\Delta A_{L.1b}=\Delta A_{L.h}+\Delta A_{L.m}+\Delta A_{L.1}$$
$$=(60.06+28.02+0.18)\times 10^4$$
$$=88.26\times 10^4(\text{kW}\cdot\text{h})$$

Allow the inlet electric supply of the 110 kV main transformers to be A_1, obtaining

$$A_1=A'_1+\Delta A_{0.1b}+\Delta A_{L.h}$$
$$=(16\ 356.06+47.57+60.06)\times 10^4$$
$$=16\ 463.69\times 10^4(\text{kW}\cdot\text{h})$$

1.3.6 *Calculation of electric energy loss of 110 kV line.*

Look for information about the LGJ185 conductor, obtaining $r_0 = 0.17$ Ω/km, $R = 12 \times 0.17 = 2.04$ Ω. $f_{P.h} = 0.823$, $\cos\varphi_{av.g} = 0.843$ and $\beta_{P.h} = 0.40$ are obtained. Use these value to obtain $C'_{11} = 1.657 \times 10^{-10}$ [1/(kW·h)] as per Table 6.2. Considering the two factors of measuring period and resistance, the following is obtained through conversion:

$$C_{11} = \left(1.657 \times 10^{-10} \times 720/7296\right) \times 2.04$$
$$= 0.334 \times 10^{-10}[1/(\mathrm{kW \cdot h})]$$

The electric energy loss of the line is calculated as per the following formula:

$$\Delta A_{11} = C_{11} A_1^2 = 0.334 \times 10^{-10} \times \left(16\,463.69 \times 10^4\right)^2$$
$$= 90.52 \times 10^4 (\mathrm{kW \cdot h})$$

The load loss of 110 kV system is $\Delta A_{L.1bl} = \Delta A_{L.h} + \Delta A_{11}$, so

$$\Delta A_{L.1bl} = (60.06 + 90.52) \times 10^4$$
$$= 150.58 \times 10^4 (\mathrm{kW \cdot h})$$

The head end electric supply of the line is

$$A_{11} = A_1 + \Delta A_{11} = (16\,463.69 + 90.52) \times 10^4$$
$$= 16\,554.21 \times 10^4 (\mathrm{kW \cdot h})$$

1.4 *Calculation of partial voltage line loss rate.*

1.4.1 *Calculation of 35 kV line loss rate.*

The total electric energy losses of the 35 kV transformers of No. 2 and No. 3 substation subsystems as well as the medium-voltage windings of the main transformers in No. 1 substation are

$$\Delta A_{35} = \Delta A_{0.35} + \Delta A_{L.35}$$
$$= (\Delta A_{0.3b} + \Delta A_{0.2b}) + (\Delta A_{L.3bl} + \Delta A_{L.2bl} + \Delta A_{L.m})$$
$$= (38.67 + 6.06) \times 10^4 + (661.18 + 8.51 + 28.02) \times 10^4$$
$$= 44.73 \times 10^4 + 697.71 \times 10^4$$
$$= 742.44 \times 10^4 (\mathrm{kW \cdot h})$$

$$\Delta A_{35}\% = \frac{\Delta A_{35}}{A_2} \times 100\% = \frac{742.44 \times 10^4}{15\,261.48 \times 10^4} \times 100\% = 4.86\%$$

Wherein

$$\Delta A_{0.35}\% = \frac{44.73 \times 10^4}{15\,261.48 \times 10^4} \times 100\% = 0.29\%$$

$$\Delta A_{L.35}\% = \frac{697.71 \times 10^4}{15\,261.48 \times 10^4} \times 100\% = 4.57\%$$

Allow the total load loss coefficient of the 35 kV system to be C_{35}, obtaining

$$C_{35}=\frac{\Delta A_{L.35}}{A_2^2}=\frac{697.1\times10^4}{(15\ 261.48\times10^4)^2}$$
$$=2.996\times10^{-10}[1/(\mathrm{kW\cdot h})]$$

The electricity line loss of the 35 kV system is

$$\Delta A_{35}=44.73\times10^4+2.996\times10^{-10}\times A_2^2$$

The line loss rate of the 35 kV system is

$$\Delta A_{35}\%=0.29\%+2.996\times10^{-10}\times A_2$$

1.4.2 *Calculation of 110 kV line loss rate.*

See below for the calculation:

$$\Delta A_{110}=\Delta A_{0.110}+\Delta A_{L.1bl}=(47.57+150.58)\times10^4$$
$$=198.15\times10^4(\mathrm{kW\cdot h})$$
$$\Delta A_{110}\%=\frac{\Delta A_{110}}{A_{11}}\times100\%$$
$$=\frac{198.15\times10^4}{16\ 554.21\times10^4}\times100\%=1.197\%$$

Wherein

$$\Delta A_{0.110}\%=\frac{47.57\times10^4}{16\ 554.21\times10^4}\times100\%=0.287\%$$
$$\Delta A_{L.110}\%=\frac{150.58\times10^4}{16\ 554.21\times10^4}\times100\%=0.910\%$$

The load loss coefficient of the 110 kV system is

$$C_{110}=\frac{\Delta A_{L.110}}{A_{11}^2}=\frac{150.58\times10^4}{(16\ 554.21\times10^4)^2}$$
$$=0.549\times10^{-10}[1/(\mathrm{kW\cdot h})]$$

The electricity line loss of the 110 kV system is

$$\Delta A_{110}=47.57\times10^4+0.549\times10^{-10}\times A_{11}^2$$

The line loss rate of the 110 kV system is

$$\Delta A_{110}\%=0.29\%+0.549\times10^{-10}\times A_{11}$$

1.4.3 *Calculation of line loss rate of the whole network.*
See below for the calculation:

$$\begin{aligned}\Delta A_{\Sigma} &= \Delta A_{0.\Sigma} + \Delta A_{L.\Sigma}\\ &= (44.73 + 47.57) \times 10^4 + (697.71 + 150.58) \times 10^4\\ &= 92.30 \times 10^4 + 848.29 \times 10^4\\ &= 940.59 \times 10^4 (\text{kW}\cdot\text{h})\end{aligned}$$

$$\begin{aligned}\Delta A_{\Sigma}\% &= \frac{\Delta A_{\Sigma}}{A_{11}} \times 100\%\\ &= \frac{940.59 \times 10^4}{16\ 554.21 \times 10^4} \times 100\% = 5.68\%\end{aligned}$$

Wherein

$$\Delta A_{0.\Sigma}\% = \frac{92.30 \times 10^4}{16\ 554.21 \times 10^4} \times 100\% = 0.56\%$$

$$\Delta A_{L.\Sigma}\% = \frac{848.29 \times 10^4}{16\ 554.21 \times 10^4} \times 100\% = 5.12\%$$

Allow the load loss coefficient of the losses of the whole network to be C_{Σ}, obtaining

$$\begin{aligned}C_{\Sigma} &= \frac{\Delta A_{L.\Sigma}}{A_{11}^2}\\ &= \frac{848.29 \times 10^4}{(16\ 544.21 \times 10^4)^2}\\ &= 3.096 \times 10^{-10} [1/(\text{kW}\cdot\text{h})]\end{aligned}$$

The electricity line loss of the whole network is

$$\Delta A_{\Sigma} = 92.30 \times 10^4 + 3.096 \times 10^{-10} \times A_{11}^2$$

The line loss rate of the whole network is

$$\Delta A_{\Sigma}\% = 0.56\% + 3.096 \times 10^{-10} \times A_{11}$$

The load loss coefficient of losses of the whole network can also be calculated as per Formula (6.17), that is

$$\begin{aligned}C_{\Sigma} &= C_{110} + C_{35}\left(\frac{A_2}{A_{11}}\right)^2 + C_{10}\left(\frac{A_3}{A_{11}}\right)^2\\ &= 0.549 \times 10^{-10} + 2.996 \times 10^{-10} \times \left(\frac{15\ 261.48 \times 10^4}{16\ 544.21 \times 10^4}\right)^2 + 0.148\ 9 \times 10^{-10} \times \left(\frac{1094.58 \times 10^4}{16\ 544.21 \times 10^4}\right)^2\\ &= 0.549 \times 10^{-10} + 2.546 \times 10^{-10} + 0.000\ 65 \times 10^{-10}\\ &= 3.096 \times 10^{-10} [1/(\text{kW}\cdot\text{h})]\end{aligned}$$

Wherein

$$C_{10} = \frac{\Delta A_{L.1}}{A_3^2} = \frac{0.178\ 5\times 10^4}{(1094.58\times 10^4)^2}$$
$$= 0.148\ 9\times 10^{-10}[1/(\text{kW}\cdot\text{h})]$$

The calculation shows that both methods derive the same C_Σ.

6.4 Determination of Voltage Level by Loss Reduction Requirement

6.4.1 Voltage Characteristics of Various Loads and Comprehensive Loads of Distribution Lines

According to relevant domestic analyses of measured data [23, 24], the voltage characteristics of power loads represented by asynchronous motors and illumination loads represented by incandescent and fluorescent lamps are generally expressed by exponential functions, and the no-load current and voltage of distribution transformers present a higher-power relation. Using formulas for these voltage characteristics, the voltage characteristic of comprehensive loads of distribution lines can be calculated based on a certain proportion of load combinations and by considering the difference in power factors. The calculation result shows that the voltage characteristic of comprehensive loads of distribution lines is also a kind of exponential function. The difference in the voltage index for the voltage characteristic of comprehensive loads of distribution lines is about 0.10–0.17 when including and excluding the no-load current of distribution transformers. See Table 6.4 for relevant formulas.

Some Chinese electricity workers once measured the voltage characteristic of distribution lines. See Table 6.5 for collected data and analysis results.

According to Table 6.5, the approximate formula of voltage characteristic of 10 kV distribution lines is

$$I_* = 1.0U_*^{0.63} \tag{6.18}$$

When the no-load current of distribution transformers is excluded, the voltage power exponent in the empirical formula of $I(U)$ characteristic of comprehensive loads should be reduced by 0.10–0.17. Therefore, to analyze how to control voltage and reduce line loss, the voltage characteristic of comprehensive loads can be approximately taken as

$$I_* = 1.0U_*^{0.5} \tag{6.19}$$

Table 6.4 Voltage characteristics of various loads and comprehensive loads of distribution lines.

Type		Voltage characteristic $I(U)$	Description
Asynchronous motor	Load rate 25%	$I_* = 1.023U_*^{0.976}$	Calculated by measured data in Reference [24]
	Load rate 50%	$I_* = 1.00U_*^{0.327}$	
Incandescent lamp		$I_* = 1.00U_*^{0.515}$	Calculated by measured data of the author
Fluorescent lamp		$I_* = 1.006U_*^{2.67}$	Calculated by measured data in Reference [24]
Distribution transformer		$I_{T.0} = 1.00U_*^{5.0}$	Calculated by measured data in Reference [23]
Distribution line	Including $I_{T.0}$	$I_* = 1.004U_*^{1.046}$	Calculated by a certain proportion of load combinations
	Excluding $I_{T.0}$	$I_* = 1.003U_*^{0.871}$	

Table 6.5 Measurement and analysis of voltage characteristic of distribution lines.

Actual measurement of line		Measured data						Approximate formula
10 kV line supplying electricity for small and medium-sized industries and urban life; at night, when $U = U_N$, $I = 472$ A. See Reference [24]	U_* I_*	0.93 0.981	0.95 0.97	1.0 1.0	1.015 1.011	1.05 1.032	1.08 1.08	$I_* = 1.01U_*^{0.634}$ Correlation coefficient $r = 0.935\,6$
Total loads of four 10 kV rural trunk lines, mainly including power loads. In the daytime, when $U = 9.9$ kV, $I = 260$ A. See Reference [23]	U_* I_*	0.934 3 0.953 8	0.959 6 0.961 5	0.984 8 0.973 1	1.0 1.0	— —	— —	$I_* = 0.991U_*^{0.638}$ Correlation coefficient $r = 0.921\,4$

6.4.2 *Control of Voltage Level and Reduction of Electric Energy Loss*

When the voltage changes within $(0.90–1.10)U_N$, the no-load electric energy loss of a high-loss distribution transformer can be calculated as per the approximate Formula (6.7), that is $\Delta A_0 = 1.103U^4_{*av}\Delta P_0 T \times 10^{-3}$. Allow $\Delta A_{0*} = \Delta A_0/\Delta A_{0.e}$, as $\Delta A_{0.N} = \Delta P_0 T \times 10^{-3}$, obtaining

$$\Delta A_0 = 1.103U^4_{*av}\Delta A_{0.N} \tag{6.20}$$

When the voltage changes or is controlled within a certain range, the load loss also changes with the voltage. Substitute Formula (6.19) into Formula (1.39) to obtain

$$\begin{aligned}
\Delta A_L &= 3I^2_{max}(U)FRT \times 10^{-3} \\
&= 3\left(1.0U_*^{0.5}I_{max.N}\right)^2 FRT \times 10^{-3} \\
&= U_*\left(3I^2_{max.N}FRT \times 10^{-3}\right) \\
&= U_*\frac{1}{(U_{av}/U_N)^2} \times \left(\frac{R}{U_N^2}P^2_{max}F \times 2T \times 10^{-3}\right)
\end{aligned}$$

So

$$\Delta A_L \approx \frac{1}{U_{*av}}\Delta A_{L.N} \tag{6.21}$$

According to Formula (6.2), the calculation formula of line loss rate is as below:

$$\Delta A\% = \frac{\Delta A_0}{A} + \frac{\Delta A_L}{A} = \Delta A_0\% + \Delta A_L\%$$

Substitute Formulas (6.20) and (6.21) into the above formula to obtain

$$\Delta A\% = 1.103U^4_{*av}\Delta A_{0.N}\% + \frac{1}{U_{*av}}\Delta A_{L.N}\%$$

Take differential of the line loss rate $\Delta A\%$ to obtain the following approximate formula

$$d(\Delta A\%) = (4\Delta A_0\% - \Delta A_L\%)\frac{\Delta U_{*av}}{U_{*av}}$$
$$= \Delta A_L\%(4\theta - 1)\delta U_{*av} \quad (6.22)$$
$$\theta = \frac{\Delta A_0\%}{\Delta A_L\%} = \frac{B}{CA^2}, \quad \delta U_{*av} = \frac{\Delta U_{*av}}{U_{av}}$$

Wherein θ – loss constituent ratio between no-load line loss rate and load line loss rate;
δU_{*av} – rate of change in voltage level.

The following conclusions can be reached according to Formula (6.22):

1. When $\theta > 0.25$ and $\delta U_{*av} < 0$, $d(\Delta A\%) < 0$. This indicates that when the no-load loss is larger than one-quarter of the load loss (the no-load loss is larger than one-fifth of the total loss), the reduction of voltage level can reduce the line loss rate.
2. When $\theta < 0.25$ and $\delta U_{*av} > 0$, $d(\Delta A\%) < 0$. This indicates that when the no-load loss is smaller than one-quarter of the load loss (the no-load loss is smaller than one-fifth of the total loss), the increase in voltage level can reduce the line loss rate.
3. When $\theta > 0.25$ and $\delta U_{*av} > 0$, or when $\theta < 0.25$ and $\delta U_{*av} < 0$, $d(\Delta A\%) > 0$. This indicates that when the no-load loss is larger than one-quarter of the load loss, the increase in voltage level can increase the line loss rate; when the no-load loss is smaller than one-quarter of the load loss, the reduction of voltage level can also increase the line loss rate.

For electrical heating lines or rectification load lines, $P = aU^2$, $I = aU$, so $I_* = U_*$. As mentioned above,

$$\Delta A_L = 3\left(I_{max.e}\frac{U}{U_N}\right)^2 FRT \times 10^{-3} = \frac{U^2}{U_N^2}\frac{R}{U_{av}^2}\dot{P}_{max}^2\dot{F} \times 2T \times 10^{-3}$$
$$= \frac{1}{U_{*av}^2} \times \frac{R}{U_N^2}\dot{P}_{max}^2\dot{F} \times 2T \times 10^{-3} = \frac{1}{U_{*av}^2}\Delta A_{L.N} \quad (6.23)$$
$$d(\Delta A\%) = (4 \times \Delta A_0\% - 2\Delta A_L\%) \times \frac{\Delta U_{*av}}{U_{*av}}$$
$$= \Delta A_L\%(4\theta - 2) \times \delta U_{*av}$$

The following similar conclusions can also be reached according to Formula (6.23):

1. When $\theta > 0.50$ and $\delta U_{*av} < 0$, $d(\Delta A\%) < 0$, indicating that when the no-load loss is larger than one-half of the load loss, the reduction of voltage level can reduce the line loss rate.
2. When $\theta < 0.50$ and $\delta U_{*av} > 0$, $d(\Delta A\%) < 0$, indicating that when the no-load loss is smaller than one-half of the load loss, the increase in voltage level can reduce the line loss rate.
3. When $\theta > 0.50$ and $\delta U_{*av} > 0$, or when $\theta < 0.50$ and $\delta U_{*av} < 0$, $d(\Delta A\%) > 0$. This indicates that when the no-load loss is larger than one-half of the load loss, the increase in voltage level can increase the line loss rate; when the no-load loss is smaller than one-half of the load loss, the reduction of voltage level can also increase the line loss rate.

In preliminary conclusion, because the no-load loss of a distribution transformer is in direct proportion to higher power of the voltage, while the voltage characteristic $I_*(U_*)$ of comprehensive loads of a distribution line is an exponential function with power exponent of 0.5–1.0, when the no-load electric energy loss accounts for 20–33% of the total loss, the reduction of operating voltage level (below the rated value) can reduce the total loss; when the no-load electric energy loss is smaller than 20–33% of the total loss, the increase in operating voltage level (above the rated value) can reduce the total loss. The former case often occurs in agricultural electric lines, while the latter case usually exists in urban distribution lines. For specific lines, the statistical calculation of line losses can be used to estimate the loss constituent ratio θ, so as to correctly control the level of operating voltage and reduce line losses. Note that the operating voltage should be adjusted within a reasonable range.

7

Analysis and Control of Line Loss Rate Indicators of Power Grids

7.1 Analysis of Line Loss Rate Composition

For regional power grids, it is necessary to analyze the composition of line loss rate by voltage class and the composition of line loss rate by category.

7.1.1 Line Loss Rates and Total Line Loss Rate of Different Voltage Grids

Regional power grids are generally composed of power transmission grids and power distribution grids, so it is necessary to analyze the relationship between the line loss rates and total line loss rate of the two relevant voltage power grids.

Assume that the line loss rates of two relevant voltage grids are $\Delta A_1\%$ and $\Delta A_2\%$, and their power sales quantities are A_1 and A_2. If the ratio of power sales quantities is $A_2/A_1 = \alpha$, when the electric supply of the lower class voltage grids is input from the higher class voltage grids, the schematic diagram of electric flows under different voltages is shown in Figure 7.1.

According to Figure 7.1, if $\Delta A_1\% = \frac{\Delta A_1}{A} \times 100\%$, and $\Delta A_2\% = \frac{\Delta A_2}{A'} \times 100\%$, then the total line loss rate is

$$A_\Sigma\% = \frac{\Delta A_1 + \Delta A_2}{A} = \Delta A_1\% + \frac{A'\Delta A_2\%}{A}$$

As $A' = A(1 - \Delta A_1\%) - A_1$, substitute it into the above formula to obtain

$$\Delta A_\Sigma\% = \Delta A_1\% + (1 - \Delta A_1\%)\Delta A_2\% - \frac{A_1}{A} \times \Delta A_2\%$$

Wherein, the third term is

$$\frac{A_1}{A}\Delta A_2\% = \frac{A_1}{(A_1 + A_2)/(1 - \Delta A_\Sigma\%)} \times \Delta A_2\%$$

$$= \frac{(1 - \Delta A_\Sigma\%)\Delta A_2\%}{1 + \alpha}$$

Line Loss Analysis and Calculation of Electric Power Systems, First Edition. Anguan Wu and Baoshan Ni.

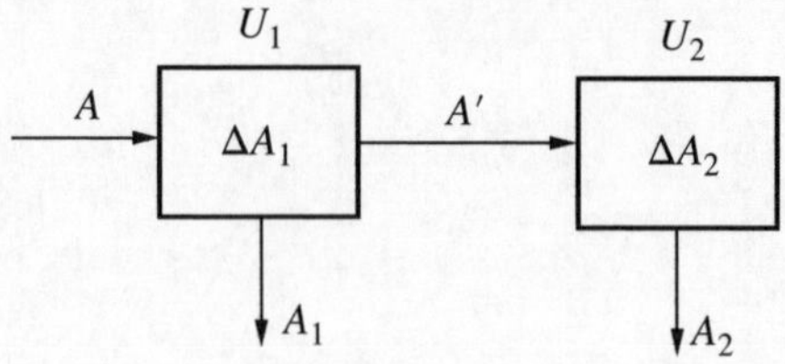

Figure 7.1 Schematic diagram of electric flows under different voltages.

Substitute it into the $\Delta A_\Sigma\%$ formula and obtain the following formula after transposition and simplification:

$$\Delta A_\Sigma\% = \frac{\Delta A_1\% + \Delta A_2\%\left(1 - \Delta A_1\% - \dfrac{1}{1+\alpha}\right)}{1 - \dfrac{\Delta A_2\%}{1+\alpha}} \tag{7.1}$$

According to Formula (7.1): when $\alpha \to 0$, $\Delta A_\Sigma\% = \Delta A_1\%$; when $\alpha \to \infty$, $A_1 \to 0$, $\Delta A_\Sigma\% = \Delta A_1\% + \Delta A_2\%(1 - \Delta A_1\%)$; when α is any other value, $\Delta A_1\% < \Delta A_\Sigma\% < [\Delta A_1\% + \Delta A_2\%(1 - \Delta A_1\%)]$; when α is the larger, the total line loss rate is closer to the upper limit of line loss rate.

7.1.2 No-load Line Loss Rate and Load Line Loss Rate

To analyze the relationship between the ratio of the no-load line loss rate to the load line loss rate of whole grids (i.e. the loss constituent ratio θ_Σ) and the line loss rates of different voltage grids, it is necessary to analyze the no-load line loss rate and the load line loss rate, respectively. For the two relevant voltage grids shown in Figure 7.1, the total load line loss rate can be expressed by

$$\begin{aligned}\Delta A_{L.\Sigma}\% &= \frac{\Delta A_{L.1} + \Delta A_{L.2}}{A} \\ &= \Delta A_{L.1}\% + \Delta A_{L.2}\%\frac{A'}{A} \\ &= \Delta A_{L.1}\% + \Delta A_{L.2}\%\left[\frac{A(1 - \Delta A_1\%) - A_1}{A}\right]\end{aligned}$$

The total line loss rate and the partial voltage line loss rate of U_1 grid can be expressed by the following two formulas:

$$\Delta A_\Sigma\% = \Delta A_{L.\Sigma}\%(1 + \theta_\Sigma), \ \Delta A_1\% = \Delta A_{L.1}\%(1 + \theta_1)$$

After derivation and simplification,

$$\begin{aligned}\Delta A_{L.\Sigma}\% &= \frac{\Delta A_{L.1}\% + \Delta A_{L.2}\%\left[1 - \Delta A_{L.1}\%(1+\theta_1) - \dfrac{1}{1+\alpha}\right]}{1 - \dfrac{1+\theta_\Sigma}{1+\alpha}\Delta A_{L.2}\%} \\ &= \frac{T}{1 - \dfrac{1+\theta_\Sigma}{1+\alpha}\Delta A_{L.2}\%}\end{aligned} \tag{7.2}$$

$$T = \Delta A_{L.1}\% + \Delta A_{L.2}\%\left[1 - \Delta A_{L.1}\%(1+\theta_1) - \frac{1}{1+\alpha}\right]$$

Wherein, T – designation of numerator polynomial.

Likewise, the formula of the total no-load line loss rate can be obtained

$$\Delta A_{0.\Sigma}\% = \frac{\Delta A_{0.1}\% + \Delta A_{0.2}\%\left[1 - \Delta A_{0.1}\%\left(\frac{1}{\theta_1}+1\right) - \frac{1}{1+\alpha}\right]}{1 - \Delta A_{0.2}\%\left(\frac{1}{\theta_\Sigma}+1\right)/(1+\alpha)} = \frac{S}{1 - \Delta A_{0.2}\%\left(\frac{1}{\theta_\Sigma}+1\right)/(1+\alpha)} \tag{7.3}$$

$$S = \Delta A_{0.1}\% + \Delta A_{0.2}\%\left[1 - \Delta A_{0.1}\%\left(\frac{1}{\theta_1}+1\right) - \frac{1}{1+\alpha}\right]$$

Wherein, S – designation of numerator polynomial.

As $\theta_\Sigma = \frac{\Delta A_{0.\Sigma}\%}{\Delta A_{L.\Sigma}\%}$,

$$\theta_\Sigma = \frac{S}{1 - \Delta A_{0.2}\%\left(\frac{1}{\theta_\Sigma}+1\right)/(1+\alpha)} \times \frac{1 - \Delta A_{L.2}\%(1+\theta_\Sigma)/(1+\alpha)}{T}$$

After simplification,

$$\theta_\Sigma = \frac{S\left(1 - \frac{\Delta A_{L.2}\%}{1+\alpha}\right) - T\frac{\Delta A_{0.2}\%}{1+\alpha}}{T + S\Delta A_{L.2}\%/(1+\alpha) - T\Delta A_{L.2}\%/(1+\alpha)} = \frac{S[(1+\alpha) - \Delta A_{L.2}\%] - T\Delta A_{0.2}\%}{T(1+\alpha) + S\Delta A_{L.2}\% - T\Delta A_{0.2}\%} \tag{7.4}$$

If the no-load and load line loss rates of different voltage grids are $\Delta A_{0.1}\%$, $\Delta A_{L.2}\%$ and $\Delta A_{L.1}\%$, $\Delta A_{0.2}\%$, then T and S of polynomials can be calculated, and θ_Σ can be calculated as per Formula (7.4). Then, the total load line loss rate $\Delta A_{L.\Sigma}\%$ and the total no-load line loss rate $\Delta A_{0.\Sigma}\%$ can be calculated as per Formulas (7.2) and (7.3). According to Formula (7.4), the loss constituent ratio θ_Σ of power grids in the entire region depends on no-load line loss rates and load line loss rates under various classes of voltages as well as the ratio of power sales quantities of different voltage grids.

Example 7.1 In Example 6.1, no-load line loss rates, load line loss rates, and total line loss rates of 35, 110 kV systems, and the entire region were calculated, as shown in Table 7.1.

The schematic diagram of electric flows under different classes of voltages is shown in Figure 7.2.

Try to use Formulas (7.2) to (7.4) to calculate line loss rates $\Delta A_{0.\Sigma}\%$ and $\Delta A_{L.\Sigma}\%$ of the entire region as well as the loss constituent ratio θ_Σ, and compare them with values given in Table 7.1.

Table 7.1 Calculation of regional line loss rates and loss constituent ratio.

Item	ΔA_0 ($\times 10^4$ kW·h)	ΔA_L ($\times 10^4$ kW·h)	A ($\times 10^4$ kW·h)	$\Delta A_0\%$	$\Delta A_L\%$	ΔA	θ
35 kV	44.73	697.71	15 261.48	0.290	4.57	4.86	0.063 46
110 kV	47.57	150.58	16 554.2	0.287	0.910	1.197	0.315 4
Entire system	92.30	848.29	16 554.2	0.560	5.12	5.68	0.109 4
Calculation by formulas				0.554	5.123	5.677	0.108 1

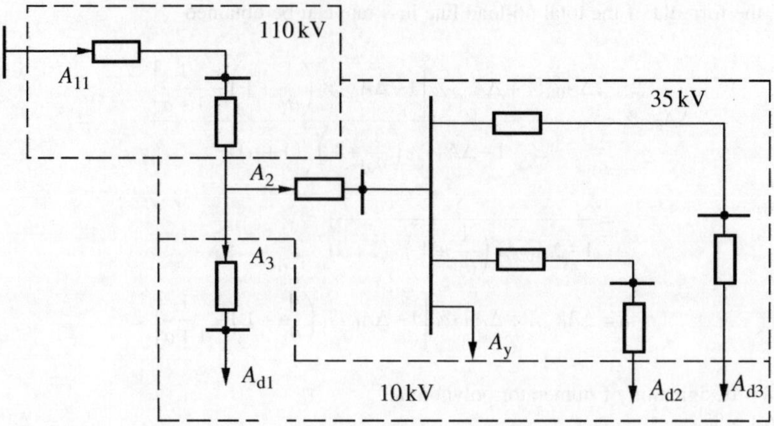

Figure 7.2 Schematic diagram of electric flows under different classes of voltages.

Solution

According to Example 6.1, the total power sales quantity of the 35 kV system is

$$\begin{aligned} A_{35} &= A_y + A_{d2} + A_{d3} \\ &= (8000 \times 0.90 \times 7296) + (510.72 + 8755.2) \times 10^4 \\ &= 14\ 519.04 \times 10^4 (\text{kW}\cdot\text{h}) \end{aligned}$$

The power sales quantity (deemed as the direct power sales quantity of a 110 kV system) of the 10 kV system of No. 1 substation is

$$A_{d1} = 1094.4 \times 10^4\ (\text{kW}\cdot\text{h})$$

The ratio between the power sales quantities of the 35 kV and 110 kV systems is

$$\alpha = \frac{14\ 519.04 \times 10^4}{1094.4 \times 10^4} = 13.27$$

According to Formula (7.1),

$$\begin{aligned} \Delta A_\Sigma\% &= \frac{1.197\% + 4.86\% \times \left[1 - 1.197\% - \dfrac{1}{1+13.27}\right]}{1 - 4.86\%/(1+13.27)} \\ &= \frac{5.658\%}{0.996\ 6} = 5.667\% \end{aligned}$$

This is basically consistent with $\Delta A_\Sigma\% = 5.68\%$, as calculated in Example 6.1. According to Formula (7.2),

$$\begin{aligned} T &= 0.91\% + 4.57\% \times \left[1 - 0.91\% \times (1 + 0.315\ 4) - \frac{1}{1+13.27}\right] \\ &= 0.91\% + 4.57\% \times 0.917\ 9 \\ &= 5.105\% \end{aligned}$$

According to Formula (7.3),

$$S = 0.287\% + 0.29\% \times \left[1 - 0.287\% \times \left(\frac{1}{0.3154} + 1\right) - \frac{1}{1 + 13.27}\right]$$
$$= 0.287\% + 0.29\% \times 0.9179 = 0.553\%$$

Substitute T and S into Formula (7.4) to obtain

$$\theta'_{\Sigma} = \{0.005\,53 \times [(1 + 13.27) - 0.045\,7] - 0.051\,05 \times 0.002\,9\} /$$
$$[0.051\,05 \times (1 + 13.27) + 0.005\,53 \times 0.045\,7 - 0.051\,05 \times 0.002\,9]$$
$$= \frac{0.078\,51}{0.728\,59} = 0.107\,76$$

Substitute this θ'_{Σ} into Formulas (7.2)~(7.3) to obtain $\Delta A_{Lz\cdot\Sigma}\%$ and $\Delta A_{0.\Sigma}\%$

$$\Delta A_{Lz.\Sigma}\% = \frac{5.105\%}{1 - \frac{1 + 0.107\,66}{1 + 13.27} \times 0.045\,7}$$
$$= \frac{5.105\%}{0.996\,45} = 5.123\%$$

$$\Delta A_{0.\Sigma}\% = \frac{0.553\%}{1 - \frac{1 + 1/0.107\,76}{1 + 13.27} \times 0.002\,9}$$
$$= \frac{0.553\%}{0.997\,9} = 0.554\%$$

So θ_{Σ} is calculated

$$\theta_{\Sigma} = \Delta A_{0.\Sigma}\% / \Delta A_{L.\Sigma}\%$$
$$= 0.554\% / 5.123\% = 0.108\,1$$

$\Delta A_{\Sigma}\%$ is also calculated

$$\Delta A_{\Sigma}\% = \Delta A_{0.\Sigma}\% + \Delta A_{L.\Sigma}\%$$
$$= 0.554\% + 5.123\% = 5.677\%$$

This is exactly the same as the result calculated by Formula (7.1).

According to the comparison of line loss rates $\Delta A_{0\cdot\Sigma}\%$, $\Delta A_{fz.\Sigma}\%$ and $\Delta A_{\Sigma}\%$ for the entire region as well as θ_{Σ} calculated by Formulas (7.1) to (7.4) with the values obtained in Example 6.1, the errors are very small, indicating that Formulas (7.1) to (7.4) are correct. These formulas are helpful to make line loss rate analysis programs of microcomputers to analyze whether partial voltage line loss rates are "consistent".

7.2 Analysis of Influence of Grid Electric Supply Structure on Line Loss Rate

7.2.1 Repeated Electric Supply Rate

For ease of analysis, the schematic diagram of electric flow of three voltage systems is shown in Figure 7.3.

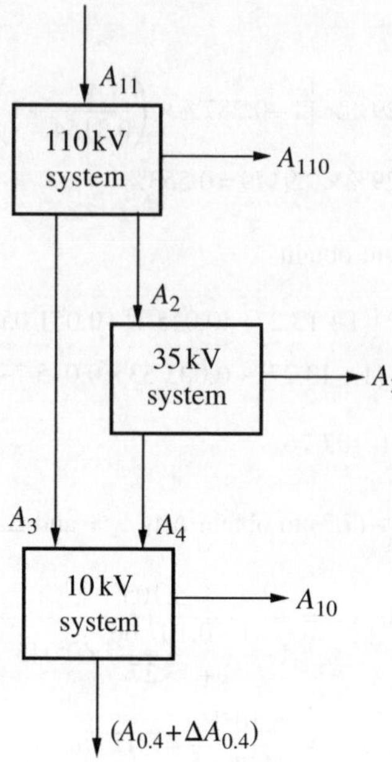

Figure 7.3 Schematic diagram of the electric flow of three voltage systems.

35 kV and 10 kV systems have no electric input from any power plant or external grids, so the following expression formula of electric quantity balance is obtained

$$A_{11}-(A_2+A_3)=A_{110}+\Delta A_{110}$$
$$A_2-A_4=A_{35}+\Delta A_{35}$$
$$A_3+A_4=A_{10}+A_{0.4}+\Delta A_{10}+\Delta A_{0.4}$$

If the repeated electric supply rate is defined as

$$\gamma=\frac{\text{electric supply of two continuous voltage drops}}{\text{sum of electric supply of two classes of voltage systems}} \tag{7.5}$$

According to Figure 7.3, $\gamma=\dfrac{A_4}{A_2+A_3}$.

If the input electric quantity of the 110 kV system is A_{11}, then the total line loss rate of 10 kV and above power grids can be calculated as per the following formula

$$\Delta A_{\Sigma}\%=[\Delta A_{110}\%A_{11}+\Delta A_{35}\%A_2+(A_3+A_4)\Delta A_{10}\%]/A_{11}$$
$$=\Delta A_{110}\%\left[1+\frac{\Delta A_{35}\%}{\Delta A_{110}\%}\cdot\frac{A_2}{A_{11}}+\frac{\Delta A_{10}\%}{\Delta A_{110}\%}\cdot\frac{A_3+A_4}{A_{11}}\right]$$

If $\dfrac{A_2}{A_{11}}=\alpha_1$, $\dfrac{A_3}{A_{11}}=\alpha_2$, $\dfrac{\Delta A_{35}\%}{\Delta A_{110}\%}=a$, $\dfrac{\Delta A_{10}\%}{\Delta A_{110}\%}=b$, then the above formula can be simplified to

$$\Delta A_{\Sigma}\%=\Delta A_{10}\%[1+(a\alpha_1+b\alpha_2)+\gamma b(\alpha_1+\alpha_2)] \tag{7.6}$$

According to Formula (7.6), the total line loss rate depends on the output electric quantities at medium voltage side and low voltage side of the 110 kV system, ratios α_1 and α_2 between such output electric quantities and input electric quantities, ratios a and b of line loss rates of various voltage systems, and the repeated electric supply rate γ. Obviously, when other ratios except γ are constant values, the smaller γ is, the smaller the total line loss rate is, so are total line losses; and vice versa.

7.2.2 Calculation of Loss Reduction Effect of Reducing Repeated Electric Supply Rate

Formula (7.6) can be used to check the total line loss rate of regional power grids and also calculate the loss reduction effect which may be achieved by reducing the repeated electric supply rate.

Example 7.2 If the line loss rate of the 10 kV system is 10% in Example 6.1, *try to use the direct calculation method and Formula (7.6) to calculate the total line loss rate of 10 kV and above systems in the entire region.* If the repeated electric supply rate is reduced by 50%, *how much is the total line loss rate reduced?*

Solution

1. *Calculation of the total line loss rate by direct method.* The total electric supply of the 10 kV system is

$$
\begin{aligned}
A_{10} &= A_3 + A_4 = 1094.58 \times 10^4 + 9265.92 \times 10^4 \\
&= 10\ 360.50 \times 10^4 (\mathrm{kW \cdot h})
\end{aligned}
$$

Wherein

$$
\begin{aligned}
A_4 &= A_{d2} + A_{d3} = (510.72 + 8755.2) \times 10^4 \\
&= 9265.92 \times 10^4 (\mathrm{kW \cdot h}) \\
\Delta A_{10} &= A_{10} \Delta A_{10}\% = 10\ 360.50 \times 10^4 \times 10\% \\
&= 1036.05 \times 10^4 (\mathrm{kW \cdot h})
\end{aligned}
$$

According to Example 6.1, the electricity losses of exceed by 10 kV systems are 940.59×10^4 kW·h, so the total electricity losses of 10 kV and above systems are

$$
\begin{aligned}
\Delta A_\Sigma &= 940.59 \times 10^4 + 1036.05 \times 10^4 \\
&= 1976.64 \times 10^4 (\mathrm{kW \cdot h})
\end{aligned}
$$

So

$$
\begin{aligned}
\Delta A_\Sigma\% &= \frac{\Delta A_\Sigma}{A_{11}} \times 100\% \\
&= \frac{1976.64 \times 10^4}{16\ 554.2 \times 10^4} \times 100\% = 11.94\%
\end{aligned}
$$

2. *Calculation of the total line loss rate by Formula* (7.6). According to the data listed in Table 7.1 and the calculation results in Example 6.1,

$$a=\frac{\Delta A_{35}\%}{\Delta A_{110}\%}=\frac{4.86\%}{1.197\%}=4.07$$

$$b=\frac{\Delta A_{10}\%}{\Delta A_{110}\%}=\frac{100\%}{1.197\%}=8.35$$

$$\alpha_1=\frac{A_2}{A_{11}}=\frac{15\,261.48\times 10^4}{16\,544.2\times 10^4}=0.922$$

$$\alpha_2=\frac{A_3}{A_{11}}=\frac{1094.58\times 10^4}{16\,544.2\times 10^4}=0.066$$

$$\gamma=\frac{A_4}{A_2+A_3}=\frac{9265.92\times 10^4}{(15261.48+1094.58)\times 10^4}=0.566\,5$$

Substitute the above values into Formula (7.6) to obtain

$$\Delta A_{\Sigma}\%=1.197\%\times[1+(4.07\times 0.922+8.35\times 0.066)+0.566\,5\times 8.35\times(0.922+0.066)]=11.94\%$$

As a result, the result calculated by Formula (7.6) is the same as that calculated by direct method.

3. *Calculation of the total line loss rate when γ is reduced by 50%.* Letters with "·" on the top refer to calculation values after γ is reduced. When γ is changed, the reduced A_4 electric quantity is transferred to A_3, so α_1 and α_2 are also changed. Then,

$$\begin{aligned}\dot{A}_4&=\dot{\gamma}(A_2+A_3)=\frac{1}{2}\times 0.566\,5\times(15\,261.48+1094.58)\times 10^4\\&=4632.85\times 10^4(\mathrm{kW\cdot h})\\ \partial(A_4)&=A_4-\dot{A}_4=(9265.92-4632.85)\times 10^4\\&=4633.07\times 10^4(\mathrm{kW\cdot h})\\ \dot{A}_3&=A_3+\partial(A_4)=(1094.58+4633.07)\times 10^4\\&=5727.65\times 10^4(\mathrm{kW\cdot h})\\ \dot{A}_2&=A_2-\partial(A_4)=(15\,261.48-4633.07)\times 10^4\\&=10\,628.41\times 10^4(\mathrm{kW\cdot h})\end{aligned}$$

So

$$\dot{\alpha}_1=\frac{\dot{A}_2}{A_{11}}=\frac{10\,628.41\times 10^4}{16\,554.2\times 10^4}=0.642$$

$$\dot{\alpha}_2=\frac{\dot{A}_3}{A_{11}}=\frac{5727.65\times 10^4}{16\,554.2\times 10^4}=0.346$$

Because the electric supply of the 35 kV system is reduced, the line loss rate of the 35 kV system is certainly reduced. The calculation formula of the electricity line losses of the 35 kV system is given in Example 6.1

$$\Delta A_{35}=44.73\times 10^4+2.996\times 10^{-10}\times A_2^2$$

After transfer of electric quantity, $\dot{A}_2$ = 10 628.41 × 10^4 kW·h. So

$$\begin{aligned}\Delta\dot{A}_{35}\%&=\frac{44.73\times 10^4}{10\,628.41\times 10^4}+2.996\times 10^{-10}\times\left(10\,628.41\times 10^4\right)\\&=0.42\%+3.18\%=3.60\%\end{aligned}$$

As A_3 is increased, low-voltage winding load losses of the main transformer of No. 1 substation are increased.

As $C_{10} = 0.148\,9 \times 10^{-10}$ 1/(kW·h) is calculated in Example 6.1, $\Delta A_{fz.d} = C_{10}\dot{A}_3^2 = 0.148\,9 \times 10^{-10} \times (5727.65 \times 10^4)^2 = 4.88 \times 10^4$ (kW·h), $\partial(\Delta A_{fz.d}) = (4.88 - 0.18) \times 10^4 = 4.7 \times 10^4$ (kW·h). Then,

$$\Delta\dot{A}_{10}\% = \frac{(1036.05 + 4.7) \times 10^4}{(10\,360.5 + 4.7) \times 10^4} \times 100\% = 10.04\%$$

So

$$\dot{a} = \frac{3.6\%}{1.197\%} = 3.01, \dot{b} = \frac{10.04\%}{1.197\%} = 8.39,$$

Substitute the values into Formula (7.6) to obtain

$$\Delta\dot{A}_\Sigma\% = 1.197\% \times [1 + (3.01 \times 0.642 + 8.39 \times 0.346) + 0.566\,5 \times 0.5 \times 8.39 \times (0.642 + 0.346)$$
$$= 9.69\%$$

The relative reduction rate of the total line loss rate is

$$\delta(\Delta A_\Sigma\%) = \left|\frac{9.69 - 11.94}{11.94}\right| \times 100\%$$
$$= 18.8\%$$

The above analysis results show that the improvement in the electric supply structure can reduce the total line loss rate, but the percentage of reduction is smaller than the percentage of reduction of the repeated electric supply rate.

It should be noted that the optimization of electric supply structure and the reduction of repeated electric supply rate are always influenced by the geographical distribution of loads and line routes, and thus should be considered in the planning of power grids and the earlier stage feasibility study of substations. To change the electric supply structure of power grids that have been put into operation, it is necessary to conduct economic and technical comparison for making the right decision.

7.3 Analysis of Power Sales Quantity Composition

Because power supply enterprises use different metering methods for consumers, such as power sales quantity without loss, or power sales quantity with loss, the comprehensive line loss rate obtained by statistics is smaller than the line loss rate with loss (see below for its definition). Some power supply enterprises undertake transit electric supply, so their comprehensive line loss rate is increased. This section will conduct a quantitative analysis of the two problems.

7.3.1 Influence of Power Sales Quantity without Loss or Power Sales Quantity with Loss on Line Loss Rate

The power sales quantity of dedicated lines supplying electricity to consumers can be measured in the substations of power supply enterprises, or measured by meters installed for consumers. But in the latter case, electricity losses in the lines should be added in the consumers' power sales quantity. In both cases, the electric energy losses of the lines should be borne by the consumers, and consumers are usually known

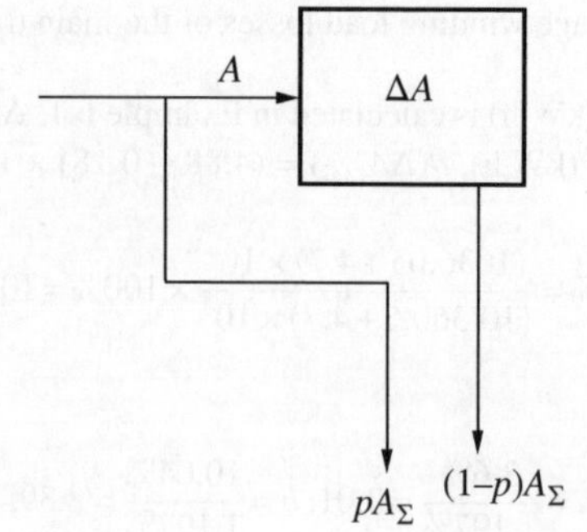

Figure 7.4 Schematic diagram of electric flow considering power sales quantity without loss.

as "consumers without loss" and "consumers with loss". In line loss management, the line loss rate including power sales quantity without loss is called a comprehensive line loss rate, and the line loss rate excluding power sales quantity without loss is called a line loss rate with loss. Obviously, the comprehensive line loss rate is smaller than the line loss rate with loss. To know the difference between the two line loss rates, it is necessary to analyze the relationship between them.

The schematic diagram of electric flow considering the power sales quantity without loss is shown in Figure 7.4.

Assume that the total electric supply with loss of a system is A; the sum of electricity losses is ΔA; the total power sales quantity is A_Σ; the ratio of the power sales quantity without loss to the total power sales quantity is p. Then, the line loss rate with loss $\Delta A\%$ meets $(1 - \Delta A\%)A = (1 - p)A_\Sigma$, so

$$A_\Sigma = \frac{(1-\Delta A\%)A}{1-p}$$

Because the comprehensive line loss rate is $\Delta A'\% = \dfrac{\Delta A}{A + pA_\Sigma}$, substitute the A_Σ formula into it to obtain

$$\Delta A'\% = \frac{\Delta A}{A + p\dfrac{(1-\Delta A\%)A}{1-p}} = \frac{\Delta A\%}{1 + \dfrac{p}{1-p}(1-\Delta A\%)} \tag{7.7}$$

When $\Delta A\% \le 10\%$, Formula (7.7) can be simplified to

$$\Delta A'\% \approx (1-p)\Delta A\% = \Delta A\% - p\Delta A\% \tag{7.8}$$

According to Formula (7.8), the difference between the line loss rate with loss and the comprehensive line loss rate is approximately equal to the product of the rate of power sales quantity without loss and the line loss rate with loss.

7.3.2 Calculation of Influence of Transit Electric Supply on Line Loss Rate

Assume that the transit electric supply of a voltage system is A_2 and the power sales quantity at current class of voltage is A_1, as shown in Figure 7.5. Assume the rate of transit electric supply is q, that is $q = \frac{A_2}{A}$. Not considering the transit electric supply, the line loss rate with loss is

$$\begin{aligned}\Delta A\% &= \Delta A_0\% + CA \\ &= \Delta A_L\% \times (1+\theta)\end{aligned}$$

When the transit electric supply A_2 passes and it is assumed that the load loss coefficient of the current class of voltage grid is constant, the comprehensive line loss rate is

$$\begin{aligned}\Delta A'\% &= \frac{\Delta A_0\% A}{A + A_2} + C(A + A_2) \\ &= \frac{\Delta A_0\%}{1+q} + CA(1+q) = \Delta A_0'\% + \Delta A_L'\%\end{aligned}$$

The loss constituent ratio considering the transit electric quantity is $\theta' = \frac{\Delta A_0'\%}{\Delta A_L'\%}$, so $\theta' = \frac{\Delta A_0\%/(1+q)}{CA(1+q)} = \frac{\theta}{(1+q)^2}$. Due to $\Delta A'\% = \Delta A'_L(1+\theta')$, and $\Delta A'_L\% = \Delta A_L\%(1+q)$,

$$\begin{aligned}\Delta A'\% &= \Delta A_L\%(1+q)\left[1 + \frac{\theta}{(1+q)^2}\right] \\ &= \Delta A_L\%\left[\frac{(1+q)^2 + \theta}{1+q}\right]\end{aligned}$$

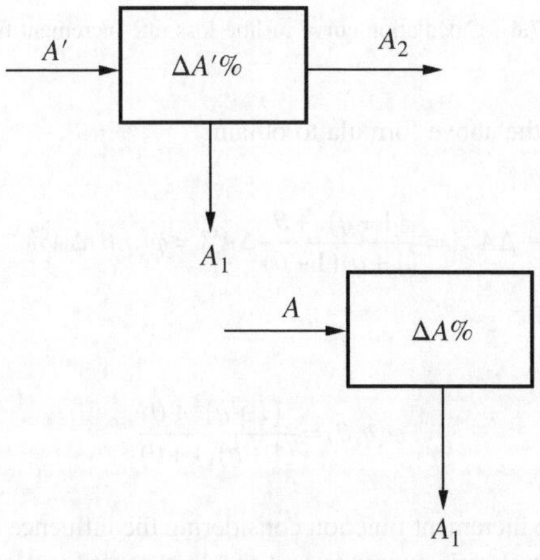

Figure 7.5 Influence of transit electric supply.

Table 7.2 Calculation table of line loss rate increment function $\varphi(q, \theta)$.

q / θ	0.04	0.06	0.08	0.10	0.12	0.14	0.16	0.18	0.20
0.10	1.032 9	1.049 4	1.066 0	1.082 6	1.099 4	1.116 1	1.132 9	1.149 8	1.166 7
0.15	1.029 8	1.044 8	1.059 9	1.075 1	1.090 4	1.105 7	1.121 1	1.136 6	1.152 2
0.20	1.026 9	1.040 6	1.054 3	1.068 2	1.082 1	1.096 2	1.110 3	1.124 6	1.138 9
0.25	1.024 3	1.036 7	1.049 2	1.061 8	1.074 6	1.087 4	1.100 4	1.113 5	1.126 7
0.30	1.021 9	1.033 1	1.044 4	1.055 9	1.067 6	1.079 4	1.091 2	1.103 3	1.115 4
0.35	1.019 7	1.029 8	1.040 1	1.050 5	1.061 1	1.071 9	1.082 8	1.093 8	1.104 9

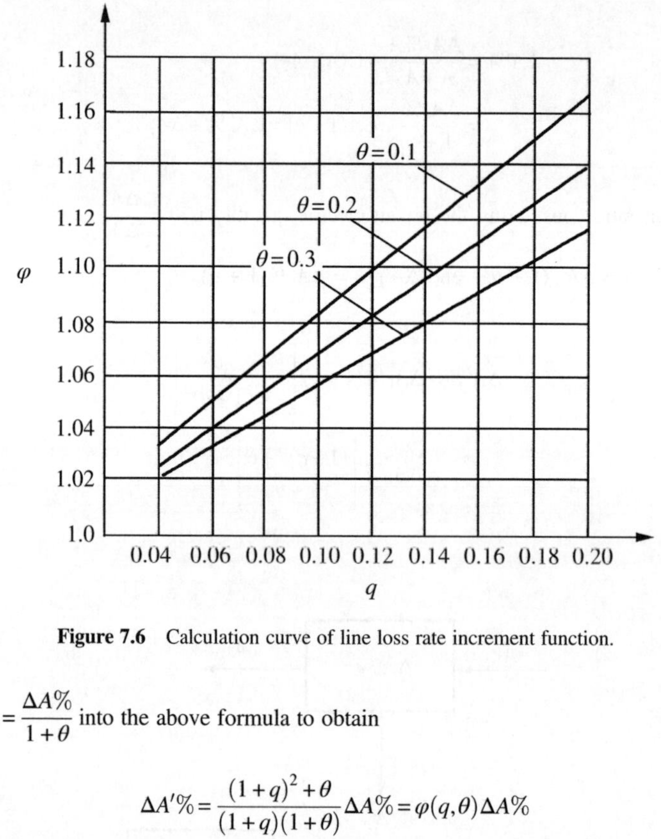

Figure 7.6 Calculation curve of line loss rate increment function.

Substitute $\Delta A_{\mathrm{L}}\% = \dfrac{\Delta A\%}{1+\theta}$ into the above formula to obtain

$$\Delta A'\% = \frac{(1+q)^2+\theta}{(1+q)(1+\theta)}\Delta A\% = \varphi(q,\theta)\Delta A\% \tag{7.9}$$

Wherein

$$\varphi(q,\theta) = \frac{(1+q)^2+\theta}{(1+q)(1+\theta)} \tag{7.10}$$

Wherein $\varphi(q, \theta)$ – line loss rate increment function considering the influence of transit electric quantity; this is related to the rate of transit electric supply q and the loss constituent ratio θ of grids without transit electric quantity.

According to Formula (7.10), the calculation cable (see Table 7.2) and calculation curve (see Figure 7.6) of the increment function $\varphi(q, \theta)$ can be made. According to Table 7.2, when q is certain, the increment function increases as θ decreases. This is because the smaller θ is, the larger the load line loss rate of the system is than the no-load line loss rate. With the transit electric supply, the load line loss rate of the system is greatly increased, while the no-load line loss rate is slightly decreased.

Example 7.3 In Example 7.1 if the 35 kV bus of the 110 kV substation has 2000 × 10^4 kW·h transit electric supply within the same measuring period, *how much is the comprehensive line loss rate of the 110 kV system increased? What is the total line loss rate of 35 kV and above systems in the entire region?*

Solution

The transit electric supply is $A_g = 2000 \times 10^4$ kW·h, and the electric supply of the original 110 kV system is $A = 16\ 554.2 \times 10^4$ kW·h, so $q = \dfrac{2000 \times 10^4}{16\ 554.2 \times 10^4} = 0.120\ 8$. According to Example 7.1, θ = 0.315 4, $\Delta A\%$ = 1.197%. Substitute the above values into Formula (7.10) to obtain $\varphi(q,\theta) = \dfrac{(1+0.120\ 8)^2 + 0.315\ 4}{(1+0.120\ 8) \times (1+0.315\ 4)} = 1.066$, indicating that the line loss rate of the 110 kV system is increased by 6.6%. $\Delta A'\% = \varphi(q, \theta) \times \Delta A\% = 1.066 \times 1.197\% = 1.276\%$. Through calculation, the increased losses of medium-voltage windings as a result of the transit electric supply are $\partial(\Delta A_{\text{L.m}}) = 7.85 \times 10^4$ kW·h. According to the output electric quantity and transit electric supply under medium and low voltages of the 110 kV substation, the input electric quantity at the 110 kV side after the transit electric supply is added is

$$A'_{110} = \frac{A_2 + A_3 + A_g + \partial(\Delta A_{\text{L.m}})}{1 - \Delta A'\%}$$

$$= \frac{(15\ 261.48 + 1094.58 + 2000 + 7.85) \times 10^4}{1 - 1.276\%}$$

$$= \frac{18\ 363.91 \times 10^4}{1 - 1.276\%} = 18\ 601.26 \times 10^4 (\text{kW} \cdot \text{h})$$

The electricity losses of the 110 kV system are

$$\Delta A'_{110} = (18\ 601.26 - 18\ 363.91) \times 10^4$$

$$= 237.35 \times 10^4 (\text{kW} \cdot \text{h})$$

According to Example 7.1, $\Delta A_{35} = (44.73 + 697.71) \times 10^4 = 742.44 \times 10^4$ (kW·h). After $\partial(\Delta A_{\text{L.m}})$ is included, the total line loss rate of 35 kV and above buses in the entire region is

$$\Delta A'_{\Sigma}\% = \frac{\Delta A'_{110} + \Delta A'_{35}}{A'_{110}} \times 100\%$$

$$= \frac{237.35 \times 10^4 + (742.44 + 7.85) \times 10^4}{18\ 601.26 \times 10^4} \times 100\%$$

$$= 5.29\%$$

The above calculation results show that although the line loss rate of the 110 kV system is increased due to the transit electric supply, the total line loss rate of the entire regions is reduced because of the large increment in the electric supply.

It should be noted that the above calculations are made under certain assumptions. Because q is generally small, they are within the allowable range of engineering calculation despite certain errors as compared with accurate calculation.

7.4 Multiple-factor Analysis of Changes in Electricity Line Losses

Based on the line loss binomial $\Delta A = B + CA^2$, this section introduces the concept of loss structure function $H(B, C, A)$ and puts forward the method for calculating changes in electricity line losses.

7.4.1 *Loss Structure Coefficient and Electricity Line Loss Increase Rate Function*

7.4.1.1 Loss Structure Coefficient

Use the line loss binomial to calculate the rate of change in electricity line losses to the electric supply:

$$\frac{\mathrm{d}(\Delta A)}{\mathrm{d}A} = 2CA$$

The electric energy loss structure coefficient H is defined as the rate of change in the electricity line losses to the electric supply per unit of line loss rate, that is

$$H = \frac{\mathrm{d}(\Delta A)/\mathrm{d}A}{\Delta A\%} = \frac{2CA}{\frac{B}{A} + CA} = \frac{2}{\theta + 1} \tag{7.11}$$

$$\theta = \frac{B}{CA^2} = \frac{\Delta A_0\%}{\Delta A_L\%}$$

Wherein θ – loss constituent ratio.

According to Formula 6.3, when the minimum condition of line loss rate is satisfied, $\theta = 1.0$. So at this time, $H = 1.0$, $\mathrm{d}(\Delta A)/\mathrm{d}A = \Delta A\%$, that is the rate of change in the electricity line losses to the electric supply is equal to the line loss rate.

In general situations, $\theta < 1$, $H > 1$, $\mathrm{d}(\Delta A)/\mathrm{d}A > \mathrm{d}A\%$, that is no-load losses are smaller than load losses, and the rate of change in the electricity line losses to the electric supply is larger than the line loss rate.

When the mode of connection and equipment conditions (i.e. grid structure) of power grids are not obviously changed, θ and H can be deemed as basically constant. According to the definition of loss structure coefficient in Formula (7.11),

$$\mathrm{d}(\Delta A) \approx H\Delta A\%\mathrm{d}A \tag{7.12}$$

So the following formula can be approximately obtained

$$\Delta A_2 - \Delta A_1 = H\Delta A\%(A_2 - A_1)$$

According to Formula (7.12), the quantity of change in electricity line losses is approximately equal to the product of the electric energy loss structure coefficient, the original line loss rate and the quantity of change in the electric supply, so this formula can be called the constraint condition of change in electricity line losses within a short period. Based on this, the line loss rate obtained through statistics within a measuring period can be checked for reasonability, and electricity line losses corresponding to the increase in the electric supply within the next measuring period can be predicted.

7.4.1.2 Electricity Line Loss Increase Rate Function

Assume that the rate of increase in electricity line losses is y and the rate of increase in the electric supply is x. According to the concept of rate of increase, $y = \mathrm{d}(\Delta A)/\Delta A$, $x = \mathrm{d}A/A$. Then,

$$y = \frac{2CA\mathrm{d}A}{B + CA^2} = \frac{2}{\theta + 1}\chi = H\chi \tag{7.13}$$

According to Formula (7.13), the rate of increase in electricity line losses is in linear function of the rate of increase in the electric supply, and the proportionality coefficient is the loss structure coefficient H.

In two continuous measuring periods, if the grid structure and load conditions are not obviously changed, then θ can be deemed as unchanged. Formula (7.13) is applicable to any system with a certain class of voltage or any regional power grids composed of several voltages, so it not only predicts the quantity of change in electricity line losses by class of voltage, but also predicts the changes in electricity line losses of power grids in the entire region.

Example 7.4 See Table 7.3 for the results of theoretical calculation of line losses in 10 kV power distribution lines located in a mountainous place in some month during 1974.

1. The electric supply of No. 1 and No. 5 lines is increased by 3000 kW·h, respectively. *Try to calculate the increase in line losses of the two lines.*
2. Calculate the electricity line loss increase rate function of 10 kV grids in the entire region. If the electric supply is increased by 10% in the entire region, *is the line loss rate increased or reduced?*

Solution

1. See Table 7.3 for the loss constituent ratios θ and loss structure coefficients H calculated by $\theta = \dfrac{B}{CA^2}$ and Formula (7.11). The increase in line losses of the two lines when the electric supply is increased by 3000 kW·h can be calculated as per Formula (7.12):

$$\text{For No. 1 line } \Delta A_2 - \Delta A_1 = 1.44 \times 0.086\,5 \times 3000 = 374 \quad (\text{kW}\cdot\text{h})$$
$$\text{For No. 5 line } \Delta A_2 - \Delta A_1 = 0.337 \times 0.236\,8 \times 3000 = 239 \quad (\text{kW}\cdot\text{h})$$

   The above results indicate that although the increase in electric supply is the same for the two lines, the increase in line losses for the line with larger θ is smaller. This is because no-load losses which account for a large proportion do not change with the increase in electric supply. When θ is small, load losses are much greater than no-load losses, and load losses caused by the increase in electric supply are largely increased.

2. The loss structure coefficient of 10 kV grids in the entire region is H = 0.69, so the electricity line loss increase rate function is $y = 0.69x$. $x = 0.10$ is given and substituted into it to obtain

$$y = 0.69 \times 0.10 = 0.069$$

   So

$$(\Delta A_2 - \Delta A_1)/\Delta A_1 = 0.069$$

   After transposition of terms,

$$\Delta A_2 = 1.069 \Delta A_1 = 1.069 \times 68\,484 = 73\,209 (\text{kW}\cdot\text{h})$$

$$\Delta A_2\% = \frac{\Delta A_2}{A_2} \times 100\% = \frac{73\,209}{1.10 \times 598\,600} \times 100\% = 11.11\% < \Delta A_1\% = 11.44\%$$

   According to the above results, although the electric supply is increased by 10% in the entire region, the line loss rate of the entire region is slightly reduced. For agricultural electric lines, the frequent case is $\Delta A_0\% > \Delta A_L\%$. At this time, the total line loss rate will be reduced when the electric supply is increased.

Table 7.3 Results of theoretical calculation of line losses of 10 kV power distribution lines.

Line No.	Electric supply (× 10^4 kW·h)	Transformer losses (W)		Line losses (W)		Total load losses (W)	Calculated line losses		Line loss rate $\Delta A\%$	Loss constituent ratio θ	Loss structure coefficient H	$(\Delta A\%)_m$ $2 \times \Delta A_L\%$
		No-load	Load	High voltage	Low voltage		ΔP (W)	ΔA (kW·h)				
1	8.11	2660	833	149	5790	6772	9432	7017	8.65	0.39	1.44	0.125
2	9.36	8360	843	2790		3633	11 993	8923	9.53	2.30	0.606	0.058
3	14.2	17 840	1361	8641		10002	27 842	20 714	14.59	1.78	0.719	0.105
4	4.96	9780	848	1748		2596	12 376	9207	18.56	3.77	0.419	0.078
5	3.03	8020	501	1122		1623	9643	7174	23.68	4.94	0.337	0.080
6	20.2	13 610	1027	5440	688	7155	20 765	15 449	7.65	1.90	0.690	0.053
Total	59.86	60 270	5413	19 890	6478	31781	92 051	68 484	11.44	1.90	0.690	0.078 9

7.4.2 Loss Structure Function and Calculation of Increase in Electricity Line Losses

When the interval between two measuring periods for comparison is very long, not only is the electric supply largely changed, but also the power grid conditions are greatly changed, so it is necessary to conduct multiple-factor analysis of changes in electricity line losses.

7.4.2.1 Loss Structure Function

Take the total derivative of Formula (6.1) to obtain

$$\begin{aligned}\mathrm{d}(\Delta A) &= \frac{\partial(\Delta A)}{\partial B}\mathrm{d}B + \frac{\partial(\Delta A)}{\partial C}\mathrm{d}C + \frac{\partial(\Delta A)}{\partial A}\mathrm{d}A \\ &= \mathrm{d}B + A^2\mathrm{d}C + 2CA\mathrm{d}A\end{aligned}$$

Connect the total derivative $\mathrm{d}(\Delta A)$ with the increase in the electric supply $\mathrm{d}A$ and the line loss rate $\Delta A\%$, and use the definition of the rate of change in the electricity line losses to the electric supply per unit of line loss rate to obtain the loss structure function

$$\begin{aligned}H(B,C,A) &= \frac{\mathrm{d}(\Delta A)/\mathrm{d}A}{\Delta A\%} \\ &= \frac{\delta B/\delta A}{\frac{1}{\theta}+1} + \frac{\delta C/\delta A}{\theta+1} + \frac{2}{\theta+1}\end{aligned} \tag{7.14}$$

$$\delta B = \mathrm{d}B/B,\quad \delta C = \mathrm{d}C/C,\quad \delta A = \mathrm{d}A/A$$

Wherein δB – relative quantity of change in no-load losses;
δC – relative quantity of change in load loss coefficient;
δA – relative quantity of change in electric supply.

When $\mathrm{d}B$ and $\mathrm{d}C$ are both equal to 0, $H(B,C,A) = \frac{2}{\theta+1}$, that is the above-mentioned loss structure coefficient H. Obviously, Formula (7.11) is a special case of Formula (7.14).

7.4.2.2 Calculation of Increase in Electricity Line Losses

Considering the influence of multiple factors, the loss structure function can be used to calculate the increase in electricity line losses, that is

$$\begin{aligned}\mathrm{d}(\Delta A) &= H(B,C,A)\Delta A\%\mathrm{d}A \\ &= \mathrm{d}B + \left(\frac{\delta C/\delta A}{\theta+1} + \frac{2}{\theta+1}\right)\Delta A\%\mathrm{d}A\end{aligned} \tag{7.15}$$

According to Formula (7.15), the increase in electricity line losses is composed of the increase in no-load losses and the increase in load losses, the latter of which depends on the change in the load loss coefficient (reflecting changes in the loss calculation resistance, load factor and power factor) and the change in the electric supply.

It is well-known that the differential dx is approximately equal to Δx only when the increase is very small as compared with the original value, so when Formula (7.15) is used to predict the increase in electricity line losses, the increase in three variables (B, C, A) must be much smaller than the original values. Otherwise, larger prediction errors will be caused.

Formula (7.15) can be used to estimate the increase in electricity line losses based on the given data of $\Delta A\%$, B, C, A within a measuring period and the predicted electric supply; it can also be used for recursive rolling prediction. The shorter the recursive period is, the more accurate the prediction is. In addition, Formula (7.15) can be used to determine the technical requirements for loss reduction measures (such as replacement of conductor to reduce resistance) according to the loss reduction electric quantity, and determine the specific requirements for operated loss reduction measures (such as improvement in load factor or power factor).

7.5 Marginal Line Loss Rate and Optimal Distribution of Increase in Electric Supply

7.5.1 Marginal Line Loss Rate

According to Example 7.4, when the increase in the electric supply is the same, the increase in line losses of lines with different loss structure coefficients varies. Therefore, for a new consumer, if lines connected to it are optional within a certain scope, the principle that the increase in line losses is the minimum with each per unit increase in the electric supply should be used for judgment. Then, the concept of marginal line loss rate is introduced.

The increase in line losses of a power grid unit with each per unit increase in the electric supply is called the marginal line loss rate, that is $(\Delta A\%)_{\mathrm{m}}$. According to Formula (7.12),

$$(\Delta A\%)_{\mathrm{m}} = \frac{\mathrm{d}(\Delta A)}{\mathrm{d}A} = H\Delta A\% \tag{7.16}$$

According to the line loss binomial in Formula (6.1),

$$(\Delta A\%)_{\mathrm{m}} = \frac{\mathrm{d}(B + CA^2)}{\mathrm{d}A} = 2CA = 2\Delta A_{\mathrm{L}}\% \tag{7.17}$$

According to Formulas (7.16) and (7.17), the marginal line loss rate is equal to the product of the loss structure coefficient and the line loss rate, and also equal to two times the load line loss rate.

According to the principle of minimum marginal line loss rate, that is the principle of minimum load line loss rate, the distribution of the increase in the electric supply can be optimized to realize the objective of minimum increase in line losses.

7.5.2 Optimal Distribution of Increase in Electric Supply

The electric supply in each region is generally increased year by year. For all lines connected to the same voltage bus, the increase in the electric supply of regional customers generally cannot be controlled. However, for new consumers for which connection lines are optional, if the electric supply is undertaken by a line with a small marginal line loss rate, the effect of a smaller increase in total electricity line losses can be achieved. Under certain conditions, the optimal calculation of the increase in electric supply can be conducted based on the increment size and to a certain degree.

The following part proves optimal distribution conditions of the increase in electric supply.

The load losses of the first line are $\Delta A_{L1} = C_1 A_1{}^2$, and the load losses of the second line are $\Delta A_{L2} = C_2 A_2{}^2$, with $2C_1A_1 < 2C_2A_2$. If the increase in the electric supply is D, try to determine the distribution conditions of the increase in electric supply under which the total increase in load line losses is minimal.

Assume that the increase in electric supply for the first line is D_1 and $D - D_1$ for the second line. Load losses of the two lines are

$$\Delta A'_{L1} = C_1\ (A_1 + D_1)^2,\ \Delta A'_{L2} = C_2\ (A_2 + D - D_1)^2$$

After expansion and addition,

$$\Delta A'_{L1} + \Delta A'_{L2} = C_1\left(A_1^2 + 2A_1D_1 + D_1^2\right) + C_2\left[A_2^2 + 2A_2(D - D_1) + D^2 - 2DD_1 + D_1^2\right]$$

If the change in total load losses before and after the increase in electric supply is $\delta(\Delta A_L)$, then

$$\begin{aligned}\delta(\Delta A_L) &= \left(\Delta A'_{L1} + \Delta A'_{L2}\right) - (\Delta A_{L1} + \Delta A_{L2})\\ &= 2C_2A_2D + C_2D^2 + 2A_1C_1D_1 - 2A_2C_2D_1 - 2C_2DD_1 + C_2D_1^2 + C_1D_1^2\end{aligned}$$

To have a minimal value of $\delta(\Delta A_L)$, calculate the derivative of the variable D_1 and make it equal to 0, that is d$(\delta\Delta A_L)/\mathrm{d}D_1 = 0$, obtaining

$$2A_1C_1 - 2A_2C_2 - 2C_2D + 2C_2D_1 + 2C_1D_1 = 0$$

then

$$2A_1C_1 - 2A_2C_2 = 2C_2(D - D_1) - 2C_1D_1$$

When $2A_1C_1 = 2A_2C_2$ and $(D - D_1)/D_1 = C_1/C_2$, the above formula is satisfied. Therefore, the optimal distribution of the increase in electric supply has two equal conditions: (i) the marginal line loss rates are equal; (ii) the distribution is in inverse proportion to the load loss coefficient. Condition (ii) is consistent with the distribution of current of DC parallel circuits in inverse proportion to resistance.

When the increase in electric supply is large, the electric supply can be increased in line with the minimum load line loss rate, until the load line loss rate of the line is increased to the subminimum load line loss rate. If there is a margin for the increase in electric supply, it is distributed in two lines whose load line loss rates are equal, and alike. Marginal line loss rates in increasing order are $(\Delta A\%)_{m1}$, $(\Delta A\%)_{m2}$, and so on. The first increase in the electric supply of the line with minimum marginal line loss rate is ∂A_{11}, and the second increase is ∂A_{12}, and so on. After the increase in the electric supply of the line with minimum marginal line loss rate, its marginal line loss rate is increased and equal to the subminimum value, so

$$2(\partial A_{11} + A_1)C_1 = 2A_2C_2$$

Then

$$\begin{aligned}\partial A_{11} &= \frac{A_2C_2 - A_1C_1}{C_1}\\ &= \frac{1}{2}\frac{(\Delta A\%)_{m2} - (\Delta A\%)_{m1}}{C_1}\end{aligned} \tag{7.18}$$

When the margin of the increase in electric supply is small, it should be distributed in two lines whose marginal line loss rates are equal in inverse proportion to the load loss coefficient, so that marginal line loss rates of the two lines are still equal after the increase in electric supply.

Example 7.5 The increase in the electric supply of the distribution system in Example 7.4 is expected to be $\partial A_\Sigma = 10 \times 10^4$ (kW·h). It can be randomly distributed in lines. *Try to do the calculation of optimal distribution.*

Solution

Arrange the marginal line loss rates of lines listed in Table 7.3 in increasing order, and list the calculation results of electric supply and load loss coefficient C in Table 7.4.

1. *First distribution.* As

$$\partial A_{11} = \frac{1}{2} \times \frac{0.058 - 0.053}{1.311\,9 \times 10^{-7}} = 1.906 \times 10^4 (\text{kW} \cdot \text{h})$$

The marginal of the increase in electric supply is

$$\begin{aligned}\partial A_{\Sigma 1} &= \partial A_\Sigma - \partial A_{11} = (10 - 1.906) \times 10^4 \\ &= 8.094 \times 10^4 (\text{kW} \cdot \text{h})\end{aligned}$$

2. *Second distribution.*

a. *First calculation.* Continue to increase the electric supply of No. 6 line, so

$$\partial A_{12} = \frac{1}{2} \times \frac{0.078 - 0.058}{1.311\,9 \times 10^{-7}} = 7.623 \times 10^4 (\text{kW} \cdot \text{h})$$

At this time, the marginal of the increase in electric supply is

$$\partial A_{\Sigma 2} = (8.094 - 7.623) \times 10^4 = 0.471 \times 10^4 (\text{kW} \cdot \text{h})$$

The marginal line loss rate of No. 6 line is $(\Delta A\%)'_{\text{ml}} = 0.078$, and the marginal line loss rate of No. 2 line is $(\Delta A\%)_{\text{m}} = 0.058$. The electric supply of No. 2 line should be increase, so

$$\begin{aligned}\partial A_{21} &= \frac{1}{2} \times \frac{0.078 - 0.058}{3.098\,3 \times 10^{-7}} \\ &= 3.228 \times 10^4 (\text{kW} \cdot \text{h})\end{aligned}$$

It can be found that $\partial A_{21} > \partial A_{\Sigma 2}$. That is, the increase in electric supply required for No. 2 line is larger than the marginal of the increase in electric supply, so the increase in electric supply should be redistributed.

b. *Second calculation.* The following two conditions should be satisfied at the same time.

$$\begin{cases} \partial A_{12} + \partial A_{21} = \partial A_{\Sigma 1} = 8.094 \times 10^4 (\text{kW} \cdot \text{h}) \\ \dfrac{\partial A_{21}}{\partial A_{12}} = \dfrac{C_1}{C_2} = \dfrac{1.311\,9 \times 10^{-7}}{3.098\,3 \times 10^{-7}} \end{cases}$$

So

$$\frac{\partial A_{\Sigma 1}}{\partial A_{12}} = \frac{C_1 + C_2}{C_2} = \frac{1.311\,9 + 3.098\,3}{3.098\,3} = 1.423$$

Table 7.4 Calculation of distribution of increase in electric supply.

Marginal line loss rate					Distribution of increase in electric supply (× 10^4 kW·h)			Calculation of line loss rate after distribution				
Increasing order	$(\Delta A\%)_m$	Original line No.	Electric supply A (× 10^4 kW·h)	Load loss coefficient C [×10^{-7} 1/(kW·h)]	First	Second	Total	$(\Delta A\%)'_m$	$\Delta A'_L\%$	$\Delta A'_0\%$	$\Delta A\%$	H
1	0.053	6	20.2	1.311 9	1.906	5.688	7.594	0.072 9	3.645	3.643	7.288	1.00
2	0.058	2	9.36	3.098 3	—	2.406	2.406	0.072 9	3.645	5.286	8.931	0.816 3
3	0.078	4	4.96	7.862 9				0.078 0	3.90	14.66	18.56	0.419

Substitute $\partial A_{\Sigma 1}$ and $\partial A_{\Sigma 1}/\partial A_{12}$ into it to obtain $(8.094 \times 10^4)/\partial A_{12} = 1.423$, so $\partial A_{12} = 5.688 \times 10^4$ (kW·h), $\partial A_{12} = 2.406 \times 10^4$ (kW·h).

After the distribution of the increase in electric supply is conducted by the above method, the marginal line loss rates can be calculated as per Formula (7.18), that is

$$
\begin{aligned}
(\Delta A\%)'_{\text{m1}} &= 2(\partial A_{11} + \partial A_{12} + A_1)C_1 \\
&= 2(1.906 + 5.688 + 20.2) \times 10^4 \times 1.311\,9 \times 10^{-7} \\
&= 0.072\,9 \\
(\Delta A\%)'_{\text{m2}} &= 2(\partial A_{21} + A_2)C_2 \\
&= 2(2.406 + 9.36) \times 10^4 \times 3.098\,3 \times 10^{-7} = 0.072\,9
\end{aligned}
$$

The calculation shows that, after the distribution of the total increase in electric supply, the current marginal line loss rate of the two lines whose original marginal line loss rates are minimum and sub-minimum have increased to the same value, but are still smaller than the marginal line loss rate of the line ranked third in the increasing order. The checking calculation indicates that the distribution of the increase in the electric supply as listed in Table 7.4 can realize the objective of a minimum increase in the total line losses.

Table 7.4 lists the calculation results of load line loss rate, no-load line loss rate, total line loss rate and loss constituent ratio of lines after the optimal distribution of the increase in electric supply. The marginal line loss rate calculated by Formula (7.16) is the same as that calculated by Formula (7.17). Example 7.5 shows that, to distribute the increase in electric supply, it is necessary to compare the margin of the increase in electric supply with the increase in electric supply distributed to a line, and to adopt two different calculation methods according to the comparison results. If simple computer programs are worked out, a microcomputer can be used to do the calculation of optimal distribution of the increase in electric supply.

8

Theoretical Calculation of Electric Energy Losses of Power Grid Units

Previous chapters have respectively explained methods of line loss calculation according to current load curve parameters and of line loss calculation considering power factor for no-branch lines, as well as the change law for electric energy losses of power grids, providing conditions for the theoretical calculation of the line losses of various power grid units. This chapter will briefly introduce the calculation of electric energy losses of overhead lines, cable lines, main transformers, and other power grid units.

8.1 Classification of Electric Energy Losses

8.1.1 Classification of Electric Energy Losses by Whether Theoretical Calculation is Feasible

1. *Unknown losses which are difficult to calculate* These losses consist of unknown management losses and unknown technical losses. Unknown technical losses include leakage losses caused by poor line insulation and electric energy losses caused by equipment grounding or short-circuit faults.
2. *Technical losses which can be calculated* These losses can be obtained by theoretical calculation, so they are also known as theoretical line losses. Technical losses include resistance heat losses, dielectric magnetization losses, dielectric polarization loss, and corona losses.

8.1.2 Classification of Calculable Technical Losses by Change Law

1. *No-load losses* These losses are independent from the current passing power grid units, but are related to the voltage withstood by the power grid units. Customarily, no-load losses are also called fixed losses.
2. *Load losses* These are in direct proportion to the square of load power or current passing the power grid units. Customarily, load losses are also called variable losses.

8.1.3 Classification of Electric Energy Losses by Different Power Grid Units

Electric energy losses can be divided into line losses, transformer losses, and other power grid unit losses by different power grid units.

Line Loss Analysis and Calculation of Electric Power Systems, First Edition. Anguan Wu and Baoshan Ni.

1. Line losses include overhead line losses and cable line losses. Overhead line losses consist of power transmission line losses, power distribution line losses, and low-voltage line losses. Cable line losses are composed of three-phase cable losses and single-phase cable losses.
2. Transformer losses include main transformer losses and distribution transformer losses. Main transformer losses consist of two-winding transformer losses and three-winding transformer losses.
3. Other power grid unit losses include reactive compensation equipment (capacitor and synchronous compensator) losses, reactor (high-voltage and low-voltage reactor) losses, transformer (current transformer and voltage transformer) losses, switch equipment (switching device and high-voltage bushing) losses, and measuring meter (indicating meter and watt-hour meter) losses.

The following sections will respectively describe the calculation methods of electric energy losses of various power grid units.

8.2 Calculation of Electric Energy Losses of Overhead Lines

8.2.1 Calculation of Corona Losses of Power Transmission Lines

As is known to all, an incomplete self-excited discharge concentrated around an electrode with a larger curvature is called a "corona". For overhead lines, when the operating electric field intensity on the conductor surface of high-voltage power transmission lines exceeds the air breakdown strength, corona discharge will occur. Corona discharge increases the active power and electric energy losses of the lines, so the calculation result of corona losses of high-voltage power transmission lines serves as one of the important bases to verify whether the conductor selection is economical and reasonable. This verification adopts not only the limit on the absolute value (kW/km) of annual average corona power loss per unit line length, but also the comparison between the corona losses and resistance heat losses of the lines. In general, corona losses should be smaller than 10% of resistance heat losses.

8.2.1.1 Calculation for Comparison Between Critical Field Intensity that Starts to Cause Corona and Conductor Surface Maximum Operating Field Intensity

When the conductor surface maximum operating field intensity is smaller than the critical field intensity that causes a corona, corona discharge can be avoided. Therefore, in the line design stage, such two intensities must be calculated according to meteorological conditions in regions where the high-voltage power transmission lines pass, as well as selected conductors and conductor arrangements.

1. *Calculation of critical field intensity that starts to cause corona*

   The Northwest Electric Power Design Institute of China and the Xi'an High Voltage Apparatus Research Institute, considering to our country's situation, put forward a corrected Peek Formula [20] used to calculate the critical field intensity E_0 (kV/cm) that starts to cause a corona, that is

$$E_0 = 30.3M\sqrt{\delta}\left(1 + \frac{0.3}{\sqrt{r_0\delta}}\right) \tag{8.1}$$

Wherein M – conductor surface roughness coefficient; when there are more than 24 conductors in the outer layer, the coefficient of 0.9 conforms to the actual situation;
r_0 – conductor radius (cm);
δ – relative air density.

It is basically consistent with the formula set forth in the former Soviet Union's corona calculation guideline.

Table 8.1 δ_h values at various altitudes.

H (m)	0	500	1000	1500	2000	2500	3000	3500
δ_h	1	0.955	0.908 5	0.865	0.824	0.784	0.745	0.708

$$E_0 = 30.1M\sqrt{\delta}\left(1 + \frac{0.299}{\sqrt{r_0}}\right) \tag{8.2}$$

With atmospheric pressure p (MPa) at temperature t (°C), the relative air density can be calculated as per the following formula:

$$\delta_t = \frac{2892p}{273 + t} \tag{8.3}$$

For high-altitude regions, the relative air density can be calculated as per the following formula:

$$\delta_h = \delta_0\left(1 + \frac{\alpha H}{T_0}\right)^{4.26} \tag{8.4}$$

Wherein δ_0 – relative air density under standard state;
H – altitude (m);
α – temperature gradient of air, about 0.006 5 °C/m;
T_0 – absolute temperature under standard state; T_0 = 273 + 20 = 293 (K).

See Table 8.1 for δ_h values under various altitudes calculated as per Formula (8.4).

2. *Calculation of conductor surface maximum field intensity E_M (kV/cm)*

For single-conductor lines,

$$E_M = 0.014\,7\frac{C_W U_m}{r} \tag{8.5}$$

For bundled-conductor lines,

$$E_M = K \times 0.014\,7\frac{C_W U_m}{nr} \tag{8.6}$$

When bundled conductors are arranged in the form of regular polygon,

$$K = 1 + \frac{2r}{a}(n-1) \times \sin\frac{\pi}{n}$$

$$C_{av} = \frac{24.1}{\tan\dfrac{D_{av}}{r_{eq}}} \tag{8.7}$$

$$D_{av} = \sqrt[3]{D_{ab}D_{bc}D_{ca}} \tag{8.8}$$

Wherein U_m – actual maximum operating voltage of a line (for a line length < 100 km, it takes 1.1 times the rated voltage; for a line length > 100 km, it takes 1.1 times the rated voltage in the former half section and 1.05 times the rated voltage in the latter half section; kV);
r – calculated radius of the conductor (cm);
C_W – working capacitance of conductors of various phase (pF/m; for side phase conductors, C_W = 1.03C_{av}; for medium phase conductors, C_W = 1.10C_{av});

Table 8.2 Mean geometrical spacing between bundled conductors (cm).

n	2	3	4	5	6
a_{pj}	40	40	44.8	50.8	56

C_{av} – average capacitance of three-phase transposition overhead power transmission lines;
D_{av} – mean geometrical distance between three-phase conductors (cm);
D_{ab}, D_{bc}, D_{ca} – distances between three-phase conductors (cm);
r_{eq} – equivalent radius of conductors of each phase (for single-phase conductors, $r_{eq} = r$; for bundled conductors, $r_{eq} = \sqrt[n]{ra_{av}^{n-1}}$; cm);
a_{av} – mean geometrical distance between bundled conductors (when the distance between two adjacent bundled conductors is $a = 40$ cm and the number of bundled conductors n varies, see Table 8.2 for a_{av});
K – coefficient used to calculate bundled conductor surface maximum field intensity;
a – split spacing between bundled conductors (cm).

3. *Comparison criteria*

According to actual comparisons between the design and operation of high-voltage power transmission lines in high-altitude regions of China, regarding the calculation as per Formulas (8.1), (8.5), or (8.6), if the conductor surface maximum operating field intensity does not exceed 85% of the critical field intensity that starts to cause a corona, that is $E_M/E_0 \le 0.85$, then no corona will occur in the line under normal weather.

8.2.1.2 Calculation of Corona Losses of Operating Lines

As there are many factors that affect corona losses, an accurate calculation formula has not yet been theoretically derived. An empirical formula derived from measured test data or general calculation curves is normally used for an approximate calculation. The former Wuhan High Voltage Research Institute collected and analyzed studies of the corona discharge mechanism of conductors and measured data from test lines of single and double bundled conductors, thereby drawing a set of curves of corona loss power ΔP_{dy} which are divided into four categories by different climatic conditions: (i) icy and snowy days, including rime, glaze, wet snow, and dry snow; (ii) rainy days, including drizzle and weather with various rainfall intensities; (iii) foggy days, including weather with various fog intensities, frosty and frosted days; and (iv) good days, other than the above three types of weather. See Appendix A for the set of curves. It should be pointed out that the above calculation curves are only applicable to single-conductor and double bundled-conductor lines, and are only for reference for triplicate and above bundled-conductor lines.

The corona loss power of each conductor per phase P_t [kW/(km·phase)] has a functional relationship with the relative air density δ_t, the calculation radius of the conductor r, and the conductor surface maximum electric field intensity E_M, that is

$$P_t = \frac{\Delta P_c}{n} = f\ (\delta_t r,\ E_M/\delta_t) \tag{8.9}$$

Wherein P_t or $\dfrac{\Delta P_c}{n}$ – corona loss power of each conductor per phase and per kilometer under various weather conditions (kW);
δ_t – relative air density, calculated as per Formula (8.3);
n – number of bundled conductors.

When three-phase conductors are horizontally arranged, the corona electricity loss ΔA_c (kW·h) within the measuring period can be calculated as per the following formula:

$$\Delta A_c = n[2\Sigma(P_{1t}T_t) + \Sigma(P_{2t}T_t)]L \tag{8.10}$$

Wherein n – number of bundled conductors;
T_t – duration of a weather condition (h);
L – line length (km);
P_{1t} – corona loss power of each conductor per side phase and per kilometer under a weather condition [with calculated $\delta_t r$ and E_M/δ_t, refer to the corresponding corona loss calculation curve to obtain P_{1t}; kW/(km·phase)];
P_{2t} – corona loss power of each conductor per medium phase and per kilometer under a weather condition [kW/(km·phase)].

8.2.1.3 Calculation of Corona Losses of DC Lines

China's first ±500 kV remote DC power transmission line, the Gezhou Dam–Nanqiao line, was formally put into operation in September 1987. In the early 1990s, Chinese scientific and technical workers made in-depth studies into corona losses of the Gezhou Dam–Nanqiao line and obtained the corrected Anneberg Formula for calculating corona power losses of monopolar lines by means of least square fitting of the measured results:

$$P = K_c UnrZ^{b(g_{max}-14)}$$

Wherein U – operating voltage of the line (kV);
n – number of bundled conductors;
r – sub-conductor radius (cm);
g_{max} – conductor surface maximum field intensity $g_0 = 14$ kV/cm is the conductor surface starting field intensity measured by some foreign researchers on lines (kV/cm);
K_c, b – empirical coefficients related to conductor surface condition, conductor structure, and weather condition.

The Gezhou Dam–Nanqiao line uses 4×LGJQ300 conductors, with a spacing between sub-conductors of 0.45 m. When there is no rain in the line, $K_c = 0.121$, $b = 0.226$; when there is rain in one-third of the line, $K_c = 0.206$, $b = 0.200$.

According to domestic and foreign measured data, grounding corona losses of bipolar lines in parallel connection are approximately two times those of monopolar lines, and operating corona losses of bipolar lines are generally four times those of monopolar lines. Corona losses in rainy days are 2.0~4.0 times those of fine days. As to the actual measurement of the +500 kV Gezhou Dam–Nanqiao line, its grounding corona losses of bipolar lines in parallel connection are 4.28 W/m. The loads in the whole line are 4.47 MW, accounting for 0.75% of monopolar operating full loads (600 MW) and 19.9% of resistance losses (22.5 MW) in the whole line. Obviously, such a ratio is much higher than the ratio of corona losses of an AC power transmission line to its resistance losses.

The actual measurement also indicates that corona losses of monopolar lines are in direct proportion to about biquadratic of operating voltage. When the lines are operating under low loads, a reduction of voltage can reduce the corona losses of DC power transmission lines and mitigate the influence on the environment.

Example 8.1 330 kV line is located in a region with an altitude of 3500 m and has a length of about 100 km, with 2×LGJQ400 conductors, $r = 1.41$ cm. According to the statistics of local weather stations, the entire year sees 7632 h fine days, 708 h rainy days, 60 h foggy days, and 360 h snowy days. *Try to calculate the corona losses of a whole year.*

Solution

According to Formula (8.6), the maximum field intensity of side conductors is $E_{M1} = 22.25$ kV/cm, and the maximum field intensity of medium-phase conductors is $E_{m2} = 23.5$ kV/cm. Refer to Table 7.1 to obtain $\delta_h = 0.708$, so $\delta_t r = 0.708 \times 1.41 = 0.998 \approx 1.0$, $E_{M1}/\delta_t = 31.42$ kV/cm, $E_{M2}/\delta_t = 33.19$ kV/cm. Refer to the corona loss calculation curves in Appendix A to obtain P_{1t} and P_{2t}. Based on the durations of four known climatic conditions, annual corona losses can be calculated, as shown in Table 8.3.

Table 8.3 Calculation table of annual corona losses.

Climatic conditions	Snowy days, t_1	Rainy days, t_2	Foggy days, t_3	Fine days, t_4	Total
P_{1t} [kW/(km·phase)]	7.5	4.0	3.0	0.33	
P_{2t} [kW/(km·phase)]	15.5	6.7	5.8	0.78	
Duration T_t (h)	360	708	60	7632	8760
$2(P_{1t}T_t)+(P_{2t}T_t)$ (kW)	10 980	10 408	708	10 990	33 086

Because the conductors are double bundled conductors, $n = 2$. Use Formula (8.10) to calculate the annual corona electricity losses of the line are

$$\Delta A_c = 2[2\Sigma(P_{1t}T_t) + \Sigma(P_{2t}T_t)]L$$
$$= 2 \times 33\ 086 \times 100$$
$$= 661.72 \times 10^4\ (\text{kW·h})$$

The corona loss power (annual average) per unit length is

$$\Delta P_c = 33\ 086/8760 = 3.78\ [\text{kW/(km·three-phase)}]$$

8.2.2 Calculation of Resistance Heat Losses of Overhead Lines

8.2.2.1 Correction of Calculation Resistance

To calculate the electric energy losses of overhead lines, the influence of the temperature rise caused by the load current and the ambient air temperature on resistance change should be considered, so the calculation resistance R (Ω) needs to be corrected as below:

$$R = R_{20}(1 + \beta_1 + \beta_2) \tag{8.11}$$

$$\beta_1 = 0.2\left(\frac{I_{rms}}{I_{ad}}\right)^2 \tag{8.12}$$

$$\beta_2 = \alpha(T_{av} - 20) \tag{8.13}$$

Wherein R_{20} – resistance of conductors per phase at 20°C (Ω);
β_1 – correction coefficient considering the influence of conductor temperature rise on resistance;
I_{ad} – allowable constant current when conductors reach the allowable temperature under the ambient air temperature of 20°C (A);
β_2 – correction coefficient considering the influence of the ambient air temperature on resistance;
T_{av} – average ambient temperature within the line loss measuring period (°C);
α – temperature coefficient of conductor resistance (for aluminum wires or steel-cored aluminum wires, $\alpha = 0.004$).

According to Formula (8.13), when the mean monthly temperature in summer exceeds 32.5 °C, the resistance heat losses corrected by considering the influence of ambient temperature on calculation resistance are 5% larger than those not corrected. When the mean monthly temperature in winter is below 7.5 °C, the resistance heat losses corrected by considering the influence of ambient temperature on calculation resistance are 5% smaller than those not corrected. Generally, when the mean monthly temperature range is 12–28 °C, the influence of ambient temperature on calculation resistance are not considered.

8.2.2.2 Calculation of Resistance Heat Losses of No-branch Lines

If operating data in the measuring period are given, Formula (8.1) can be used to calculate resistance heat losses ΔA (kW·h), that is

$$\Delta A = 3I_{rms}^2 RT \times 10^{-3}$$

Wherein I_{rms} – rms current in the measuring period (A);
R – calculation resistance of no-branch lines, which is corrected by considering all factors (Ω);
T – measuring period (h).

When the rms current is calculated based on current data recorded in typical days, it should be corrected according to Formula (2.2) with the square of the ratio between average daily electric supplies within the measuring period and electric supplies of typical days.

If the minimum current, maximum current, and load factor within the measuring period are given, the loss factor F can be calculated as per the loss factor Formula (2.28), and the resistance heat losses ΔA (kW·h) can be calculated according to Formula (1.39), that is

$$\Delta A = 3I_{max}^2 FRT \times 10^{-3}$$

If the active electricity, reactive electricity, average voltage, maximum current, and minimum current within the measuring period are given, the average current I_{av} within the measuring period can be calculated first, and then the load factor f and minimum load rate β are calculated. Next, the loss factor is calculated as per Formula (2.28) and the resistance heat losses are finally calculated as per Formula (2.4):

$$\Delta A = 3I_{av}^2 \frac{F}{f^2} RT \times 10^{-3}$$

$$I_{av} = \frac{\sqrt{A_P^2 + A_Q^2}}{\sqrt{3} U_{av} T}$$

$$f = I_{av} / I_{max}$$

Wherein A_P, A_Q – active electricity (kW·h) and reactive electricity (kvar·h) passing no-branch lines within the measuring period;
U_{av} – average voltage within the measuring period (kV);
T – duration of the measuring period (h);
f – current load factor within the measuring period, which is calculated with I_{av} and I_{max}.

For the calculation of electric energy losses of no-branch lines during the planning and design stages of power grids, refer to the method introduced in Section 5.3 of Chapter 5.

8.2.2.3 Calculation of Resistance Heat Losses of Multi-branch Distribution Lines

Various methods introduced in Chapter 9 can be used, but the calculation should be based on the calculation resistance corrected according to Formula (8.11) by considering various factors.

8.2.3 Calculation of Electric Energy Losses of Low-voltage Lines

Low-voltage lines are widely distributed and lack complete and accurate line parameters and load information, so it is very difficult to calculate the electric energy losses of low-voltage lines in a detailed and accurate manner. An approximate simplified calculation method is generally used. Section 9.5 in Chapter 9 introduces a method of approximately calculating the line loss rate of low-voltage lines by the voltage loss rate at a time of maximum load.

The approximate calculation method introduced below considers four factors, namely, the number of directions of low-voltage lines in a distribution transformer, the power supply structure of low-voltage lines, the imbalance of current in all line directions, and the distribution of line current in each direction.

1. Low-voltage lines probably do not supply electricity in one direction. In cities the most common case is that low-voltage lines supply electricity in two directions, followed by three or four directions. The number of directions of power supply is N, $1 \le N \le 4$ in general.
2. The structure of low-voltage lines also affects line losses. For a single-phase two-wire system line, phase conductors and neutral conductors generally have the same sectional area and equal current, so electric energy losses of such a single-phase line are two times those of a conductor. If the structure of a low-voltage line is a three-phase four-wire system, then neutral conductors have a smaller sectional area and smaller current than phase conductors, so electric energy losses of such line are 3.5 times those of a conductor.
3. In fact, current at the start end of each direction in a low-voltage line is not equal. To calculate the electric energy losses of the low-voltage line in a distribution transformer, line losses can be first calculated based on the assumption that current at the start end of each direction in the line is equal. But it needs to be multiplied with the correction coefficient K_1 considering the imbalance of current distribution at the start end of each direction in the line. First of all, the imbalance coefficient K_{ubl} of current at the start end of each power supply direction of the low-voltage line is defined as the ratio between the maximum difference and the average, that is

$$K_{ubl} = \frac{I_{max} - I_{min}}{I_{av}} \tag{8.14}$$

Wherein I_{max}, I_{min}, I_{av} – maximum value, minimum value and average of the current in the start end of each power supply direction of the low-voltage line (A).

The relation between the correction coefficient K_1 and the imbalance coefficient K_{ubl} can be identified by analytical methods. Assume that the shape of the current load curve in each power supply direction is the same, and that the sum of rms current in each direction of the low-voltage line within the measuring period is equal to the sum of rms current in the low-voltage side of the distribution transformer; also assume that the resistance in each power supply direction of the low-voltage line is the same. For example, as for a single-phase line, the unbalanced current distribution set randomly in each power supply direction is used to calculate K_{ubl} and K_1, and then the analytical method or curve fitting method is used to identify the relational expression between K_1 and K_{ubl}.

a. If the number of power supply directions of the low-voltage line is $N = 2$, and assuming that the rms current at the start end of Line A and Line B is respectively $I_{rms.A}$ and $I_{rms.B}$, and $I_{rms.B} = a$(A), $I_{rms.A} = na$(A), then use Formula (8.14) to calculate K_{ubl}, that is

$$K_{ubl} = \frac{(n-1)a}{0.5(n+1)a} = \frac{2(n-1)}{n+1}$$

According to the definition of the correction coefficient K_1,

$$(I_{rms.B}^2 R + I_{rms.A}^2 R) = K_1 \times 2\left(\frac{I_{rms.B} + I_{rms.A}}{2}\right)^2 R$$

so

$$K_1 = \frac{(I^2_{\text{rms.B}} + I^2_{\text{rms.A}})}{2\left[\frac{1}{2}(I_{\text{rms.B}} + I_{\text{rms.A}})\right]^2} = \frac{a^2 + n^2a^2}{\frac{1}{2}(n+1)^2a^2} = \frac{2(n^2+1)}{(n+1)^2}$$

As $K_{\text{ubl}} = \dfrac{2(n-1)}{n+1}$, so $n = \dfrac{2+K_{\text{ubl}}}{2-K_{\text{ubl}}}$, and substitute it into the above formula and obtain the following formula after simplification:

$$K_1 = 1 + 0.25K^2_{\text{ubl}} \tag{8.15}$$

b. If the number of power supply directions of the low-voltage line is N = 3, and assuming that the rms current at the start end in the three directions is respectively $I_{\text{rms.A}}$, $I_{\text{rms.B}}$, $I_{\text{rms.C}}$, and $I_{\text{rms.C}}$ = a(A), $I_{\text{rms.B}}$ = ma(A), $I_{\text{rms.A}}$ = na(A), then the calculation formulas of K_{ubl} and K_1 are obtained:

$$\begin{cases} K_{\text{ubl}} = \dfrac{3(n-1)}{n+m+1} \\ K_1 = \dfrac{3(n^2+m^2+1)}{(n+m+1)^2} \end{cases}$$

When n and m are different values, corresponding K_{ubl} and K_1 values are obtained as shown in Table 8.4.

According to the K_{ubl} and K_1 data listed in Table 8.4, the curve fitting method is used to obtain the following approximate formula:

$$K_1 = 1.067 - 0.113K_{\text{ubl}} + 0.236K^2_{\text{ubl}} \text{ (correlation coefficient } r = 0.98) \tag{8.16}$$

Figure 8.1 shows 17 data points and curve fitting.

4. The electric energy losses of a low-voltage line are also related to the load distribution in each direction of the line. If the electric energy losses are calculated based on the current at the start end in each direction of the low-voltage line and its resistance, the calculated result should be multiplied with the correction coefficient K_2. Section 9.4 of Chapter 9 will compare the electric energy losses of multi-branch lines in a case of uneven load distribution with those in cases of even load distribution or loads concentrating at the tail end. According to the actual situation of a low-voltage grid, distribution transformers are equipped in the center of loads, and loads in the start-end section of low-voltage lines are generally larger than those in the tail-end section, so it is advisable to take the correction coefficient K_2 as 0.25–0.30.

Table 8.4 Calculation table of K_{ubl} and K_1.

n	1	3	5	7	9	2	3	5	7
m	1	1	1	1	1	2	2	2	2
K_{ubl}	0	1.20	1.714	2.0	2.182	0.60	1.0	1.50	1.80
K_1	1	1.32	1.653	1.889	2.058	1.08	1.167	1.406	1.62
n	9	3	5	7	9	11	13	15	
m	2	3	3	3	3	3	3	3	
K_{ubl}	2.0	0.857	1.333	1.636	1.846	2.0	2.118	2.211	
K_1	1.792	1.163	1.296	1.463	1.615	1.747	1.858	1.953	

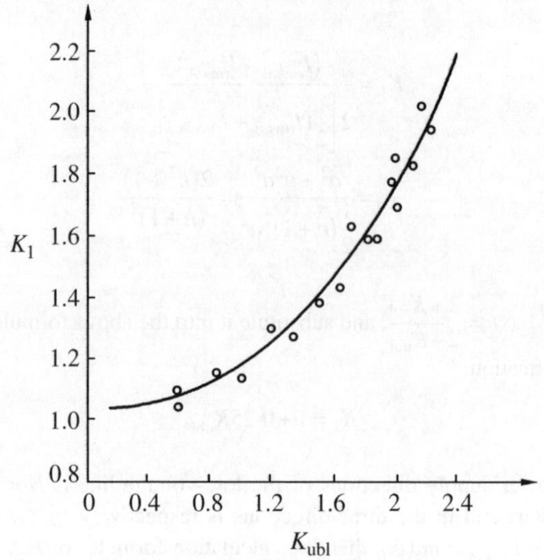

Figure 8.1 $K_1(K_{ubl})$ fitting curve.

According to the above description, the electricity losses ΔA_{eq} (kW·h) of low-voltage lines in a distribution transformer within the measuring period can be calculated as per the following formula:

$$\Delta A_{eq} = MNK_1K_2\left(\frac{I_{rms.bd}}{N}\right)^2\left(\frac{R_{av}}{N}\right)T\times 10^{-3} \tag{8.17}$$

Wherein M – constant considering the structure of low-voltage line (for single-phase two-wire system, $M = 2$; for three-phase four-wire system, $M = 3.5$);

N – number of power supply directions of low-voltage line;

K_1 – correction coefficient considering unequal current in the start end of each power supply direction, which can be calculated as per Formulas (8.15) or (8.16);

K_2 – correction coefficient considering uneven load distribution in each power supply direction of the line, generally in the range 0.25–0.30;

$I_{rms.bd}$ – rms current in the low-voltage side of the distribution transformer within the measuring period, which can be obtained by multiplying the rms current in the high-voltage side of the distribution transformer within the measuring period by the ratio of transformation (A);

R_{av} – average of total resistance of per phase conductor of the low-voltage line, which is obtained based on the statistics of distribution transformers of the same capacity and the same load type (Ω).

Due to the large quantity of common distribution transformers, it is impossible to calculate electric energy losses of the low-voltage line in each distribution transformer by the above method. To regularly conduct theoretical calculation of line losses in the whole region, first of all, select several kinds of representative distribution transformers for detailed calculation of low-voltage lines. Then, compare the total low-voltage rms current of various kinds of distribution transformers with the low-voltage rms current of representative distribution transformers and calculate the ratio between the two. Next, multiply the square of this ratio by the losses of low-voltage lines of representative distribution transformers to obtain electric energy losses of low-voltage lines of the same kind of distribution transformers. Add up the electric energy losses of low-voltage lines of several kinds of distribution transformers to obtain the total electric energy losses of low-voltage lines in the whole region.

8.3 Calculation of Electric Energy Losses of Cable Lines

Due to complex structures, power cables have resistance heat losses in conductors, dielectric losses in insulating layers, and sheath losses and armor losses in sheathed, armored, and reinforced layers. The calculation of no-load losses and load losses of cable lines will be introduced below.

8.3.1 Calculation of No-load Losses (Dielectric Losses in Insulating Layers) of Cable Lines

When the operating voltage of oil impregnated paper insulated cables exceeds 30 kV and the operating voltage of PVC insulated cables exceeds 6 kV, that is when the operating voltage of cables is high, the dielectric losses in insulating layers are of significance. Dielectric losses are independent of the cable load, but are related to the cable operating voltage, so dielectric losses are regarded as a kind of no-load losses. The dielectric losses ΔA_0 (kW·h/km) per unit length of per phase cable are calculated as per the following formula[1]:

$$\Delta A_0 = 3U_{\text{ph.av}}^2 \omega C_0 \tan\delta T \times 10^{-9} \tag{8.18}$$

$$C_0 = \frac{\varepsilon_r \times 10^6}{18\ln\dfrac{r_{21}}{r_1}} \tag{8.19}$$

Wherein $U_{\text{ph.av}}$ – average phase voltage of cable lines within the measuring period (kV);

ω – AC angular frequency (when AC frequency is 50 Hz, $\omega = 2\pi f = 2\pi \times 50 \approx 314$);

C_0 – operating capacitance per unit length of cable, which can be found in the product category of cables (pF/km);

ε_r – relative dielectric constant of insulating material (for oil impregnated paper insulated cables, $\varepsilon_r = 3.55$)

r_1, r_{21} – outside radius of cable core and outside radius of insulating layer (cm);

$\tan\delta$ – tangent of dielectric loss angle of insulating material in cable, which is related to the material, structure and rated voltage of cable. See Table 8.5 for ε_r and $\tan\delta$ values of common insulating materials used in cables.

According to the factory technical parameters of 630 mm^2 copper cables, calculations as per Formulas (1.6) and (1.7) indicate that the percentage of dielectric losses of cross-linked cables in resistance losses changes significantly with the operating voltage. The percentage of 6–10 kV cross-linked cables is below 1×10^{-4}, approximately 5.3×10^{-4} for 35 kV cables, around 15×10^{-4} for 66 kV cables, and up to 69×10^{-4} for 110 kV cables. This percentage is dramatically reduced as the quality of insulating materials is improved. For urban power grids with intensive new and old high-voltage cable, no-load losses of cables should be particularly considered for the calculation and analysis of line losses and should not be mistakenly included in "unknown losses" due to omission. Refer to Appendix A and see Example 2-1 in Reference [65] for these calculations.

8.3.2 Calculation of Load Losses of Cable Lines

8.3.2.1 Calculation of Resistance Heat Losses of Cable Cores

Because each phase of the cores of three-phase cables is close, an obvious skin effect and proximity effect occur so that the AC resistance of cores is increased. If the AC resistance considering the temperature influence around the cores is R', then

$$\begin{aligned} R' &= R_{20}(1+\beta_2) \\ \beta_2 &= \alpha_{20}(\theta_c - 20) \end{aligned} \tag{8.20}$$

[1] In Formula (1.7), $\varepsilon = \varepsilon_0 \varepsilon_r$. Substitute $\varepsilon_0 = 8.86 \times 10^{-12}$ F/m into it and divide by the length l to obtain Formula (8.19).

Table 8.5 ε_r and $\tan\delta$ values of common insulating materials used in cables.

Cable type	ε_r [a]	Tanδ [a]
Viscous impregnated, pre-impregnated, or non-draining insulated cable	4	0.01
Low-pressure oil-filled cable	3.3	0.004
High-pressure oil-filled cable	3.5	0.004 5[b]
Steel pipe type oil-filled cable	3.7	0.0045
Pressure cable	3.5	0.0045
Gas-filled cable	3.4	0.0045
Butyl rubber insulated cable	4.0	0.05
Ethylene propylene rubber (EPR) cable	0	0.003
PVC insulated cable	5.0	0.07
Polyethylene insulated cable	2.3	0.0005
Cross-linked polyethylene insulated cable	2.3	0.0005

[a] 20 °C, 50 Hz.

[b] This value is greater than that of high-pressure oil-filled cable, which is attributable to the different paper used, not the different pressure.

Wherein R_{20} – DC resistance of cores at 20 °C (Ω);
β_2 – correction coefficient considering the influence of temperature around cores on resistance;
α_{20} – temperature coefficient of resistance of core conductor material at 20 °C (for copper, α_{20} = 0.003 93 1/°C; for aluminum, α_{20} = 0. 004 03 1/°C);
θ_c – core temperature (°C).

Considering the skin effect and proximity effect, the AC resistance of three-phase AC cable is calculated as per the following formula:

$$R = R'(1 + K_3 + K_4) \tag{8.21}$$

$$K_3 = \frac{X_3^2}{192 + 0.8X_3^2} \tag{8.22}$$

$$K_4 = \frac{0.66X_4^2}{192 + 0.8X_4^2}\left(\frac{D_c}{S}\right)^2 \times \left[0.312\left(\frac{D_c}{S}\right)^2 + \frac{1.18}{\frac{X_4^2}{192 + 0.8X_4^2} + 0.27}\right] \tag{8.23}$$

$$X_3 = \frac{8\pi f}{R'} \times 10^{-9}$$

$$X_4 = k_p \frac{8\pi f}{R'} \times 10^{-9}$$

Wherein K_3 – coefficient of skin effect;
K_4 – coefficient of proximity effect;
X_3 – calculation coefficient;
X_4 – calculation coefficient;
k_p – coefficient changing with core structure (for dry impregnated paper insulated cables, k_p = 0.80; for non-dry impregnated paper insulated cables, k_p = 1.0);
D_c – outer diameter of core (cm);
S – spacing between cores (cm).

According to the calculation result for 10 kV three-core sector-shaped cables, the sum of K_3 and K_4 coefficients of both copper cables with a sectional area of 240 mm^2 and above and aluminum cables with a sectional area of 400 mm^2 and above exceeds 5%, so the influences of skin effect and proximity effect must be considered so as to avoid a smaller calculation of electric energy losses.

8.3.2.2 Calculation of Cable Sheath Losses

The quantity of losses in the metal sheaths of cables depends on the cable structure and the sheath connection. Sheath losses are in direct proportion to the square of current in the cores, so the ratio between sheath losses and core losses is a constant, that is $\Delta A_h = \lambda_1 \Delta A$. λ_1 is called the loss coefficient of the metal sheaths of cables.

1. The coefficient of sheath losses of 10 kV three-phase integrated steel tape armored cables can be calculated as per the following formula:

$$\lambda_1 = \left[1 + \left(\frac{D_h}{D_k}\right)^2 \frac{1}{1+\frac{D_k}{\mu\Delta_k}}\right]^2 \times 0.94\frac{R_h}{R}\left(\frac{2r_1+t}{D_h}\right)^2 \frac{1}{1+\left(1.59R_h \times 10^8/f\right)^2} \tag{8.24}$$

$$\Delta_k = \frac{A}{\pi D_k}$$

Wherein D_h – average diameter of metal sheath (cm);
D_k – average diameter of armor (cm);
r_1 – radius of circumference surrounding three-phase sector-shaped core (cm);
t – insulation thickness between cores (cm);
μ – average permeance coefficient of steel tape, $\mu \approx 300$;
Δ_k – equivalent thickness of armor (cm);
A – sectional area of armor (cm^2).

According to the calculation, λ_1 is generally smaller than 1% for common 10 kV steel tape armored cables, so it can be ignored.

2. The the coefficient of sheath losses is calculated when single-core cables are laid with both ends of the sheathes being directly interconnected, not transposed, and linearly and equidistantly arranged. In some special cases (such as submarine cable lines), metal sheaths cannot be cross-connected and can only be directly connected in both ends. In such cases, wherein a closed loop is formed within the sheathes, induced electromotive forces in metal sheaths will produce a sheath current whose direction is opposite to the core current, thereby leading to larger sheath losses. In this case, the per phase equivalent coefficient of the sheath losses of three-phase cables can be calculated as per the following formula:

$$\lambda_1 = \frac{1}{2}\frac{R_h}{R}\left[\frac{(X_h+X_m)^2}{R_h^2+(X_h+X_m)^2} + \frac{\left(X_h-\frac{1}{3}X_m\right)^2}{R_h^2+\left(X_h-\frac{1}{3}X_m\right)^2}\right] \tag{8.25}$$

$$X_h = \left(2\omega \ln\frac{2S}{D_h}\right) \times 10^{-9},\ X_m = 2\omega \ln 2 \times 10^{-9} \tag{8.26}$$

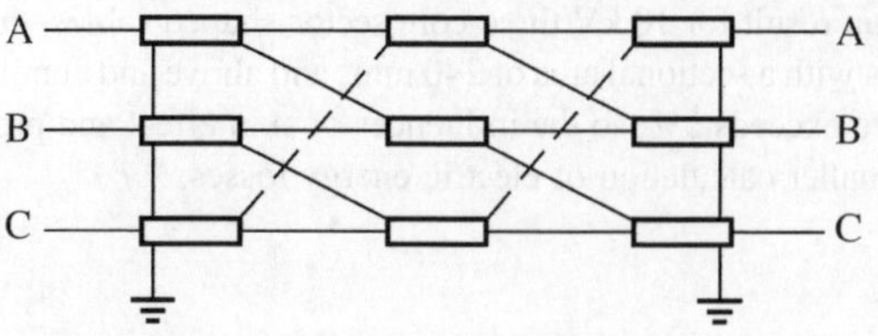

Figure 8.2 Non-transposed and cross-connected cables.

Wherein R – effective resistance per unit length of core [Ω/cm; calculated as per Formula (8.23)];
R_h – effective resistance per unit length of sheath (Ω/cm);
X_h – reactance per unit length of sheath (Ω/cm);
S – center distance between two cables (cm);
D_h – diameter of metal sheath (cm);
ω – AC angular frequency;
X_m – calculation reactance per unit length (when the AC frequency is 50 Hz, $X_m = 4.35 \times 10^{-7}$ Ω/cm).

According to the calculation, when both ends of sheaths of single-core cables are directly interconnected, the sheath losses may be very large.

3. When the sheaths of single-core cables are single-point interconnected or cross-connected, and are laid in the form of a linear and equidistant arrangement (see Figure 8.2), the cable sheath losses are very small.
   Current-circulation losses λ'_1 are generally ignored and the coefficient of eddy-current losses λ''_1 [21] is only calculated as per the following formula:

$$\lambda''_1 = A_1 \frac{R_h}{R} \frac{(D_h/2S)^2\left[1 + A_2(D_h/2S)^2\right]}{R\left(R_h \times 10^9/\omega\right)^2 + (1/S)(2S/D_n)}$$

Wherein R_h – resistance per unit length of metal sheath (Ω/cm);
R – resistance per unit length of conductor (Ω/cm);
D_h – average diameter of metal sheath (cm);
S – spacing between center axes of cable conductors (cm);
A_1, A_2 – calculation coefficients related to the way of laying of cables (when cables are arranged in a regular triangle manner, $A_1 = 3$, $A_2 = 0.417$; when cables are arranged equidistantly and in parallel, $A_1 = 1.5$, $A_2 = 0.27$ for side phase, and $A_1 = 6$, $A_2 = 0.083$ for medium-phase).

For example for 110 kV and 240 mm^2 oil-filled cables, when the phase spacing is 40 cm, $\lambda''_1 = 2.12\%$ in the case of regular triangle arrangement of three-phase cables; side phase $\lambda''_1 = 1.04\%$ and medium phase $\lambda''_1 = 4.14\%$ in the case of horizontal arrangement. According to the comparison, when the phase spacing is the same, the eddy-current losses of sheaths arranged horizontally are approximately 2.5 times those of sheaths arranged in the form of a regular triangle.

4. The coefficient of sheath losses of split-phase lead-sheathed armored cables can be calculated as per the following formula:

$$\lambda_1 = \frac{R_h}{R} \cdot \frac{1.7X_h^2}{R_h^2 + X_h^2} \tag{8.27}$$

8.3.2.3 Calculation of Cable Armored Losses

Losses caused by hysteresis and eddy current in armored layers and reinforced layers of cables are known as armored losses, ΔA_k, which are in direct proportion to the square of current in the cores, that is $\Delta A_k = \lambda_2 \Delta A$. λ_2 is called the coefficient of cable armored losses.

For split-phase lead-sheathed wire-armored cables, λ_2 can be calculated as per the following formula:

$$\lambda_2 = 1.23(1-\lambda_1)\frac{R_k}{R}\cdot\frac{2C}{D_k}\cdot\frac{1}{\left(44R_k\times10^6/f\right)^2+1} \tag{8.28}$$

Wherein λ_1 – coefficient of sheath losses calculated as per Formula (8.27);
R_k – resistance per unit length of armor (Ω/cm);
D_k – average diameter of armor (cm);
C – distance between cable center and core center (cm).

According to the calculations of Formulas (8.27) and (8.28), the sheath losses and armored losses of split-phase lead-sheathed cables are large, and the sum of the two may reach over 15% of core losses. Therefore, such two losses must be considered for the calculation of line losses in these cable lines.

8.3.2.4 Other Losses in Cable Lines

1. *Additional losses caused by unequal lengths of cross-connected sections.* Single-core high-voltage cables use the cross-connected grounding mode which requires equal lengths of three sections in a unit system. However, due to the limit on setting of junction manhole in engineering practice, sections may have unequal lengths. At this time, additional losses of metal sheathes can be calculated as per the following formula [66]

$$\Delta P_s = \Delta W_s(\Delta L/L)^2$$

Wherein ΔP_s – additional loss rate of cable metal sheath (%);
ΔW_s – ratio of current-circulation losses of metal sheaths to cable core losses when both ends of metal sheaths are completed grounded;
ΔL – difference between maximum length and minimum length of three sections of the unit system (m);
L – sum of lengths of three sections of the unit system (m).

The writers of Reference [66] found during their surveys that ΔP_s of a 66 kV cable line with $L = 770$ m and $\Delta L = 160$ m was about 4.32% of the current-circulation losses as calculated as per the above formula. Because the current-circulation losses of the cable line were small after cross-connection, unequal lengths of section did lead to additional losses which, however, were very small and thus generally can be ignored.

2. *Eddy-current losses of brackets.* According to Japan's relevant literature dated July 1983, when the operating current of AC single-phase large cross-section cables reaches 1450 A, losses in steel cable brackets will be produced due to an eddy effect, which are 160 W/m for three-phase triangular configuration and 530 W/m for split-phase (horizontal) configuration, approximately accounting for 20–70% of cable losses. To avoid such bracket losses, non-magnetic material such as stainless steel, glass steel, or aluminum alloy have been used at both home and abroad to produce brackets of large cross-section single-core cables.

8.4 Calculation of Electric Energy Losses of Main Transformers

8.4.1 Active Power Losses of Main Transformers

8.4.1.1 No-load Losses of Main Transformers

When the secondary winding of a main transformer is in open circuit and a rated voltage of a rated frequency is applied in its primary winding, the current passing the primary winding is called the no-load current. The active component of the no-load current is called the loss current, and the active power supplied by the loss current are called no-load losses, which are generally expressed by ΔP_0. No-load losses ignoring resistance losses of primary winding are also called iron core losses. In the calculation of line losses, iron core losses are usually regarded as equal to no-load losses ΔP_0(W), which can be calculated as per the following formula:

$$\Delta P_0 = K_0 p_t G_t \tag{8.29}$$

Wherein K_0 – processing increase coefficient, taking 1.2–1.5;
p_t – unit loss of magnetic flux density of iron core (W/kg);
G_t – mass of iron core (kg).

8.4.1.2 Load losses of Main Transformers

When the secondary winding of a transformer is in short circuit and a rated current passes the secondary winding, the voltage applied in the primary winding is called the short-circuit voltage or impedance voltage of the transformer. Losses produced at this time are called rated load losses, which are usually expressed by ΔP_{LN}, and are divided into resistance losses and additional losses of the winding. Additional losses include eddy-current losses of the winding, current-circulation losses of shunt winding, structure losses, and lead losses. Considering the additional losses, load losses can be calculated based on 1.05–1.15 times resistance losses.

Resistance losses ΔP_r(W), also known as copper losses or aluminum losses, can be calculated as per the following formula:

$$\Delta P_r = K_r j^2 G \tag{8.30}$$

Wherein K_r – coefficient (2.4 for copper wires; 13.22 for aluminum wires);
j – current density of winding conductors at 75 °C (A/mm^2);
G – total mass of winding conductors (kg).

8.4.2 Calculation of Electric Energy Losses of Main Transformers

If, within the measuring period T, the average voltage at the primary side of the transformer is U_{av} and the rms current is I_{rms}, then the electric energy losses ΔA (kW·h) of the three-winding transformer are calculated as per the following formula:

$$\Delta A = \Delta A_0 + \Delta A_L = \left[\Delta P_0 \left(\frac{U_{av}}{U_f}\right)^2 + \Delta P_{LN}\left(\frac{I_{rms}}{I_N}\right)^2\right] T \times 10^{-3} \tag{8.31}$$

Wherein ΔP_0 – rated no-load losses of the transformer, when rated voltage is applied in the transformer which is operating in the tapping point (W);
U_f – voltage of operating tapping point of the transformer (kV);
ΔP_{LN} – rated load losses of the transformer (W);
I_N – rated current at primary side of the transformer (A).

If the maximum load current I_{max}, minimum load current I_{min}, and average current (obtained from readings of active and reactive watt-hour meters) within the measuring period are given, the loss factor can also be calculated with the load factor f and the minimum load rate as per Formula (2.26), and then the electric energy losses ΔA (kW·h) of the main transformer can be calculated as per the following formula:

$$\Delta A = \left[\Delta P_0 \left(\frac{U_{av}}{U_f}\right)^2 + \Delta P_{LN}\left(\frac{I_{max}}{I_N}\right)^2 F\right] T \times 10^{-3} \tag{8.32}$$

For a three-winding transformer, electricity losses in each winding should be calculated and are added up to obtain the load electricity losses of the three-winding transformer. Subscripts *h*, *m*, and *l* are used below to represent windings at high-voltage, medium-voltage, and low-voltage sides. When 100% rated current passes each winding, if the rated load losses among windings are known, then the rated load losses of each winding can be calculated as per the following formula:

$$\Delta P_{\mathrm{hN}} = \frac{1}{2}(\Delta P_{\mathrm{h.m}} + \Delta P_{\mathrm{h.l}} - \Delta P_{\mathrm{m.l}}) \tag{8.33}$$

$$\Delta P_{\mathrm{mN}} = \frac{1}{2}(\Delta P_{\mathrm{h.m}} + \Delta P_{\mathrm{m.l}} - \Delta P_{\mathrm{h.l}}) \tag{8.34}$$

$$\Delta P_{\mathrm{lN}} = \frac{1}{2}(\Delta P_{\mathrm{h.l}} + \Delta P_{\mathrm{m.l}} - \Delta P_{\mathrm{h.m}}) \tag{8.35}$$

Wherein $\Delta P_{\mathrm{h.m}}$, $\Delta P_{\mathrm{h.l}}$, $\Delta P_{\mathrm{m.l}}$ – rated load losses of high voltage–medium voltage, high voltage–low voltage, and medium voltage–low voltage windings, which are included to the rated capacity of the main transformer.

If the rms current passing each winding within the measuring period is calculated according to measured load records, that is $I_{1\mathrm{rms}}$, $I_{2\mathrm{rms}}$, $I_{3\mathrm{rms}}$, and the rms current included in the high-voltage side winding is $I_{1\mathrm{rms}}$, $I'_{2\mathrm{rms}}$, $I'_{3\mathrm{rms}}$, then the electricity losses of each winding are

$$\Delta A_1 = \Delta P_{\mathrm{hN}} \left(\frac{I_{1\mathrm{rms}}}{I_{1\mathrm{N}}}\right)^2 T \times 10^{-3} \tag{8.36}$$

$$\Delta A_2 = \Delta P_{\mathrm{mN}} \left(\frac{I'_{2\mathrm{rms}}}{I_{1\mathrm{N}}}\right)^2 T \times 10^{-3} \tag{8.37}$$

$$\Delta A_3 = \Delta P_{\mathrm{lN}} \left(\frac{I'_{3\mathrm{rms}}}{I_{1\mathrm{N}}}\right)^2 T \times 10^{-3} \tag{8.38}$$

No-load electric energy losses of the three-winding transformer are calculated in the same way as that of the two-winding transformer, so electric energy losses of the three-winding transformer can be calculated as per the following formula:

$$\Delta A = \Delta P_0 \left(\frac{U_{\mathrm{av}}}{U_{\mathrm{f}}}\right)^2 T \times 10^{-3} + \Delta A_1 + \Delta A_2 + \Delta A_3 \tag{8.39}$$

Wherein ΔA_1, ΔA_2, ΔA_3 – electric energy losses of each winding, which are calculated as per Formulas 8.36 to (8.38).

8.4.3 *Calculation of Electric Energy Losses of Main Transformers in Parallel Operation*

When two main transformers are in parallel operation, if the total rms current at the primary side or the secondary side is given, then the rms current passing each transformer can be determined by the following approximate formula:

$$\left.\begin{aligned} I_{\mathrm{rms1}} &= \frac{\chi_2}{\chi_1 + \chi_2} I_{\mathrm{rms}} \\ I_{\mathrm{rms2}} &= \frac{\chi_1}{\chi_1 + \chi_2} I_{\mathrm{rms}} \end{aligned}\right\}$$

$$\begin{aligned} \chi_1 &= \frac{\Delta u_{\mathrm{k1}} U_{\mathrm{N}}^2 \times 10}{S_{\mathrm{N1}}} \\ \chi_2 &= \frac{\Delta u_{\mathrm{k2}} U_{\mathrm{N}}^2 \times 10}{S_{\mathrm{N2}}} \end{aligned} \tag{8.40}$$

Wherein χ_1, χ_2 – reactance of two transformers (Ω);
Δu_{k1}, Δu_{k2} – percentage of short-circuit voltage of two transformers;
U_N – rated voltage of transformer (kV);
S_N – rated capacity of transformer (kVA).

The total electric energy losses ΔA_Σ (kW·h) of the two transformers can be calculated as per the following formula:

$$\Delta A_\Sigma = \left[(\Delta P_{01} + \Delta P_{02})\left(\frac{U_{av}}{U_f}\right)^2 + \Delta P_{LN}\left(\frac{I_{rms1}}{I_{N1}}\right)^2 + \Delta P_{LN2}\left(\frac{I_{rms2}}{I_{N2}}\right)\right] T \times 10^{-3} \tag{8.41}$$

Wherein I_{N1}, I_{N2} – rated current of two transformers (A).

8.5 Calculation of Electric Energy Losses of Other Electrical Equipment

8.5.1 Shunt Capacitors

Electric energy losses of shunt capacitors can be determined by the tangent measurement of loss angle ($\tan\delta$) of the capacitor, and this measurement is one of the items required to be tested in the factory test of the capacitor. For 1 kV and below non-self-healing shunt capacitors, Reference [67] stipulates that $\tan\delta$ values of capacitors under voltages of $(0.9–1.1)U_N$ and frequencies of $(0.8–1.2)f_N$ should be measured before and after the thermal stability test, and that $\tan\delta$ measurements should not guarantee values given by manufacturers under the test temperature and voltage or values agreed by manufacturers and purchasers.

Reference [68] specifies that measurement of the $\tan\delta$ value of a capacitor, as one of the routine test items, should be conducted under 0.9–1.1 times the rated voltage and in a method that can eliminate errors caused by harmonics. The accuracy of the measuring method and the relationship between it and the measurements under the rated voltage and rated frequency should be identified. For multiphase capacitors, the measuring voltage should be adjusted so that each unit can be subject to 0.9–1.1 times the rated voltage. The old version of this standard (dated 2001) stipulates that the requirement for capacitor losses (or $\tan\delta$) should be discussed and determined by manufacturers and purchasers, while the new version [68] (dated 2010) explicitly stipulates that the $\tan\delta$ of all film dielectric capacitors should not exceed 0.000 5. If there is a case for a lower requirement than this, the manufacturers and purchasers should have a discussion to determine it. According to the comparison, given the requirement for loss reduction and energy conservation in electric power systems, the new version puts forward higher requirements for the standard of energy consumption of widely used shunt capacitors having a rated voltage above 1000 V.

The new version [68] also stipulates that losses caused by all accessories such as external fuse protectors and reactors should be included when calculating the total losses of a bank of capacitors. As agreed, manufacturers should provide curve graphs or data tables indicating the changes in capacitance and $\tan\delta$ with temperature under rated capacity steady-state conditions. Such provisions of the new version provide basic conditions for the correct calculation of the electric energy losses of shunt capacitors considering various influence factors.

Therefore, the electricity losses ΔA (kW·h) of shunt capacitors within the measuring period T can be calculated as per the following formula:

$$\Delta A = Q_C \tan\delta T \times 10^{-3} = U_1^2 \omega C \tan\delta T \times 10^{-3} \tag{8.42}$$

Wherein Q_C – average of total capacity of shunt capacitors put into operation within the measuring period (kvar);
U_1 – line voltage (kV);
C – capacitance corresponding to the average of total capacity of capacitors put into operation (μF).

8.5.2 Shunt Reactors and Series Current-limiting Reactors

8.5.2.1 Shunt Reactors

Shunt reactors are used for ultra-high voltage remote power transmission lines to absorb the capacitive charging power of line capacitance, so as to stabilize the system voltage and improve system stability.

Similar to transformers operating with no loads, the power losses of shunt reactors depend on the operating voltage. Using factory test data [i.e. the power losses ΔP_{re} (three-phase average) of reactors under the rated voltage] provided by manufacturers, the average voltage within the operating period and the operating period T can be used to calculate electric energy losses as per the following formula:

$$\Delta A = 3\Delta P_{\mathrm{re}}\left(\frac{U_{\mathrm{av}}}{U_{\mathrm{N}}}\right)^2 T\times 10^{-3} \tag{8.43}$$

Wherein U_{av}, U_{N} – average voltage of reactors within the operating period and the rated voltage of the system to which reactors are connected (kV).

8.5.2.2 Series Current-limiting Reactors

Series current-limiting reactors are generally installed in loops of substation transformers, buses, power transmission lines, or high-voltage distribution lines, and current passing the reactors within the measuring period is directly or indirectly measured and recorded. Therefore, the electric energy losses of a series current-limiting reactor can be calculated as per the following formula:

$$\Delta A = 3\left(\frac{I_{\mathrm{rms}}}{I_{\mathrm{N}}}\right)^2 \Delta P_{\mathrm{rel}} T\times 10^{-3} \tag{8.44}$$

Wherein I_{rms} – rms current passing the reactor (A);
I_{N} – rated current of the reactor (A);
ΔP_{rel} – power losses of the single-phase reactor at 75 °C (W; this can be found in factory data; if no data can be found, refer to [69]).

8.5.3 Synchronous Compensator

The losses of a synchronous compensator are losses of the main engine and can be generally divided into basic losses, additional losses, and mechanical losses. Basic losses include basic iron losses, basic copper losses, and excitation losses. Additional losses include rotor surface losses, additional losses produced in the stator by higher harmonic in the rotor magnetic field, pulsating losses in stator teeth, additional copper losses in stator winding conductors, and losses produced in the rotor surface by harmonic magnetic potential and flux of the stator. The former three losses exist when the synchronous compensator is operating with no loads and can be regarded as no-load losses, while the latter two losses are in direct proportion to the square of stator current and can be regarded as load losses. Mechanical losses include mechanical friction losses of the bearing and electric brush, and friction losses of ventilation; they are basically independent of the loads of the synchronous compensator and account for over half of the total losses.

According to the power loss percentage $\Delta p\%$ (provided by the manufacturer) of a synchronous compensator under rated loads and the ratio between no-load losses and load losses, the electric energy losses ΔA (kW·h) of the synchronous compensator within the measuring period can be calculated as per the following formula:

$$\Delta A = \Delta p\% S_{\mathrm{N}} T\left[k + (1-k)\left(\frac{S_{\mathrm{rms}}}{S_{\mathrm{N}}}\right)^2\right] \tag{8.45}$$

$$S_{rms} = \sqrt{\frac{0.2}{f} + 0.8S_{av}} \tag{8.46}$$

Wherein S_N – rated capacity of the synchronous compensator (kVA);
S_{rms} – rms loads of the synchronous compensator with the operating period T (kVA);
k – ratio of no-load losses to load losses; $k > 0.5$ in general;
T – operating period of the synchronous compensator (h).

The synchronous compensator should be equipped with a low power factor watt-hour meter that can measure consumed electrical energy for the purpose of making a statistical collection of electricity losses of the synchronous compensator. This watt-hour meter allows recording readings at the beginning and end of operation, thus determining the electric energy consumed by the synchronous compensator within the operating period.

8.5.4 *Watt-hour Meter and Other Instruments*

Electric energy losses in the current coils of watt-hour meters are generally much smaller than those in voltage coils and iron cores, so the upper limit of no-load losses can be used for calculating the electricity losses of watt-hour meters, as a step of considering ignored load losses.

The no-load power losses of a single-phase watt-hour meter generally do not exceed 1.5 W, so the monthly electricity losses of a single-phase watt-hour meter can be taken as 1 kW·h. The monthly losses of a three-phase watt-hour meter with two units can be taken as 2 kW·h. If the number of single- and three-phase watt-hour meters is given, then the total electricity losses of watt-hour meters within a measuring period can be obtained.

The power consumed by the voltage coil of various measuring instruments or electromagnetic relays is generally not more than 5 W; and the power consumed by the voltage coil of some automatic recording instruments may reach 6 or 13 W. The power consumed by each current coil under full loads is generally not more than 1 W; and the power consumed by the current coil of some automatic recording instruments may reach 6 W.

On average, the power losses in each current transformer are not more than 20–50 W. The losses of a voltage transformer are mainly no-load losses and depend on the model and capacity of the voltage transformer. For a general electromagnetic voltage transformer, its no-load losses approximately account for 10% of its capacity. A theoretical calculation of the line losses of some power grids indicates that the total electricity losses in the current transformer, voltage transformer, electromagnetic relay, and measuring instrument only account for 0.1–0.4% of total line losses and thus can be ignored in general. In addition, the electricity used by a substation is mainly for the inside and outside lighting of the substation, cooling the main transformer (fan, oil-submerged pump), auxiliary engine, synchronous compensator and AC + DC operation, control and protection, and the power supply of the signal power source. Some substations use electric boilers with a larger power consumption, leading to a higher electricity usage. Watt-hour meters should be installed to measure the electricity required for operation, and an electricity rating should be established, depending on the scale of the substation, and be strictly assessed.

9

Calculation of Electric Energy Losses of Multi-branch Lines

Power transmission lines connected with a number of concentrated loads, 6–10 kV distribution lines connected with many high-voltage customers and dedicated or common distribution transformers, and 380/220 V low-voltage distribution lines connected with a lot of low-voltage power and lighting customers are all multi-branch lines. Because of the different shapes of load curves, power factors and voltages at different load points, it is very difficult to accurately calculate the electric energy losses of multi-branch lines. Various simplified methods have to be used for the theoretical calculation of line losses of distribution lines with many branch lines. These methods need less original operating data and less calculation workload while producing highly accurate calculation results. This chapter will introduce several methods for calculating the electric energy losses of multi-branch lines.

9.1 Basic Method for Calculating Electric Energy Losses of Multi-branch Lines

9.1.1 Weighted Average Method

A planned or designed 35 kV power transmission line is always connected with several concentrated loads, as shown in Figure 9.1.

Electric energy losses of the whole line can be calculated as per the following formula:

$$\Delta A = \Delta A_3 + \Delta A_2 + \Delta A_1$$
$$= \left[\left(\frac{S_{3\max}}{U_{\rm d}}\right)^2 r_3F_3 + \left(\frac{S_{2\max}}{U_{\rm c}}\right)^2 r_2F_2 + \left(\frac{S_{1\max}}{U_{\rm b}}\right)^2 r_1F_1\right] T\times 10^{-3}$$

$$S_{3\max} = S_{\rm d,max},\ S_{2\max} = S_{3\max} + \Delta S_3 + S_{\rm c,max}$$

$$S_{1\max} = S_{2\max} + \Delta S_2 + S_{\rm b,max}$$

$$U_{\rm c} = U_{\rm d} + \Delta U_3,\ U_{\rm b} = U_{\rm c} + \Delta U_2,\ U_{\rm a} = U_{\rm b} + \Delta U_1$$

$$\Delta U = \frac{P_{\max} r + Q_{\max} \chi}{U}$$

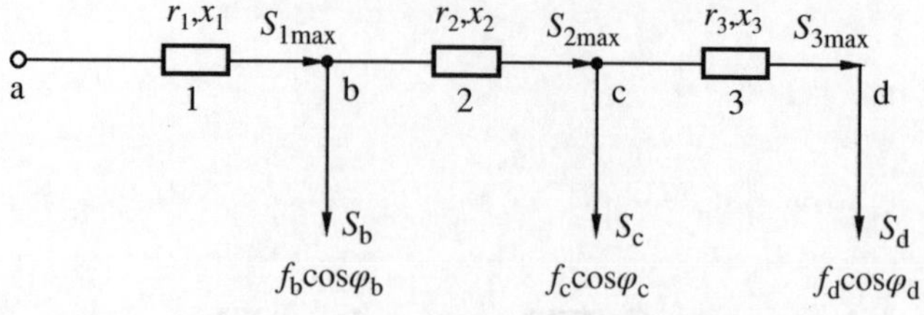

Figure 9.1 Line connected with three loads.

Wherein S_{3max}, S_{2max}, S_{1max} – maximum values of input value at nodes a, b, and c;
ΔS_2, ΔS_3 – power losses in sections 2 and 3;
U_d, U_c, U_b – voltages at nodes d, c, and b (if U_d or U_a is given, they can be obtained with voltage loss);
ΔU – voltage loss (without considering the transverse component of the voltage losses in each section, ΔU can be calculated based on the maximum load rate passing each section and the impedance of each section);
r_3, r_2, r_1 – resistance of each section (Ω);
x_3, x_2, x_1 – reactance of each section (Ω);
F_3, F_2, F_1 – loss factor of each section [calculated by formulas listed in Table 4.4 (Chapter 4) and Formula (4.13) based on the load factor and power factor of each section];
T – calculation period (h).

To derive the load curves and operating parameters as a result of the superposition of several concentrated loads, concepts of the weighted average load factor [Formula (9.1)] and the weighted average power factor [Formula (9.2)] can be quoted, that is

$$f_{we} = \Sigma(P_{max.i} f_i)/\Sigma P_{max.i} \tag{9.1}$$

$$\cos\varphi_{we} = \Sigma(S_{max.i} \cos\varphi_i)/\Sigma S_{max.i} \tag{9.2}$$

For the multi-branch line shown in Figure 9.1,

$$f_{we2} = \frac{P_{max.d} f_d + P_{max.c} f_c}{P_{max.d} + P_{max.c}}$$

$$f_{we1} = \frac{P_{max.d} f_d + P_{max.c} f_c + P_{max.b} f_b}{P_{max.d} + P_{max.c} + P_{max.b}}$$

$$\cos\varphi_{we2} = \frac{\cos\varphi_d S_{max.d} + \cos\varphi_c S_{max.c}}{S_{max.d} + S_{max.c}}$$

$$\cos\varphi_{we1} = \frac{\cos\varphi_d S_{max.d} + \cos\varphi_c S_{max.c} + \cos\varphi_b S_{max.b}}{S_{max.d} + S_{max.c} + S_{max.b}}$$

The loss factors F_3, F_2, and F_1 can be found according to the calculation curves of electric energy losses shown in Figure 5.4 based on the three sets of parameters of (f_d, $\cos\varphi_d$), (f_{we2}, $\cos\varphi_{we2}$), and (f_{we1}, $\cos\varphi_{we1}$).

According to Formulas (9.1) and (9.2), the weighted average method is on the premise that the shapes of load curves of all branch loads are the same. Actually, the superposition of load curves with different shapes causes a staggered peak, so f_{we} calculated by Formula (9.1) is smaller than the actual operating value. The weighted average method may produce larger electric energy losses of multi-branch lines.

9.1.2 Point by Point Section Simplification Method

To calculate electric energy losses of a 6–10 kV high-voltage distribution line, the following original data should be generally collected:

1. Single line connection diagram of the distribution line, indicating the conductor model and length of each section, and the capacities of all connected distribution transformers.
2. Records of monthly active electricity and reactive electricity at the start end and high-voltage customers of the distribution line.
3. Records of 24 h current or active electricity and reactive electricity at the start end and high-voltage customers of the distribution line on a representative day.
4. Classification of common distribution transformers, and actual measuring records of current of each type of distribution transformer on a representative day.
5. Voltage curve at the start end of the distribution line on a representative day.

To simplify the calculation, the point by point section simplification method needs the following two basic assumptions:

1. The power factor at each load point is approximately equal to the power factor at the start end.
2. The influence of the voltage change along the multi-branch line on electric energy losses is ignored.

If the average voltage at the start end on a representative day is regarded as U_{av}, and the monthly average power factor is regarded as $\cos\varphi_{\mathrm{av}}$, then

$$U_{\mathrm{av}} = \Sigma U_i/24$$

$$\cos\varphi_{\mathrm{av}} = A_P\Big/\sqrt{A_P^2 + A_Q^2}$$

Wherein A_P, A_Q – monthly active electricity (kW·h) and reactive electricity (kvar·h).

Due to assumption (1), for each load point,

$$I_{\mathrm{av}.i} = \frac{A_{P.i}}{T}\Big/\left(\sqrt{3}\cos\varphi_{\mathrm{av}} U_{\mathrm{av}}\right)$$

Due to assumption (2),

$$\begin{aligned}\Sigma I_{\mathrm{av}.i} &= \Sigma\left[\frac{A_{P.i}}{T}\Big/\left(\sqrt{3}\cos\varphi_{\mathrm{av}} U_{\mathrm{av}}\right)\right] \\ &= (\Sigma A_{P.i})\frac{1}{T}\Big/\left(\sqrt{3}\cos\varphi_{\mathrm{av}} U_{\mathrm{av}}\right) \\ &= \frac{A_{P.0}}{T}\Big/\left(\sqrt{3}\cos\varphi_{\mathrm{av}} U_{\mathrm{av}}\right) = I_{\mathrm{av}.0}\end{aligned}$$

Wherein $A_{P.0}$, $I_{\mathrm{av}.0}$ – monthly active electricity (kW·h) and monthly average current (A) at the start end of the line.

The above formula indicates that, because of the above two basic assumptions, the average current at the start end of the line is equal to the sum of average current at all load points.

Because of the difference in the shapes of load curves of 6–10 kV customers connected in the 6–10 kV distribution line and in the shapes of load curves of various types of common distribution transformers, loss factors of different sections of the line are also different.

If the rms current of each section of the line is used to calculate line losses, the form coefficient of each section must be calculated after the distribution of average current is obtained. As there are many load points in the distribution line, the maximum load duration of each section load curve is very short due to staggering peak of loads, so Formula (2.16) can be selected to calculate the form coefficient of each section, that is

$$\begin{aligned} K &= \sqrt{F/f} = \sqrt{0.2f + 0.8f^2}/f \\ &= \sqrt{0.2/f + 0.8} \end{aligned} \tag{9.3}$$

According to Formula (9.3), the form coefficient of each section of the 6–10 kV multi-branch distribution line is deemed as only relevant to the load factor. Therefore, the average current and load rates of a load point and the section behind the point can be used to calculate the weighted average load factor $f_{we.i}$ of the section before the point as per Formula (9.1), and then $f_{we.i}$ is substituted into Formula (9.3) to obtain the form coefficient K_i.

The point by point section simplification method follows the steps below to calculate electric energy losses of the distribution line:

1. Determine the number of sections in the line and the resistance of each section, and draw the single line diagram used for calculating line losses.
2. Calculate the average voltage U_{av} within the measuring period based on the voltage records at the start end of the line on the representative day.
3. Calculate the monthly average current at the start end and high-voltage customers of the line based on the monthly active electricity and reactive electricity at the start end and high-voltage customers, that is

$$I_{av.0} = \sqrt{A_P^2 + A_Q^2}/(\sqrt{3}U_{av}T)$$

$$I_{av.m} = \sqrt{A_{P.m}^2 + A_{Q.m}^2}/(\sqrt{3}U_{av}T)$$

   Wherein subscript 0, m – start end of the line, and number of each high-voltage customer.
4. Calculate the monthly average current of each common distribution transformer by means of capacity assignment, that is

$$i_{av} = (I_{av.0} - \Sigma I_{av.n})/\Sigma W_n$$

$$I_{av.n} = W_n i_{av}$$

   Wherein i_{av} – monthly average current assigned to each common distribution transformer per kVA rated capacity (A/kVA);
   W_n – rated capacity of each common distribution transformer (kVA);
   $I_{av.n}$ – monthly average current assigned to each common distribution transformer (A).

   Do an algebraic addition section by section from the tail ends of main line and branch line to the start end, and calculate the average current of each section. Mark them on the single line diagram.
5. Calculate the load factor f_n of each common distribution transformer according to typical surveys or representative day records of each common distribution transformer; calculate the load factor f_m of each high-voltage customer based on load records of the customer on the representative day; calculate the weighted average load factor $f_{we.i}$ of each section according to the distribution of average current and load factor at all load points as per Formula (9.1); calculate the form coefficient K_i of each section as per Formula (9.3); and finally obtain the rms current of each section $I_{rms.i} = K_i I_{av.i}$. Fill in the data in the calculation table.
6. Calculate the monthly electric energy losses of the whole line by means of the point by point section simplification method and accumulation:

$$\Delta A_1 = 3(\Sigma I_{rms.i}^2 R_i)T \times 10^{-3} \tag{9.4}$$

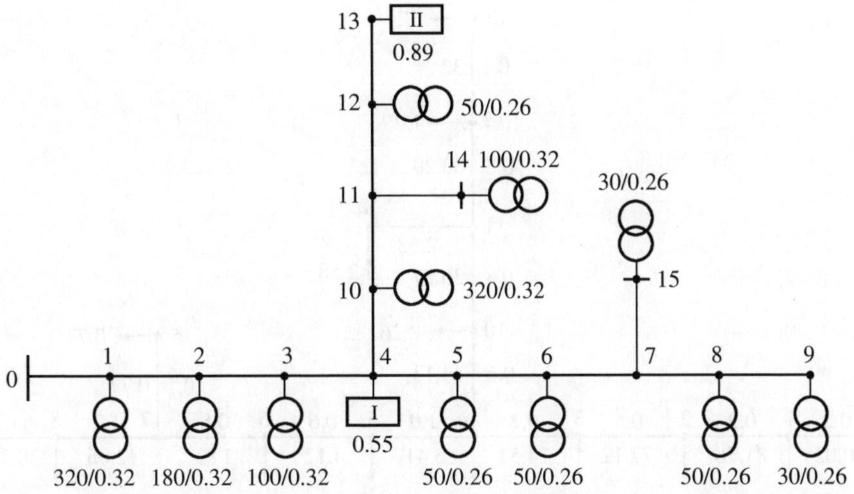

Figure 9.2 Connection diagram of high-voltage distribution line [I and II are high-voltage customers; the numerator in each number beside a transformer is the rated capacity (kVA), and the denominator is the load factor].

Example 9.1 There is a 10 kV high-voltage distribution line whose connection is shown in Figure 9.2.

Known data for calculating line losses of this line include: (i) conductor model and length of each section of the line, (ii) capacity of each common distribution transformer in the line, (iii) active electricity and reactive electricity at the start end of the line and two high-voltage customers (I and II) within the measuring period (one month), (iv) load factor of the two high-voltage customers and common distribution transformers which are calculated based on load data on the representative day, and (v) average voltage at the start end of the line during the measuring period. *Try to calculate the monthly theoretical electricity line losses of this line and its line loss rate.*

Solution

1. *Determine the number of sections in the line and the resistance of each section.*
   The whole line is divided into 15 sections according to the distribution of load points. Calculate the resistance of each section based on the conductor model and length of each section. The resistance values are marked in the single line diagram for calculation, as shown in Figure 9.3.
2. *Determine the distribution of average current.*
   The monthly average voltage at the start end of the line is $U_{av} = 10$ kV, and the monthly active electricity and reactive electricity are 94.6×10^4 kW·h and 60.46×10^4 kvar·h, respectively. For high-voltage customer I, the monthly electric supply is 28.54×10^4 kW·h and $\cos\varphi_{av} = 0.92$. For high-voltage customer II, the monthly electric supply is 34.52×10^4 kW·h and $\cos\varphi_{av} = 0.86$. According to the above known data, calculate the monthly average current at the start end of the line and the two high-voltage customers:

$$I_{av.0} = \left(\sqrt{94.6^2 + 60.46^2}\right) \times 10^4 / \left(\sqrt{3} \times 10 \times 720\right)$$

$$= 112.27 \times 10^4 / 1.247 \times 10^4 = 90.03(\text{A})$$

$$I_{av.I} = 28.5 \times 10^4 / \left(\sqrt{3} \times 10 \times 0.92 \times 720\right) = 25(\text{A})$$

$$I_{av.II} = 34.52 \times 10^4 / \left(\sqrt{3} \times 10 \times 0.86 \times 720\right) = 32(\text{A})$$

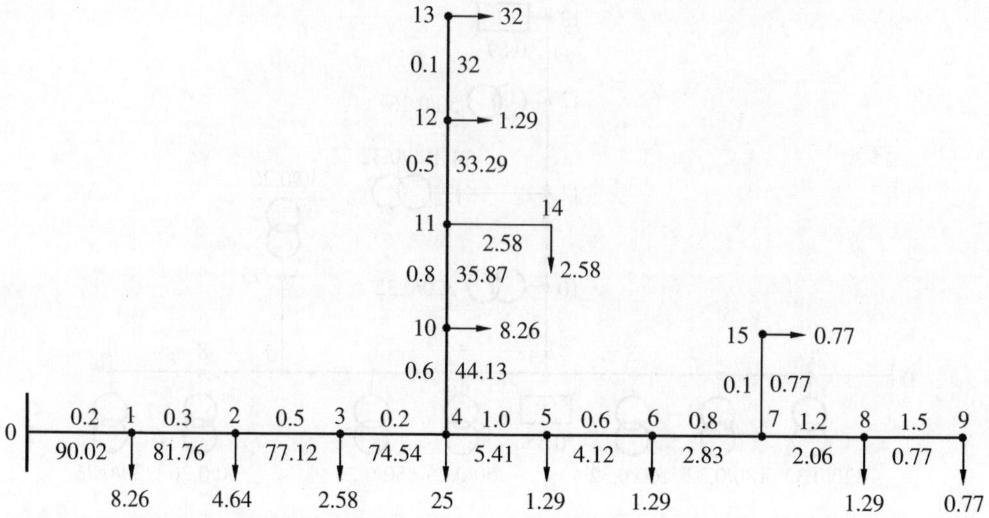

Figure 9.3 Single line diagram for calculating electricity line losses of the high-voltage distribution line [The number under the arrow is the average current (A) at each load point; the number above a section (left) is the resistance (Ω) of the section; the number under a section (right) is the average current (A) of the section].

The average current of each common distribution transformer can be assigned by capacity, that is

$$i_{av} = \frac{90.03-(25+32)}{1280} = 0.025\,8(\text{A/kVA})$$

The average current of each common distribution transformer is calculated as per $I_{av.n} = W_n i_{av}$. Do a point by point calculation from the load points of the tail ends of main line and branch line to the start end, to obtain the average current of each section. The results are marked in Figure 9.3.

3. *Calculate the weighted average load rate of each section.*
   It is known that 30 kVA and 50 kVA common distribution transformers supply loads only for lighting, with the load factor f_1 = 0.26, and distribution transformers with other capacities supply urban common loads, with the load factor of f_2 = 0.32; high-voltage customer I is a two-shift company, with the load factorof f_{I} = 0.55; and high-voltage customer II is a three-shift company, with the load factor of f_{II} = 0.89. The load factor of all load points are marked in the single line diagram. According to the distribution of average current shown in Figure 9.3, calculate the weighted average loadfactor $f_{we.i}$ and the form coefficient K_i of each section. See columns 4 and 5 of Table 9.1 for the calculation results.
4. *Calculate electric energy losses.*
   Calculate $I^2_{rms.i}R_i$ section by section for the 15 sections. According to Table 9.1, $\sum I^2_{rms.i}R_i = 11764.36$ (W), so the monthly electricity losses are

$$\Delta A_1 = 3\times 11\,764.36\times 720\times 10^{-3}$$
$$= 25\,411(\text{kW·h})$$

The monthly line loss rate of the line is

$$\Delta A_1\% = \frac{25\,411}{94.6\times 10^4}\times 10\% = 2.69\%$$

Table 9.1 Calculation table of line losses of high-voltage distribution line through point by point section simplification.

Section numbers	Section resistance (Ω)	Average current for section calculation (A)	Weighted average load factor $f_{we.i}$	Form coefficient of section K_i	rms current of section $I_{rms.i}$ (A)	Power losses of section $I^2_{rms.i}R_i$ (W)	Calculation of weighted average load factor $f_{we.i}$
0–1	0.2	90.02	0.582	1.069	96.23	1852.04	$(0.609 \cdot 81.76 + 0.32 \cdot 8.26)/(81.76 + 8.26) = 0.582$
1–2	0.3	81.76	0.609	1.062	86.83	2226.83	$(0.626 \cdot 77.12 + 0.32 \cdot 4.64)/(77.12 + 4.64) = 0.609$
2–3	0.5	77.12	0.626	1.058	81.59	3328.46	$(0.637 \cdot 74.54 + 0.32 \cdot 2.58)/(74.54 + 2.58) = 0.626$
3–4	0.2	74.54	0.637	1.055	78.64	1236.85	$(0.26 \cdot 5.41 + 0.732 \cdot 44.13 + 0.55 \cdot 25)/74.54 = 0.637$
4–5	1.0	5.41	0.26	1.253	6.78	45.97	
5–6	0.6	4.12	0.26	1.253	5.16	15.98	
6–7	0.8	2.83	0.26	1.253	3.55	10.08	
7–8	1.2	2.06	0.26	1.253	2.58	7.99	
8–9	1.5	0.77	0.26	1.253	0.96	1.38	
7–15	0.1	0.77	0.26	1.253	0.96	0.09	
4–10	0.6	44.13	0.732	1.036	45.72	1254.9	$(0.827 \cdot 35.87 + 0.32 \cdot 8.26)/(35.87 + 8.26) = 0.732$
10–11	0.8	35.87	0.827	1.02 1	36.62	1072.82	$(0.866 \cdot 33.29 + 0.32 \cdot 2.58)/(33.29 + 2.58) = 0.827$
11–12	0.5	33.29	0.866	1.0 15	33.79	570.88	$(0.89 \cdot 32 + 0.26 \cdot 1.29)/(32 + 1.29) = 0.866$
12–13	0.1	32.0	0.89	1.0 12	32.38	104.85	
11–14	0.1	2.58	0.32	1.194	3.08	0.95	

$\sum I^2_{rms.i}R_i = 11\,764.36(\text{W})$

According to Example 9.1, because high-voltage customers and various common distribution transformers have different shapes of load curves and load factors, the weighted average method can be used to calculate the load rate and form a coefficient for each section after the distribution of average current is obtained, thereby obtaining the distribution of rms current and finally calculating the electric energy losses of the whole line. This point by point section simplification method can be called a double current (average current and rms current) distribution simplification method.

9.2 Equivalent Resistance Method and Calculation of Electric Energy Losses of Distribution Transformers

9.2.1 Equivalent Resistance of a Line

The electric energy losses of the multi-branch line shown in Figure 9.4 are

$$\begin{aligned}\Delta A_1 &= 3\left(I^2_{\text{rms.0}}R_0 + \Sigma I^2_{\text{rms.}i}R_i\right)T \times 10^{-3} \\ &= 3I^2_{\text{rms.0}}\left[R_0 + \Sigma\left(\frac{I_{\text{rms.}i}}{I_{\text{rms.0}}}\right)^2 R_i\right]T \times 10^{-3}\end{aligned} \tag{9.5}$$

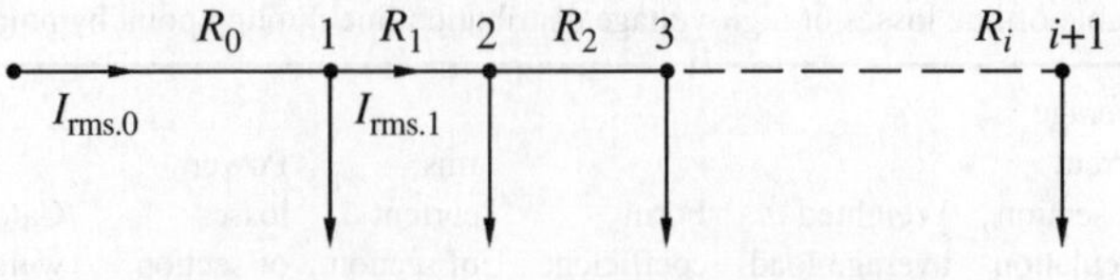

Figure 9.4 Schematic diagram of multi-branch line.

The concept and calculation expression of equivalent resistance can be obtained from Formula (9.5):

$$R_{eql} = R_0 + \Sigma\left(\frac{I_{rms.i}}{I_{rms.0}}\right)^2 R_i \tag{9.6}$$

Wherein R_0, R_i – resistances of the start-end section and section i (Ω).

Due to $I_{rms.i} = K_i I_{av.i} = K_i \dfrac{A_i/T}{\sqrt{3}U_i \cos\varphi_i}$, substitute this into Formula (9.6) to obtain

$$R_{eql} = R_0 + \Sigma D_i^2 R_i \tag{9.7}$$

$$D_i = \frac{K_i}{K_0}\frac{A_i}{A_0}\frac{U_0}{U_i}\frac{\cos\varphi_0}{\cos\varphi_i} = \frac{I_{rms.i}}{I_{rms.0}} \tag{9.8}$$

Wherein D_i – conversion coefficient of section resistance.

Example 9.2 Use the calculation results in Table 9.1.

1. *Try to calculate the equivalent resistances of Sections 4-13 and 4-9 in [Example 9.1], and compare them with the actual resistances.*
2. *Calculate the equivalent resistance of the whole line.*

Solution

1. According to Table 9.1, the rms current at the start end of Section 4-13 is $I_{rms.0} = 45.72$ A, and power losses of this section are $\sum I^2_{rms.i}R_i$ = (1254.19 + 1072.82 + 570.88 + 104.85 + 0.95) = 3003.69 (W). Then, the equivalent resistance of Section 4-13 can be obtained as per Formula (9.5):

$$R_{eq14.13} = \frac{3003.69}{45.72^2} = 1.437(\Omega)$$

The actual resistance of the section is $R_{4.13}$ = 2.0 Ω, and the ratio is $R_{eql4.13}/R_{4.13}$ = 1.437/2 = 0.72.

2. As for Section 4-9, according to Table 9.1,

$$I_{rms.0} = 6.78\text{A}$$

$$\Sigma I^2_{rms.i}R_i = (45.97 + 15.98 + 10.08 + 7.99 + 1.38 + 0.09) = 81.49(\text{W})$$

So, the equivalent resistance is obtained

$$R_{eql4.9} = \frac{81.49}{6.78^2} = 1.77(\Omega)$$

The actual resistance of the section is $R_{4.9}$ = 5.1 Ω, and the ratio is $R_{eql4.9}/R_{4.9}$ = 1.77/5.1 = 0.35.

The comparison shows that, if loads tend to be distributed at the tail end of the line, the ratio R_{eql}/R is close to one; if loads tend to be distributed evenly, the ratio is close to one-third.

3. The rms current and power losses at the start end of the whole line can be obtained according to Table 9.1, so the equivalent resistance of the whole line can be calculated as per Formula (9.5):

$$R_{\text{eql}} = \frac{11\,764.36}{(96.23)^2} = 1.27(\Omega)$$

Replace Sections 4-13 and 4-9 with equivalent resistances and use Formula (9.6) to calculate the equivalent resistance of the whole line, that is

$$\begin{aligned}
R_{\text{eql}} &= R_0 + \Sigma\left(\frac{I_{\text{rms}.i}}{I_{\text{rms}.0}}\right)^2 R_i \\
&= 0.2 + \left(\frac{86.83}{96.23}\right)^2 \times 0.3 + \left(\frac{81.59}{96.23}\right)^2 \times 0.5 + \left(\frac{78.64}{96.23}\right)^2 \\
&\quad \times 0.2 + \left(\frac{45.72}{96.23}\right)^2 \times 1.437 + \left(\frac{6.78}{96.23}\right)^2 \times 1.77 \\
&= 0.2 + 0.244 + 0.359 + 0.134 + 0.324 + 0.009 \\
&= 1.27\ (\Omega)
\end{aligned}$$

The result is exactly the same as that calculated by Formula (9.5).

9.2.2 Calculation of Electric Energy Losses of Distribution Transformers

It is widely known that the electric energy losses of a single distribution transformer can be calculated by the following formula:

$$\Delta A_{\text{T0}} = \Delta A_{\text{T0.0}} + \Delta A_{\text{T0.L}} = \left[\Delta P_0 + \Delta P_{\text{LN}}\left(\frac{I_{\text{rms}}}{I_{\text{N}}}\right)^2\right] T \times 10^{-3} \tag{9.9}$$

Wherein ΔP_0, ΔP_{LN} – no-load active power losses and rated load power losses of the distribution transformer (W);

I_{rms} – rms current of the distribution transformer within the measuring period (A);

I_{N} – rated current of the distribution transformer (A).

If a multi-branch line is connected with several distribution transformers, then the electric energy losses of all distribution transformers can be calculated as per the following formula:

$$\begin{aligned}
\Delta A_{\text{T}} &= \Delta A_{\text{T0}} + \Delta A_{\text{TL}} \\
&= \left[\Sigma \Delta P_{0.n} + \Sigma\left(\Delta P_{\text{LN}.n} \frac{I^2_{\text{rms}.n}}{I^2_{\text{N}.n}}\right)\right] T \times 10^{-3}
\end{aligned} \tag{9.10}$$

Wherein ΔA_{To}, ΔA_{TL} – no-load and load electric energy losses of common distribution transformers.

9.2.3 Equivalent Resistance and Equal Resistance of Common Distribution Transformers

Similar to the definition of the equivalent resistance of the line, the equivalent resistance of all common distribution transformer windings in the line can be defined as

$$R_{\text{eq.T}} = \frac{\Delta A_{\text{TL}} \times 10^3}{3(\Sigma I_{\text{rms}.n})^2 T} \tag{9.11}$$

Substitute ΔA_{TL} in Formula (9.10) into Formula (9.11) to obtain

$$R_{eq.T} = \frac{\Sigma \Delta P_{L.n}}{3(\Sigma I_{rms.n})^2} \tag{9.12}$$

$$\Delta P_{L.n} = \Delta P_{LN.n} \frac{I_{rms.n}^2}{I_{N.n}^2}$$

Wherein $\Delta P_{L.n}$ – actual load power losses of each common distribution transformer (W).

To measure the utilization of the rated capacities of all common distribution transformers, an equivalent distribution transformer with constant total capacities and total rated load losses can be used to replace all common distribution transformers, and the resistance of this equivalent distribution transformer is called the equal resistance of all common distribution transformers, that is

$$R_{eq} = \frac{\Sigma \Delta P_{LN.n}}{3(\Sigma I_{N.n})^2} \tag{9.13}$$

Allow the total utilization ratio of common distribution transformers to be $\alpha = \frac{\Sigma I_{rms.n}}{\Sigma I_{N.n}}$, and Formulas (9.12) and (9.13) can be used to obtain

$$\alpha = \sqrt{\frac{\Sigma \Delta P_{L.n}}{\Sigma \Delta P_{LN.n}} \frac{R_{eq}}{R_{eq.T}}} \tag{9.14}$$

Example 9.3 As stated in Example 9.1, the rated capacity, no-load losses and rated load losses of each common distribution transformer are given, as shown in columns 1, 3 and 4 of Table 9.2.

The load factor of various common distribution transformers are marked in Figure 9.2. *Try to calculate:*

1. *Monthly electric energy losses of all common distribution transformers.*
2. *Equivalent resistance, equal resistance and total utilization ratio of common distribution transformers.*
3. *Total monthly electric energy losses and line loss rate of the line.*

Solution

1. List the average current of each common distribution transformer as calculated in Example 9.1 in column 6 of Table 9.2.

According to Example 9.1, the loads rates of two types of common distribution transformers are $f_1 = 0.32, f_2 = 0.26$, with corresponding form coefficients of load curves of $K_1 = 1.194$, $K_2 = 1.253$ which are listed in column 7 of Table 9.2. Due to $I_{rms.n} = K_i I_{av.n}$, calculate $I_{rms.n}$ in column 8. According to $\Delta P_{L.n} = \Delta P_{LN.n} \left(\frac{I_{rms.n}}{I_{N.n}}\right)^2$, calculate the load losses of various distribution transformers, which are listed in column 9 of Table 9.2.

The monthly electric energy losses of all common distribution transformers are

$$\begin{aligned}\Delta A_T &= (\Sigma \Delta P_{0.n} + \Sigma \Delta P_{L.n}) T \times 10^{-3} \\ &= (8880 + 8211) \times 720 \times 10^{-3} \\ &= 12\,305.5(\text{kW·h})\end{aligned}$$

Wherein, no-load electric energy losses are $\Delta A_{T0} = 6393.6$ kW·h, and load electric energy losses are $\Delta A_{TL} = 5911.9$ kW·h. The ratio between them is $\Delta A_{T0}/\Delta A_{TL} = 6393.6/5911.9 = 1.082$.

Table 9.2 Calculation table of electric energy losses of common distribution transformers.

Capacity of distribution transformer (kVA)	Quantity	No-load losses ΔP_0 (W)	Load losses ΔP_{LN} (W)	Rated current at high-voltage side (A)	Average current for calculation as assigned by capacity $I_{av.n}$ (A)	Form coefficient K_n	rms current for calculation $I_{rms.n}$ (A)	Load losses of each distribution transformer $\Delta P_{L.n}$ (W)	Remarks
1	2	3	4	5	6	7	8	9	10
320	2	1900	6200	18.5	8.26	1.194	9.86	1761	
180	1	1200	4000	10.4	4.64	1.194	5.54	1135	
100	2	730	2400	5.8	2.58	1.194	3.08	677	
50	4	440	1325	2.88	1.29	1.253	1.62	419	
30	2	330	850	1.73	0.77	1.253	0.96	262	
Total	11	8880	28 200	73.98	33.02	1.206	39.82	8211	$K_{dx}=\dfrac{39.82}{33.02}=1.206$

2. According to Formula (9.12), $R_{eq}.T=\dfrac{8211}{3\times 39.82^2}=1.73\,(\Omega)$; according to Formula (9.13), $R_{eq}=\dfrac{28\,200}{3\times 73.98^2}=1.72\,(\Omega)$; according to Formula (9.14), $\alpha=\sqrt{\dfrac{8211}{28\,200}\times\dfrac{1.72}{1.73}}=0.538$.
3. As $\Delta A_1=25\,411$ kW·h is calculated in Example 9.1, and $\Delta A_b=12\,305.5$ kW·h is calculated in this example, the electric energy losses and line loss rate of the whole line are

$$\Delta A=\Delta A_1+\Delta A_b$$
$$=25\,411+12\,305.5=37\,716.5\,(\text{kW·h})$$
$$\Delta A\%=\frac{37\,716.5}{94.6\times 10^4}\times 100\%=3.99\%$$
$$\Delta A_0\%=\frac{6393.6}{94.6\times 10^4}\times 100\%=0.68\%$$

9.2.4 Calculation of Electric Energy Losses by Equivalent Resistance Method

If the electric quantity and power factor at the start end of the line change, the line losses of the whole line also change. If the range of change in the electric quantity is close at all load points on the whole line, then it can be assumed that the equivalent resistance of the line and the total equivalent resistance of common distribution transformers remain unchanged, so the electric energy losses of the whole line can be calculated as per the following formula:

$$\Delta A=\Delta A_{T0}+\Delta A_1+\Delta A_{TL}=\left[\Sigma\Delta P_{0.n}+3I_{rms.0}^2R_{eq.1}+3(\Sigma I_{rms.n})^2R_{eq.T}\right]T\times 10^{-3} \quad (9.15)$$

Wherein $\sum I_{rms.n}$ – sum of rms current of n common distribution transformers (A).

Example 9.4 The line connection and the number and capacities of common distribution transformers in Example 9.1 remain unchanged; the monthly electric supply at the start end is increased by 20% and the average power factor at the start end remains unchanged; the electric supply of high-voltage customer I is increased by 13% and the electric supply of high-voltage customer II is increased by 16%, but their average power factors remain the same. *Calculate the monthly line loss rate of the line and its range of change.*

Solution

Symbols with ['] on the top right corner indicate various calculation items under new conditions.

1. The form coefficient and power factor at the start end of the line remain unchanged, so the percentage of increase in its average current and rms current is the same as the percentage of increase in the electric supply, so

$$I'_{rms.0} = 1.2I_{rms.0} = 1.2\times 96.23 = 115.48\ (A)$$

$$I'_{av.0} = 1.2\times 90.02 = 108.02\ (A)$$

For high-voltage customer I, $I'_{av.I} = 1.13 \times 25 = 28.25$ (A); for high-voltage customer II, $I'_{av.II} = 1.16 \times 32 = 37.12$ (A).

2. The sum of average current of common distribution transformers can be calculated as per the following formula:

$$\Sigma I'_{av.Tn} = 108.02 - (28.25 + 37.12) = 42.65(A)$$

Suppose the load rates of common distribution transformers remain the same, so the form coefficient of the equivalent distribution transformer also remains the same, that is $K'_{eq} = K_{eq} = 1.206$ (see Table 9.2 for the calculation table of K_{eq}). Then, the sum of rms current of common distribution transformers can be calculated, that is

$$\Sigma I'_{rms.Tn} = K'_{eq}\Sigma I'_{av.Tn}$$
$$= 1.206 \times 42.65 = 51.44(A)$$

3. As $R_{eq.l} = 1.27\ \Omega$ and $R_{eq.T} = 1.73\ \Omega$ are calculated, the monthly electricity line losses and line loss rate after the electric supply is increased can be obtained as per Formula (9.15):

$$\Delta A' = \left(8880 + 3\times 115.48^2\times 1.27 + 3\times 51.44^2\times 1.73\right)\times 720\times 10^{-3}$$
$$= 52\ 864(kW\cdot h)$$

$$\Delta A'\% = \frac{52\ 864}{1.2\left(94.6\times 10^4\right)}\times 100\% = 4.66\%$$

Wherein

$$\Delta A'\% = \frac{8.88\times 720}{1.2\left(94.6\times 10^4\right)} = 0.56\%$$

So

$$\Delta A'_L\% = 4.66\% - 0.56\% = 3.90\%$$

According to Example 9.3, $\Delta A\% = 3.99\%$, so $\Delta A_f\% = 3.99\% - 0.68\% = 3.31\%$. Only from the perspective of the line loss rate of loads, $\frac{\Delta A'_L\%}{\Delta A_L\%} = \frac{3.90\%}{3.31\%} = 1.18$, indicating that the range of increase in the line loss rate of loads is smaller than the range of increase in the electric supply at the start end of the line. The relationship between the change in the line loss rate and the change in the electric supply was discussed in Chapter 7.

9.3 Double Component Balance Method

Section 9.1 mentioned two basic assumptions of the point by point section simplification method; and one of the assumptions is that the power factor at each load point is approximately equal to the power factor at the start end. Based on this assumption, the vector addition of current becomes an algebraic addition and thus the calculation can be simplified. In the practice of line loss management, to make the calculation of line losses more accurate, the balance method of active and reactive components of current can be used to calculate line losses. This method leads to a lower calculation workload and produces calculation results that are very beneficial to the analysis of line losses.

The double component method only needs the following three kinds of information: (i) load curves of active power and reactive power of high-voltage customers on a representative day, which are used for calculating f_a and f_r, (ii) no-load current of common distribution transformers, and (iii) reactive load factor of various types of common distribution transformers, which are obtained through typical surveys.

The double component balance method follows the steps below to calculate line losses:

1. Calculate the active components and reactive components of the monthly average current according to the monthly active electricity and reactive electricity as well as the average operating voltages at the start end of the line and high-voltage customers.
2. Calculate the distribution of active components of the monthly average current of the whole line according to the point by point section simplification method in Section 9.1.
3. Calculate the no-load exciting current $I_{0.n}$ of the common distribution transformers, and then calculate the increased monthly average reactive current (A/kVA) of each distribution transformer by capacity, that is

$$i_{sv.Q}=\frac{I_{av.Q.0}-\Sigma I_{av.Q.m}-\Sigma I_{0.n}}{\Sigma W_n}$$

Wherein $I_{av.Q.0}$ – reactive component of monthly average current at start end of the line (A);
$\sum I_{av.Q.m}$ – sum of reactive components of monthly average current of high-voltage customers (A);
$\sum I_{0.n}$ – sum of no-load exciting current of common distribution transformers (A), wherein the influence of active component in no-load current is ignored.

The reactive component of monthly average current of each common distribution transformer can be calculated as per the following formula:

$$I_{av.Q.n}=I_{0.n}+i_{av.Q}W_n$$

With the calculated reactive component of monthly average current of each load point, the distribution of reactive components of monthly average current of the whole line can be obtained.

4. Use the weighted average method to calculate the load factor of each section according to the active load factor and reactive load factor of each load point, and then calculate the form coefficients of the active and reactive components of each section, thereby calculating the active and reactive components of the monthly rms current of each section and finally calculating the power losses and electric energy losses of each section.
5. Calculate the active component $I_{rms.P.n}$ and the reactive component $I_{rms.Q.n}$ of the monthly rms current of a common distribution transformer according to the active and reactive components of its monthly average current and its form coefficients $K_{P.n}$ and $K_{Q.n}$, and calculate the load losses of each common distribution transformer as per the following formula:

$$\Delta P_{L.n}=\Delta P_{LN.n}\left(\frac{I^2_{rms.P.n}+I^2_{rms.Q.n}}{I^2_{N.n}}\right) \tag{9.16}$$

Example 9.5 Assume that the line connection and the number and capacities of common distribution transformers in Example 9.1 remain unchanged, and that the monthly electric supplies and monthly average power factors at the start end of the line and high-voltage customers also remain unchanged. The calculation according to records of high-voltage customers on the representative day produces $f_{P.I}=0.553, f_{Q.I}=0.62, f_{P.II}=0.89, f_{Q.II}=0.92$. See Table 9.3 for the no-load current of common distribution transformers.

Conduct typical surveys by active and reactive watt-hour meters to obtain load curve parameters of two types of common distribution transformers: $f_{P1}=0.24, f_{Q1}=0.30, f_{P2}=0.30, f_{Q2}=0.35$. *Try to use the double component balance method to calculate monthly electric energy losses of common distribution transformers and the line via tables.*

Table 9.3 Calculation table of electric energy losses of common distribution transformers.

Capacity of distribution transformer S_n (kVA)	Quantity	No-load losses ΔP_0 (W)	Load losses ΔP_{LN} (W)	Rated current I_n (A)	Distribution of reactive components of monthly average current: I_0 (A)	$i_{av.Q}S_n$ (A)	$I_{av.Q}$ (A)	Active component of monthly average current $I_{av.P}$ (A)	Form coefficient: K_P	K_Q	rms current: $I_{rms.P}$ (A)	$I_{rms.Q}$ (A)	Load losses $\Delta P_{L.n}$ (W)
1	2	3	4	5	6	7	8 = 6+7	9	10	11	12 = 9 × 10	13 = 8 × 11	14
320	2	1900	6200	18.5	1.30	4.23	5.53	6.34	1.211	1.171	7.68	6.48	1829
180	1	1200	4000	10.4	0.73	2.38	3.11	3.56	1.211	1.171	4.31	3.64	1177
100	2	730	2400	5.8	0.43	1.32	1.75	1.98	1.211	1.171	2.40	2.05	711
50	4	440	1325	2.88	0.23	0.66	0.89	0.99	1.278	1.211	1.265	1.078	441
30	2	330	850	1.73	0.16	0.40	0.56	0.59	1.278	1.211	0.75	0.678	290
Total[a]	11	8880											8601

[a] The factor of quantity has been considered for the numbers in the "Total" row.

Solution

1. Calculate the monthly average current at the start end of the line and high-voltage customers. The *monthly* average current at the start end of the line is

$$I_{av.P.0} = \frac{A_P}{\sqrt{3}U_{av}T} = \frac{94.6\times10^4\times10^3}{\sqrt{3}\times10^4\times720} = 75.86\,(A)$$

$$I_{av.Q.0} = \frac{A_Q}{\sqrt{3}U_{av}T} = \frac{60.46\times10^7}{\sqrt{3}\times10^4\times720} = 48.48\,(A)$$

High-voltage customer I

$$I_{av.P.I} = I_{av.I}\cos\varphi_I = 25\times0.92 = 23(A)$$

$$I_{av.Q.I} = I_{av.I}\sin\varphi_I = 25\times0.392 = 9.80(A)$$

High-voltage customer II

$$I_{av.P.II} = 32\times0.86 = 27.52\ (A)$$

$$I_{av.Q.II} = 32\times0.51 = 16.33\ (A)$$

Calculate form coefficients of two high-voltage customers according to the given load factor:

$$K_{PI} = \sqrt{0.2/0.553+0.8} = 1.078$$

$$K_{QI} = \sqrt{0.2/0.62+0.8} = 1.060$$

$$K_{PII} = \sqrt{0.2/0.89+0.8} = 1.012$$

$$K_{QII} = \sqrt{0.2/0.92+0.8} = 1.009$$

Likewise, calculate form coefficients of two types of distribution transformers:

$$K_{P1} = 1.278,\quad K_{Q1} = 1.211$$

$$K_{P2} = 1.211,\quad K_{Q2} = 1.171$$

2. Current of common distribution transformers is distributed by capacity for calculating their load electric energy losses. The active component and reactive component of monthly average current of a common distribution transformer per unit capacity are respectively calculated:

$$i_{av.P} = \frac{75.86-(23+27.52)}{1280} = 0.019\,80(A/kVA)$$

$$i_{av.Q} = \frac{48.48-5.43-(9.80+16.33)}{1280} = 0.013\,22(A/kVA)$$

Wherein $\sum I_{0.n} = 5.43$ A is the sum of no-load current and calculated by column 6 of Table 9.3. Use $i_{av.P}$, $I_{0.n}$ and $i_{av.Q}$ to calculate the active component and reactive component of monthly average current of each distribution transformer, as shown in columns 9 and 7 of Table 9.3.

Use the calculated K as well as $I_{av.P}$ and $I_{av.Q}$ to calculate the active component and reactive component of rms current, as shown in columns 12 and 13 of Table 9.3. Then, use Formula (9.16) to calculate load losses ΔP_{Ln} of each distribution transformer, as shown in column 14 of Table 9.3.

3. Calculate the distribution of active components and reactive components of average current and rms current of the line, for the purpose of electric energy losses of each section. Mark active components and

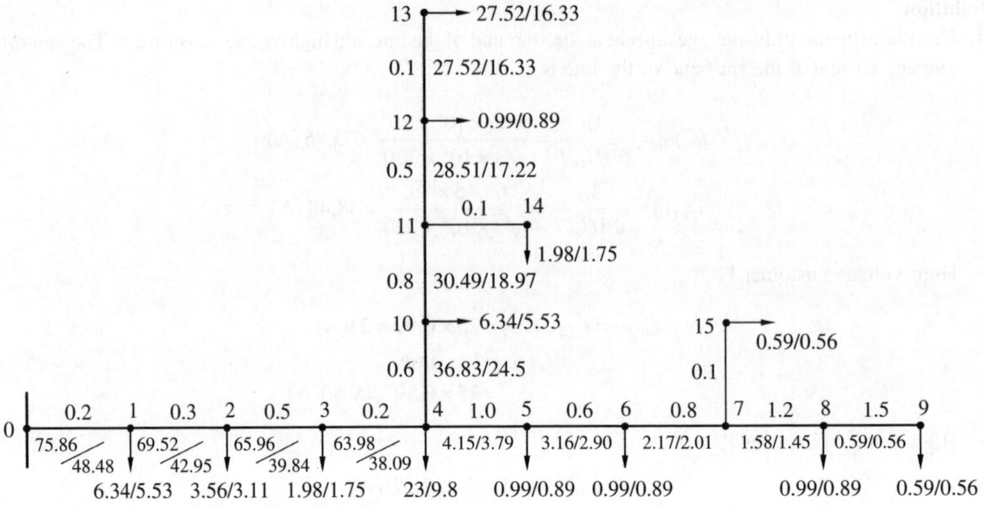

Figure 9.5 Single line diagram for calculating electric energy losses of high-voltage distribution line by double component balance method. Fractions under single line (right) and load point: the numerator is the active component of the monthly average current, and the denominator is the reactive component of the monthly average current.

reactive components of monthly average current of high-voltage customers I and II and all common distribution transformers in the single line diagram of Figure 9.5, and use the method of accumulation to obtain the distribution of active components and reactive components of monthly average current, as listed in columns 3 and 4 of Table 9.4.

Fill the load rates and form coefficients of Sections 4–9, Sections 11–14, and Sections 12, 13 in columns 5–8 of Table 9.4, and the weighted average load rates of other sections are calculated. The calculation processes are listed in columns 13 and 14 of Table 9.4. Use the weighted average load rate to calculate form coefficients K_{Pi} and K_{Qi}, and use K_{Pi}, K_{Qi} and $I_{av.Pi}$, $I_{av.Qi}$ to calculate the active component and reactive component of rms current of each section, thereby finally calculating power losses ($I^2_{rms.i}R_i$) of each section as listed in column 11 of Table 9.4. Column 12 of Table 9.4 lists losses caused by the reactive component of monthly rms current for analysis.

4. Calculate electric energy losses of the whole line. Use the no-load losses and load losses of common distribution transformers in Table 9.3 as well as load losses of each section in Table 9.4 to calculate the monthly electric energy losses of the whole line:

$$\begin{aligned}\Delta A &= \Delta A_1 + \Delta A_T\\ &= \left[3\Sigma I^2_{rms.i}R_i + (\Sigma\Delta P_{0n} + \Sigma\Delta P_{Ln})\right]T\times 10^{-3}\\ &= [3\times 11\,740.7 + (8880 + 8601)]\times 720\times 10^{-3}\\ &= 25\,360 + 12\,586\\ &= 37\,946(\text{kW·h})\end{aligned}$$

The line loss rate is

$$\Delta A\% = \frac{37\,946}{94.6\times 10^4}\times 100\% = 4.01\%$$

Compared with ΔA_1 obtained in Example 9.3, the load losses of the line are slightly larger as calculated by the point by point section simplification method and under the assumption of same power factors at all load points.

Table 9.4 Table of point by point section calculation of line losses by double component balance method. The columns are numbered 1–14, for ease of reference in the text.

Section numbers	Section resistance R_i (Ω)	Average current (A)		Weighted average load rate		Form coefficient		rms current (A)		Power losses (W)		Calculation of weighted average load rate	
		$I_{av.Pi}$	$I_{av.Qi}$	$f_{we.Pi}$	$f_{we.Qi}$	K_{Pi}	k_{Qi}	$I_{rms.Pi}$	$I_{rms.Qi}$	$I^2_{rms.i}R_i$	$I^2_{rms.Qi}R_i$	$f_{we.Pi}$	$f_{we.Qi}$
1	2	3	4	5	6	7	8	9	10	11	12	13	14
0–1	0.2	75.86	48.48	0.587	0.592	1.068	1.067	81.02	51 73	1848.0	535.2	(69.52 × 0.613 + 6.34 × 0.30) / (69.52 + 6.34)	(42.95 × 0.623 + 5.53 · 0.35) / (42.95 + 5.53)
1–2	0.3	69.52	42.95	0.613	0.623	1.061	1.059	73.76	45.48	2252.7	620.5	(65.96 × 0.63 + 3.56 × 0.30) / (65.96 + 3.56)	(39.84 × 0.644 + 3.11 × 0.35) / (39.84 + 3.11)
2–3	0.5	65.96	39.84	0.630	0.644	1.057	1.054	69.72	41.99	312.0	881.6	(63.98 × 0.64 + 1.98 × 0.30) / (63.98 + 7.98)	(38.09 × 0.658 + 1.75 × 0.35) / (38.09 + 1.75)
3–4	0.2	63.98	38.09	0.640	0.658	1.055	1.051	67.50	40.03	1231 8	320.5	(4.15 × 0.24 + 36.83 × 0.739 + 23 × 0.553) / (4.15 + 36.83 + 23)	(3.79 × 0.3 + 24.5 × 0.728 + 9.8 × 0.62) / (3.79 + 24.5 + 9.8)
4–5	1.0	4.15	3.79	0.24	0.30	1.278	1.211	5.30	4.59	49.2	21.1		
5–6	0.6	3.16	2.90	0.24	0.30	1.278	1.211	4.04	3.51	17.2	7.4		
6–7	0.8	2.17	2.01	0.24	0.30	1.278	1.211	2.77	2.43	10.8	4.7		
7–8	1.2	1.58	1.45	0.24	0.30	1.278	1.211	2.02	1.76	8.6	3.7		
8–9	1.5	0.59	0.56	0.24	0.30	1.278	1.211	0.75	0.68	1.5	0.7		
7–15	0.1	0.59	0.56	0.24	0.30	1.278	1.211	0.75	0.68	0.15	0.05		
4–10	0.6	36.83	24.5	0.739	0.728	1.035	1.037	38.12	25.41	1259.3	387.4	(30.49 × 0.83 + 6.34 × 0.30) / (30.49 + 6.34)	(18.97 × 0.838 + 5.53 × 0.35) / (18.97 + 5.53)
10–11	0.8	30.49	18.97	0.830	0.838	1.020	1.019	31.10	19.33	1072.7	298.9	(28.51 – 0.867 + 1.98 × 0.30) / (28.51 + 1.98)	(17.22 × 0.888+ 1.75 × 0.35) / (17.22 + 1.75)
11–12	0.5	28.51	17.22	0.867	0.888	1.015	1.013	28.94	17.44	570.9	152.1	(27.52 × 0.89 + 0.99 × 0.24) / (27.52 + 0.99)	(16.33 × 0 92 + 0.89 × 0.30) / (16.33 + 0.89)
12–13	0.1	27.52	16.33	0.890	0.920	1.012	1.009	27.85	16.48	104.8	27.2		
11–14	0.1	1.98	1.75	0.30	0.35	1.211	1.171	2.40	2.05	1.0	0.4		
Total	$\sum I^2_{rms.i}R_i = 11\,740.7$ (W), $\sum I^2_{rms.Qi}R_i = 3261.5$(W), $\sum I^2_{rms.Pi}R_i = 8479.2$(W)												

5. According to column 12 of Table 9.4, losses caused by the reactive component of monthly rms current are most significant for Sections 0–4 and Sections 4–12. If capacitors are considered for dispersion compensation, they can be installed near the tail end of Section 4-13 and at Node 4. According to the last row "Total" of Table 9.4, losses caused by reactive components of rms current approximately account for 30% of total losses. This indicates that reactive dispersion compensation may obtain the largest loss reduction effect, reducing the line loss rate by 30%. As a result, the double component balance method can produce new information for selecting loss reduction measures.

9.4 Dispersion Coefficient Method

When the distribution of loads along a multi-branch line follows a certain rule, the electric energy losses of the whole line can be obtained by mathematical analysis. If the distribution of loads in an actual line is close to a typical distribution, then the electric energy losses of such a typical distribution can be regarded as an approximate value of actual losses of this line. This is the essence of the dispersion coefficient method.

9.4.1 Calculation of Power Losses of Typically Distributed Loads

Power losses of several types of typically distributed loads are analyzed first.

1. Loads are concentrated in the tail end of the line. In this case, the total power losses of three phases are $\Delta P = 3I^2R$.
2. Loads are distributed uniformly. As shown in Figure 9.6, three-phase loads are distributed uniformly along the line. If the current at the start end of the line is I, the resistance per unit length of the line being r_0, and the length of the line being L, then the current at each branch point is $i_x = I/L$, and the sum of branch current at the position which is x away from the start end is $I_x = (L-x)I/L$, $\mathrm{d}(\Delta P) = I^2{}_x\mathrm{d}R = I^2{}_x r_0\mathrm{d}x$. Power losses of the whole line can be calculated as per the following formula:

$$\Delta P = 3\int_0^L I_\chi^2 r_0 \mathrm{d}\chi = 3I^2 r_0 \int_0^L \left(1-\frac{\chi}{L}\right)^2 \mathrm{d}\chi$$
$$= I^2 r_0 L = I^2 R$$

3. Loads are distributed in a linear decreasing manner. As shown in Figure 9.7, three-phase loads are distributed along the line in a linear decreasing manner.

   If the current at the start end of the line is I and the branch current at the start end is i_0, then the branch current at the position which is x away from the start end is $i_x = i_0(1-x/L)$. As the sum of branch current is I, the following formula can be obtained:

$$I = \int_0^L i_\chi \mathrm{d}\chi = \int_0^L i_0(1-\chi/L)\mathrm{d}\chi = \frac{1}{2} i_0 L$$

So

$$i_0 = 2I/L,\ i_\chi = \frac{2I}{L}\left(1-\frac{\chi}{L}\right)$$

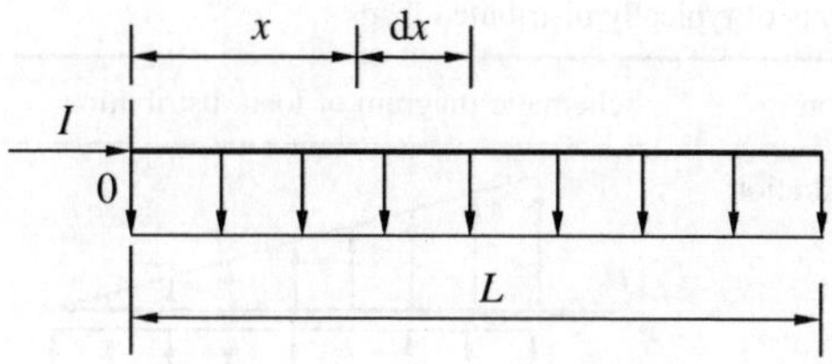

Figure 9.6 Uniform distribution.

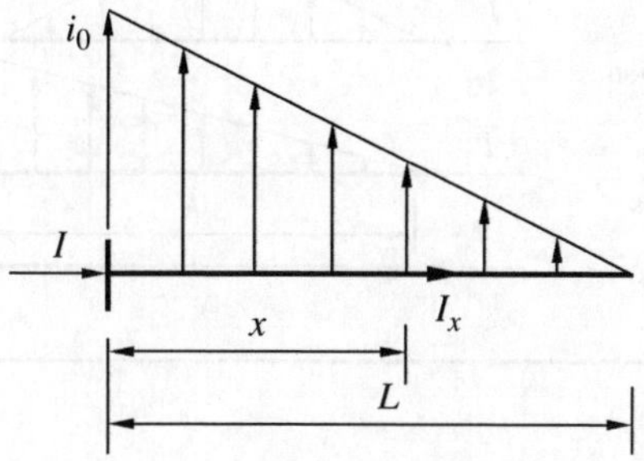

Figure 9.7 Linear decreasing distribution.

The sum of branch current I_x at the position which is x away from the start end can be calculated as per the following formula:

$$I_\chi = I - \int_0^\chi i_\chi \mathrm{d}\chi = I - \int_0^\chi \frac{2I}{L}\left(1-\frac{\chi}{L}\right)\mathrm{d}\chi = I\left(1-\frac{\chi}{L}\right)^2$$

The power losses of the whole line are

$$\Delta P = 3\int_0^L I_\chi^2 r_0 \mathrm{d}\chi = 3\int_0^L I^2\left(1-\frac{\chi}{L}\right)^4 r_0 \mathrm{d}\chi$$

$$= 3I^2 r_0 \times \frac{L}{5}\left(1-\frac{\chi}{L}\right)^5\Bigg|_L^0 = 0.60I^2R$$

4. Other typical distributions. For linear increasing distribution, ΔP = 1.60I2R; for loads heavily distributed in the middle, $\Delta P = 1.14I^2R$. See Table 9.5 for schematic diagrams of the above two distribution.

9.4.2 Dispersion Coefficient

Dispersion coefficient refers to the ratio G' between the power losses of a multi-branch line and the power losses when loads are concentrated at the tail end of the line (or the ratio G between the power losses of a multi-branch line and the power losses when loads are distributed uniformly). Dispersion coefficients G' and G are listed in Table 9.5.

Table 9.5 Dispersion coefficients of typically distributed loads.

No.	Type of load distribution	Schematic diagram of load distribution	ΔP	G'	G
1	Linear decreasing distribution	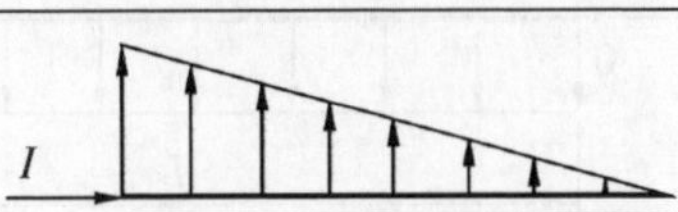	$0.60I^2R$	0.20	0.60
2	Uniform distribution	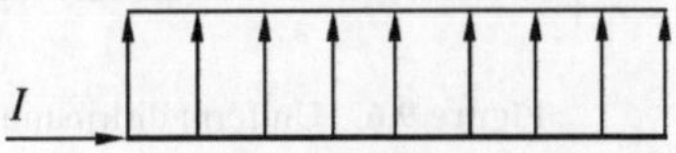	I^2R	0.333	1.0
3	Middle heavy distribution	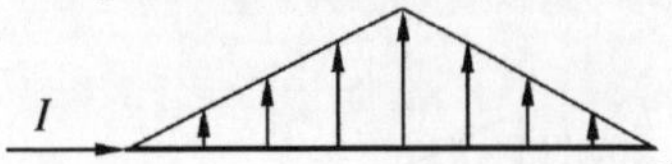	$1.14I^2R$	0.380	1.14
4	Linear increasing distribution	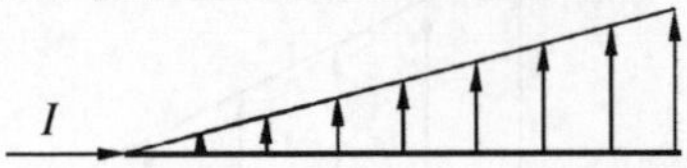	$1.60I^2R$	0.533	1.60
5	Tail end concentrated loads	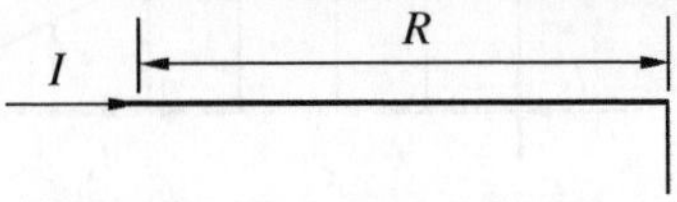	$3I^2R$	1.0	3.0

9.4.3 Conversion of Length Under Different Sectional Areas of Conductors

The above calculation of dispersion coefficients is on the premise that the resistance r_0 per unit length of the whole line remains unchanged, that is sectional areas of conductors are the same. However, sectional areas of conductors in each section of a multi-branch line are actually unlikely to be the same, so the concept of length conversion is put forward.

If a line consists of three sections whose sectional areas of conductors are respectively S_1, S_2, and S_3, and whose lengths are respectively L_1, L_2, and L_3. L_2 and L_3 are converted to the length equivalent to that of S_1, so

$$L_2' = L_2 \frac{S_1}{S_2}$$

$$L_3' = L_3 \frac{S_1}{S_2}$$

If L_k is the length conversion coefficient, then $L_k = \frac{L_\Sigma'}{L_\Sigma}$. If $\frac{S_1}{S_2} > 1$, $\frac{S_1}{S_2} > 1$, then $L_k = \frac{L_1 + L_2' + L_3'}{L_1 + L_2 + L_3} > 1.0$.

After the length conversion, power losses of a multi-branch line in which loads are distributed differently can be calculated as per the following formula:

$$\begin{aligned}\Delta P &= GI^2(r_0 L_k L_\Sigma) \\ &= GL_k I^2 r_0 L_\Sigma \\ &= GI^2R'\end{aligned} \tag{9.17}$$

Wherein R' – calculation resistance after length conversion.

9.4.4 Calculation of Power Losses of Complexly Distributed Loads

Any actual load distribution of a multi-branch line falls in any combinations of several typical distributions listed in Table 9.5. The combination of uniform distribution with tail end concentrated loads is to be analyzed below as an example, as shown in Figure 9.8.

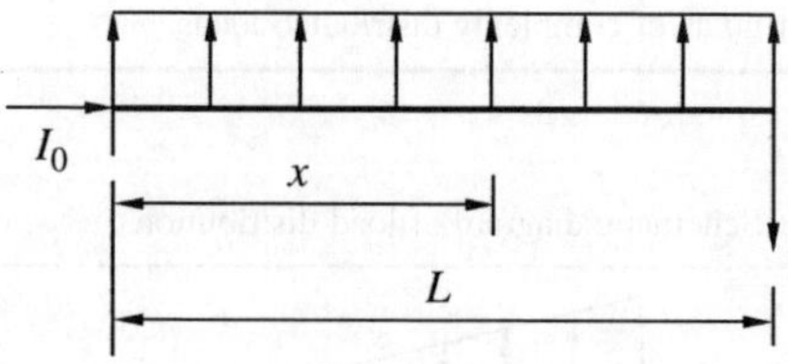

Figure 9.8 Uniform distribution with tail end concentrated loads.

If the ratio between the concentrated loads at the tail end of the line and the loads at the start end of the line is $a = I_1/I_0$, then the branch current at the position which is x away from the start end is $i_\chi = \frac{I_0 - I_1}{L} = I_0\frac{1-a}{L}$. The sum of branch current at the position which is x away from the start end is $I_\chi = (L-\chi)i_\chi + aI_0 = I_0\left[1-(1-a)\frac{\chi}{L}\right]$.

The power losses of the whole line can still be calculated as per $\Delta P = 3\int_0^L I_\chi^2 r_0 \mathrm{d}\chi$, that is

$$\begin{aligned}\Delta P &= 3\int_0^L I_0^2\left[1-(1-a)\frac{\chi}{L}\right]^2 r_0 \mathrm{d}\chi \\ &= 3I_0^2 r_0\left[L-(1-a)L+\frac{(1-a)^2}{3}L\right] \\ &= (1+a+a^2)I_0^2 r_0 L \\ &= GI_0^2 R \\ G &= 1+a+a^2\end{aligned}$$

Wherein G – dispersion coefficient of uniform distribution with tail end concentrated loads. If $a = 0$, then $G = 1$, indicating uniform distribution of loads along the line; if $a = 1$, then $G = 3$, indicating concentrated loads at the tail end.

The derivation of several other cases is ignored, and dispersion coefficient formulas are directly listed in Table 9.6.

According to Table 9.6, when a is large, tail end concentrated loads have a larger influence on power losses, and G values of various complex distributions approach 3; when a is small, distributed loads have a larger influence on power losses, and the dispersion coefficient is significantly increased, with the load distribution tending to be on the tail end.

9.4.5 Calculation of Electric Energy Losses by Dispersion Coefficient Method

The dispersion coefficient method is a simplified calculation method based on three assumptions: (i) shapes of load curves of differently distributed loads are the same, (ii) power factors of loads are the same, and (iii) changes in the voltages along the line are ignored. Because this method calculates power losses and electric energy losses by comparing different types of load distributions, the accuracy of calculation results is lower than that of the point by point section simplification method. The calculation steps of this method are as follows:

1. Divide a multi-branch line into several sections with different types of load distributions according to the distribution of average current within the measuring period, and calculate the dispersion coefficients of these sections.

Table 9.6 Dispersion coefficient formulas of complexly distributed loads.

No.	Type of load distribution	Schematic diagram of load distribution	Formula $G=\varphi(a)$	G_1 (a = 0.80)	G_2 (a = 0.4)
1	Decreasing distribution with tail end concentrated loads	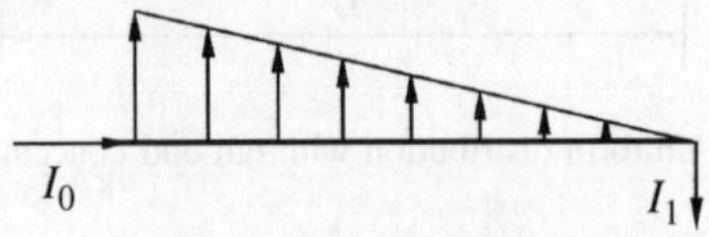	$\frac{(3+4a+8a^2)}{5}$	2.264	1.176
2	Uniform distribution with tail end concentrated loads	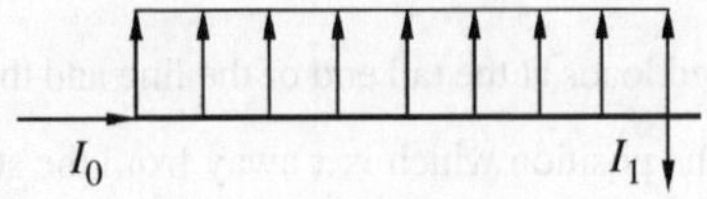	$(1+a+a^2)$	2.44	1.56
3	Middle heavy distribution with tail end concentrated loads	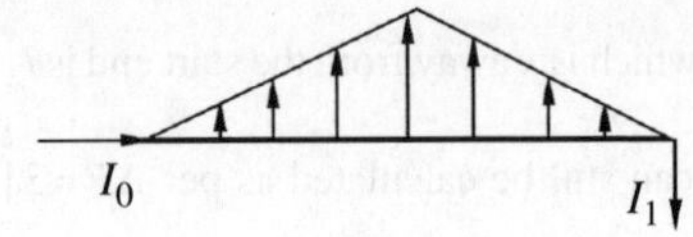	$\frac{(23+14a+23a^2)}{20}$	2.446	1.614
4	Increasing distribution with tail end concentrated loads	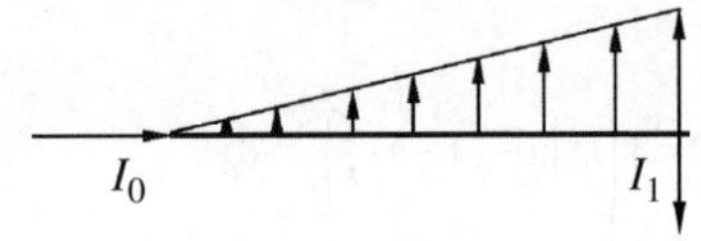	$\frac{(8+4a+3a^2)}{5}$	2.624	2.016

2. Calculate the rms current at the start end of each section by the form coefficient of the load curve at the start end of the section. Use the dispersion coefficient for each type of distribution to calculate electric energy losses of each section, thereby obtaining the electric energy losses of the whole line, that is

$$\Delta A=\Sigma G_i(K_0 I_{av.0.i})^2 R_i T\times 10^{-3} \tag{9.18}$$

Wherein G_i – dispersion coefficient of a section;
K_0 – form coefficient at the start end of the line;
$I_{av.0.i}$ – average current at the start end of a section (A);
R_i – resistance of a section (Ω).

Example 9.6 If the line connection and the conditions of high-voltage customers and common distribution transformers in Example 9.1 remain unchanged, *then try to use the dispersion coefficient method to calculate the monthly electric energy losses of the line.*

Solution

1. *Determine the type of load distribution of each section.*
   According to the distribution of monthly average current shown in Figure 9.3, Section 0-4 is the No. 1 section, belonging to the type of decreasing distribution with tail end concentrated loads; Section 4-9 is the No. 2 section, approximately belonging to the type of uniform distribution; Section 4-13 is the No. 3 section, belonging to the type of decreasing distribution with tail end concentrated loads. See Figure 9.9 for the schematic diagram of load distributions.
2. *Calculate dispersion coefficients.*
   According to Figure 9.3, $I_{av.0.1}$ = 90.02 A, $I_{av.0.2}$ = 5.41 A, $I_{av.0.3}$ = 44.13 A, $I_{av.I}$ = 25 A, so the ratio can be obtained as $a_1=\frac{5.41+44.13+25}{90.02}=0.828$. The dispersion coefficient of No. 1 section is

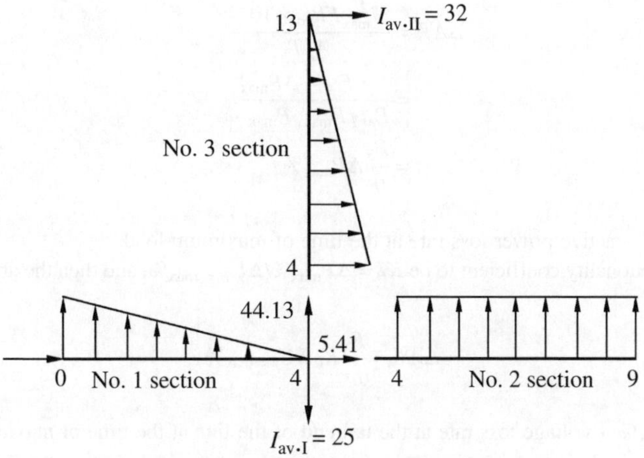

Figure 9.9 Schematic diagram of load distributions.

$G_1 = (3 + 4 \times 0.828 + 8 \times 0.828^2)/5 = 2.359$; No. 2 section has a uniform distribution with $G_2 = 1.0$; No. 3 section has concentrated loads at the tail end, $I_{av.II} = 32$ A, so $a_3 = 32/44.13 = 0.725$, obtaining $G_3 = (3 + 4 \times 0.725 + 8 \times 0.725^2)/5 = 2.021$.

3. *Calculate electric energy losses.*

According to Table 9.1 and Figure 9.3, $K_0 = 1.069$, $R_1 = 1.2\ \Omega$, $R_2 = 5.1\ \Omega$, $R_3 = 2\ \Omega$. According to Formula (9.18),

$$\begin{aligned}\Delta A &= \left[G_1(K_0 I_{av.0.1})^2 R_1 + G_2(K_0 I_{av.0.2})^2 R_2 + G_3(K_0 I_{av.0.3})^2 R_3\right] T \times 10^{-3} \\ &= \left[2.359 \times (1.069 \times 90.02)^2 \times 1.2 + 1.0 \times (1.069 \times 5.41)^2 \times 5.1 + 2.021 \times (1.069 \times 44.13)^2 \times 2\right] \times 720 \times 10^{-3} \\ &= (26\,214.6 + 170.6 + 8995.4) \times 0.72 \\ &= 25\,474(\text{kW·h})\end{aligned}$$

Compared with the calculation result in Example 9.1, the calculation error is

$$\begin{aligned}\delta(\Delta A)\% &= \frac{25\,474 - 25\,411}{25\,411} \times 100\% \\ &= 0.25\%\end{aligned}$$

According to the calculation in Example 9.6, as long as the type of load distribution is correctly determined, the accuracy of electric energy losses of a multi-branch line as calculated by the dispersion coefficient method can be satisfactory.

9.5 Calculation of Electric Energy Losses of Multi-branch Lines by Voltage Drop Method

9.5.1 Calculation of Line Loss Rate by Proportionality Coefficient Method

When the load curve shapes for all load points of a multi-branch line are the same, the electric energy losses of the line are calculated with the maximum current I_{max} at the start end of the line and equivalence resistance R_{eq}, that is $\Delta A = 3I^2_{max} F R_{eq} T \times 10^{-3}$, and the line loss rate is $\Delta A\% = \Delta A/A$, so

$$\Delta A\% = \frac{3I_{\max}^2 F R_{\text{eq}} \times 10^{-3}}{A/T} = \frac{F}{P_{\text{av}}/P_{\max}} \frac{\Delta P_{\max}}{P_{\max}} = \frac{F_S}{f_P} \Delta P_{\max}\%$$

Wherein $\Delta P_{\max}\%$ – active power loss rate at the time of maximum load.

Allow the proportionality coefficient to be $K_P = \Delta P_{\max}\%/\Delta U_{P=\max}\%$, and then the above formula can be rewritten to

$$\Delta A\% = \frac{F_S}{f_P} K_P \Delta U_{P=\max}\% \tag{9.19}$$

Wherein $\Delta U_{P=\max}\%$ – voltage loss rate at the tail end of the line at the time of maximum load.

When loads are concentrated at the tail end of the line, $\Delta P_{\max}\% = 3I^2_{\max}R/(\sqrt{3}I_{\max}U\cos\varphi)$, while $\Delta U_{P=\max}\% = \sqrt{3}(I_{\max}U\cos\varphi R + I_{\max}U\sin\varphi X)/U^2 = \sqrt{3}I_{\max}R\cos\varphi(1 + \tan\varphi\frac{X}{R})/U$, so the proportionality coefficient K_P can be obtained

$$K_P = \frac{\Delta P_{\max}\%}{\Delta U_{P=\max}\%} = \frac{\sqrt{3}I_{\max}R}{U\cos\varphi} \bullet \frac{U}{\sqrt{3}I_{\max}R\cos\varphi\left(1+\tan\varphi\frac{X}{R}\right)} = \frac{1}{\cos^2\varphi\left(1+\tan\varphi\frac{X}{R}\right)}$$

Wherein $\cos\varphi$, $\tan\varphi$ – power factor and corresponding tangent value at the time of maximum load;
X, R – reactance and resistance of the line (Ω).

The analysis of several typically distributed loads can lead to K_P formulas, as shown in Table 9.7.

If the five distributions of No. 2 to No. 6 in Table 9.7 account for the same percentage, then the average K_P of multi-branch lines within a region can be calculated as per $K_{P.\text{av}} = \sum K_P/n$.

Substitute $K_{P.\text{av}} \approx 0.75/[\cos^2\varphi(1+ \tan\varphi\frac{X}{R})]$ into Formula (9.19) to obtain

$$\Delta A\% = (F_S/f_P)\frac{0.75}{\cos^2\varphi(1+\tan\varphi X/R)}\Delta U_{P=\max}\%$$

Given the large quantity of loads in the multi-branch line, the load combined effect leads to extremely short maximum load duration. So allow $F_P = 0.2f_P + 0.8f^2{}_P$, and substitute this F into the above formula. Allow $F_S \approx F_P$ to obtain

$$\Delta A\% = \frac{(0.15+0.60f)\Delta U_{P=\max}\%}{\cos^2\varphi(1+\tan\varphi X/R)} \tag{9.20}$$

For a 10 kV distribution line, X/R generally ranges within 0.37–1.60 and changes with the conductor model and spacing. Use Formula (9.20) to estimate the line loss rate of the line based on the parameter $\cos\varphi$ at the time of maximum load and the measurement of $\Delta U_{P=\max}\%$.

For a 0.38/0.22 kV low-voltage line, X is so small that the influence of reactance can be ignored; for a pure lighting line, $f \approx 0.26$; for a low-voltage line for combined lighting and power, $f = 0.32$. Then, calculation formulas of the line loss rate can be respectively obtained:

$$\left.\begin{aligned} \Delta A\% &\approx \frac{0.30\Delta U_{P=\max}\%}{\cos^2\varphi} \quad \text{(lighting line)} \\ \Delta A\% &\approx \frac{0.35\Delta U_{P=\max}\%}{\cos^2\varphi} \quad \text{(combined load line)} \end{aligned}\right\} \tag{9.21}$$

Because the power factor of a low-voltage line at the time of maximum load is generally not large, this factor must be considered. When the power factor the low-voltage line is small, $\Delta A\%$ may be greater than $\Delta U_{P=\max}\%$.

9.5.2 *Calculation of Line Losses by Voltage Drop Measurements*

9.5.2.1 Calculation of Voltage Loss

It is known that the longitudinal component of voltage drop is calculated as per the following formula:

$$\Delta U = \frac{PR+QX}{U} = \sqrt{3}IR\cos\varphi\left(1+\tan\varphi\frac{X}{R}\right)$$

For a multi-branch line, when the power factors of all sections of the line are the same and the ratios of X/R are close, the calculation formula of voltage loss at the tail end of the line can be obtained, that is $\Sigma\Delta U_i = \sqrt{3}\cos\varphi\left(1+\tan\varphi\frac{\text{X}}{\text{F}}\right)\Sigma I_i R_i$, so

$$\Sigma I_i R_i = \frac{\Sigma\Delta U_i}{\sqrt{3}\cos\varphi(1+\tan\varphi X/R)} \tag{9.22}$$

9.5.2.2 Bias Ratio and Voltage Loss

The degree of deviation of the load center of distributed loads to the geometric center of the line is called the bias ratio, which can be calculated as per the following formula:

$$\delta = 1 - \frac{\Sigma I_{\text{rms}.i}R_i/I_{\text{rms}.0}}{0.5R} = 1 - \frac{R_{\text{bj}}}{0.5R} \tag{9.23}$$

Wherein R_{bj} – power supply radius calculated by the distribution of rms current (Ω);
R – actual resistance of main line (Ω).

Bias ratios δ of various typical load distributions as calculated by Formula (9.23) are listed in Table 9.7.

Substitute Formula (9.22) into Formula (9.23) to obtain the calculation formula of δ:

$$\delta = 1 - \frac{2\Sigma\Delta U_i}{\sqrt{3}\cos\varphi(1+\tan\varphi X/R)I_{\text{rms}.0}R} \tag{9.24}$$

Wherein $\sum\Delta U_i$ – voltage loss at the tail end of the line when the rms current at the start end of the line is $I_{\text{rms}.0}$ (V);
$\cos\varphi$, $\tan\varphi$ – power factor and corresponding tangent value when the load current at the start end of the line is $I_{\text{rms}.0}$.

Table 9.7 K_P formulas and eccentricity ratio δ of typically distributed loads.

No.	Type of load distribution	ΔP	$\sum I_i R_i$	K_P formula	G	δ
1	Start end concentrated loads	0	0	—	0	1
2	Linear decreasing distribution	$0.60I^2R$	$\frac{1}{3}IR$	$\frac{0.60}{\cos^2\varphi\left(1+\tan\varphi\frac{X}{R}\right)}$	0.60	0.333
3	Uniform distribution	I^2R	$\frac{1}{2}IR$	$\frac{0.667}{\cos^2\varphi\left(1+\tan\varphi\frac{X}{R}\right)}$	1.0	0
4	Middle heavy distribution	$1.14I^2R$	$\frac{1}{2}IR$	$\frac{0.766}{\cos^2\varphi\left(1+\tan\varphi\frac{X}{R}\right)}$	1.14	0
5	Linear increasing distribution	$1.60I^2R$	$\frac{2}{3}IR$	$\frac{0.80}{\cos^2\varphi\left(1+\tan\varphi\frac{X}{R}\right)}$	1.60	−0.333
6	Tail end concentrated loads	$3I^2R$	IR	$\frac{1.0}{\cos^2\varphi\left(1+\tan\varphi\frac{X}{R}\right)}$	3.0	−1

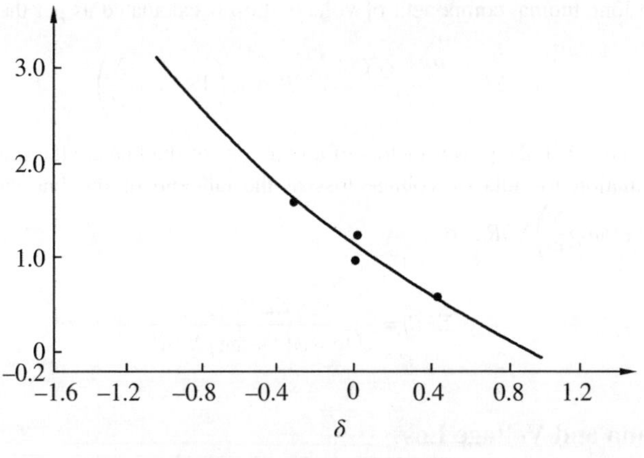

Figure 9.10 $G(\delta)$ fitting curve.

9.5.2.3 Relational Expression Between Dispersion Coefficient and Bias Ratio

Use the data fitting technique to obtain the following approximate formula based on the six (G, δ) data sets listed in Table 9.7:

$$\begin{aligned} G &= 1.061\,28 - 1.500\,15\delta + 0.437\,633\delta^2 \\ &\approx 1.06 - 1.50\delta + 0.438\delta^2 \end{aligned} \tag{9.25}$$

The $G(\delta)$ fitting curve is shown in Figure 9.10.

9.5.2.4 Calculation of Electric Energy Losses

According to the definition of dispersion coefficient and the selected criteria for comparison, the calculation formula of electric energy losses for a type of load distribution can be obtained:

$$\Delta A = G\Delta A_{jy} = GI_{rms.0}^2 RT \times 10^{-3} \tag{9.26}$$

Wherein ΔA_{jy} – electric energy losses for uniform distribution of loads.

Substitute Formula (9.24) into Formula (9.25), and then substitute the result into Formula (9.26) to obtain

$$\Delta A=\left[\frac{0.72\Sigma\Delta U_i I_{\text{rms.0}}}{\cos\varphi(1+\tan\varphi X/R)}+\frac{0.584(\Sigma\Delta U_i)^2/R}{\cos^2\varphi(1+\tan\varphi X/R)^2}\right]T\times10^{-3} \tag{9.27}$$

According to Formula (9.27), if the parameter *X/R* of the multi-branch line, the rms current $I_{\text{rms.0}}$ at the start end of the line and the power factor and tangent value at the corresponding time are given, the electric energy losses of the main line can be directly calculated according to actual measurements of voltage loss of the main line and via conversion by the current and the rms current at the start end of the line at the time of measuring voltage loss.

Example 9.7 In Example 9.5, Sections 0–4–10–13 are regarded as the main line, and the resistance is $R = 3.3\ \Omega$. According to Table 9.4, $I_{\text{rms.0}} = 96.13$ A, $\sum I_{\text{rms.}i}R_i = 178.56$ V, $\tan\varphi = 0.67$, $\cos\varphi = 0.83$ are calculated. The conductor model is LGJ50, with $X/R = 0.615$. *Try to use the formula of voltage drop method to calculate electric energy losses of the main line and compare the result with that in Example* 9.1.

Solution

$$\cos\varphi\left(1+\tan\varphi\frac{X}{R}\right)=0.83\times(1+0.67\times0.615)=1.172$$

According to Formula (9.22),

$$\begin{aligned}\Sigma\Delta U_i&=\sqrt{3}\cos\varphi\left(1+\tan\varphi\frac{X}{R}\right)\Sigma I_{\text{rms.}i}R_i\\&=\sqrt{3}\times1.172\times178.56\\&=326.46(\text{V})\end{aligned}$$

Substitute this into Formula (9.27) to obtain

$$\begin{aligned}\Delta A&=\left[0.72\times\frac{362.46\times96.13}{1.172}+0.584\times\frac{362.46^2/3.3}{1.172^2}\right]\times720\times10^{-3}\\&=27\,599\ (\text{kW}\cdot\text{h})\end{aligned}$$

The electric energy losses of the main line as calculated by the point by point section simplification method in Example 9.1 are

$$\begin{aligned}\Delta A'&=3\times11682\times720\times10^{-3}\\&=25\,233\ (\text{kW}\cdot\text{h})\end{aligned}$$

According to the calculation results of the above two methods, the result calculated by the voltage drop method is larger and the error is

$$\begin{aligned}\delta(\Delta A)\%&=\frac{\Delta A-\Delta A'}{\Delta A'}\times100\%\\&=\frac{27\,599-25\,233}{25\,233}\times100\%\\&=9.38\%\end{aligned}$$

Table 9.8 Comparison table of calculation methods of line losses of multi-branch lines.

No.	Name of method	Necessary assumptions	Calculation workload	Accuracy of calculation result	Occasion applicable	Example number
1	Dispersion coefficient method	Same shapes of load curves at all load points Same power factors at all load points Regardless of change in voltage along the line	Small	Not high; possible positive or negative error as compared with calculation result of method 5	Too many load points, and load distribution following an obvious rule	Example 9.6
2	Point by point section simplification method (double current distribution method)	Same power factors at all load points Regardless of change in voltage along the line	Large	High; larger calculation result in general	Not too many load points, and load distribution not following an obvious rule	Example 9.1
3	Equivalent resistance method	Same power factors at all load points Regardless of change in voltage along the line	Small	Low	Line connection remains unchanged, while electric quantity (current) at start end of the line changes	Example 9.4
4	Voltage drop method	Same power factors at all load points Same shapes of load curves at all load points	Small	Low; larger calculation result in general	Calculation of electric energy losses of high- and low-voltage distribution main lines with many load points	Example 9.7
5	Double component balance method	Regardless of change in voltage along the line (use the average voltage at the start end of the line for calculation)	Maximum	Highest	Necessary determination of technical measures of loss reduction or calculation of line compensation effect	Example 9.5

9.6 Comparison and Selection of Calculation Methods of Electric Energy Losses of Multi-branch Lines

This chapter introduces a total of five calculation methods for the electric energy losses of multi-branch lines, and they are listed in Table 9.8 for comparison.

The use of a general-purpose tester and data acquisition unit for distribution transformers facilitates the acquisition of operating parameters of multi-branch lines and distribution transformers, and in combination with the wide application of microcomputers, the double component balance method with large calculation workload but high accuracy will be applied more often; while the voltage drop method needs less data and can be applied in some cases.

9.7 Calculation of Loss Reduction Benefits after Connection of Distributed Resources to System

Referring to American and German standards [70, 71], the State Grid Corporation of China formulated its enterprise standard in 2010 [72]. Distributed resources in this Standard refer to small resources connected to power grids of 35 kV and below, including synchronous motors, induction motors, and current transformers.

This Standard [72] stipulates three quantification principles for the connection to the electric power system:

1. The total capacity of distributed resources should not exceed 25% of the maximum load in the power supply area served by the higher level transformer in principle; following this principle, electric power generated by distributed resources can be balanced within the current level of distribution area, therefore preventing any big influence of reverse flow on the higher level power grids.
2. The ratio between the short-circuit current at the point connecting distributed resources and the rated current of distributed resources should not be lower than 10; following this principle, the influence of power fluctuation of distributed resources on the voltage at the point connecting distributed resources can be reduced.
3. Distributed resources of 200 kW and below are connected to 380 V power grids; distributed resources of 200 kW and above are connected to 10(6) kV and above power grids.

In distributed resources, wind power generation, photovoltaic power generation, and biomass power generation are intermittent. In other words, within a complete measuring period (a day or a year), there is both a generation period and a consumption period as small loads. Therefore, to analyze the loss reduction benefits after distributed resources are connected to the system, line losses in both the generation period and the consumption period should be calculated.

9.7.1 Calculation of Loss Reduction Benefits During Generation Period of Distributed Resources

Distributed resources supply electricity to medium- and low-voltage customers nearby, eliminating the accumulation of line losses caused by step-down. Thus, before calculating the loss reduction benefits, it is important to know the line loss rates of power grids under different voltage classes.

From 500 kV high-voltage power grids to 380 V low-voltage customers, the longest step-down path is shown in Figure 9.11. In the figure, 0 is the generator bus; 1# to 6# are respectively 500 to 0.38 kV buses. If the line loss rates of power grids under different voltage classes are known, the line losses caused by step-down from a network under a certain voltage class to the consumption point can be calculated.

According to the results of line loss theoretical calculations of all regional and provincial enterprises, as summarized by the State Grid in 2005 [53], under large load conditions, the change in the line loss rate of each voltage class network varies, as do the average values, as shown in Table 9.9. Formula (3.15) can be used to calculate the expected value $\Delta A_e\%$ of the theoretical line loss rate of each network, as listed in the fourth row of the table.

To calculate the loss reduction benefits after distributed resources are connected to the system, it is necessary to compare the accumulated line loss rate from the power source point to the point connecting those distributed resources with the line loss rate of the network in which the connection point is located. Assume that the power source point is a large power plant supplying electricity to the 500 kV network, and the connection point of photovoltaic power generation is the 0.38 kV low-voltage system. Values A with different

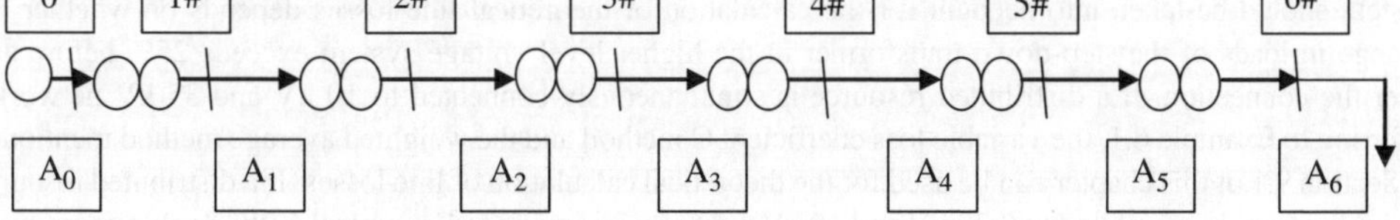

Figure 9.11 Schematic diagram of step-down network.

Table 9.9 Change in line loss rate of each voltage class network and expected line loss rate (Unit: %).

Voltage class	500	220	110	35	10	0.38
$\Delta A_b\%/\Delta A_a\%$	3.15/0.26	6.96/0.65	2.84/0.64	3.12/0.56	5.70/2.27	11.93/5.43
$\Delta A_m\%$	1.5	2.0	1.5	1.5	4.0	8.0
$\Delta A_e\%$	1.568	2.60	1.58	1.613	3.995	8.226

subscripts refer to the power supply (sales) quantities of each network. According to the power supply (sales) quantities and the basic relationship between line loss rates:

$$\begin{aligned} A_0 &= A_1/(1-\Delta A_{500}\%) = A_2/(1-\Delta A_{220}\%)/(1-\Delta A_{500}\%) \\ &= A_3/(1-\Delta A_{110}\%)/(1-\Delta A_{220}\%)/(1-\Delta A_{500}\%) \\ &= A_4/(1-\Delta A_{35}\%)/(1-\Delta A_{110}\%)/(1-\Delta A_{220}\%)/(1-\Delta A_{500}\%) \\ &= A_5/(1-\Delta A_{10}\%)/(1-\Delta A_{35}\%)/(1-\Delta A_{110}\%)/(1-\Delta A_{220}\%)/(1-\Delta A_{500}\%) \\ &= A_6/[(1-\Delta A_{0.38}\%)(1-\Delta A_{10}\%)(1-\Delta A_{35}\%)(1-\Delta A_{110}\%)(1-\Delta A_{220}\%)(1-\Delta A_{500}\%)] \end{aligned} \tag{9.28}$$

Substitute data in the fourth row of Table 9.9 to the above formula to obtain

$$\begin{aligned} A_0 &= A_6/0.81976 = 1.223A_6; \\ \Delta A_\Sigma\% &= (A_0 - A_6)/A_0 \\ &= 0.233/1.233 = 0.1890 = 18.90\%; \end{aligned}$$

According to the comparison between the above six-level continuous step-down power supply system and the 0.38 V network which provides electricity nearby, the loss reduction benefits from planned photovoltaic power generation can be obtained by multiplying the decrease value in the line loss rate by the on-grid energy of photovoltaic power generation A_g:

$$\begin{aligned} \partial(\Delta A_{0.38}) &= (\Delta A_\Sigma\% - \Delta A_{0.38}\%)A_g \\ &= (18.90\% - 8.23\%)A_g = 10.67\%A_g; \end{aligned}$$

The loss reduction benefits of distributed resources connected to the 10 kV or 35 kV network can be calculated in the similar way. If the connection point of a wind power plant is 35 kV, the original power supply mode has two types, 500–220–110–35 kV or 500–220–35 kV. Due to many step-down levels in the former case, the loss reduction benefits are higher provided that connection capacity is the same. For a specific distributed resource, the power supply mode of the network connecting the distributed resource and possible line loss rates of all voltage class networks in the planning period should be considered for calculating loss reduction benefits after the distributed resource is connected to the system. As the line loss rate varies with the electric supply, the line loss rate in the planning period should be calculated based on the expected electric supply. See Chapter 13 for details.

When multiple distributed resources are connected to two voltage systems, whether the higher level voltage system should be taken into account for the calculation of theoretical line losses depends on whether the change in loads of the step-down transformer in the higher level voltage system exceeds 25% before and after the connection. If a distributed resource is simultaneously connected to 10 kV and 35 kV networks, referring to Example 6.1, the variable loss coefficient C method and the weighted average method mentioned in Section 9.1 of this chapter can be used for the theoretical calculation of line losses. If a distributed resource is simultaneously connected to 0.38 kV and 10 kV networks, referring to Example 9.5, the double component balance method can be used for the theoretical calculation of line losses. During the above calculation, if the generated output and the loads in a nearby area reach a balance, then the new load or energy distribution should be determined before the theoretical calculation of line losses.

9.7.2 Calculation of Line Loss Change During Consumption Period of Distributed Resources

During the consumption period, distributed resources become power consumption loads, and methods mentioned in this chapter can be selected to calculate line losses. If a synchronous motor is in phasing operation during this period, the double component balance method can better show the effect of loss reduction by reactive nearby compensation. If the active load of a distributed resource is small and the original reactive load at the connection is large, the effect of loss reduction by reactive compensation is obvious. It may be possible that line losses of the voltage system at the connection point during the consumption period are decreased.

9.7.3 Calculation of Loss Reduction Benefits During the Full Period of Distributed Resources

When integrating line loss changes of the network connecting a distributed resource during the generation and consumption periods, the loss reduction benefits during the full period can be calculated, as shown in the formula below:

$$\begin{aligned}\partial\Delta A_{\Sigma} &= \partial\Delta A_g - \partial\Delta A_C \\ &= \left(\Delta A_{\Sigma}\% - \Delta A_g\%\right)A_g - \left[(\Delta A_2\% - \Delta A_1\%)A_0 + \Delta A_2\% A_C\right]\end{aligned} \tag{9.29}$$

Wherein, the former part refers to reduced electricity losses during the generation period. $\Delta A_1\%$ and $\Delta A_2\%$ in square brackets respectively refer to the line loss rates of the system at the connection point before and after the consumption electricity A_C of the distributed resource is added. A_0 is the power supply electricity of the original system.

9.7.4 Benefit Evaluation of Distributed Energy System

The distributed energy system includes not only core distributed resources, but also heating, refrigeration and other energy systems. Evaluation of the benefits of a distributed energy system requires a consideration of the generation benefits during the peak period, the benefits of utilizing common power grids during the valley period, and the comprehensive benefits from cogeneration cooling, heating, and power. The combined heating and power system [56] for a gas turbine at Shanghai Pudong International Airport is composed of one 10.5 kV, 4.0 MW gas turbine generator unit, one 9.7 t/h heat recovery boiler, several centrifugal refrigeration units, and potassium bromide heating and cooling units. The cooling in summer lasts five months and the heating in winter lasts three months. The electricity price is high during the peak period of the system, and the gas turbine generator unit generates electricity for use by the airport itself. The price is low during the night valley period, and the gas turbine generator unit uses common electricity. The generator unit is operating for 4200 h/year. Given depreciation, the total investment in the project is RMB 30 million, and the payback time is about eight years. This shows that the benefits of the distributed energy system are significant.

10

Calculation of High-voltage Power Grid Losses

It is generally recognized that the calculation of losses in an electric power system has two purposes: one is to determine the difference between theoretical losses and actual losses, to identify the reason for this difference, and to propose loss reduction measures; the other is to estimate loss indexes according to expected load level and generated output arrangement, and to provide a basis for planning departments to formulate various plans. This chapter will introduce various methods for a theoretical calculation of losses, as well as comparing and selecting these methods according to the purpose of the theoretical calculation.

10.1 Characteristics and Requirements of Loss Calculation

10.1.1 Classification of High-voltage Power Grids

Power grids above 35 kV can be classified into the following three categories according to the classification needs of the line losses under assessment:

1. The first category is regional high-voltage power grids, including tie lines among provincial regions and access system lines of regional large-scale power plants, whose voltage class is generally 220–500 kV.
2. The second category is regional power grids, including various voltage classes of substations within cities, user lines and substations, as well as a few 110–220 kV power plants.
3. The third category is prefectural power grids, including some 35–110 kV substations, user lines and substations, as well as incoming and outgoing lines of a few 35 kV small hydroelectric power plants and small thermal power plants; in some developed regions, the maximum voltage of prefectural power grids can be up to 220 kV.

As a matter of convenience, the first and second categories are collectively referred to as regional power grids, and the third category is called prefectural power grids.

Line Loss Analysis and Calculation of Electric Power Systems, First Edition. Anguan Wu and Baoshan Ni.

10.1.2 Characteristics of Regional Power Grids and Requirements of Loss Calculation

There are many power plants, substations, and lines in regional power grids. Although the loss rate is not high, the absolute value of electricity losses is large due to wide coverage and large load electricity. The loss calculation should provide decision reference for the operation, transformation, and development of power grids.

Regional power grids also feature a coexistence of industrial loads with a high load factor and commercial, municipal and residential loads with a low load factor, so overall loads of regional power grids have good integrity and a relatively high load factor. The air conditioning loads of some Chinese cities in summer skyrocket with the rising temperature, and the maximum loads of regional power grids are concentrated within a certain period, so the theoretical calculation of line losses of regional power grids once a year is always conducted based on the load curve of a typical day in a month with maximum loads. The loss calculation of regional power grids involves many loss calculation units whose operation modes vary significantly over a long period, so the result calculated by a load flow calculation is generally the power loss within a time section. To grasp the change law in the losses of regional power grids, it is important to explore the relationship between the result of a load flow calculation at a finite time section and the losses during a long period. This is a difficult problem when calculating the line losses of regional power grids.

No-load losses account for a small percentage in the structure of losses of regional power grids, so to maintain a high voltage level is helpful to reduce losses. To identify a reasonable voltage level within a period with low load level and to find loss reduction measures, it is necessary to increase the loss calculations within the period with low loads, and comprehensively to analyze the result of load flow calculations at each time section. This is another problem to be considered in when calculating the losses of regional power grids.

10.1.3 Characteristics of Prefectural Power Grids and Requirements of Loss Calculation

Agricultural loads account for a large proportion in most prefectural power grids with obvious seasonal characteristics. To comprehensively consider loss reduction measures, it is necessary to calculate losses by season, so it is better to select a loss calculation method which has an acceptable calculation workload and produces reliable results.

The proportion of no-load losses of main transformers and distribution transformers in prefectural power grids is larger than that in regional power grids, and the economical operation of main transformers may be a preferential loss reduction measure; the influence of no-load losses of distribution transformers should be taken into account when the voltage level of prefectural power grids is controlled to reduce losses, which is different from regional power grids.

Although a certain number of substations are arranged in prefectural power grids for concentrated compensation, the overall operating power factor of prefectural power grids is smaller than that of regional power grids due to the low number of cable lines and the lower self-generating reactive power of the lines, so it is more imperative to look for reactive optimization approaches through the loss calculation. This results in a special requirement for the expression of loss calculation results of prefectural power grids.

10.2 Real-time Loss Measuring Method for High-voltage Power Grids

According to Reference [30], a computerized load flow calculation method is generally adopted to calculate the electric energy losses of high-voltage power grids of 35 kV and above, and on-line computation should be conducted if conditions permit. As a result of the large calculation workload and the volume of operating data required for the calculation of regional power grid losses, the work is often undertaken by professionals of dispatching departments.

Provincial and local dispatching departments in China generally established supervisory control and data acquisition (SCADA) systems in the 1990s. Through years of practical operation, some dispatching operators find that the power difference or electricity difference between both ends of grid units within a short period can be obtained through certain processing of real-time data collected by the system, and that real-time losses can be obtained by the storage and accumulation of such difference values required for the loss calculation period. In the case of distorted or missing data, the arithmetic operation of long-period measured data cannot produce final loss results, so the real-time calculation of losses, similar to other advanced applications of a dispatching automation system, requires continuous, complete, and correct real-time data.

10.2.1 Function and Method of State Estimation

10.2.1.1 Function of State Estimation

State estimation means real-time data processing of telemetering quantity and telemetering information at a time section of the system. During the operation of an electric power system, a piece of complete measuring information should include node voltage amplitude, phase angle, power, and power angle of all power plants and substations within the dispatching jurisdiction; due to limitations on investment in tele-control equipment and channels, measuring information collected by the dispatching center is always not complete, and the part of the system where measuring points are not arranged cannot be measured. In addition, remote measuring information may be subject to small interferences in various links of transmission, resulting in random errors, and sometimes subject to big interferences (random failure of equipment), leading to big errors and bad data. At this time, the partial measuring information received by the dispatching center cannot reflect the true conditions of the system operation, which is therefore difficult to judge. Overall, the tele-control information collected by the dispatching center can be called *raw data* and this is always incomplete measuring set whose elements are not entirely accurate.

State estimation is a method used for learning the comprehensive operating conditions of the system under incomplete measuring conditions. It is a kind of digital filtering and able to predict the state variables of the system by a mathematical method provided that measuring points are arranged reasonably and have a certain redundancy (the number of measuring points is greater than the number of state variables of the system), thereby obtaining comprehensive operating data for the system, meaning a complete measuring is obtained. If the measuring redundancy is large enough, it is possible to detect a sampling for bad data, and even identify any measuring point with bad data and the degree of badness and thereby correct the bad data.

10.2.1.2 Method of State Estimation

The difference between the state estimation algorithm and the load flow calculation is that the number of measuring points is greater than the number of system state variables, that is the number of equations is greater than the number of variables to be calculated. Such equations are called statically indeterminate equations in mathematics (in mechanical structure they are called statically indeterminate structures). Due to errors in measuring, these equations are mutually incompatible and cannot derive definite solutions. State estimation adopts optimized estimation methods, and the most commonly used is the least square method which aims to minimize the sum of squares of errors in all measuring points, thereby obtaining a set of optimal estimates of system state variables. Results obtained from the state estimation are more accurate than measuring values directly collected and reduce measuring errors.

10.2.1.3 State Estimation Based on SCADA System

Currently, SCADA systems in Chinese dispatching centers of regional power grids update data every 3 s and use two saving modes at 5 min and 1 h. Effective state estimation is realized in the saving time (5 min), which means detecting bad data and supplementing missing data, and is limited by the grid size and computer operation speed. In recent years, progress has been made in the quick least square estimation and quick PQ decomposition state estimation used in the calculation of state estimation [38]. For a power grid with seven load points and one equivalent generator, when the measuring redundancy (the ratio between the number of measuring vector dimensions m and the number of state dimensions n) is 1.40, if the voltage difference is used as convergence criterion and the given error is 10^{-4}~10^{-6} of the reference voltage, then convergence can be done through six to eight load flow calculations by the quick PQ decomposition state estimation method, which needs a total time of about 5.6 ms. According to such calculations, if the number of nodes increases by n times, then the number of required convergence iterations is generally less than n times. If the number of grid nodes reaches the general size of a real prefectural grid, the number of required iterations is nearly 50 and the total time is generally not more than 50.0 ms. Therefore, under the existing sampling and saving conditions of a SCADA system, the quick state estimation algorithm can produce complete and correct system state data, thus realizing the real-time calculation of losses.

10.2.2 Real-time Calculation of Losses by State Estimation Combined with Excel

At the beginning of the twenty-first century, some Chinese regions where the level of application of SCADA systems was high started to explore the application of state estimation in the real-time measuring of losses [39], by following the three steps below:

1. Modify state estimation software to allow output to include power at both sides of each branch.
2. Establish a custom ACCESS database which includes the following data fields: (i) name of branch (line or transformer), (ii) start time of loss calculation, (iii) end time of loss calculation, (iv) transmitted active power, (v) last transmitted active power, (vi) calculated line losses, and (vii) last calculated line losses. The seventh field acts as alternative line losses in the case of a failure in the divergence of state estimation caused by too many errors in the telemetering data.
3. Send line losses calculated in the ACCESS database to the analysis process of line losses for further analysis. For the purpose of layered and divisional analysis of losses, conduct classified statistics of lines and main transformers, and adopt an Excel form with strong functions to convert power to electric quantities at the time of sampling for accumulation. Import relevant fields of the Excel form and place them adjacent to accumulated losses, so as to conveniently and automatically calculate and display losses. See Table 10.1 for the form of real-time calculation table of losses and the samples.

According to Table 10.1, when SCADA data is very complete and correct, the accuracy of real-time calculation of losses is very high, and the maintenance of simultaneity and integrity of information for a long

Table 10.1 Form of real-time calculation table of losses and samples.

Name of equipment	Voltage class	Electric quantity (10^4 kW·h)	Electricity losses (10^4 kW·h)	Loss rate
No. 2 transformer of XX substation	220 kV	2770.656	1.047 63	
	110 kV	645.972 37	0.223 39	
	35 kV	2123.364 97	3.412 01	
	Subtotal	2770.656 00 (electric supply)	4.683 03	0.169%

period needs to be guaranteed by other technical means and reliable and effective state estimation. One of the effective technical means to guarantee the measuring simultaneity is a global positioning system (GPS) with an operating frequency of 1.5 GHz and the maximum error in pulse per second is only 10 ns (10^{-8} s); GPS can realize synchronous sampling of operating electrical quantities of power plants and substations in power grids and mark *time tags* in all collected data. In a GPS message of measuring data, a time tag is used as a field to transmit the name of sampling quantity and measuring value to the ACCESS database, thus realizing an arithmetic operation of simultaneous power values and completing the real-time calculation of losses.

10.2.3 Typical Day Method Based on Actual Load Measurement and State Estimation

To understand the maximum loads and distribution in a region and to reasonably arrange a power source construction and power transmission and transformation projects for the next year, prefectural power grid enterprises have conducted test work on large loads in recent years. Therefore, some prefectural dispatching departments use actual measurement data of tests on large loads, the state estimation method, and load curves on the typical day to calculate losses by the typical day (representative day) method. The typical day is a day when the operation mode, daily electric supply, load level, temperature conditions, and other factors affecting losses are all typical. However, the operation mode, distribution of electric quantities, and temperature conditions are not always typical during the measuring period of large loads, so various factors need to be considered to adjust loss calculation parameters and to select a specific calculation method [40].

10.2.3.1 Selection of Power Method and Electric Quantity Method

1. *Calculation of losses by power method.*
   The current and voltage of each branch are obtained through grid analysis and load flow calculation of the active and reactive power at the generation side and load side and the bus voltage data for one hour. Considering power grid parameters, active and reactive power losses of each branch can be calculated and constitute electric energy losses within the hour. Such electric energy losses accumulated within 24 h are losses on the representative day. This method has two advantages: one is that reactive power losses are obtained at the same time so that the distribution of reactive power and its voltage conditions can be easily analyzed; the other is that, thanks to real-time load flow records, calculated losses of each branch can be easily checked to avoid excessive distortion of loss calculation due to any change in the grid connection. The disadvantage of this method is that measured distribution of electric quantities for system large loads is not used, and electricity losses cannot be directly related to electric quantities, thus affecting the reliability of the calculation result of the loss rate on the typical day.
2. *Calculation of losses by electric quantity method.*
   To maintain the distribution of electric quantities on the representative day, and reflect the typicality of output load curves of each load and power plant, the concept of output distribution coefficient is introduced. Assume that there are n loads; the maximum value of a typical load curve is $p_{i.\max}$; the load per hour is p_{ij} (j = 1–24); the daily electric quantity is A_i; the total electric quantity of the whole grid is A_Σ. The output distribution coefficient of load i can be calculated as per the following formula:

$$K_{ij} = \frac{A_\Sigma}{\sum_{i=1}^{n} A_i} \frac{p_{ij}}{p_{i.\max}} = k_A f_{ij} \tag{10.1}$$

Wherein $f_{ij} = p_{ij}/p_{i.\max}$ – instantaneous (hourly) load factor of load i;

$k_A = A_\Sigma / \sum_{i=1}^{n} A_i$ – ratio between the total electric quantity of the whole grid on the representative day and the sum of electric quantities of n loads on the typical load curve day; also called the electric quantity ratio coefficient.

As for the calculation of losses by the representative day electric quantity method, loads at each load node for 24 h can be calculated as per the following formula:

$$\bar{p}_{ij} = k_{ij} p_{ij} \quad (i = 1 \sim n,\ j = 1 \sim 24) \tag{10.2}$$

10.2.3.2 Real-time Connection Analysis and Supplement to Grid Structure Parameters

The typical day method needs 24 h load flow calculation, but the connection of the main grid may vary within 24 h. The theoretical calculation of losses should reflect loss conditions under the most frequent connection of the main grid, so the grid connection adopted in the typical day method should be analyzed in real time, including processing switch information in real time depending on the operation mode, and automatically dividing any new grid connection formed by nodes of power plants and substations. Generally, there are three steps as follows:

1. *Connection analysis of power plants and substations.* According to the pre-stored switch information table and the switch switching table collected by the SCADA system, connect nodes at both sides of closed switches based on the principle of same-name connection. Connected nodes are simplified to a node, and substations are finally simplified into several nodes.
2. *System grid analysis.* According to the pre-stored branch information table and the branch switching table collected by the SCADA system, connect nodes at both sides of switched-in branches based on the principle of same-name connection. Connected nodes are simplified to a subsystem. The result of the system grid analysis is to divide electrically connected nodes into a subsystem.
3. *Formation of node name.* Unified node names are formed for simplified nodes collected from the above two steps in the whole grid and serve as the basis of grid structure parameters and operating data.

The load flow calculation results can lead to power losses and the theoretical calculation of line losses results in energy losses, so the latter is related to the energy loss parameter of concentrated compensation capacitors in substations, that is tangent of dielectric loss angle (tanδ). As a result, the energy loss parameter of each concentrated compensation point in the whole grid should be supplemented for the loss calculation, so as to ensure the integrity of grid structure parameters for the loss calculation and prevent missing of any item for the energy loss calculation.

10.2.3.3 Adjustment of Measuring Point Configuration and Analysis of Loss Calculation Results

For a large system with many nodes, measuring systems should be divided by subsystem, and the balance node of each subsystem should be analyzed and determined. Input power of this node is to be determined. When the state estimation provides incomplete or wrong data, causing insufficient input power in more than one node of a subsystem, the load flow of this subsystem cannot be calculated and its nodes should be reconfigured.

In the loss calculation, the 24 h load flow calculation results can be compared with the load flows recorded in real time. If they are close to each other under maximum, minimum and normal operation modes, then the calculation results of the typical day method can be deemed as having higher reliability; if the difference is larger under an operation mode, the real-time grid connection analysis should be conducted again, and the calculation results under actual loads and those under load distribution are compared to analyze the reason for the larger difference. Then, the connection or any measuring point is adjusted or the output distribution coefficient is changed for supplementary calculation.

The typical day method requires the analysis of typicality of grid connection. This shows that the 24 h load flow calculation on the typical day to reflect the influence of three operation modes on losses does have difficulties and deficiencies. If sections of the three different modes can be defined, the load flow calculation in three periods (less than or equal to 24 h) is not even difficult; however, here is the second problem that how the loss calculation results under the three modes can be integrated into losses within a long period. These two problems will be analyzed in Section 10.4 of this chapter.

10.2.4 Comprehensive Analysis Method of Losses Based on Real-time System Data

With the gradual improvement and practical application of the real-time dispatching system, management information system, and electric energy metering system, several theoretical line loss values may be obtained. So some authors proposed to utilize the dispatching automation system and management information system to generate the theoretical line loss value based on load flow, the theoretical line loss value based on state estimation, and the statistical line loss value based on electric energy metering system, and to use a mathematical analysis method to integrate the three values [41], thus a more reliable theoretical line loss value is expected to be obtained.

10.2.4.1 Three Theoretical Line Loss Rates Based on Real-time System Data

1. *Real-time line loss calculation based on load flow.* Conduct load flow calculation of real-time data of the SCADA system, and store integral of current square ($\int i^2 \mathrm{d}t$) of each loss calculation unit and the integral of input and output power of each bus by month, day, and hour. Invoke equipment parameters in the management information system (MIS), and calculate line and transformer losses (including no-load losses). Summarize the total electric supply, electricity line losses and line loss rate of the whole system, as well as the electric supply, electricity line losses, and line loss rate under various voltage classes.
2. *Theoretical line loss calculation based on state estimation.* State estimation is one of the advanced application software in the energy management system (EMS). As mentioned in Section 10.2.4.2 below, state estimation is able to identify and eliminate bad data in the real-time data of the SCADA system, and calculate measuring point data which is not provided by the measuring system. The theoretical line loss value based on state estimation can be obtained by the same method in (1) above. In the case of non-convergence of load flow or bad data, the data of state estimation can be referenced. If the accuracy of data obtained by state estimation is not high, the data of the load flow calculation can be used for check. Therefore, the methods in (1) and (2) are complementary.
3. *Statistical line loss calculation based on electric energy metering system.* On-line statistics can produce real-time system statistical line losses according to the data of remote meter reading system. Any error in the metering system leads to the fluctuation in the line loss rate (see Section 2.5 in Chapter 2). Due to management line losses and other management reasons, the error in the statistical line loss rate obtained here may be larger than that in (1) or (2) above[1].

---

[1] According to materials of the relevant R&D institution, when the real-time metering system is operating stably, the error in the statistical line loss rate of high-voltage lines is very small.

10.2.4.2 Calculation of Theoretical Line Loss Rate by three Value Comprehensive Method

The three value comprehensive line loss rate is X which can be calculated as per the following formula:

$$X = W_1X_i + W_2X_2 + W_3X_3 \tag{10.3}$$

Wherein X_1, X_2, X_3 – theoretical line loss rate based on load flow, theoretical line loss rate based on state estimation, and statistical line loss rate based on electric energy metering system;
W_1, W_2, W_3 – weights of three line loss rates, which can be determined by the following conditions:

$$\left.\begin{array}{l}\text{Objective function } a^2W_1^2 + b^2W_2^2 + c^2W_3^2 = \min \\ \text{Constraint condition } W_1 + W_2 + W_3 = 1, \text{ and } W_1, W_2, W_3 \geq 0\end{array}\right\} \tag{10.4}$$

Wherein a, b, c are the differences (relative value) between three line loss rates on the last day and those on the current day, respectively. After the weights W_1, W_2, W_3 are obtained, the comprehensive line loss rate X can be calculated as per Formula (10.3), that is the theoretical line loss rate whose reliability is higher than any line loss rate. In the actual example provided in Reference [41], the six-day comprehensive line loss rate calculated by this method acts as the accurate value, and the line loss rates from three sources are compared with this accurate value to calculate to the sum of mean square errors of six days. It turns out that the total mean square error in the line loss rate calculated by load flow method is the lowest, successively followed that in the statistical line loss rate and that in the line loss rate calculated by state estimation. This indicates that the loss rate calculated by load flow method is perhaps the most close to the actual value under normal circumstances.

10.2.4.3 Comparison with other Real-time Loss Calculation Methods

This method makes the most of the information of various kinds of real-time systems and comprehensively produces highly reliable real-time losses based on the calculation of three theoretical line loss rates. Such real-time losses are evaluated as sufficiently reliable from the perspective of mathematical treatment. Looking from the method itself, the analysis of its final result is hard to be connected with real-time information on which each specific method is based, and it is difficult to analyze the influence of grid operating parameters and connection conditions on line losses. In conclusion, this method is effective to the short-term theoretical line loss calculation which is for the purpose of check, but not applicable to the long-term theoretical calculation of line losses.

10.3 Equivalent Node Power Method for Calculation of High-voltage Power Grid Losses

The equivalent node power method for the calculation of high-voltage power grid losses was put forward in Reference [29] by Professor Yang Xiutai of Chongqing University in the 1980s. Over the past 20 years, this method has been widely used in electric power systems.

10.3.1 Equivalent Node Power and its Distribution

10.3.1.1 Two-dimensional Array for Calculation of Losses of Power Grids

Assume that a power grid has n loss calculation units; the calculation period is T; the time interval between measurements is t_0. Losses of the main grid within the measuring period T can be obtained by the accumulation of electric energy losses of each unit within $i(T/t_0)$ intervals in T, or obtained by the accumulation of the sum of electric energy losses of all units within each measuring interval in T. The expression by two-dimensional array is

	Unit (n) ⟶			
	ΔA_{11}	ΔA_{12}	$\cdots$	ΔA_{1j}
Time (i) ↓	$\vdots$			
	ΔA_{i1}	ΔA_{i2}	$\cdots$	ΔA_{ij}

Losses of the power grid within T can be calculated as per the following formula:

$$\Delta A_{\Sigma}\big|_0^T = \sum_{j=1}^{n}\left[\sum_{i=1}^{i=T/t_0}(\Delta A_{ij})\right] = \sum_{i=1}^{i=T/t_0}\left[\sum_{j=1}^{n}(\Delta A_{ij})\right] \tag{10.5}$$

10.3.1.2 Equivalent Power at Each Load Point

The calculation of losses with equivalent node power (rms average power) within the measuring period can ensure the equivalence of losses of each unit within T, thus ensuring the equivalence of calculation of losses of the whole power grid. At this time, the power distribution of all load points is the distribution of equivalent node power with constant losses of the whole power grid.

10.3.1.3 Consideration of No-load Losses of Transformers

To include no-load losses of transformers in the calculation of losses of the power grid, exciting power of transformers can be treated like grounding branches in the grid in the form of constant exciting admittance, that is

$$Y_{\mathrm{T}} = G - \mathrm{j}B_{\mathrm{T}} = \frac{\Delta P_0}{1000(U')^2} - \mathrm{j}\frac{i_0\%S_{\mathrm{N}}}{100(U')^2} \tag{10.6}$$

When high-voltage side loads and no-load losses of a transformer are located in the same node, a small impedance branch and a node can be added to separate them, as shown in Figure 10.1, wherein R_{T}, X_{T}, ΔP_{T}, ΔQ_{T} are transformer winding impedances and rated no-load losses, respectively.

10.3.2 Relationship between Power Losses and Electric Energy Losses Under Distribution of Equivalent Node Power

10.3.2.1 Calculation of Grid Electric Energy Losses

If the calculation period of electric energy losses is one day (i.e. 24 h) then grid electric energy losses can be calculated by the following two methods:

1. Multiply the 24 h average power losses of each loss calculation unit with time and then accumulate them, that is

$$\Delta A = \sum_{i=1}^{n}(\Delta P_{ij})_j T$$

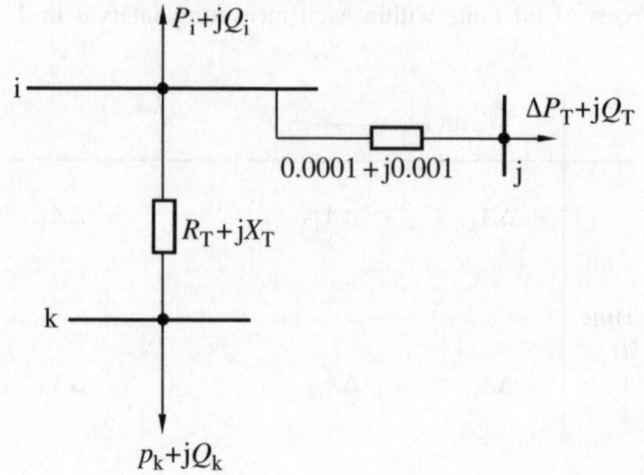

Figure 10.1 Node connection of transformer loads and no-load losses.

2. Accumulate grid electric energy losses within each hour, that is

$$\Delta A=\sum_{i=1}^{24}(\Delta A_{ij})=\sum_{i=1}^{24}\left(\sum_{j=1}^{n}\Delta P_{av}T\right)$$

According to the first method, the grid electric quantity is

$$A=\left(\sum_{j=1}^{n}P_{av}\right)T+\Delta A=\sum_{j=1}^{n}(P_{av}T)+\sum_{j=1}^{n}(\Delta P_{av})_jT=\left[\sum_{j=1}^{n}(P_{av}+\Delta P_{av})\right]T$$

That is, grid electric quantity = (daily average power of all grid loads + daily average power losses) × loss calculation period.

So the line loss rate is

$$\Delta A\%=\sum_{j=1}^{n}(\Delta P_{av})_jT/T\left[\sum_{j=1}^{n}(P_{av}+\Delta P_{av})\right]=\sum_{j=1}^{n}(\Delta P_{av})_j\Big/\sum_{j=1}^{n}(P_{av}+\Delta P_{av})_j \tag{10.7}$$

10.3.2.2 Power Loss Value under Distribution of Grid Equivalent Node Power

The power loss rate under the distribution of grid equivalent node power is the ratio between the sum of grid power losses and the sum of grid equivalent average power under the distribution of equivalent node power, that is

$$\Delta P_{eq}\%=\sum_{j=1}^{n}(\Delta P_{eq})_j\Big/\sum_{j=1}^{n}(P_{eq.j}+\Delta P_{eq.j}) \tag{10.8}$$

10.3.2.3 Ratio between Electric Energy Loss Rate and Power Loss Rate Under Distribution of Equivalent Node Power

Under the above definition of the power loss rate, the relationship between the electric energy loss rate and the power loss rate under the distribution of equivalent power can be obtained, that is

$$\Delta A\%/\Delta P_{\text{av}}\% = \frac{\sum_{j=1}^{n}(\Delta P_{\text{av}})_j \sum_{j=1}^{n}(P_{\text{eq}.j}+\Delta P_{\text{eq}.j})}{\sum_{j=1}^{n}(P_{\text{av}}+\Delta P_{\text{av}})_j \sum_{j=1}^{n}(\Delta P_{\text{eq}})_j} \approx \frac{\sum_{j=1}^{n}(P_{\text{eq}.j}+\Delta P_{\text{eq}.j})}{\sum_{j=1}^{n}(P_{\text{av}}+\Delta P_{\text{av}})_j} \tag{10.9}$$

When the loss power is very small, the above formula can be approximated to

$$\Delta A\%/\Delta P_{\text{eq}}\% \approx \sum_{j=1}^{n}(P_{\text{eq}.j}) \Big/ \sum_{j=1}^{n} P_{\text{av}} = K_{\Sigma} \tag{10.10}$$

According to Formula (10.10), when the grid power loss rate is small, the ratio between the loss rate and the equivalent node power loss rate is approximately equal to the form coefficient of the total power curve of the whole grid.

To understand the principle of the equivalent power method, the example below is analyzed.

Example 10.1 A distribution line is connected with two pure resistive loads, and the voltage remains unchanged at 10 kV within the measuring period. The step type load curves of two loads, the load curve of the start end of the line, and the single line diagram for loss calculation are shown in Figure 10.2.

Try to calculate the daily average power loss rate, weighted average power loss rate and line loss rate within the measuring period of the line, and analyze the relationship among the three values.

Solution

1. *Calculation of daily average power losses.*

   According to Figure 10.2a, $R_1 = 2\ \Omega$, $R_2 = 1\ \Omega$ are marked, and $U = 10$ kV is given. Subscripts I and II are used to indicate two sections of the distribution line. The average power losses within the period t can be obtained:

$$\begin{aligned}
(\Delta P_{\text{av}})_{\text{II}} &= \sqrt{3}\left(I_{\text{II.t1}}^2 R_2 \times 0.5t + I_{\text{II.t2}}^2 R_2 \times 0.5t\right)/t = \sqrt{3}\left(6^2 R_2 \times 0.5 + 2^2 R_2 \times 0.5\right)\\
&= \sqrt{3} \times 20 \times R_2 = 20\sqrt{3} \times 10^{-3}(\text{kW})\\
(\Delta P_{\text{av}})_{\text{I}} &= \sqrt{3}\left(I_{\text{I.t1}}^2 \times R_1 \times 0.5t + I_{\text{I.t2}}^2 \times R_1 \times 0.5t\right)/t\\
&= \sqrt{3}\left(8^2 \times R_1 \times 0.5 + 6^2 \times R_1 \times 0.5\right)\\
&= \sqrt{3} \times 50 \times R_1 = 100\sqrt{3} \times 10^{-3}(\text{kW})\\
(\Delta P_{\text{av}})_{\Sigma} &= (\Delta P_{\text{av}})_{\text{I}} + (\Delta P_{\text{av}})_{\text{II}} = 120\sqrt{3} \times 10^{-3}(\text{kW})
\end{aligned}$$

2. *Calculation of power losses under the distribution of rms current.*

   The rms current of unit II is

$$I_{\text{rms.II}} = \sqrt{\left(6^2 \times 0.5t + 2^2 \times 0.5t\right)} = \sqrt{18+2} = 4.472(\text{A})$$

   The rms current of load i_1 is

$$I_{\text{rms.}i} = \sqrt{\left(2^2 \times 0.5t + 4^2 \times 0.5t\right)/t} = \sqrt{2+8} = 3.162(\text{A})$$

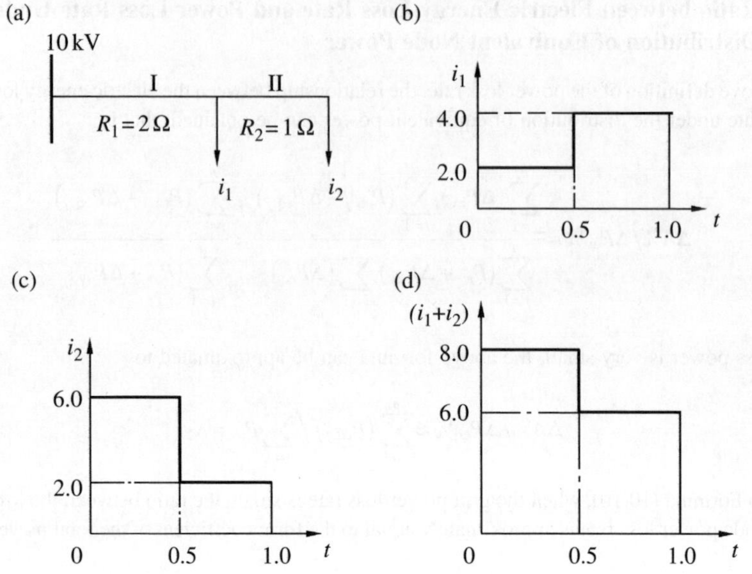

Figure 10.2 Calculation of line loss rate of distributed loads: (a) single line diagram; (b) i_1 load curve; (c) i_2 load curve; (d) start end load curve.

The distribution of rms current is

$$I_{\text{rms.I}}=I_{\text{rms.II}}+I_{\text{rms.}i}=4.472+3.162=7.634(\text{A})$$

The actual rms current in section I is

$$I_{\text{rms.I}}=\sqrt{0.5\times 8^2+0.5\times 6^2}=\sqrt{50}=7.07(\text{A})$$

$$\Sigma(\Delta P_{\text{eq}})=\sqrt{3}(I^2_{\text{rms.I}}R_1+I^2_{\text{rms.II}}R_2)=\sqrt{3}[7.634^2R_1+4.472^2R_2]$$
$$=136.56\sqrt{3}\times 10^{-3}(\text{kW})$$

So

$$\Sigma(\Delta P_{\text{eq}})>(\Delta P_{\text{av}})_{\Sigma}$$

3. *Calculation of the line loss rate.*
   The electric energy losses are

$$\Delta A=(\Delta P_{\text{av}})_{\Sigma}t=120\sqrt{3}\times 10^{-3}\times 24=4.988\ (\text{kW}\cdot\text{h})$$

The electric supply is

$$A=A_0+\Delta A=\sqrt{3}U(I_{1t}\bullet 0.5t+I_{2t}\bullet 0.5t)+\Delta A$$
$$=\sqrt{3}\times 10\times(8\times 0.5\times 24+6\times 0.5\times 24)+4.988$$
$$=2909.85+4.988=2914.83\ (\text{kW}\cdot\text{h})$$

The line loss rate is

$$\Delta A\%=\Delta A/(A_0+\Delta A)=4.988/2914.75=0.171\,1\%$$

4. *Calculation of power losses under the distribution of rms power.*
The rms power at the start end of the line is

$$P_{\text{rms.I}} = \sqrt{3}UI_{\text{rms.I}} = \sqrt{3}\times 10\times 7.634 = 76.34\sqrt{3}(\text{kW})$$

The power loss rate under the distribution of equivalent power is

$$\Delta P_{\text{eq}}\% = \Sigma(\Delta P_{\text{eq}})/\left[P_{\text{rms.I}} + \Sigma(\Delta P_{\text{eq}})\right]$$
$$= 0.136\ 6\sqrt{3}/(76.34 + 0.136\ 6)\sqrt{3} = 0.178\ 6\%$$

5. *Calculation of form coefficient and error analysis.* Use the start end rms current and the average current to calculate the form coefficient of total grid loads, that is $K_{\Sigma} = 7.634/7.0 = 1.0906$, and use approximate Formula (10.10) to calculate the line loss rate as below:

$$\Delta A'\% = K_{\Sigma}\Delta P_{\text{eq}}\% = 1.090\ 6\times 0.178\ 6\% = 0.194\ 8\%$$

Compared with the calculated line loss rate $\Delta A\% = 0.171\ 1\%$, this line loss rate is larger and the error is $\delta(\Delta A\%) = \dfrac{0.194\ 8 - 0.1711}{0.1711} = 13.83\%$.

According to Example 10.1, in fact except load branches, there is no distribution of rms current in the whole grid. As for the equivalent node power method, equivalent power of load points forms the distribution of grid rms current, and the power losses and the form coefficient are used to calculate the loss rate under the actual current distribution. When the shapes of load curves at load points are different, the arithmetic sum of rms current at load points of each branch is certainly greater than the actual rms current of the main line (in this example, $7.634A > 7.07A$), so the equivalent node power method is bound to produce larger calculation results of losses. Under normal circumstances, $\sum(\Delta P_{\text{eq}}) > \sum(\Delta P_{\text{av}})$. Two approximations in Formulas (10.9) and (10.10) have opposite influences on the error in the loss calculation, so the larger result of losses calculated by the equivalent node power method can be eased. This is an inherent problem of the equivalent node power method, which has been proved in Example 10.1.

10.3.3 Analysis of Equivalent Node Power Method

The equivalent node power method creates a virtual distribution of equivalent node power and uses the load flow calculation to obtain power losses $\Delta P_{\text{eq}}\%$ which are multiplied by the form coefficient of grid loads to obtain approximate losses $\Delta A\%$. The principle of the method is clear, and the degree of approximation depends on actual losses and the difference in shape of the load curves of all load points. Compared with other methods, this method can consider the load characteristics of all load points, and the total load level can be compared with the maximum, minimum, and normal operation modes to determine the mode to which losses are close. When there are many nodes and the difference in shape of the load curves of all load points is not large, the result of losses calculated by this method is satisfactory. Due to the above advantages, this method has been widely applied for a long time.

The disadvantage of this method is that the reliability of the calculated larger loss rate is not sufficiently demonstrated. Some scholars once used this method to calculate losses in actual grids with several typical load curves, indicating that calculation errors could be accepted [29]. However, they did not use the international general standard multi-node grid and various typical load curves for verification calculation, so whether the larger calculation errors are acceptable in a wide range needs to be further studied.

10.4 Calculation of Losses of High-voltage Power Grids Based on Power Losses under Three Modes

Section 3.4 of Chapter 3 introduced the three-mode division of long-period load curves, which provides conditions for the transition from the calculation of power losses under three modes to the calculation of loss rate within the full period.

Because the long-period three-mode section division is transited from the "multiplying probability" division in loss curves to the division in power curves based on the condition that energy losses are constant within the full period, typical load flow calculations are conducted in three sections under three modes, to obtain respective average power losses under typical daily loads. Obviously, the three kinds of daily average power losses can represent energy losses within the three mode periods. Line loss rates under the three modes are integrated to obtain the loss rate within the full period, which is the basic concept of calculation of losses of high-voltage power grids based on power losses under three modes to be explained in this section.

The long-period (annual or quarterly) theoretical loss calculation which serves as a check follows the steps below:

1. Use the long-period electric quantity to obtain the average loads, and use the measured maximum loads to obtain the long-period active load factor f_P and the minimum load rate β_P, thereby obtaining the change index $1/\lambda$ [as per Formula (3.8)]. Use Formulas (3.22) and (3.25) to obtain the power $P_{\text{s.m}}$ and $P_{\text{m.b}}$ (per unit value)[2] at three-mode division points.
2. Allow the average power of three-mode sections to be P_{1*}, P_{2*}, P_{3*}, and then $P_{1*} = (\beta_P + P_{\text{s.m}})/2$, $P_{2*} = f_P$, $P_{2*} = (P_{\text{m.b}} + 1.0)/2$; the subscript $*$ refers to the per unit value.
3. Select typical daily load curves of the three modes, and allow their daily maximum loads are respectively $P_{1\max}$, $P_{2\max}$ and $P_{3\max}$. Multiply the actual maximum load within the full period by the per unit value of average power. Compare the daily maximum loads with the product, and take $(P_{\Sigma}.\max P_{1*})/P_{1\max}$, $(P_{\Sigma.\max} P_{2*})/P_{2\max}$ and $(P_{\Sigma.\max} P_{3*})/P_{3\max}$ as proportionality coefficients for adjustment, thus obtaining the new distributions of three-mode typical daily loads.
4. Conduct 24 h load flow calculation of the adjusted distributions of three-mode typical daily loads, to obtain daily average power losses $\Delta P_1\%$, $\Delta P_2\%$ and $\Delta P_3\%$, respectively.
5. Use typical daily grid load curves under the three modes to calculate form coefficients K_1, K_2 and K_3, and use Formula (10.10) to calculate loss rates $\Delta A_1\%$, $\Delta A_2\%$, and $\Delta A_3\%$ of the three-mode sections.
6. As combinations of low load factor coefficients occur in most long periods, according to Table 4.4, obtain the ratio among electric quantities in the three-mode sections A_1:A_2:A_3, and calculate the ratio of electric quantities in the three-mode sections to the total electric quantity α_1:α_2:α_3. According to calculations under various $1/\lambda$ values in Table 3.4: α_1:α_2 = 1:2.56~1:3.0; α_1:α_3 = 1:0.36~1:0.522. For the four situations where $1/\lambda$ values are smaller or larger, that is 1.0, 2.0, 5.0, 6.0, take the occurrence weight as 0.10. For the three situations where $1/\lambda$ values are 3.0, 4.0, 4.5, take the occurrence weight as 0.20. Obtain the weighted average of the ratio as α_1:α_2:α_3 = 1:2.83:0.458; so $\alpha_1/(\alpha_1 + \alpha_2 + \alpha_3) = 0.233$.
7. Calculate the loss rate within the full period as per the following formula:

$$\Delta A_{\Sigma}\% = \Delta A_{\Sigma}/A = (\Delta A_1\% A_1 + \Delta A_2\% A_2 + \Delta A_3\% A_3)/(A_1 + A_2 + A_3)$$

Substitute the weighted average ratio into this to obtain

$$\Delta A_{\Sigma}\% = 0.233\Delta A_1\% + 0.660\Delta A_2\% + 0.107\Delta A_3\% \tag{10.11}$$

According to Formula (10.11), as the situations of the three-mode sections are different, they have different degrees of influence on the full-period loss rate: the normal mode has the highest influence, successively followed by the minimum mode and the maximum mode.

[2] For concise expression, subscripts s, m, and b of the three section power points of division refer to minimum, normal, and maximum in this chapter, which correspond to the subscripts x, zh, and d in Chapter 3.

As for the loss calculation which is for the purpose of check, this method has the following characteristics:

1. The division of three-mode sections in a long-period load duration curve has been sufficiently demonstrated in Section 3.4 of Chapter 3, and the foundation of this method is reliable.
2. There is no special difficulty in selecting typical load curves in the long period by minimum, normal and maximum modes, and the calculated three form coefficients have stable characteristics.
3. The three-period division corresponds to the three modes, and the average power losses of the three sections fluctuate in a small range.
4. The line loss rate is large under the maximum mode, small under the minimum mode, and average under the normal mode. The long-term loss rate is calculated based on the principle of electric quantity balance in the long period and deemed highly reliable in the sense of statistics and probability.
5. This method adopts the concept of "output distribution coefficient" in the typical day method, and uses the average power of the three-mode sections to adjust the distribution of typical daily loads under the three modes. However, it adopts typical days for three modes, which is more representative than one typical day. This is an improvement in the method. This method also uses approximate Formulas (10.9) and (10.10) of the average power loss rate and loss rate in the equivalent node power method, and solves the transition from the power loss rate under three modes to the loss rate. Therefore, this method is a new method for calculating losses of high-voltage lines by making use of advantages of other methods and adopting the principle of "multiplying probability" to resolve the difficulty in the three-mode section division.
6. This method uses the weighted average method on the ratio among electric quantities of three-mode sections to obtain Formula (10.11). The above steps can be followed to estimate the possible range of errors caused by the weighted average method: $\delta(A_2)\% = -5.67$ to $+10.55\%$, $\delta(A_3)\% = -13.9$ to $+27.2\%$. To avoid larger errors in total grid losses, directly use $1/\lambda$ values in the combination of low load factor parameters; look for corresponding $\alpha_1{:}\alpha_2{:}\alpha_3$ ratios in Table 4.4; and modify coefficients before $\Delta A_2\%$ and $\Delta A_3\%$ in Formula (10.11), thereby obtaining total grid losses with smaller errors.

The predicted calculation of the loss rate by the above method has the following features:

1. Predict the electric quantity according to the normal operation mode, and look for the ratio among electric quantities of the three sections in Table 3.4 based on the predicted load change index $1/\lambda$. Determine the electric quantities of the minimum mode section and the maximum mode section, and compare them with the predicted total electric quantity for necessary adjustment.
2. Consider separate or comprehensive measures under three modes, such as "valley filling" under the minimum mode, or system input in pumped storage power stations. These measures can improve load curve shapes and power distribution under the minimum mode and maximum mode and reduce the loss rate. The predicted electric quantities can be used to estimate the effect of loss reduction.
3. In the middle and later time of the long period, comprehensively consider positive and negative condition changes according to the loss rates under three modes and the total loss rate for the preceding period, and conduct load flow calculations under the three modes in the middle and later time, thereby obtaining the predicted total loss rate, so as to provide calculation conditions for the controllable and controlled line loss rate in the full period.

10.5 Calculation and Analysis of Samples

10.5.1 Verification of Loss Calculation of Standard Power Grid with 39 Nodes

10.5.1.1 Connection and Parameters of Standard Power Grid with 10 Generators and 39 Nodes

1. The connection of the standard power grid with 10 generator and 39 nodes is shown in Figure 10.3.
2. Line parameters and transformer parameters are listed in Table 10.2 and Table 10.3, respectively.
3. The bus generated output and load data are shown in Table 10.4.

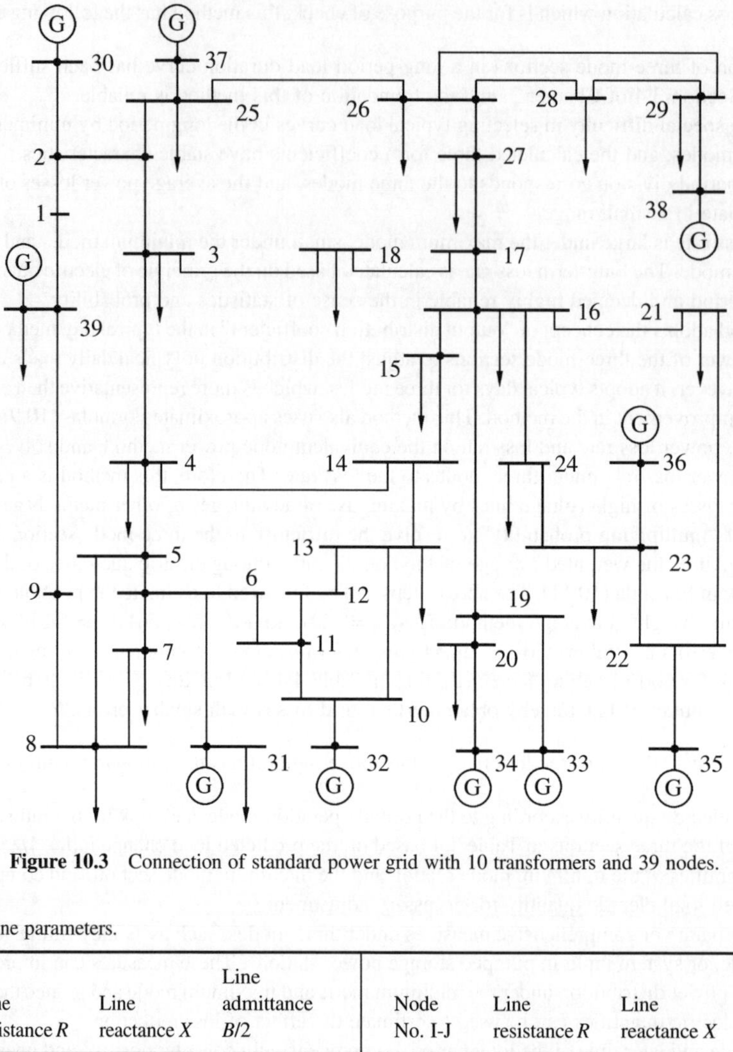

Figure 10.3 Connection of standard power grid with 10 transformers and 39 nodes.

Table 10.2 Line parameters.

Node No. I-J	Line resistance R	Line reactance X	Line admittance $B/2$	Node No. I-J	Line resistance R	Line reactance X	Line admittance $B/2$
1–2	0.0035	0.0411	0.349 35	3–18	0.0011	0.0133	0.1069
1–39	0.001	0.025	0.375	4–5	0.0008	0.0128	0.0671
2–3	0.0013	0.0151	0.1286	4–14	0.0008	0.0129	0.0691
2–25	0.007	0.0086	0.073	5–6	0.0002	0.0026	0.0217
3–4	0.0013	0.0213	0.1107	5–8	0.0008	0.0112	0.0738
6–7	0.0006	0.0092	0.0565	16–21	0.0008	0.0135	0.1274
6–11	0.0007	0.0082	0.069 45	16–24	0.0003	0.0059	0.034
7–8	0.0004	0.0046	0.039	17–18	0.0007	0.0082	0.065 95
8–9	0.0023	0.0363	0.1902	17–27	0.0013	0.0173	0.1608
9–39	0.001	0.025	0.60	21–22	0.0008	0.014	0.128 25
10–11	0.0004	0.0046	0.036 45	22–23	0.0006	0.0096	0.0923
10–13	0.0004	0.0043	0.036 45	23–24	0.0022	0.035	0.1805
13–14	0.0009	0.0101	0.086 15	25–26	0.0032	0.0323	0.2565
14–15	0.0018	0.0217	0.1830	26–27	0.0014	0.0147	0.1198
15–16	0.0009	0.0094	0.0855	26–28	0.0043	0.0474	0.3901
16–17	0.0007	0.0089	0.0671	26–29	0.0057	0.0625	0.5145
16–19	0.0016	0.0195	0.1520	28–29	0.0014	0.0151	0.1245

Table 10.3 Transformer parameters.

Node No. I-J	Transformer resistance R	Transformer reactance X	Node No. I-J	No-load active power P_0	No-load reactive power Q_0
11–12	0.0016	0.0435	12–40	0.0037	0.024
13–12	0.0016	0.0435	6–41	0.005	0.064
31–6	0.0006	0.0250	10–42	0.0072	0.0384
32–10	0.0007	0.020	19–43	0.0072	0.0384
33–19	0.0007	0.0142	20–44	0.006 25	0.0984
34–20	0.0009	0.0180	22–45	0.0072	0.0384
35–22	0.0007	0.0143	23–46	0.005	0.064
36–23	0.0005	0.0272	25–47	0.005	0.064
37–25	0.0006	0.0232	2–48	0.0034	0.0432
30–2	0.0018	0.0181	29–49	0.0102	0.120
38–29	0.0008	0.0156			
20–19	0.0007	0.0138			

Note: The transformer no-load power branch node 40–49 is not marked in Figure 10.3.

Table 10.4 Bus generated output and load data.

Bus No.	Active output P_G	Reactive output Q_G	Bus No.	(Type)	Active load P_L	Reactive load Q_L
			3	(1)	3.22	0.024
			4	(2)	5.0	1.84
			7	(4)	2.338	0.84
			8	(6)	5.22	1.76
			12		0.085	0.88
			15	(2)	3.2	1.53
			16	(1)	3.294	0.323
			18	(3)	1.58	0.30
30	2.5	1.451	20	(3)	6.80	1.03
31	5.63	2.055	31	(2)	0.092	0.046
32	6.5	2.057	21	(4)	2.74	1.15
33	6.32	1.091	23	(5)	2.475	0.846
34	5.08	1.67	24	(5)	3.086	−0.922
35	6.5	2.113	25	(3)	2.24	0.472
36	5.6	1.005	26	(1)	1.39	0.17
37	5.4	0.007	27	(4)	2.81	0.755
38	8.3	0.228	28	(1)	2.06	0.276
39	10.0	0.86	29	(1)	2.835	0.269
			39	(6)	11.04	2.50

10.5.1.2 Selection of 24 h Load Curves

Load curves listed on pages 11 and 222 of Reference [29] are various load curves used in verification examples, and their 24 h per unit values are shown in Table 10.5. See Table 10.5 for the type of loads at each node.

10.5.1.3 Reactive Load Values for Load Flow Calculation

Based on the power factor cosφ calculated by basic load flow in Table 10.5, when per unit value of active load is lower than 0.20, the power factor is also decreased. The decreased values of the power factor Δcosφ when loads of the type (1) to (6) load curves are low are listed in the last row of Table 10.5.

Table 10.5 Load curves of all load points (per unit values).

Type / Hour	(1)	(2)	(3)	(4)	(5)	(6)
1	0.1254	0.050	0.3713	0.5850	0.800	0.830
2	0.1254	0.050	0.3713	1.00	0.800	0.830
3	0.1254	0.050	0.3653	0.5850	0.800	0.830
4	0.1254	0.050	0.3653	0.4830	0.800	0.830
5	0.1254	0.100	0.3653	1.00	0.800	0.840
6	0.1254	0.100	0.5449	0.7551	0.800	0.840
7	0.1254	0.400	0.5449	0.3061	0.8957	0.850
8	0.1254	0.950	0.5449	0.4898	0.8957	0.850
9	0.1254	1.00	0.6467	0.4898	1.00	0.950
10	0.5017	0.960	0.8204	0.9184	0.9565	1.00
11	0.5017	0.920	1.00	0.7483	1.00	0.900
12	1.00	0.650	0.3953	0.4558	0.7826	0.850
13	1.00	0.900	0.3593	1.00	0.800	0.850
14	0.1254	0.960	0.5449	0.7483	0.800	0.950
15	0.1254	0.850	0.7246	0.5986	1.00	0.900
16	0.1254	0.750	1.00	0.2517	1.00	0.870
17	0.1627	0.650	1.00	0.8367	1.00	0.900
18	0.6610	0.100	1.00	0.6667	0.800	0.920
19	1.00	0.100	0.4551	1.00	0.800	0.900
20	0.1254	0.100	0.7246	0.5034	0.9217	0.900
21	0.2847	0.050	0.7246	0.7483	0.9130	0.900
22	0.6407	0.050	0.5569	0.7483	0.9130	0.900
23	0.6271	0.050	0.5449	1.00	0.800	0.900
24	0.1254	0.050	0.4491	0.4082	0.7913	0.850
$\Delta\cos\varphi$	0.05	0.08	0.04	0.03	0.02	0.02

10.5.1.4 Results of 24 h Load Flow Calculation for 10 Generators and 39 Nodes

The PSASP program of the China Electric Power Research Institute is used to conduct the 24 h load flow calculation for the standard test grid with 10 generators and 39 nodes, and the results are shown in Table 10.6.

1. Calculation of parameters of generated output curve

$$f_G = \frac{\Sigma P_G/t}{P_{G.max}} = \frac{93\ 567.4/24}{5250.1} = 0.742\ 58$$

$$\beta_G = P_{G.min}/P_{G.max} = 2778.3/5250.1 = 0.529\ 2$$

The load change index is

$$\frac{1}{\lambda_G} = \frac{1-f_G}{f_G-\beta_G}$$

2. Calculation of three-mode power points of division. Use Formulas (3.24) and (3.25) to calculate the three-mode power points of division:

$$P_{s.m*} = 0.007\ 7\times 1.206\ 4^2 - 0.119\ 4\times 1.206\ 4 + 0.758\ 5 = 0.625\ 7$$

$$P_{s.m} = 0.625\ 7\times 5250.1 = 3284.99\ \text{(MW)}$$

$$P_{m.b*} = 0.004\times 1.206\ 4^2 - 0.050\ 4\times 1.206\ 4 + 1.007\ 1 = 0.952\ 1$$

$$P_{m.b} = 0.952\ 1\times 5250.1 = 4998.62\ \text{(MW)}$$

Table 10.6 Results of 24 h load flow calculation for 10 generators and 39 nodes.

Hour	Load P_L (MW)	Generated output P_G (MW)	ΔP (MW)	ΔP%	Number of iterations
1	2749.4	2778.3	28.9	1.04	6
2	3078.8	3106.4	27.6	0.89	6
3	2863.3	2887.4	24.1	0.83	6
4	2782.2	2806.9	24.7	0.88	7
5	3250.5	3272.6	22.1	0.68	6
6	3246.8	3266.1	19.3	0.59	6
7	3208.6	3228.2	19.6	0.61	6
8	3831.2	3885.2	54.0	1.39	6
9	4179.9	4235.8	55.9	1.32	6
10	5191.1	5250.1	59.0	1.12	6
11	5095.0	5149.3	54.3	1.05	5
12	4394.0	4433.2	39.2	0.88	6
13	5043.0	5091.9	48.9	0.96	5
14	4133.5	4168.7	35.2	0.84	7
15	4144.0	4174.4	30.4	0.73	6
16	4029.4	4058.6	29.2	0.72	6
17	4506.4	4537.2	30.8	0.68	7
18	4475.8	4504.3	28.5	0.63	8
19	4562.7	4598.2	35.5	0.77	7
20	3379.8	3402.7	22.9	0.67	6
21	3756.1	3776.4	20.3	0.54	6
22	3951.0	3984.0	33.0	0.83	6
23	4055.2	4092.1	36.9	0.90	5
24	2839.7	2879.4	39.7	1.38	8
Total	92 747.4	93 567.4	820.0	0.876	

According to Table (3.4), when $1/\lambda = 1.0$ and $1/\lambda = 2.0$, the ratio between the electric quantities of three-mode sections α_1:α_2:α_3 is 1:2.56:0.522 and 1:2.79:0.507, respectively. Use the interpolation method to calculate α_1:α_2:α_3 = 1:2.6074:0.5189 in this example, so the electric quantity of the minimum mode section is

$$A_s = \Sigma P_{G\cdot t} \times 1.0(\text{h})/(\alpha_1 + \alpha_2 + \alpha_3) = 93\ 567.4/(1 + 2.607\ 4 + 0.518\ 9)$$
$$= 22\ 675.86\ (\text{MW}\cdot\text{h})$$

So

$$A_m = 2.607\ 4A_s = 59\ 125.04\ (\text{MW}\cdot\text{h})$$
$$A_b = 0.518\ 9A_s = 11\ 766.50\ (\text{MW}\cdot\text{h})$$

3. Calculation of average power and power loss rates of three sections. The average power under the minimum mode is

$$P_{s.av} = (\beta P_{max} + P_{s.m})/2 = (2778.3 + 3284.99)/2 = 3031.65\ (\text{MW})$$

According to Table 10.7, this is close to the generated output at hour 2.

Take the power loss rate $\Delta P\%$ = 0.89% at hour 2 to obtain

$$\Delta P_s\% = (3031.65/3106.4)^2 \times 0.89\% = 0.848\%$$

Table 10.7 Total loads and measured loss power over 24 h during 2004 in a province.

Hour	10	11	12	13	14	15	16	17	18	19	20	21	Total
Total load (10^4 kW)	2400	2349.3	2146.48	2239.44	2334.08	2366.20	2391.55	2374.65	2222.54	2290.14	2374.65	2323.94	—
Loss (10^4 kW)	25.0	24.5	17.5	20.0	23.0	24.0	25.0	24.75	19.75	21.75	24.5	22.5	—
Hour	22	23	24	1	2	3	4	5	6	7	8	9	—
Total load (10^4 kW)	2222.54	2095.77	2121.12	1952.11	1918.31	1892.96	1901.41	1935.21	2053.52	2104.22	2205.63	2290.14	52 505.91
Loss (10^4 kW)	19.50	16.50	13.0	13.50	12.50	12.525	12.525	13.0	15.50	16.50	18.5	21.50	457.30

For the normal mode, $P_{m.av} = f_G P_{G.max} = 0.742\ 58 \times 5250.1 = 3898.6$ (MW). According to Table 10.7, generated output values at four time points are close to $P_{m.av}$: hour 8/3885.2 MW, hour 16/4058.6 MW, hour 22/3984 MW, hour 23/4092 MW. Calculate average power values and weighted average values of power loss rates at the four time points, to obtain $P'_{m.av} = 4004.98$ MW, $\Delta P_{av}\% = 0.952\ 9\%$. Use the two values to calculate the power loss rate under the normal mode:

$$\Delta P_m\% = (3898.6/4004.98)^2 \times 0.9529\% = 0.903\%$$

The average power of the maximum mode section is the average values of $P_{m.b}$ and P_{max}, that is $P_{b.av} = (1 + 0.9521)/2 \times 5250.1 = 5124.36$ (MW). According to Table 10.7, the above value is close to the generated output at hour 13, and the power loss rate at this time point can be used to obtain

$$\Delta P_b\% = (5124.36/5091.9)^2 \times 0.96\% = 0.972\%$$

4. Calculation of loss factors F and form coefficients K of three-mode sections. For long-period regional power grids, use the three-mode section division method to calculate losses, and use the three-mode typical load curves to calculate the loss factor F, thus obtaining the form coefficient K. This example is only for the purpose of verifying the correctness of the method, and can only use f and β parameters of the three sections to calculate F and K. Taking the normal mode as example, its calculation process is listed:

$$\beta_m = 0.625\ 7/0.952\ 1 = 0.657\ 2$$

$$f_m = 3898.6/4998.62 = 0.780\ 0$$

$$F_m = 0.639(0.78)^2 + 0.361(0.78 \times 0.657\ 2 + 0.78 - 0.657\ 2) = 0.618\ 2$$

$$K_m = \sqrt{F_m}/f_m = \sqrt{0.618\ 2}/0.78 = 0.786/0.78 = 1.008$$

In the same method, $K_s = 1.001$ and $K_b = 1.000\ 1$ can be obtained.

5. Calculation of electric energy losses. The sum of electric energy losses of the three-mode sections is

$$\begin{aligned}\Delta A_\Sigma &= \sum_{i=1}^{3}(K_i \Delta P_i\%)A_i \\ &= 1.001 \times 0.848\% \times 22\ 675.86 + 1.008 \times 0.903\% \\ &\quad \times 59\ 125.04 + 1.000\ 1 \times 0.972\% \times 11\ 766.5 \\ &= 192.48 + 538.17 + 114.38 \\ &= 845.035\ (\text{MW}\cdot\text{h})\end{aligned}$$

The line loss rate is

$$\Delta A_\Sigma\% = \Delta A_\Sigma/\Sigma P_G t = (845.035/93\ 567.4) \times 100\% = 0.903\%$$

10.5.1.5 Comparison and Conclusion of Electric Energy Loss Rate (Line Loss Rate)

According to the 24 h load flow calculation, $\Delta P_\Sigma = 820$ MW, $\Delta P_\Sigma\% = 820/93\ 567.4 \times 100\% = 0.876\%$

According to the generated output load curve, $f_G = 0.742\ 58$, $\beta_G = 0.529\ 2$, so the loss factor is

$$F_G = 0.639(0.742\ 58)^2 + 0.361(0.742\ 58 \times 0.529\ 2 + 0.742\ 58 - 0.529\ 2) = 0.571\ 26$$

The form coefficient is $K_G = \sqrt{F_G}/f_G = \sqrt{0.571\ 26}/0.742\ 58 = 1.017\ 8$, so the actual line loss rate is

$$\Delta A_\Sigma\% = K_G \Delta P_G\% = 1.017\ 8 \times 0.876\% = 0.892\%$$

Compared with the three-mode section division method, the relative error of this method is

$$\delta(\Delta A_\Sigma)\% = (0.903 - 0.892)/0.892 \times 100\% = 1.23\%$$

According to the above calculations, the results of the 24 h load flow calculation of the world's commonly used standard test grid with 10 generators and 39 nodes indicate that the error of the loss calculation by the three-mode section division method is within the allowable range. This method can be applied in the operation and management of power grids, and experience can be accumulated for improvement.

10.5.2 Three-mode Calculation Based on Total Loads and Measured Loss Power over 24 h in one Province During 2004

Reference [42] studied the effect of secondary voltage control. This example introduced total loads and measured loss power from the reference, as shown in Table 10.7, to verify the correctness of the loss calculation by the three-mode section division method.

1. Calculation of parameters of the total load curve. Accumulated daily electric quantities of 24 h fiducial point loads are $A = \sum_1^{24} P_i t = 52\ 505.91 \times 10^4$ (kW·h); $P_{av} = A/24 = 2187.75 \times 10^4$ (kW), $P_{max} = 2400 \times 10^4$ kW, $f = 2187.75/2400 = 0.9116$; $\beta = P_{min}/P_{max} = 1892.96/2400 = 0.788\ 7$, $\frac{1}{\lambda} = (1-f)/(f-\beta) = (1-0.911\ 6)/(0.911\ 6 - 0.788\ 7) = 0.719\ 3$; the daily power loss rate is $\Delta P\% = (\sum \Delta P_t)/A = 457.3/525\ 05.91 = 0.871\%$.
2. Calculation of three-mode power points of division. As $f = 0.9116$, use Formulas (3.22) and (3.23) for the combination of high load factor to obtain

$$P_{s.m*} = 0.008 \times 0.719\ 3^2 - 0.126\ 6 \times 0.719\ 3 + 0.904\ 3 = 0.817\ 4$$
$$P_{s.m} = P_{s.m.*} P_{max} = 0.817\ 4 \times 2400 \times 10^4 = 1961.76 \times 10^4\,(\text{kW})$$
$$P_{m.b*} = 0.005\ 2 \times 0.719\ 3^2 - 0.045\ 6 \times 0.719\ 3 + 0.991\ 8 = 0.961\ 7$$
$$P_{m.b} = P_{m.b*} P_{max} = 0.961\ 7 \times 2400 \times 10^4 = 2308.08 \times 10^4\,(\text{MW})$$

3. Calculation of average loads of each section
   For the minimum mode section:

$$P_{s.av} = (\beta + P_{s.m})/2 \cdot P_{max} = (0.788\ 7 + 0.817\ 4)/2 \times 2400 \times 10^4 = 1927.32 \times 10^4\,(\text{kW})$$

   For the maximum mode section:

$$P_{b.av} = (P_{m.1} + 1)/2 \cdot P_{max} = (0.961\ 7 + 1.0)/2 \times 2400 = 2354.04 \times 10^4\,(\text{kW})$$

4. Conversion of power loss rates corresponding to average loads of three modes. Select the time points at which measured fiducial point power values are most close to the average power values under the three modes, and convert to corresponding power loss rates based on the square of power.
   a. $P_{s.av}$ under the minimum mode is most close to the power at hour 2, and the conversion of power loss rate is as follows:

$$\Delta P_s\% = \Delta P_{2'}\%(P_{s.av}/P_{2'})^2 = (12.5/1918.31)\ (1927.32/1918.31)^2 = 0.658\%$$

b. P_{av} under the normal mode is most close to the power at hour 12, so

$$\Delta P_m\% = \Delta P_{12^\circ}\%(P_{av}/P_{12^\circ})^2 = (17.5/2146.48)\,(2187.75/2146.48)^2 = 0.847\%$$

c. $P_{max.av}$ under the maximum mode is most close to the power at hour 15, so

$$\Delta P_b\% = \Delta P_{15^\circ}\% \times \left(\frac{P_{b.av}}{P_{15^\circ}}\right)^2 = \frac{24}{2366.2} \times \left(\frac{2354.04}{2366.2}\right)^2 = 1.004\%$$

5. Calculation of electric quantity ratio of three modes. According to the integral operation from the load curve (Rossander Formula), for the combination of high load rate parameters, when $1/\lambda = 0.80$, the electric quantity ratio of three modes is $\alpha_1{:}\alpha_2{:}\alpha_3$ = 1:2.803:1.628; when $1/\lambda$ = 0.60, $\alpha_1{:}\alpha_2{:}\alpha_3$ = 1:3.63:0.978. According to calculation by the interpolation method, when $1/\lambda$ = 0.719 3, the electric quantity ratio of three modes is $\alpha_1{:}\alpha_2{:}\alpha_3$ = 1:3.30:1.366.
6. Calculation of electric energy losses of three-mode sections
   For the minimum mode section:

$$\Delta A_s = \frac{A_\Sigma \cdot \alpha_1}{\alpha_1+\alpha_2+\alpha_3}\Delta P_s\% = \frac{52\,505.91\times 10^4}{1+3.30+1.366}\times 0.658\%$$
$$= 60.98\times 10^4(\text{kW·h})$$

   For the normal mode section:

$$\Delta A_m = \frac{A_\Sigma \cdot \alpha_2}{\alpha_1+\alpha_2+\alpha_3}\Delta P_m\% = \frac{52\,505.91\times 10^4\times 3.30}{1+3.30+1.366}\times 0.847\%$$
$$= 259.02\times 10^4(\text{kW·h})$$

   For the maximum mode section:

$$\Delta A_b = \frac{A_\Sigma \cdot \alpha_3}{\alpha_1+\alpha_2+\alpha_3}\Delta P_b\% = \frac{52\,505.91\times 10^4\times 1.366}{1+3.30+1.366}\times 1.004\%$$
$$= 127.091\times 10^4(\text{kW·h})$$

7. Calculation of total losses according to the full-period loss rate. The full-period power loss rate is $\Delta P_\Sigma\%$ = 457.30/52 505.91 = 0.871%, and calculate the loss factor with f and β as per Formula (2.28):

$$F = 0.639f^2 + 0.361(f + f\beta - \beta)$$
$$= 0.639\times 0.911\,6^2 + 0.361\times(0.911\,6 + 0.911\,6\times 0.788\,7 - 0.788\,7)$$
$$= 0.834\,9$$

   The form coefficient of the total load curve is $K = \sqrt{F}/f = \sqrt{0.834\,9}/0.911\,6 = 1.002$
   Use Formula (10.10) to calculate the full-period electric energy loss rate:

$$\Delta A_\Sigma\% = K\Delta P_\Sigma\% = 1.002\times 0.871\% = 0.873\%$$
$$\Delta A_\Sigma = A\Delta A_\Sigma\% = 52\,505.91\times 0.873\% = 458.39\times 10^4(\text{kW·h})$$

8. Calculation of method error. There is an error between the electric energy losses of the three-mode sections and the calculation result from the measured loss rate, as follows:

$$\Delta A'_\Sigma = \Delta A_s + \Delta A_m + \Delta A_b = (60.98 + 259.02 + 127.091)\times 10^4$$
$$= 447.091\times 10^4(\text{kW·h})$$

The relative error is

$$\delta\Delta A\Sigma = \left(\Delta A_\Sigma - \Delta A'_\Sigma\right)/\Delta A_\Sigma \times 100\%$$
$$= (458.39 - 447.091)/458.39 \times 100\% = 2.46\%$$

If the form coefficient of three modes is greater than 1.0, the relative error in the calculation of electric energy losses may be reduced.

Example 10.2 Assume that the daily load curve in Table 10.2 lasts for 10 days. Due to the continuous adjustment of loads, the electric quantity in the large load period of a further 10 days is increased by 5%, while the maximum load is reduced by 1%. *Try to calculate the loss rate of those further 10 days. If the adjusted loads last for 20 days, what is the total loss rate for a total of 30 days? What is the effect of loss reduction of the continuously adjusted loads?*

Solution

1. *Calculation of loss rate within the second period of 10 days.*
   1.1 *Parameter calculation*

$$P_{\max 1} = 0.99P_{\max 0} = 0.99 \times \left(2400 \times 10^4\right) = 2376 \times 10^4 (\mathrm{kW})$$

The daily electric quantity is

$$A_{\Sigma 1} = 0.05 \times \left[A_{\Sigma 0}\cdot\alpha_3/(\alpha_1+\alpha_2+\alpha_3)\right] + A_{\Sigma 0}$$
$$= \left\{\left(0.05 \times \frac{1.366}{5.666} + 1\right) \times 52\,505.91\right\} \times 10^4$$
$$= 53\,138.84 \times 10^4 (\mathrm{kW{\cdot}h})$$

$$P_{\mathrm{av}} = A_{\Sigma 1}/24 = 53\,138.84 \times 10^4/24$$
$$= 2214.12 \times 10^4 (\mathrm{kW})$$

$$f = P_{\mathrm{av}}/P_{\max 1} = 2214.12/2376 = 0.931\,9$$

$$\beta = P_{\min 0}/P_{\max 1} = \frac{1892.86}{2376} = 0.796\,7$$

$$\frac{1}{\lambda} = \frac{1-f}{f-\beta} = \frac{1-0.931\,9}{0.931\,9-0.796\,7} = 0.503\,7$$

   1.2 *Calculation of three section power points of division.* According to Formulas (3.22) and (3.23),

$$P_{\mathrm{s.m}*} = 0.008 \times 0.503\,7^2 - 0.126\,6 \times 0.503\,7 + 0.904\,3 = 0.842\,6$$

$$P_{\mathrm{s.m}} = 0.842\,6 \times 2376 \times 10^4 = 2002.02 \times 10^4 (\mathrm{kW})$$

$$P_{\mathrm{m.b}*} = 0.005\,2 \times 0.503\,7^2 - 0.045\,6 \times 0.503\,7 + 0.991\,8 = 0.970\,2$$

$$P_{\mathrm{m.b}} = 0.970\,2 \times 2376 \times 10^4 = 2305.2 \times 10^4 (\mathrm{MW})$$

   1.3 *Conversion of power loss rate.* After the average power of the three sections is obtained, take the power loss rate corresponding to the fiducial point power to which the average power is close for conversion: $P_{s.av} = (1892.96 + 2002.02) \times 10^4/2 = 1947.49 \times 10^4$ (kW). The power at hour 1 is close:

$$\Delta P_s\% = (\Delta P\%)_{1^\circ}(P_{s.av}/P_{1^\circ})^2 = (13.5/1952.11)\ (1947.49/1952.11)^2$$
$$= 0.692\% \times 0.995\ 2 = 0.689\%$$

$P_m = P_{av} = 2214.12 \times 10^4$ (kW), and the power at hour 18 is close:

$$\Delta P_m\% = (\Delta P\%)_{15^\circ}(P_m/P_{15^\circ})^2 = (19.75/2222.54)\ (2214.12/2222.54)^2$$
$$= 0.889\% \times 0.992\ 4 = 0.882\%$$

$P_{b.av} = (2305.20 + 2376) \times 10^4/2 = 2340.6 \times 10^4$ (kW), and the power at hour 14 is close:

$$\Delta P_b\% = (\Delta P\%)_{14^\circ}(P_b/P_{14^\circ})^2 = (23/2334.08)\ (2340.6/2334.08)^2$$
$$= 0.985\% \times 1.006 = 0.991\%$$

1.4 *Calculation of electric quantities of three sections.* According to integral operation by the load curve (Rossander Formula), when $1/\lambda = 0.40$, $\alpha_1{:}\alpha_2{:}\alpha_3 = 1{:}3.66{:}1.05$; when $1/\lambda = 0.60$, $\alpha_1{:}\alpha_2{:}\alpha_3 = 1{:}3.63{:}0.978$. According to calculation by the interpolation method, when $1/\lambda = 0.503\ 7$, $\alpha_1{:}\alpha_2{:}\alpha_3 = 1{:}3.646{:}1.015$. The total daily electric quantity after load adjustment is calculated as $A_{\Sigma 1} = 53\ 138.84 \times 10^4$kW·h, and then the electric quantities of three sections within the second period of 10 days can be obtained:

$$A_s = A_{\Sigma 1}/(\alpha_1 + \alpha_2 + \alpha_3) = 53\ 138.84 \times 10^4/(1 + 3.646 + 1.015)$$
$$= 9386.829 \times 10^4 (\text{kW·h})$$
$$A_m = \alpha_2 A_x = 3.646 \times 9386.829 \times 10^4 = 34\ 224.379 \times 10^4 (\text{kW·h})$$
$$A_b = \alpha_3 A_x = 1.015 \times 9386.829 \times 10^4 = 9527.63 \times 10^4 (\text{kW·h})$$

1.5 *Calculation of electric energy losses and loss rate within the second period of 10 days.* Due to limits on calculation conditions, based on an approximate calculation, the loss rates of three sections are regarded as the power loss rate, and the total electric energy losses are calculated as per the following formula:

$$\Delta A_\Sigma = \Delta A_s\% A_s + \Delta A_m\% A_m + \Delta A_b\% A_b$$
$$= 0.689\% \times 9386.829 \times 10^4 + 0.882\% \times 34\ 224.379 \times 10^4 + 0.991\% \times 9527.63 \times 10^4$$
$$= (64.675 + 301.859 + 94.42) \times 10^4$$
$$= 460.953 \times 10^4 (\text{kW·h})$$
$$\Delta A_\Sigma\% = \Delta A_\Sigma / A_{\Sigma 1} = 460.953/53\ 138.84 = 0.867\%$$

Calculations in this example show that the three-mode section division method can be used for calculating the effect of any loss reduction measure taken in a mode section, such as a time-sharing electricity price. Therefore, it can be extended into the calculation of revenue and expenditure changes caused by changes in electricity purchase and the sales quantities of different mode sections. In short, the three-mode section division method provides a new calculation and analysis method for the lean management of power grid enterprises.

Compared with the original 10 days, the increase in the daily electric energy losses is

$$\Delta A_{\Sigma 1} - \Delta A_{\Sigma 0} = (460.953 - 458.39) \times 10^4 = 2.56 \times 10^4 (\text{kW·h})$$

1.6 *Calculation of loss reduction effect.* If no measure of load adjustment is taken, the electricity losses of the large load section are increased in the function of the square of electric supply, that is

$$\Delta A_{b1}-\Delta A_{b0}=\left(1.05^2-1\right)\times\Delta A_{b0}=0.1205\times 127.091\times 10^4$$
$$=13.027\times 10^4(\text{kW·h})$$

After the measure of load adjustment is taken, the actual electric energy losses are reduced by

$$\delta\Delta A_{\Sigma}=(13.027-2.56)\times 10^4=10.47\times 10^4(\text{kW·h})$$

The ratio of loss reduction is

$$(\delta\Delta A_{\Sigma})\%=[10.47/(460.953+13.027)]\times 100\%=2.22\%$$

2. *Calculation of losses for 30 days in a month.*
   2.1 *Calculation of electric energy losses in 30 days.* The electricity losses in the latter 20 days are $\Delta A_{\Sigma2}$ = 20 × 460.953 × 10^4 = 921.91 × 10^5 (kW·h), and the total electricity losses in 30 days are $\Delta A_{\Sigma3}$ = (921.91 + 447.091) × 10^5 = 1369.0 × 10^5 (kW·h)
   2.2 *The total electric supply in 30 days.* This is $\Delta A_{\Sigma3}$ = (20 × 53 138.84 + 10 × 52 505.91) × 10^4 = 158 783.59 × 10^5 (kW·h)
   2.3 *The total loss rate in 30 days.* This is $\Delta A_{\Sigma3}\%$ = ($\Delta A_{\Sigma3}/A_{\Sigma3}$) × 100% = (1369.0/158 783.59) × 100% = 0.862%
   2.4 *Calculation of loss reduction effect.* If no measure of load adjustment is taken, the total loss rate in 30 days is

$$\Delta A'_{\Sigma1}\%=[20\times(460.953+13.027)+10\times 447.091]\times 10^4/\left(158\ 783.59\times 10^5\right)$$
$$=0.878\%$$

After a measurement of the loss reduction is taken, the total loss rate in the whole month is reduced by

$$\delta(\Delta A_{\Sigma\%})=[(0.878-0.862)/0.878]\times 100\%=1.82\%$$

11

Analysis and Calculation of Loss Allocation

The traditional integrated electric power industry structure of power generation, transmission, distribution, and sales has been replaced by a market-oriented reform through the implementation of a separation between power plants and power grids and the bidding on-grid of power plants. Therefore, many practical problems arise from the planning, dispatching, production, and operation of electric power systems need to be further studied. Loss allocation is one of these problems and has been of wide public concern in recent years.

Although many loss allocation methods have been put forward, many domestic scholars think that so far there is no best method that is widely applicable and accepted. On the basis of extensive comparison, some scholars propose that: (i) the average loss allocation method is recommended for the distribution network given its complexity; (ii) the marginal loss coefficient method is recommended for the transmission network; (iii) it is recommended that the transmission network can be divided into several regions after grids are connected across the whole country, wherein the node loss coefficient method that is adjusted in real time is used within a region, while the regional marginal loss coefficient method is used between regions; (iv) the power flow tracing method is recommended for the loss allocation of independent power producers [43]. The authors are in favor of the selection of different loss allocation methods as appropriate.

This chapter explores the nature of loss allocation by model analysis, summarizes four contents that are helpful to the calculation of loss allocation, introduces the application of the regional marginal loss coefficient method, explains the process of inter-provincial loss allocation within regional power grids and gives calculation examples, and finally analyzes the calculation of loss allocation under a complex trading setup.

11.1 Occurrence of Loss Allocation Problem and Possible Solutions

Before the electric power market-oriented reform, power supply enterprises purchased electricity from power plants which were under the same jurisdiction of the provincial electric power bureau as power supply enterprises, and they supplied electricity to and charged all customers. Electric energy losses in substations and lines were all included in the cost of power supply. After payment of electricity purchase expenses to the provincial electric power bureau, internal profits deducting fixed expenses maintained the normal operation of power supply enterprises. Therefore, power supply enterprises independently bore electricity line losses and expenses before the market-oriented reform, and there was no loss allocation problem.

Line Loss Analysis and Calculation of Electric Power Systems, First Edition. Anguan Wu and Baoshan Ni.

After the electric power market-oriented reform, power grid enterprises were no longer the only electricity purchasers. In the double load power supply model shown in Figure 11.1, direct supply customer *b* (indicated by subscript b) of a power plant only has a supply–demand relationship with the power plant. At this time, however, load *b* (indicated by subscript b) passes line I, resulting in variable losses which need to be reasonably allocated between the power grid enterprise (representing the ordinary customer *a*, indicated by subscript a) and the power plant (representing the direct supply customer *b*).

11.1.1 Analysis of Double Load Power Supply Model

If the power factor of customers *a* and *b* is constant, that is 1.0, then $i_I = i_a + i_b$. The change of i_I is shown in the dot-dashed line in Figure 11.1b. The power loss calculation formula can only derive power losses ΔP_a and ΔP_{ab} in section I which are, respectively, under situations where only load *a* is included and where customers *a* and *b* are both included, as shown in Figure 11.1c.

The general formula for calculating power losses in common section I is $\Delta P = 3i^2_I R_I = 6i^2_I$. According to Figure 11.1b indicating load changes of the three sections with only numbers (omitting current unit), the curve i_a is (2, 3, 4), curve i_b (6, 4, 2), and curve I_{ab} (8, 7, 6). Use the above formula to calculate ΔP_a and ΔP_{ab} of each section, as shown in Figure 11.1c. In addition to the electric energy losses in the dedicated line II established by itself, customer *b* must bear the increased electric energy losses after its load passes section I. The increased electric energy losses can be calculated based on numbers in Figure 11.1c, as follows:

$$\begin{aligned}\partial \Delta A_b &= \Delta A_{ab} - \Delta A_a \\ &= (384+294+216)\times 8-(24+54+96)\times 8 \\ &= 7152-1392=5760=5.76\ (\text{kW}\cdot\text{h})\end{aligned}$$

Now assume that terminal A of the common section I is not connected with load *b* but connected with load *c*, as shown in the dotted line in Figure 11.1a. Its load curve i_c (2, 4, 6) is shown in the short-dashed line in Figure 11.1b. At this time, the current in section I is i_{ac} (4, 7, 10), as shown in the multidot-dashed line in Figure 11.1b. The corresponding power loss curve ΔP_{ac} is shown in Figure 11.1c. Because i_c and i_a change in the same trend, with a superposition of peak values, the maximum ΔP_{ac} is much larger than the maximum ΔP_{ab}. Likewise, the increased electric energy losses in the common section I after load *c* is added are calculated:

$$\begin{aligned}\partial \Delta A_c &= \Delta A_{ac} - \Delta A_a \\ &= (96+294+600)\times 8-(24+54+96)\times 8 \\ &= 7290-1392=6528=6.528\ (\text{kW}\cdot\text{h})\end{aligned}$$

According to the comparison between double load *ab* and double load *ac*, although direct supply customers *b* and *c* have the same daily electric supply, the increase in electric energy losses allocated to direct supply customers may vary greatly due to their different load curves.

The following conclusions can be drawn from the above analysis of the double load power supply model:

1. Increased losses resulted from the high-voltage direct supply customer in the common line (section I in Figure 11.1a) should be borne by itself, and depend on the size and change of the original common load *a* as well as the size and change of the direct supply load itself.
2. Because load curves change differently, and differences in power losses under minimum, normal, and maximum modes divided by total loads vary, power losses cannot serve as the basis of loss allocation.
3. The area under the power loss curve is the quantity of electric energy losses. The quantity of electric energy is a scalar quantity. Differences in electric energy losses under the three modes are additive and the sum is

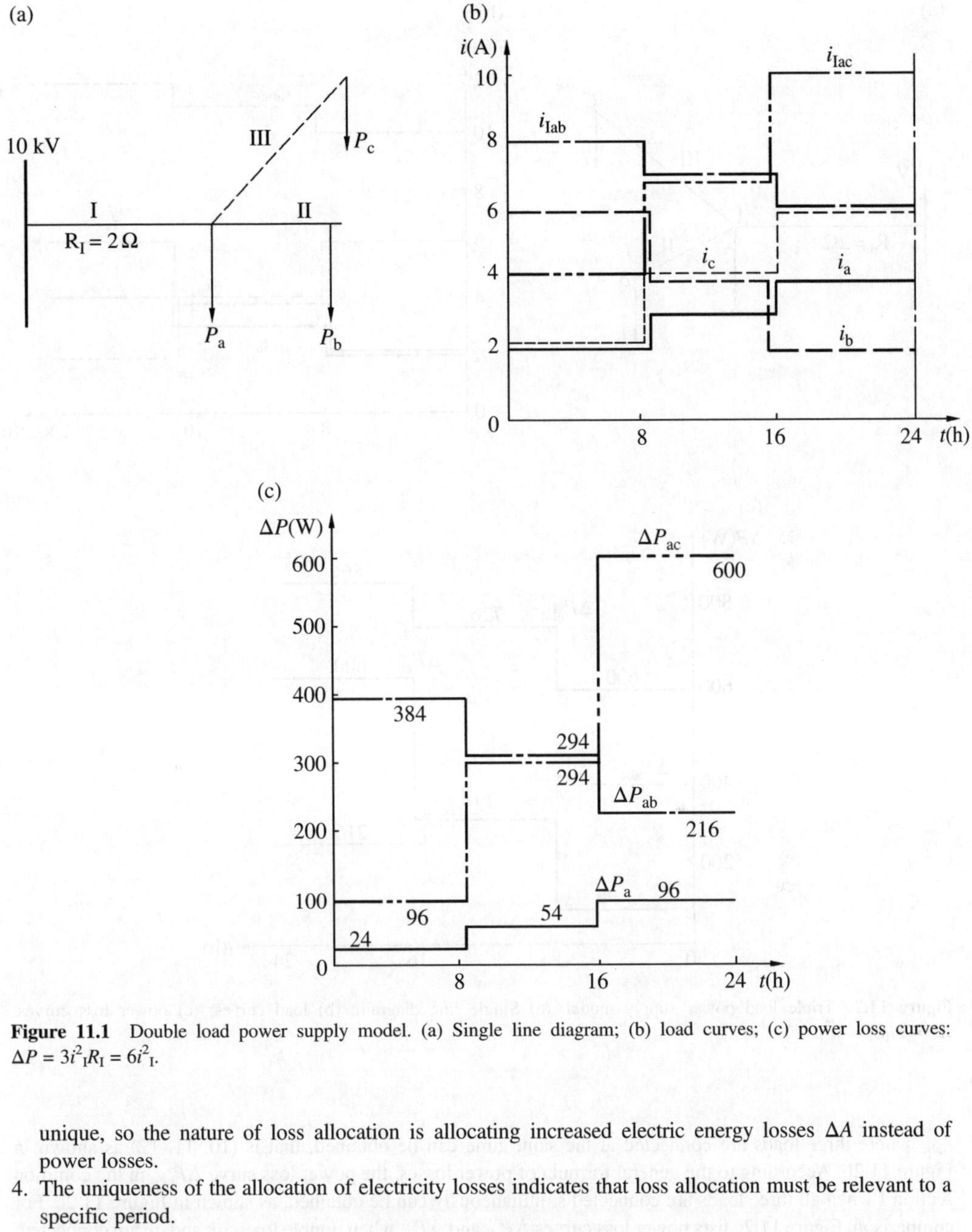

Figure 11.1 Double load power supply model. (a) Single line diagram; (b) load curves; (c) power loss curves: $\Delta P = 3i^2_I R_I = 6i^2_I$.

unique, so the nature of loss allocation is allocating increased electric energy losses ΔA instead of power losses.

4. The uniqueness of the allocation of electricity losses indicates that loss allocation must be relevant to a specific period.

11.1.2 Analysis of Triple Load Power Supply Model

When customers a, b, and c in Figure 11.1a are connected at the same time, they constitute a triple load power supply model, as shown in Figure 11.2a. Likewise, assume that the load power factor of loads a, b, and c is constant, that is 1.0. According to the load curves i_a (2, 3, 4), i_b (6, 4, 2), and i_c (2, 4, 6) shown n Figure 11.1b,

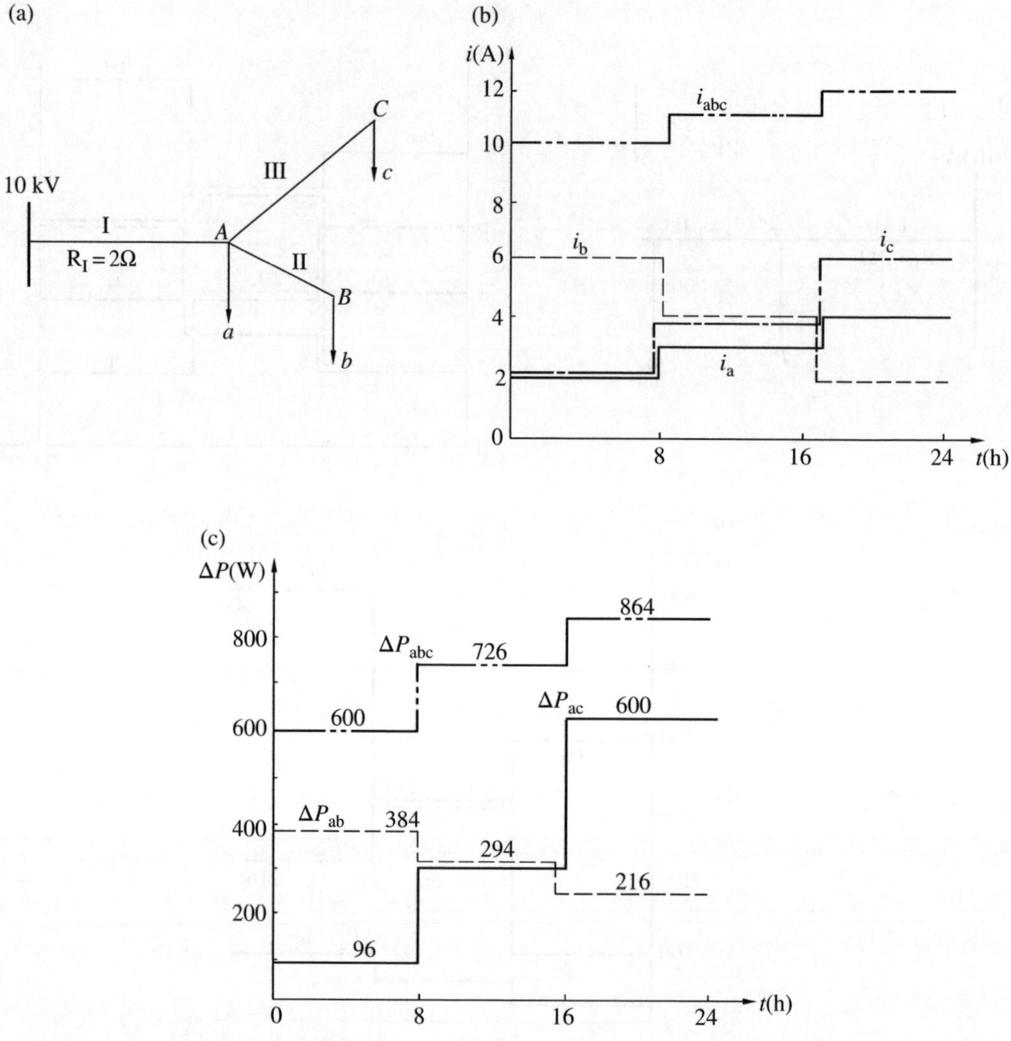

Figure 11.2 Triple load power supply model. (a) Single line diagram; (b) load curves; (c) power loss curves: $\Delta P = 3i_I R_I = 6i^2_I$.

i_{Iabc} where three loads are connected at the same time can be obtained, that is (10, 11, 12) as shown in Figure 11.2b. According to the general formula of power losses, the power loss curve ΔP_{abc} in the common section I when all three loads are connected simultaneously can be obtained, as shown in Figure 11.2c. For comparison, Figure 11.2c lists power loss curves ΔP_{ab} and ΔP_{ac} when double loads *ab* and *ac* are connected.

According to Figure 11.2c, due to the connection of the third load, electric energy losses in the common section I are greatly increased, that is ΔA_{abc} = 8 (600 + 726 + 864) = 17520 (W·h). The loss allocation to customers *b* and *c* can be discussed as below:

1. In the first situation, assume that customers *b* and *c* are connected simultaneously and share the difference between losses when three customers are connected and losses when only load *a* is connected according to

the ratio of electric supply. Because customers *b* and c have the same daily electric supply, the increased losses shared by them are

$$\partial\Delta A_b=\partial\Delta A_c=\frac{1}{2}(\Delta A_{abc}-\Delta A_a)=\frac{1}{2}(17\,520-1392)=8064=8.064\ (\text{kW}\cdot\text{h})$$

Compared with the situation where only customers *a* and *b* are connected, the losses allocated to customer *b* are increased from 5.76 to 8.064 kW·h, with an increase of $\delta\Delta A_b=(8.064-5.76)/5.76=40\%$. This value can be regarded as a subsidy proportion of customer *b* to customer *c*; on the contrary, it can be deemed that customer *c* provides subsidy for customer *b*, thus the percentage of increased losses allocated to it is increased by $\delta\Delta A_c=(8.064-6.528)/6.528=23.5\%$. This shows that two simultaneously connected customers can both claim that one provides subsidy for the other one, as is known as "cross-subsidization". The ratio of subsidization varies with customers' load curves, that is, the three-mode sections have different ratios of cross-subsidization.

2. If customer *b* is first connected to the system, then there are two possible allocations. The first is that customer *c* bears all increased losses in section I after it is connected, that is

$$\partial\Delta A_{c1}=\Delta A_{abc}-\Delta A_{ab}=17\,520-7152=10\,368=10.368\ (\text{kW}\cdot\text{h})$$

Customer *c* bears too many increased losses. From the perspective of rights of flow in the common section I, apparently there is a big difference in the loss allocation if the sequence of connection is different, and this is not fair.

The second allocation is that increased losses in section I after customer *c* is connected are shared by customers *b* and *c* according to the ratio of electric supply. This is called the average loss rate method. At this time, the respective losses borne by the three customers are

$$\Delta A_a=(24+54+96)\times 8=1392=1.392\ (\text{kW}\cdot\text{h})$$
$$\partial\Delta A_{abc}=\Delta A_{abc}-\Delta A_a=17\,520-1392=16\,128=16.128\ (\text{kW}\cdot\text{h})$$

$$\begin{aligned}\partial\Delta A_{bI}&=(\Delta A_{ab}-\Delta A_a)+\frac{1}{2}[\partial\Delta A_{abc}-(\partial\Delta A_b+\partial\Delta A_c)]\\&=5760+\frac{1}{2}[16\,128-(5760+6528)]=7680=7.68\ (\text{kW}\cdot\text{h})\end{aligned}$$

$$\begin{aligned}\partial\Delta A_{cI}&=(\Delta A_{ac}-\Delta A_a)+\frac{1}{2}[\partial\Delta A_{abc}-(\partial\Delta A_b+\partial\Delta A_c)]\\&=6528+1920=8448=8.448\ (\text{kW}\cdot\text{h})\end{aligned}$$

According to a check calculation,

$$\Delta A_a+\partial\Delta A_{bI}+\partial\Delta A_{cI}=1.392+7.68+8.448=17.52\ (\text{kW}\cdot\text{h})=\Delta A_{abc}$$

The sum of increased losses in section I that are borne by three customers is equal to the actual losses in section I, indicating that this allocation is balanced.

3. There are two possible sequences of connection for customers *b* and *c*. More customers lead to more combinations of connection sequence. To consider reasonable loss allocation under multiple combinations, the frequency of occurrence should be used as a weight to average different increased losses. This is the Shapley method, to be introduced later in this chapter.
4. Because load curves change differently, the loss allocation based on the three-mode section division of total loads should differentiate and its comprehensive result should be more reasonable and fair. Therefore, in the calculation of loss allocation, it is very necessary to do the respective calculations based on the three-mode section division of the total loads.

5. According to Figure 11.2a, sections II and III may have different impedances, so the marginal line loss rates (the ratio between the increased line losses caused by increase in unit electric supply and the increase in unit electric supply) measured from the multi-branch node A to load points B and C are different. Therefore, the marginal line loss rate based on the grid topological structure, electrical parameters and load distribution can also serve as an economic signal for a fair and reasonable allocation of increased losses.

11.1.3 Possible Solutions to Loss Allocation

The simple analysis of power supply models shows the complexity and difficulty of loss allocation, and also indicates possible solutions.

1. The ranking of several loads in the same node branch according to the quantity of marginal line loss rates in different directions at the node load side is unique.
2. The allocation of increased losses to the common branch on the left of node (power source side) and to the selected branch on the right side (load side) with maximum marginal line loss rate is reasonable, because this allocation leads to smaller energy losses in the branch at the load side and meets the requirement of economic efficiency of the entire grid.
3. To allocate increased losses in a region where the transmission price varies, such as in provincial grids, it is necessary to multiply the marginal line loss rate by the transmission price to obtain the marginal loss electricity price. The reduction of expenses incurred by increased losses in the entire grid is used as the only reasonable economic signal, that is, increased losses are allocated in the decreasing order of marginal loss electricity price, to realize fair, reasonable, and unique allocation.

11.2 Theoretical Preparation for Loss Allocation

11.2.1 Three-mode Section Division of Active Load Duration Curve

Section 3.4 of Chapter 3, on the premise of multiplying probability assumption, conducted a linkage analysis of load loss curves and active load duration curves within a full period, and obtained Formulas (3.22) to (3.25) for calculating the three-mode section power points of division under minimum, normal, and maximum loads, as well as the electric quantity ratio α_1:α_2:α_3 of three-mode sections (see Table 3.4). This provides basic conditions for the calculation of loss allocation in the three-mode periods by using marginal line loss rates and peak, valley and basis electricity prices under the three modes, provided load curve parameters are given.

11.2.2 Calculation of Influence of Transit Electric Supply on Electricity Line Losses

Section 7.2 of Chapter 7 explained the influence of transit electric supply on the line loss rate. Assume that the ratio between the transit electric supply and the original regional electric supply is q, and the increased line loss rate due to the transit electric supply is $\Delta A'\%$. The ratio between $\Delta A'\%$ and the original line loss rate ΔA is shown in Formula (7.10), that is

$$\Delta A'\%/\Delta A\% = \varphi(q,\theta) = \left[\left(1+q^2\right)+\theta\right]/[(1+q)\ (1+\theta)]$$

$$\theta = B/CA^2 = \text{no-load electricity losses/load electricity losses}$$

Wherein θ – loss constituent ratio.

If the increased losses as a result of the added transit electric supply are $\partial\Delta A$, then

$$\begin{aligned}\partial\Delta A &= \Delta A'\% \times (1+q)A_1 - \Delta A\% A_1\\ &= \{\Delta A\%[(1+q^2)+\theta]/[(1+q)(1+\theta)]\}(1+q)A_1 - \Delta A\% A_1\\ &= \left[\frac{(1+q^2)+\theta}{1+\theta} - 1\right]\Delta A\% A_1\\ &= \frac{q^2}{1+\theta}\Delta A\% A_1 = q^2 \Delta A_L\% A_1\end{aligned} \tag{11.1}$$

$$\Delta A_L\% = \Delta A\%/(H\theta)$$

Wherein $\Delta A_L\%$ – original load line loss rate in the electric supply region.

According to Formula (11.1), when the ratio between the transit electric supply and the electric supply in a region is q, the increased losses in this region are q^2 times the load electricity losses under the original electric supply. Formula (11.1) provides a theoretical basis for the calculation of increased losses in loss allocation.

11.2.3 Calculation of Marginal Line Loss Rate

Section 7.5 of Chapter 7 put forward the concept of marginal line loss rate, and demonstrated the relationship between the marginal line loss rate, the load loss coefficient, and the electric supply when no-load losses are constant, as well as the relational expression between the marginal line loss rate and the load line loss rate [Formula (7.17)], that is

$$(\Delta A\%)_m = 2CA = 2\Delta A_L\%$$

Section 6.5 of Chapter 6 discussed the calculation of load loss coefficient of equivalence C_{eq} of multiple units in adjacent voltage buses according to the network shown in Figure 6.2, and gave Formula (6.17), that is

$$C_{eq} = C_b + \sum_{i=1}^{n} C_i \left(\frac{A_i}{A_0}\right)^2$$

According to the above, on the premise that the grid topological relation is given, as long as the distribution of electric supply and other operating parameters within a period are known, Formula (7.17) can be used to calculate the marginal line loss rates at all nodes according to Formula (6.14) for calculating the load loss coefficient and Formula (6.17) for calculating the multi-branch loss coefficient of equivalence. Therefore, the load loss coefficient and its equivalence inclusion as well as the concept and calculation formula of marginal line loss rate put forward in this book provide the theoretical basis and calculation conditions for the allocation of increased losses by the regional marginal loss coefficient method and the node loss coefficient method.

11.2.4 Calculation of Optimal Distribution of Increased Electric Supply

Section 7.5 of Chapter 7 demonstrated that the increased electric supply should be distributed with priority to the line with the minimal marginal line loss rate to an optional extent, and that the remaining increased electric supply should be distributed to two lines with the same marginal line loss rate in inverse proportion to the load loss coefficient, so that the electricity losses of the entire grid are minimal. This provides a similar principle for the allocation of increased losses – making grid losses tend to the minimum. If the transmission price varies, the principle of making grid loss expenses tend to the minimum should be followed.

11.3 Analysis and Calculation of Allocation of Increased Losses in Regional Power Grids

According to literature [44], against the backdrop of current electric market orientation in China, the loss allocation in regional power grids can be divided into two types: (i) the allocation of losses in the main part of the regional power grids to all provincial power grids, and (ii) the allocation of increased losses caused by power transmission and reception in inter-provincial power grids. The two types of allocation of increased losses are challenged by different conditions and problems to be solved, and therefore require different methods and calculation processes.

11.3.1 Allocation of Losses in the Main Part of Regional Power Grids to Provincial Power Grids

Because provincial power grid enterprises are managed and assessed in regional power grids, the dispatching management department of regional power grids must allocate losses in the main part of the regional power grids to all provincial power grids, which is necessary for formulating an annual plan (ex-ante measurement) and annual review (ex-post allocation). This type of loss allocation generally adopts the increased loss allocation method based on load flow calculation. The specific way is to decompose the union set of main grids into several subsets of provincial grids, to use the load flow calculation to calculate power losses before and after the decomposition, and to introduce the correction coefficient and allocate losses in the main part of the regional power grids according to the calculated power losses of each subset of provincial grids. The authors have improved this method according to the three-mode section division, and the calculation process is shown in Figure 11.3.

11.3.2 Allocation of Increased Losses Caused by Power Transmission and Reception in Inter-Provincial Power Grids

The allocation of increased losses caused by power transmission and reception in inter-provincial power grids can be analyzed from ex-post allocation. Because provincial power grids have different marginal line loss rates, power source compositions, grid conditions, and load distribution conditions, as well as various transmission prices, it is necessary to use the marginal loss electricity price (i.e. the product of marginal line loss rate and transmission price of provincial power grids) as the economic signal of loss allocation. Thus, the regional marginal loss electricity price method can be expanded based on the regional marginal loss coefficient method, and the steps below can be followed to calculate the allocation of inter-provincial losses.

Step 1: Sort provinces that transmit electricity in decreasing order by marginal loss electricity price (m_{m}). Allocate losses to several provinces (b and c) that accept electricity transmitted by the province (a) which has the maximum m_{m}.

Step 2: Sort several provinces that accept electricity transmitted by the same province in the decreasing order by m_{m}, and allocate electricity losses to the power transmitting province and these power accepting provinces one after another. If $m_{\mathrm{m}}^{\mathrm{b}} > m_{\mathrm{m}}^{\mathrm{c}}$, first allocate losses to province a and province b. If the exchange electricity between provinces a and b is A_{ab}, then the increased losses as a result of power transmission of province a to province b can be calculated as per the following formula:

$$\Delta(\Delta A_{\mathrm{ab}}) = q^2 \Delta A_{\mathrm{L.a1}} = (A_{\mathrm{ab}}/A_{\mathrm{a1}})^2 \Delta A_{\mathrm{L.a1}}$$

Step 3: Compare marginal loss electricity prices of power output and input provinces. If $m_{\mathrm{m}}^{\mathrm{b}} > m_{\mathrm{m}}^{\mathrm{a}}$, the power input province b bears all the increased losses based on the principle of economy. After the loss allocation in provinces a and b, calculate new electricity losses for each province, and deduct the constant

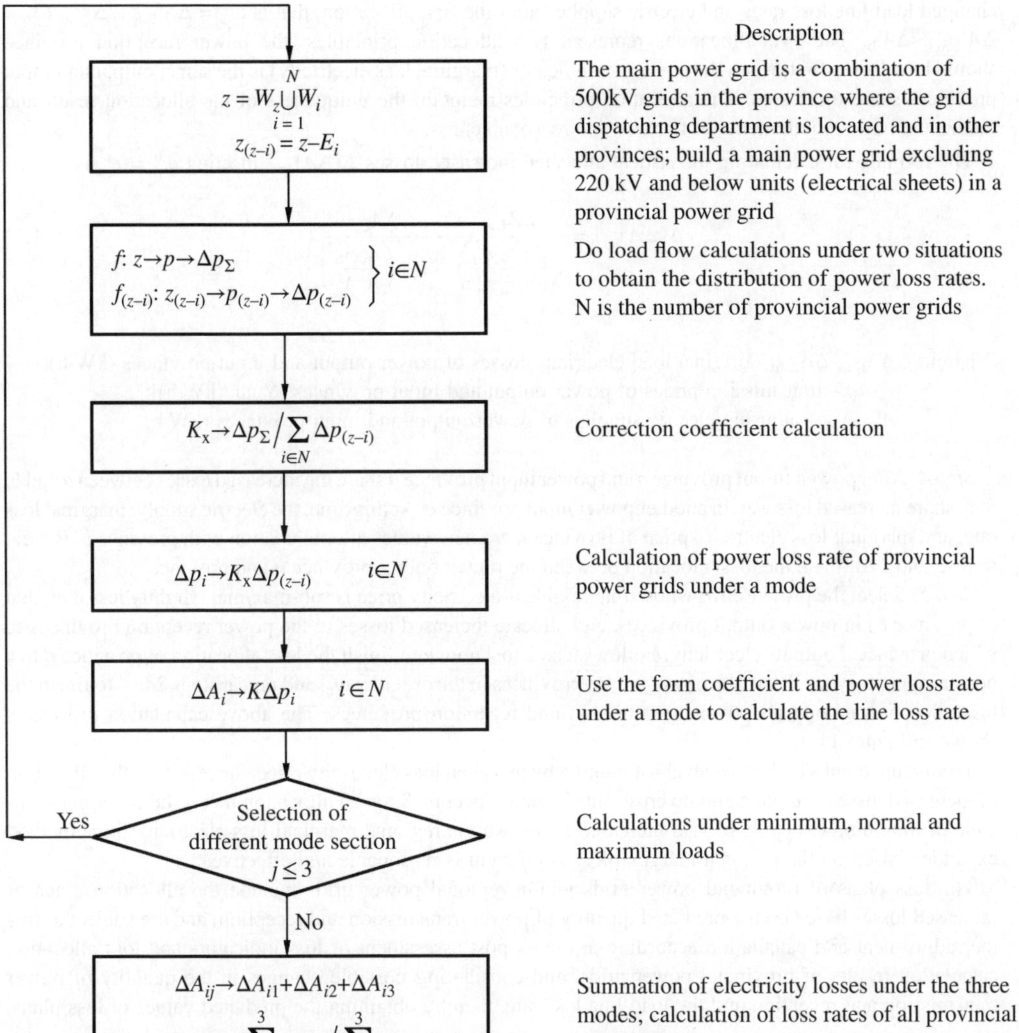

Figure 11.3 Calculation process for the allocation of losses in main part of the regional power grids to provincial power grids.

no-load electricity losses to obtain load losses, thereby calculating the new load line loss rate and obtaining a new marginal line loss rate and marginal loss electricity price. If $m_m^b < m_m^a$, then calculate the critical allocated electricity losses that make $m_m^{b'} = m_m^a$, that is $\Delta(\Delta A)_{ap.0}$, and compare it with the actual increased losses $\Delta(\Delta A)_{ab}$. If $\Delta(\Delta A)_{ab} < \Delta(\Delta A)_{ap.0}$, then province b should bear all the increased losses. If $\Delta(\Delta A)_{ab} > \Delta(\Delta A)_{ap.0}$, then province b should bear the increased losses twice. After the first allocation of the increased losses $\Delta(\Delta A)_{ab1}$, make $m_m^{b'} = m_m^{a'}$; for the second allocation, provinces a and b share the remaining increased losses $[\Delta(\Delta A)_{ab} - \Delta(\Delta A)_{ab1}]$ in inverse proportion to the product of the load loss coefficients and transmission prices of the two provinces. Note that, at this time, the load loss coefficients should be calculated by the changed load line loss rates and electric supplies after the first allocation, that is $C'_a = \Delta A'_{L.a}\%/\Delta A'_a$, $C'_b = \Delta A'_{L.b}\%/\Delta A'_b$. The two allocations represent two allocation principles: the power reception province should bear losses first; when the economic indicator (marginal loss electricity) is the same, output and input provinces share the losses. The allocation principles maintain the uniqueness of the allocation result and realize the basic requirements of fairness and reasonableness.

The formula for calculating the critical allocated increased losses $\Delta(\Delta A)_{ap.0}$ meeting $m_m^{b'} = m_m^{a'}$ is

$$\Delta(\Delta A)_{ap.0} = \frac{\Delta A_{L.a}}{1+\left(\frac{r_b}{r_a}\frac{A'_a}{A'_b}\right)} - \frac{\Delta A_{L.b}}{1+\left(\frac{r_a}{r_b}\frac{A'_b}{A'_a}\right)} \tag{11.2}$$

Wherein $\Delta A_{L.a}$, $\Delta A_{L.b}$ – original load electricity losses of power output and input provinces (kW·h);
r_a, r_b – transmission prices of power output and input provinces [Yuan/(kW·h)];
A'_a, A'_b – adjusted electric supplies of power output and input provinces (kW·h).

Step 4: After power output province *a* and power input province *b* share the increased losses between *a* and *b*, they share increased losses with another power input province *c*. At this time, the electric supply, marginal loss rate, and marginal loss electricity price of province *a* are new values after allocation with province *b*. Repeat steps 2 and 3 to finish the loss allocation between the power output province *a* and province *c*.

Step 5: Select the province (*d*) whose marginal loss electricity price is sub-maximal (slightly less than that of province *a*) in power output provinces, and allocate increased losses to the power reception provinces to which province *d* outputs electricity. Follow steps 2 to 4 above to finish the loss allocation of province *d* to a power input province. Rank the power output provinces in the order by m_{bj} and repeat steps 2 to 4 to finish the loss allocation among all power transmission and reception provinces. The above calculation process is shown in Figure 11.4.

To sum up, thanks to the two levels of ranking by marginal loss electricity price, the result for the allocation of increased losses is unique and no cross-subsidization occurs. Such an allocation meets the economic principle of minimum expenses of grid increased losses, so the regional marginal loss electricity price method expanded based on the regional marginal loss coefficient is reasonable and effective.

The loss plans of provincial power grids within regional power grids include the allocation plans of increased losses based on the predicted quantity of power transmission and reception, and are subject to further adjustment and calculation according to the ex-post assessment of loss indicators and loss allocation calculation results of provincial power grids, and considering possible changes in the quantity of power transmission and reception and the load line loss rate, thereby obtaining the predicted values of loss plans.

Example 11.1 This example is formulated based on the 2005 loss data of four eastern provinces within a regional power grid [44]. The following information is given: parameters f_P and β_P of the annual load curves of the four provinces; maximum active output, annual maximum loads, and annual electric supplies; power losses ΔP and power loss rates $\Delta P\%$ calculated by load flow calculations of the entire grid and all provincial grids under the maximum mode of the 500 kV system; the quantities of annual exchange electricity of all provinces; and transmission prices estimated by on-grid prices. See Tables 11.1 and 11.2 for details.

1. Try to calculate the annual electric supplies of three-mode sections in the entire grid and in each province.
2. If typical load curve parameters $f_{P.i}$ and $\beta_{P.i}$ under the three modes are given for the entire grid and each province, calculate the form coefficients K of typical load curves under each mode.
3. Based on the assumption that power losses are in direct proportion to the square of electric supply output, calculate the power losses ΔP_i and power loss rates $\Delta P_i\%$ of the entire grid and all provincial grids under normal mode and minimum mode.

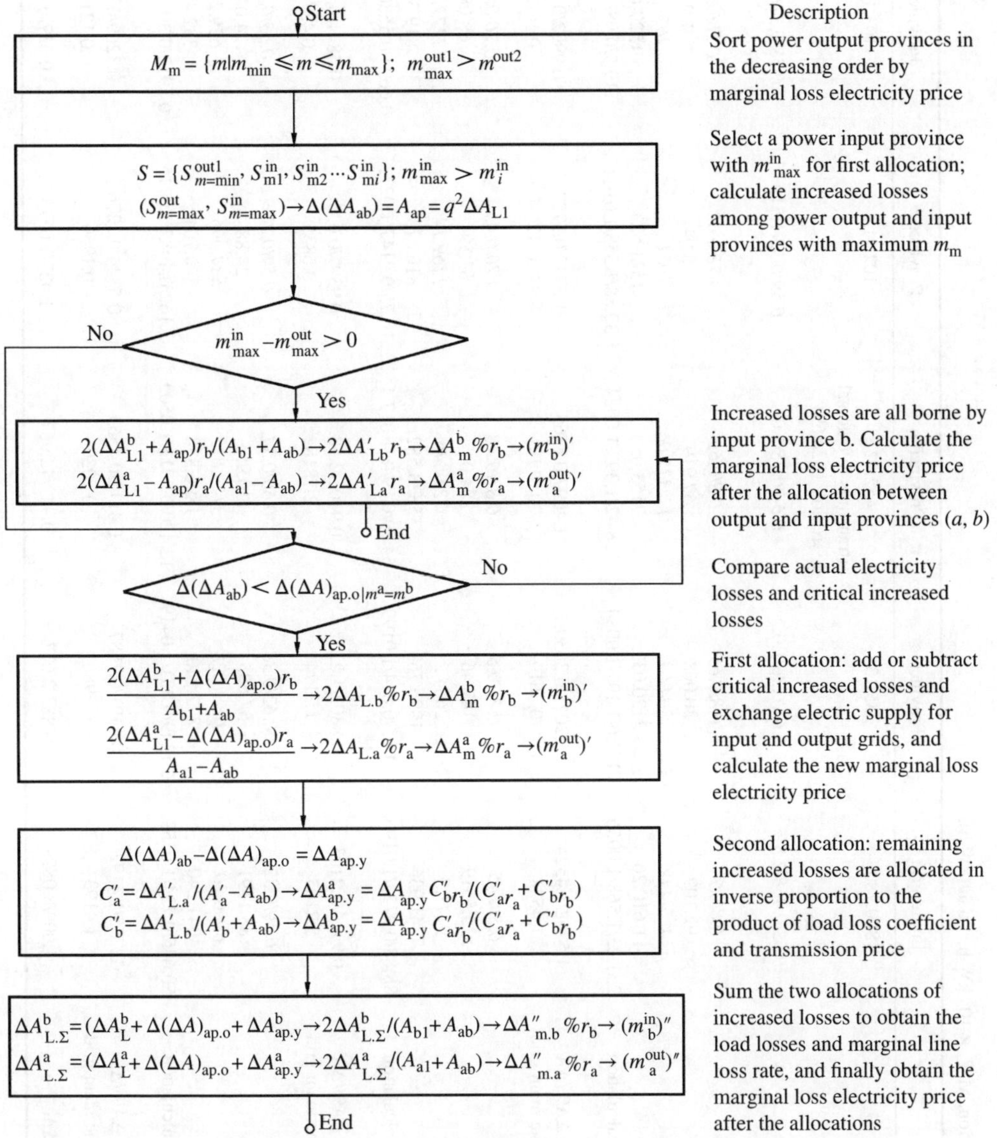

Figure 11.4 Calculation process of regional marginal loss electricity price method (main part).

4. According to power loss rates $\Delta P_i\%$, form coefficients K_i and section electric supply under the three modes, calculate loss rates $\Delta A_i\%$ and electricity losses ΔA_i of the entire grid and all provincial grids under the three modes.
5. Calculate loss allocation correction coefficients K_x under the three modes, and allocate 500 kV losses.
6. Calculate marginal loss electricity prices of all provincial grids, and use the regional marginal loss electricity price method to calculate the allocation of increased losses caused by power transmission and reception among the provincial grids.

Table 11.1 Calculation table of loss allocation, MW, × 10^8 kW·h, Yuan/(kW·h).

Scope of data		Entire grid	Province c	Province a	Province b	Province d
Annual electric supply		800.780	235.133	452.921 (transmits to external region 182.569)	102.414	103.12
Annual load curve parameters $f_P/\beta_P/\frac{1}{\lambda}$		*0.62/0.468/*2.5	*0.68/0.36/*1.0	*0.64/0.28/*1.0	*0.59/0.32/*1.519	*0.58/0.31/*1.556
Maximum mode	Typical load curve $fp/\beta p$	*0.8/0.7*	*0.8/0.65*	*0.8/0.65*	*0.78/0.62*	*0.78/0.62*
	Section electric supply	92.109	30.068	57.919	10.94	1.0
	Electric supply output	*14 744.06*	*3947.3*	*11 335.06*	*1981.55*	*2029.7*
	Average active load	11 610.26	3120.032	7153.050	1521.523	1527.64
	Representative day load flow calculation results $\Delta p/\Delta p\%/k$	231.23/1.568/1.0056	47.26/1.197/1.0084	158.42/1.398/1.0084	30.88/1.558/1.0104	19.9/0.98/1.0104
	Loss rate and electricity losses $\Delta A\%/\Delta A$	1.5768/1.4524	1.2071/0.3630	1.4097/0.8165	1.5742/0.1722	0.9902/0.1089
	Electricity losses allocated in the main grid	1.4524	0.3610	0.8119	0.1712	0.1083
Normal mode	Typical load curve $fp/\beta p$	*0.8/0.45*	*0.75/0.5*	*0.81/0.48*	*0.70/0.45*	*0.70/0.45*
	Section electric supply	525.5519	147.46	284.047	67.594	68.08
	Electric supply output	*9141.3*	*2684.2*	*7254.46*	*1169.14*	*1177.2*
	Average active load	7241.936	1996.763	5823.552	810.873	819.36
	Representative day load flow calculation results $\Delta p/\Delta p\%/k$	88.88/0.9723/1.0196	21.85/0.814/1.0199	64.89/0.8948/1.017 1	10.75/0.9147/1.0273	0.6694/0.5686/1.027 3
	Loss rate and electricity losses $\Delta A\%/\Delta A$	0.9914/5.2103	0.8302/1.2242	0.9101/2.5851	0.9397/0.6352	0.5841/0.3974
	Electricity losses allocated in the main grid	5.2103	1.3174	2.7818	0.6835	0.4276
Minimum mode	Typical load curve $fp/\beta p$	*0.59/0.32*	*0.62/0.38*	*0.60/0.34*	*0.54/0.28*	*0.53/0.27*
	Section electric supply	183.1191	57.605	110.955	23.88	24.04
	Electric supply output	*6900.20*	*1421.028*	*3173.817*	*634.096*	*629.21*
	Average active load	4041.240	877.240	1896.838	340.704	332.47
	Representative day load flow calculation results $\Delta p/\Delta p\%/k$	50.645/0.734/1.0558	6.1249/0.431/1.0419	12.420/0.3913/1.0509	3.162/0.4987/1.0715	0.1912/0.3039/1.0757
	Loss rate and electricity losses $\Delta A\%/\Delta A$	0.775/1.4192	0.4491/0.2587	0.4112/0.4563	0.5344/0.1276	0.3269/0.0786
	Electricity losses allocated in the main grid	1.4193	0.3986	0.7030	0.1966	0.1211
Annual loss rate and electricity losses $\Delta A\%/\Delta A$		1.009/8.082	0.883/2.077	0.949/4.2967	1.027/1.0513	0.637/0.657

Table 11.2 Calculation of loss allocation for inter-provincial power transmission and reception, 10^8 kW·h, Yuan/(kW·h).

Calculation item		Province *c*	Province *a*	Province *b*	Province *d*
Main grid loss rate (%)/electricity losses		0.883/2.077	0.949/4.2967	1.027/1.051 3	0.637 0/0.657 0
Annual transmitted electricity			Province *b*: 2.89/ Province *c*: 49.07		Province *b*: 1.90
Annual received electricity		Province *a*: 49.07	Province *d*: 1.90	Province *a*: 2.89	
Marginal line loss rate/no-load losses ($\Delta A_m\%/\Delta A_0$)		1.466/0.3527	1.598/0.6794	1.754/0.1536	0.974/0.1547
Transmission price *r* (1.0 is the on-grid price)		0.320	0.329	0.333	0.346
Marginal loss electricity price $m_m = 2CAr$ ($\times 10^{-3}$)		4.691 2	5.2574	5.840 8	3.370
First allocation	Rate of transit electric supply *q*/increased losses $q^2\Delta A_{1L}$		$0.0107/4.14\times 10^{-4}$		
	Electricity losses/electric supply after allocation		4.2963/267.462	1.0517/ 105.304	
	$\Delta A\%/\Delta A_L\%$ after allocation		1.606/1.3523	0.9987/0.8528	
	Marginal loss electricity price m_m ($\times 10^{-3}$) after allocation		8.898↑	5.68↓	
Second allocation	Rate of transit electric supply *q*/increased losses $q^2\Delta A_{1L}$		0.1815/0.1192		
	Electricity losses/electric supply after allocation	2.1962/ 284.203	4.1771/218.392		
	$\Delta A\%/\Delta A_L\%$ after allocation	0.773/0.649	1.913/1.602		
	Marginal loss electricity price m_m ($\times 10^{-3}$) after allocation	4.151↓	10.541↑		
Third allocation	Rate of transit electric supply *q*/increased losses $q^2\Delta A_{1L}$				$0.0184/1.71\times 10^{-4}$
	Electricity losses/electric supply after allocation		4.1772/220.292		0.973 83/101.22
	$\Delta A\%/\Delta A_L\%$ after allocation		1.896/1.5878		0.9621/0.8093
	Marginal loss electricity price m_m ($\times 10^{-3}$) after allocation		10.448↓		5.600↑

Solution

Data in italics in Table 11.1 are known, and the others are calculation results.

1. Calculate the annual electric supply based on the maximum electric supply output, annual load factor and the number of annual hours, that is $A_y = 8760 P_{max} f_P$; calculate the annual load change index $1/\lambda$ according to the given f_P and β_P; according to f_P and $1/\lambda$, refer to Table 3.4 for the electric quantity ratios $\alpha_1/\alpha_2/\alpha_3$ of the annual load curves under minimum, normal, and maximum modes. For the entire grid, due to $1/\lambda = 2.5$, refer to Table 3.4 to obtain $\alpha_1/\alpha_2/\alpha_3$ = 1.0/2.87/0.503, thus the electric supply in the maximum mode section is

$$\begin{aligned} A_{max} &= (A_{y\alpha 3})/(\alpha_1+\alpha_2+\alpha_3) \\ &= (800.780\times 10^8 \times 0.503)/(1+2.87+0.503) \\ &= 92.109\times 10^8 \ (\text{kW}\cdot\text{h}) \end{aligned}$$

In the same way, electric supplies in the normal mode section and the minimum mode section are

$$A_{\mathrm{m}}=(A_{\mathrm{y}}\alpha_2)/(\alpha_1+\alpha_2+\alpha_3),\quad A_{\mathrm{s}}=(A_{\mathrm{y}}\alpha_1)/(\alpha_1+\alpha_2+\alpha_3)$$

See Table 7.5 for the calculation results.

2. Calculate the loss factor F_P according to typical load curve parameters f_P and β_P under each mode, that is Formula (2.26)

$$F_P=0.639f_P^2+0.361(f_P+f_P\beta_P-\beta_P)$$

Calculate the form coefficient of typical load curve under each mode according to the loss factor F_P and the load factor f_P, that is Formula (1.41)

$$K_{\mathrm{n}.i}=\sqrt{F_P}/f_P$$

3. Calculate the electric supply output under normal and minimum modes according to the electric supply output under the maximum mode and annual load curve parameters f_P and β_P, that is

$$P_{\mathrm{y.m}}=f_P P_{\mathrm{y.max}},\quad P_{\mathrm{y.s}}=\beta_P P_{\mathrm{y.max}}$$

Calculate the power losses under normal and minimum modes based on the assumption that the power losses are in direct proportion to the square of electric supply output (assume that the power factor is constant)

$$\Delta P_{\mathrm{m}}=(P_{\mathrm{y.m}}/P_{\mathrm{y.max}})^2\Delta P_{\mathrm{max}},\quad \Delta P_{\mathrm{s}}=(P_{\mathrm{y.s}}/P_{\mathrm{y.max}})^2\Delta P_{\mathrm{max}}$$

Then, calculate the power loss rates under normal and minimum modes

$$\Delta P_{\mathrm{m}}\%=\Delta P_{\mathrm{m}}/P_{\mathrm{y.m}},\quad \Delta P_{\mathrm{s}}\%=\Delta P_{\mathrm{s}}/P_{\mathrm{y.s}}$$

If there are load flow calculation conditions, correct 24 h output and loads on a typical day according to the ratio between electric supply output and maximum loads on a typical day under the two modes; do the 24 h load flow calculation to obtain the daily average power losses ΔP_{av}, thus finally obtaining the power loss rate $\Delta P\%$.

4. Calculate the loss rate according to the form coefficient and power loss rate, that is $\Delta P_i\% K_{n.i}=\Delta A_i\%$. Calculate the electricity losses according to the loss rates and section electric supplies under the three modes, that is $\Delta A_i=\Delta A_i\% A_i$.
5. Calculate the loss allocation correction coefficient by maintaining allocation balance of total electricity losses, that is $K_{xi}=\Delta A_\Sigma/\Sigma\Delta A_n$. The calculation result under the maximum mode is

$$\Sigma\Delta A_n=0.363+0.8165+0.1722+0.1089=1.4606\times10^8\,(\mathrm{kW\cdot h})$$

$$\Delta A_\Sigma=1.4524\times10^8\,\mathrm{kW\cdot h},\ \text{so}\ K_{\mathrm{x.b}}=1.4524/1.4606=0.99439$$

Allocate losses in the 500 kV main grid according to this allocation coefficient, and allocate losses under the minimum mode and normal mode after the correction coefficient is calculated. See Table 11.1 for the results. The loss allocation correction coefficient under the normal mode is $K_{\mathrm{x.m}}=1.0761$, and $K_{\mathrm{x.s}}=1.5406$ under the minimum mode. This shows that correction coefficients under the three modes vary greatly, which is attributable to different load curves of provincial grids under the three modes. As a result, the calculation of loss allocation under a single mode instead of under the full period will lead to large errors.

6. Based on relevant references, to calculate the allocation of increased losses caused by inter-provincial power transmission and reception, the no-load line loss rate of provincial grids is taken as $\Delta A_0\% = 0.15\%$. If the details of main equipment parameters in all provincial grids are given, $\Delta A_0\%$ can be calculated according to actual data and may vary for different provincial grids. As the load line loss rate is $\Delta A_L\% = \Delta A\% - \Delta A_L\%$, the marginal loss electricity price of each provincial grid is $m_{m.n} = 2\Delta A_{L.n}\% r_n$. As for the calculation, the quantity of electricity transmitted to the external region should be deducted from the electric supply of province a.

The allocation of increased losses caused by inter-provincial power transmission and reception follows the process in Figure 11.4. For output provinces a and d, due to $m_{m.a} > m_{m.d}$, province a is first selected for the loss allocation with input provinces b and c. For input provinces b and c, due to $m_{m.b} > m_{m.c}$, increased losses are first allocated between provinces a and b, and then allocated between provinces a and c. After the loss allocation for output province a with input provinces b and c, the third allocation is conducted between output province d and input province a. Load losses after each round of allocation should be calculated, thereby obtaining the load line loss rate and finally calculating the marginal loss electricity price changed after allocations. According to Table 11.2, power reception provinces bear the increased losses, but due to a more obvious increase in the electric supply, m_m values of power reception provinces are reduced, but rise for power transmission provinces.

11.4 Calculation of Loss Allocation Under Complex Trading Setup

A pilot project of direct electricity purchase by large customers was first implemented in China in 2004, marking the start of opening the power sales market. The electric power market-oriented reform in China has lagged behind developed countries for over 10 years. It is imperative for electricity workers to accumulate information, study problems, and find easy and reasonable loss allocation methods in this pilot project, in order to make new efforts to develop from primary power sales marketization to bilateral trading.

This section starts with an analysis of loss allocation under the situation where a power plant directly supplies electricity to several large customers ("one to many"), it explains the loss calculation under the situation where several power plants directly supply electricity to several large customers ("many to many"), and it introduces the marginal nodal price method and the marginal loss coefficient-based generation quantity multiplier method (GMM) used in multilateral trading in developed countries.

11.4.1 Loss Allocation for Pilot Project of Direct Electricity Purchase by Large Customers Under "One to Many" Model

11.4.1.1 Policies and Regulations Concerning the Pilot Project

At the end of June 2009, the State Electricity Regulatory Commission issued Reference [73]. The new policy clarified the voluntary principle of direct electricity purchases and the nature of a direct electricity purchase which is only limited to production electricity; and it specified the standards for a transmission–distribution price which is based on the average transmission–distribution price (not including the bulk sale price) of power grid enterprises minus the price difference in voltage classes. In this, the 110 kV (66 kV) transmission–distribution price is reduced by 10%, and 220 kV by 20%. It also stipulated that, in the case of an excessive quantity of electricity being consumed by direct purchase customers, a grid electric supply is implemented based on 110% of the catalogue price; in the case of an excessive quantity of electricity being generated by power plants, the excessive quantity of electricity consumed by the power grid enterprises is implemented based on 90% of the approved price.

According to Reference [74] issued by the State Electricity Regulatory Commission and the State Administration for Industry and Commerce, the transmission–distribution price should follow the approval documents of the relevant state authority; the power transmission and distribution loss rate of grid access energy should be as specified by the state, and the allocation of power transmission and distribution losses should be borne by the consumer, the power generation enterprise, and the power grid enterprise in a proportion agreed between them. According to the national pilot policy and the contract template, in terms of the transmission–distribution price and loss allocation during the pilot project, the efforts of all three parties are needed to accumulate information and data and to carry out in-depth analysis and research, so as to make the calculation of a transmission–distribution price and the loss allocation increasingly fair and reasonable, and to create conditions for a successful transition to bilateral trading under full power sales marketization.

11.4.1.2 Loss Increment Method Based on Load Flow Calculation

Similar to the allocation of losses in the 500 kV main grid of regional grids to all provincial grids, a large customer "one to many" direct supply loss allocation can adopt the loss increment method based on a load flow calculation. Because the direct supply pilot project uses "time of use" electricity prices of four periods, namely peak, rush-hour, normal, and valley, ex-post load flow calculations need to be conducted respectively for two situations of including and excluding some direct supply customers, according to the average load and the distribution of corresponding provincial grid loads in each period, thereby obtaining power losses in the four periods and two situations. If there are n direct supply customers, the difference in results (power losses) of n times of two types of load flow calculations can be used to obtain the correct coefficient $K_x = \Sigma\Delta P_\Sigma / \Sigma\Delta P_{(z.i)}$, $K_x \le 1.0$ in general. Then the corrected power losses allocated to a large customer within a period is obtained, that is $\Delta P_i = K_x \Delta P_{(z.i)}$. Considering the form coefficient K of a typical load curve under a period, the electric energy loss rate (line loss rate) can be obtained, that is $\Delta A_i\% = K\Delta P_i\%$. Next, the electric energy losses within a period can be obtained, that is $(\Delta A_i)_j = \Delta A_i\% A_j$, wherein A_j refers to the monthly subtotal electric quantity in peak, rush-hour, normal, and valley periods. Due to different electricity prices in various periods, the actual monthly loss electricity fee allocated to a direct supply customer can be calculated as per the following formula:

$$C_i = \sum_{j=1}^{\Delta} C_{0j}(\Delta A_i)_j \tag{11.3}$$

Wherein C_{0j} – unit electricity price in each period.

The calculation to allocate electric energy losses is shown in Figure 11.3.

11.4.2 *Shapley Method of "Many to Many" Loss Allocation*

11.4.2.1 Game Theory and Shapley Value

As a branch of modern mathematics, game theory is specifically designed to study how two or more individuals with a conflict of interest make their optimized decisions through interaction. Game theory has many branches, and a cooperative game of n players is one of them. Cooperative game provides an effective mathematical model and addresses the problem of allocating benefits among many stakeholders involved in a concerted action. The group of players and characteristic function are two basic elements of the cooperative game.

The group of players is composed of all independent stakeholders who affect the result of the problem. If the problem involves n stakeholders, the set of integers $I = \{1, 2, \ldots, n\}$ is used to represent the group of players. Any possible group of players S is called a coalition; the characteristic function $v(S)$ is a real function defined within the group of players I. When many players utilize a resource together, it is necessary to allocate incurred expenses or benefits among all the players, which is also called the solution to cooperative game.

In 1953, when studying cooperative game, the American scholar Shapley proposed that each player can get paid reasonably before playing a strategy, and that in the strategy where the characteristic function is v, then $x_i(v)$ expected by player i should meet three axioms: (i) group efficiency, $\sum_{i \in N} x_i = v(N)$; (ii) symmetry, where the number of players is irrelevant; (iii) additivity, where for any two strategies u and v, $x_i(u+v) = x_i(u) + x_i(v)$.

Shapley proved that only one function can satisfy the above three axioms, and the function value is the payoff X_i of player i, known as the Shapley value and calculated as per the following formula:

$$X_i = \sum_{s \in N} \frac{(|S|-1)!(n-|S|)!}{n!}[v(s) - u(s - \{i\})] \tag{11.4}$$

Wherein the summation involves all coalitions S including player i;
$|S|$ is the number of players in these coalitions S;
$v(S)$ is the characteristic function;
$S - \{i\}$ means deducting player i in coalitions;
$[v(S) - v(S - \{i\})]$ considers different effects after and before adding player i.

According to Formula (11.4), the quantity allocated to player i is actually the average of its marginal contributions to all possible coalitions. Therefore, the Shapley value comprehensively considers the influence of each player, and the allocation result is fair and reasonable.

11.4.2.2 Analysis of Loss Allocation by Shapley Value

The load of a new direct supply large customer can be regarded as a transaction, and the grid which it will enter forms a coalition. Losses to be borne by each transaction depend on the quantity of increased losses in the coalition after it enters the coalition. Obviously, the increased losses vary greatly when the transaction enters the coalition either at the earliest point or at the latest point. They are much larger in the latter case than in the former case, because increased losses in the latter case include not only losses caused by the single action of the transaction, but also cross losses caused by the interaction of the transaction with the existing transactions already in the coalition. Therefore, the Shapley value-based loss allocation method equally treats any possible order of joining transactions and gives each the same weight, so that the allocation result is fair and acceptable to each transaction. If the characteristic function $v(S)$ is replaced by the loss function P, then

$$X_i = \sum_{S} \frac{(|S|-1)!(n-|S|)!}{n!}[P(S) - P(S - \{i\})] \tag{11.5}$$

Wherein i is a transaction involved in the loss allocation;
X_i is losses allocated to transaction i;
S is the coalition including transaction i;
$|S|$ is the number of transactions in the coalition;
n is the total number of transactions involved in the loss allocation;
P is the loss function, and obviously $P(0) = 0$ when there is no transaction;

$[P(S) - P(S - \{i\})]$ is increased losses in the coalition S after transaction i joins the coalition, that is marginal losses of coalition S; $n!$ is the arrangement in any possible order of transactions joining the large coalition (including all transactions).

To further understand the rationality of the Shapley value, the following analysis can be conducted: given that the arrangement number of the condition where transaction i is the last to join coalition S is $(|S|-1)!$, and that the arrangement number of the condition where transaction i joins coalition S earlier than any other transaction in the coalition is $(n-|S|)!$, the total number of arrangements that satisfies the Shapley value is $(|S|-1)!(n-|S|)!$. Obviously, $(|S|-1)!(n-|S|)!/n!$ represents a weight, indicating the coalition marginal losses which transaction i should bear. These results are added up to obtain the total losses which should be allocated to transaction i based on the Shapley value calculation.

11.4.2.3 Calculation Column for Loss Allocation of Five-Node Grid by Shapley Method

Power producer G1 and customer 3, and power producer G2 and customers 4 and 5 signed direct power supply contracts, respectively. The system connection is shown in Figure 11.5; the grid parameters are listed in Table 11.3; and the condition of trading power are shown in Tables 11.4 and 11.5 [45].

According to Tables 11.4 and 11.5, there are three concurrent transactions in the market. The set of all players is $N=\{1, 2, 3\}$. Each transaction combination is a coalition, and there are seven transaction combinations. The optimal power flow method is used to calculate losses under each transaction combination, and the results are shown in Table 11.5. According to the calculation results of losses, the losses of a trading coalition of any two transactions are larger than the losses of any of the two transactions which act singly, and the losses of a trading coalition of three transactions are larger than the losses of a trading coalition of any two transactions plus the losses of another transaction which acts singly.

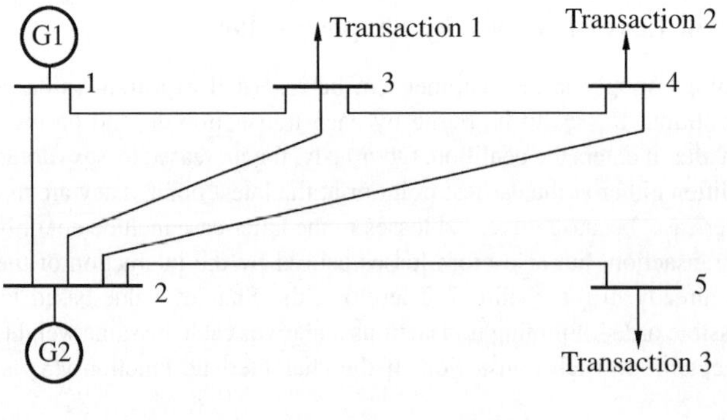

Figure 11.5 Five-node system connection.

Table 11.3 Parameters of five-node system line.

Start node	End node	R	X	B	Rated value (MVA)
1	2	0.02	0.06	0.0	200
1	3	0.08	0.24	0.0	200
2	3	0.06	0.18	0.0	80
2	4	0.02	0.18	0.0	200
2	5	0.08	0.12	0.0	200
3	4	0.01	0.03	0.0	200
4	5	0.03	0.24	0.0	200

Note: Parameters in the table are per unit values.

Table 11.4 Trading data of five-node system.

Transaction number.	Customer	Power producer	P (MW)	Q (Mvar)
1	3	1	67.5	41.83
2	4	2	60.0	37.18
3	5	2	90.0	55.77

Table 11.5 Trading losses of five-node system.

Transaction combination	Active losses (MW)	Loss rate (%)	Reactive losses (Mvar)	Transaction combination	Active losses (MW)	Loss rate (%)	Reactive losses (Mvar)
1	2.053 704	3.04	6.760 428	1 and 3	8.829 733	5.606	19.325 456
2	1.792 876	2.988	6.353 393	2 and 3	8.826 877	5.885	19.118 354
3	5.945 731	6.06	12.472 08	1, 2, and 3	13.734 904	6.315	33.532 002
1 and 2	3.532 309	2.77	13.653 32				

The losses allocated to transaction 1 between power producer 1 and customer 3 can be obtained as per Formula (11.4):

$$\begin{aligned}\chi_1 &= \frac{(3-1)!(1-1)!}{3!}[P(1)-P(1-\{1\})] + \frac{(3-2)!(2-1)!}{3!}[P(1.2)-P(2)] \\ &+ \frac{(3-2)!(2-1)!}{3!}[P(1.3)-P(3)] + \frac{(3-3)!(3-1)!}{3!}[P(1.2.3)-P(2.3)] \\ &= \frac{1}{3}(2.053\,704-0) + \frac{1}{6}(3.532\,309-1.792\,876) \\ &+ \frac{1}{6}(8.829\,733-5.945\,731) + \frac{1}{3}(13.734\,904-8.826\,877) = 3.091\,149(\text{MW})\end{aligned}$$

The losses allocated to transactions 2 and 3 can also be obtained, that is the Shapley value

$$\begin{aligned}\chi_2 &= \frac{1}{3}P(2) + \frac{1}{6}[P(2.1)-P(1)] + \frac{1}{6}[P(2.3)-P(3)] + \frac{1}{3}[P(1.2.3)-P(1.3)] \\ &= \frac{1}{3}\times 1.792\,876 + \frac{1}{6}(3.532\,309-2.053\,704) + \frac{1}{6}(8.826\,877-5.945\,731) \\ &+ \frac{1}{3}(13.734\,904-8.829\,733) \\ &= 2.959\,307\,5\ (\text{MW})\end{aligned}$$

$$\begin{aligned}\chi_3 &= \frac{1}{3}P(3) + \frac{1}{6}[P(3.1)-P(1)] + \frac{1}{6}[P(3.2)-P(2)] + \frac{1}{3}[P(1.2.3)-P(1.2)] \\ &= \frac{1}{3}\times 5.945731 + \frac{1}{6}(8.829\,733-2.053\,704) + \frac{1}{6}(8.826\,877-1.792\,876) \\ &+ \frac{1}{3}(13.734\,904-3.532\,309) \\ &= 7.684\,446\,7\ (\text{MW})\end{aligned}$$

The check calculation shows $x_1 + x_2 + x_3 = 13.734\ 904 = P$ (1.2.3), indicating that the entire allocation is break-even. The allocation results may not seem very reasonable from the customers' perspective, but in terms of any possible transaction combination, the allocation is fair and reasonable.

11.4.2.4 Transition from Allocation of Power Losses to Electric Energy Losses

Because the direct supply pilot project uses time of use electricity prices for four periods to calculate loss allocation, namely peak, rush-hour, normal, and valley, the load distribution used for the load flow calculation should be determined for each period before the load flow calculation. For peak loads, the load flow distribution at the time of maximum load of the entire grid can be directly used. For rush-hour, normal, and valley loads, the method of three-mode section division of active load duration curve in Section 3.4 of Chapter 3 can be used to calculate the electric quantities in the three sections; and the average loads of these three sections are calculated according to the duration of the three sections (see Table 3.4). Corresponding load flow calculations are conducted with real-time operating data records. With the load flow calculation results under loads for the four periods, the Shapley method can be used for the allocation of power losses. Form coefficients under average loads in rush-hour, normal, and valley periods can be calculated by typical load curves under the three sections. No load curve corresponds to the peak load which is peak value, so the form coefficient under peak load is 1.0. Through this transition, the allocation of power losses is changed to the allocation of electric energy losses. The allocation of power losses under peak load meets the needs of the electrical capacity charge calculation, and the calculation and allocation of electric energy losses in each different section satisfies the final requirement for calculating the allocation of loss expenses under time of use electricity price conditions. The interval during which the above calculations are conducted depends on the degree of load changes of direct supply large customers and can be determined in the direct supply contract.

11.4.3 Marginal Loss Coefficient-Based GMM Method

Under the complex trading manner, the electricity market in California adopts the loss coefficient-based GMM method [43].

Two assumptions are made in the calculation of marginal line loss rate: (i) generated output of other nodes remains unchanged; (ii) each node is allocated with a "relaxed load", that is system loads are increased (or reduced) as a whole. At this time, λ_{MLR} is defined as a unit change in the generated output of a node, while the generated output of other nodes remains unchanged. However, when the loads of all nodes change by the proportion which they account for in the total system loads, the total system losses also change, so λ_{MLR} is actually a sensitivity coefficient, that is

$$\lambda_{\text{MLR}.i} = \mathrm{d}(\Delta P_{\Sigma})/\mathrm{d}p_i \tag{11.6}$$

If losses allocated to each node are $\lambda_{\text{MLR}.i}G_i$, due to the above two assumptions, the total losses after allocation may be greater than the actual losses, leading to excessive loss recovery costs and excessive generated output, so the correction coefficient α is introduced and calculated as per the following formula:

$$\alpha = \sum_{i=1}^{n} \Delta P_i \Bigg/ \left(\sum_{i=1}^{n} \lambda_{\text{mLr}.i} G_i \right), \alpha < 1.0 \tag{11.7}$$

After correction, the losses allocated to a generation node are $\alpha\lambda_{\text{MLR}.i}G_i$, and the active power actually transmitted to loads is

$$\sum_{i=1}^{n}(G_i - \alpha\lambda_{\text{MLR}.i}G_i) = \sum_{i=1}^{n}(1-\alpha\lambda_{\text{MLR}.i})G_i \tag{11.8}$$

Wherein $(1 - \alpha\lambda_{\text{MLR}.i}) = \lambda_{\text{GMM}.i}$ is called generation multiplier; and Formula (11.8) is changed to

$$\lambda_{\text{GMM}.i}G_i = \sum_{i=1}^{n} G_i - \sum_{i=1}^{n} \Delta P_i = \sum_{i=1}^{n} L_i \tag{11.9}$$

According to Formula (11.9), the product of generation multiplier and generated output is the active power transmitted to loads, which is the origin of the name "generation quantity multiplier method" (GMM).

In actual operation, λ_{GMM} is calculated based on the system generation plan ($\sum G_i$) and the load plan ($\sum L_i$) per hour in advance, so $\sum G_i\lambda_{\text{GMM}.i}$ (effective generation plan) = $\sum L_i$ (load plan).

According to the GMM method, the cost recovery of losses is basically two times the actual losses, and the ratio is 1.80 according to the operation experience of the California electricity market. Note that this book has demonstrated that the marginal line loss rate is two times the load line loss rate, that is $(\Delta A\%)_{\text{m}} = 2\Delta A_{\text{f2}}\%$ [Formula (7.17)], which is exactly consistent with the above opinion.

It is believed that the marginal loss coefficient-based GMM method is a widely applicable loss allocation calculation method under the complex trading setup. The key to this method is a perfect technical support system, real-time calculation (one bidding interval lagging behind) by independent system operators (ISO) according to the bidding interval of on-grid bidding, and detailed records. Based on capacity price and time of use price (real-time price), electricity fees are calculated off-line according to the charging interval specified in the contract, the actually supplied peak, rush-hour, normal, and valley loads and electricity, and the loss allocation results.

Example 11.2 According to the trading data and losses of the five-node system in Tables 11.3 to 11.5, try to use the GMM method to calculate losses allocated to each transaction.

Solution

According to the active power losses of transactions 1, 2, and 3 in Table 11.5 and the corresponding generated output in Table 11.4, the loss rates of transactions 1, 2, and 3 are respectively obtained, namely 3.04, 2.988, and 6.06%. According to California's experience that the marginal loss rate is 1.8 times the loss rate, the marginal loss rates of transactions 1, 2, and 3 are respectively obtained, that is

$$\lambda_{\text{MLR1}} = 1.8 \times 0.0304 = 0.0547$$
$$\lambda_{\text{MLR2}} = 1.8 \times 0.029\,88 = 0.053\,69$$
$$\lambda_{\text{MLR3}} = 1.8 \times 0.0606 = 0.109\,08$$

Then

$$\sum_{i=1}^{2}\lambda_{\text{MLR}i}G_i = \lambda_{\text{MLR1}}G_1 + \lambda_{\text{MLR2}}G_2 + \lambda_{\text{MLR3}}G_3$$
$$= 0.054\,7 \times 67.5 + 0.053\,69 \times 60 + 0.109\,08 \times 90 = 16.730\,85\ (\text{MW})$$

So the correction coefficient is

$$\alpha = \Sigma\Delta P_i/(\Sigma\lambda_{\text{MLR}i}G_i) = 13.734\,904/16.730\,85 = 0.820\,933$$

The coefficients of generated energy are

$$\lambda_{\text{GMM1}} = (1-\alpha\lambda_{\text{MLR1}}) = 1-0.820\,933\times 0.054\,7 = 0.955\,095$$
$$\lambda_{\text{GMM2}} = (1-\alpha\lambda_{\text{MLR2}}) = 1-0.820\,933\times 0.053\,69 = 0.955\,924\,1$$
$$\lambda_{\text{GMM3}} = (1-\alpha\lambda_{\text{MLR3}}) = 1-0.820\,933\times 0.109\,08 = 0.910\,452\,6$$

Finally, the allocated losses are

$$\chi_1 = G_1(1-\lambda_{\text{GMM1}}) = 67.5(1-0.955\,095) = 3.0311\ (\text{MW})$$
$$\chi_2 = G_2(1-\lambda_{\text{GMM2}}) = 60(1-0.955\,924\,1) = 2.6445\ (\text{MW})$$
$$\chi_3 = G_3(1-\lambda_{\text{GMM3}}) = 90(1-0.910\,452\,6) = 8.059\,26\ (\text{MW})$$

The allocation results calculated by the Shapley method are $x_1 = 3.091\,149$, $x_2 = 2.959\,308$, $x_3 = 7.684\,447$. Compared with the results of the GMM method, the relative errors are $\delta x_1 = 1.94\%$, $\delta x_2 = 10.636\%$, $\delta x_3 = 4.88\%$.

According to the calculation in Reference [46], for the same trading data of five-node system, the maximum relative error between the allocation results calculated by the kernel method of cooperative game and the allocation results calculated by the Shapley method is about 9%.

According to the calculations and comparisons in this example, different methods may lead to different loss allocation results, and the difference may reach 10% or more. This still far from satisfies the requirement of unique result, fairness, and rationality expected to be achieved for the loss allocation. The authors think regulations can be established like "calculation guidelines" to determine the range of application of a certain method, so as to meet the requirements of fair and reasonable loss allocation in the progress of electric power marketization. Although the allocation result is not unique, it has its relative rationality in a certain historical stage. As long as we insist on accumulating experience in calculation and analysis, and conducting in-depth studies and exploring new solutions to any problem found in practice, complicated technical and economic problems in loss allocation can be resolved with our spirally escalated understanding of such problems.

12

Technical Measures for the Reduction of Line Losses

There are roughly two types of technical measures for the reduction of line losses. One type is the transformation of power grids, where line losses are reduced as the transmission capability and voltage quality of power grids are improved. This type of measures needs certain investments and should be demonstrated for its rationality according to technical and economic analysis. The other type of measures requires no investment, and it is only necessary to improve the operation and management of the power grids to reduce line losses. Power grid operation and management authorities should attach importance to this type of measures and implement them in their daily operation and line loss management. Some major technical measures for loss reduction are introduced below.

12.1 Selection of Reasonable Connection Mode and Operation Mode

When the connection mode and operation mode of all the parts of a power grid are reasonable, this affects not only the safety and quality of the electric supply, but also the quantity of line losses. There are the following loss reduction measures in this regard.

12.1.1 Introduction of High-voltage Grids to Large Cities or Load Centers

With urban development and increasing loads, the original 35 kV and 6–10 kV high-voltage distribution grids are carrying heavier and heavier loads. The load losses in electricity losses are in direct proportion to the square of loads, therefore, if the state of a long-distance electric supply by low-voltage class grids is sustained, the voltage quality cannot be guaranteed, and the electricity losses will reach an extent that is not acceptable. The transformation of such power grids by introducing 110 or 220 kV connection mode is one of the effective measures for loss reduction.

Line Loss Analysis and Calculation of Electric Power Systems, First Edition. Anguan Wu and Baoshan Ni.

12.1.2 Stepping up of Power Grid Voltage, Simplification of Voltage Class, and reduction of Repeated Substation Capacity

The load power losses ΔP (kW) of a power grid unit (line or transformer) are

$$\begin{aligned}\Delta P &= 3I^2 R\times 10^{-3} = \frac{S^2}{U^2}R\times 10^{-3}\\ &= \frac{P^2}{U^2\cos^2\varphi}R\times 10^{-3} = \frac{P^2+Q^2}{U^2}R\times 10^{-3}\end{aligned} \tag{12.1}$$

Wherein I – current passing the unit (A);
R – resistance of the unit (Ω);
S, P, Q – apparent power (kVA), active power (kW) and reactive power (kvar) passing the unit;
U – power grid voltage applied in the unit (kV).

According to Formula (12.1), on the condition that the load power is constant, if the power grid voltage is increased, then the current passing the power grid unit will be reduced, and load losses are also reduced. Therefore, stepping up the power grid voltage is a very effective measure for reducing line losses. This measure can be combined with the transformation of old power grids by reducing the voltage class and simplifying the connection of power grids, thereby meeting the needs of load growth and reducing the line losses of power grids.

Line losses can also be reduced by improving the electric supply structure and reducing repeated substation capacity. According to the calculation in Example 7.2 of Chapter 7, the total grid line losses can be significantly reduced by increasing the electric quantity transmitted by a three-winding transformer 110/35/10 kV to the 10 kV grids through direct step-down, reducing the electric quantity transmitted by 110/35 and 35/10 kV to the 10 kV grids through two continuous steps-down, and reducing the repeated electric supply rate.

12.1.3 Reasonable Determination of Closed Loop Operation or Open Loop Operation of Loop Net, or Change of Break Points of Loop Net

In the loop net, the power distribution (not considering active power and reactive power losses in all line sections) is called the approximate power distribution. The power distribution by the relationship of impedance of all sections is called the natural power distribution; and the power distribution by the relationship of the resistance of all sections is called the economic power distribution, where the active power losses in the loop net are minimal. If the power grids are uniform, that is X/R of each section is a constant, then natural power distribution and economic power distribution are consistent. If the loop net is more non-uniform, the difference between natural power distribution and economic power distribution is larger and the difference in active power losses is greater. In a loop net connected with transformers under different voltage classes, because the ratio between the reactance and resistance of transformers is larger than that of lines, the non-uniform degree of power grids is increased.

To reduce line losses, the first thing is to study which is more reasonable between closed loop operation and open loop operation. In the electric power system, because sometimes the breaker capacity is insufficient in closed loop operation or the configuration of relay protection is complicated, the loop net is always in open loop operation, and some lines are in the live-line hot spare state. After closed loop operation, power starts to flow in the original spare lines, and power losses seem to increase. As power in other sections is also changed, however, the power losses seem to be smaller than those under open loop operation, which needs to be determined through calculation and comparison.

When the shapes of load curves for all loads are basically the same, it is only necessary to compare power losses under different break point plans or under closed loop and open loop operations. If the shapes of load curves of different substations vary greatly, it is necessary to compare the electric energy losses under different operation modes to determine which operation mode is more economical and reasonable. This is because it is totally possible that electric energy losses are smaller within a period under one operation mode but are smaller within another period under the other operation mode.

The example below explains the rationality of closed loop or open loop of the loop net.

Example 12.1 As shown in Figure 12.1, the impedance of sections in No. 1 and No. 2 lines of the distribution grid as well as the maximum current of two customers are marked.

The quarterly load curve parameters of the two customers are $f_1 = 0.6$, $\beta_1 = 0.3$ and $f_2 = 0.9$, $\beta_2 = 0.6$, respectively. *Try to calculate the difference in electric energy losses under open loop and closed loop operations within the measuring period (T = 2208 h).*

Solution

When breaker QF is open, $I_1 = 150$ A, $I_2 = 150$ A. Substitute f_1, β_1 and f_2, β_2 into Formula (2.28) respectively to obtain $F_1 = 0.403$, $F_2 = 0.821$.

So the total electric energy losses are

$$\begin{aligned}\Delta A &= 3\times\left[(2.29+3.44)\times 50^2\times 0.403+(2.87+1.47)\times 150^2\times 0.821\right]\times 2208\times 10^{-3}\\&=257.83\times 2208=56.93\times 10^4\ (\text{kW·h})\end{aligned}$$

When breaker QF is closed, the current in No. 1 and No. 2 lines is respectively

$$\begin{aligned}|\dot{I}_1| &= \left|(150+50)\times\frac{Z_3+Z_4}{Z_1+Z_2+Z_3+Z_4}\right|\\&=\left|200\times\frac{4.34+\text{j}8.29}{10.07+\text{j}20.16}\right|\\&=|200\times 0.415/-1.1^\circ|\approx 83(\text{A})\end{aligned}$$

$$\begin{aligned}|\dot{I}_2| &= \left|(150+50)\times\frac{Z_1+Z_2}{Z_1+Z_2+Z_3+Z_4}\right|\\&=\left|200\times\frac{5.75+\text{j}11.87}{10.07+\text{j}20.16}\right|\\&=|200\times 0.585/0.75^\circ|\approx 117(\text{A})\end{aligned}$$

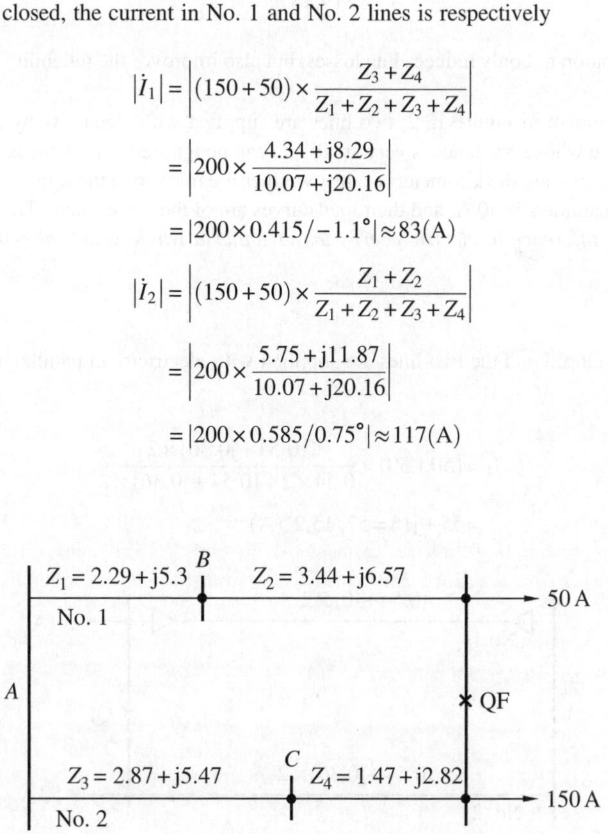

Figure 12.1 Comparison between closed loop and open loop operations.

Assume that the maximum current and minimum current of the two customers occur at the same time. The load rate and minimum load rate under the total current are

$$f=\frac{50\times0.60+150\times0.90}{50+150}=0.825$$

$$\beta=\frac{50\times0.30+150\times0.60}{50+150}=0.525$$

According to Formula (2.28), $F = 0.7$, so the total electric energy losses are

$$\begin{aligned}\Delta A'&=3\times\left[(2.29+3.44)\times83^2+(2.87+1.47)\times117^2\right]\times0.70\times2208\times10^{-3}\\&=207.66\times2208=45.85\times10^4\ (\text{kW}\cdot\text{h})\end{aligned}$$

The proportion of reduced electric energy losses under closed loop operation is

$$\begin{aligned}\delta(\Delta A)&=\frac{\Delta A-\Delta A'}{\Delta A}\times100\%\\&=\frac{56.93-45.85}{56.93}\times100\%\\&=19.46\%\end{aligned}$$

The closed loop operation not only reduces line losses, but also improves the reliability of the electric supply.

Example 12.2 As shown in Figure 12.2, two lines are supplied with electricity by the same substation. No. 1 line is a cable whose reactance is very small and can be ignored. No. 2 line is an overhead line. The impedance per unit length and the kilometers of the two lines are marked in the figure. The maximum current for each of the two customers is 50 A, and their load curves are of the same shape. The loss factor is $F = 0.7$. *Try to calculate the difference in electric energy losses under a whole year's closed loop and open loop operations.*

Solution

When breaker QF is closed and the two lines are supplied with electricity in parallel, the current in the two lines is respectively

$$\begin{aligned}I_1&=(50+50)\times\frac{(0.54+\text{j}0.36)\times2}{0.54\times2+(0.54+\text{j}0.36)\times2}\\&=55+\text{j}15=57/15.2^\circ(\text{A})\end{aligned}$$

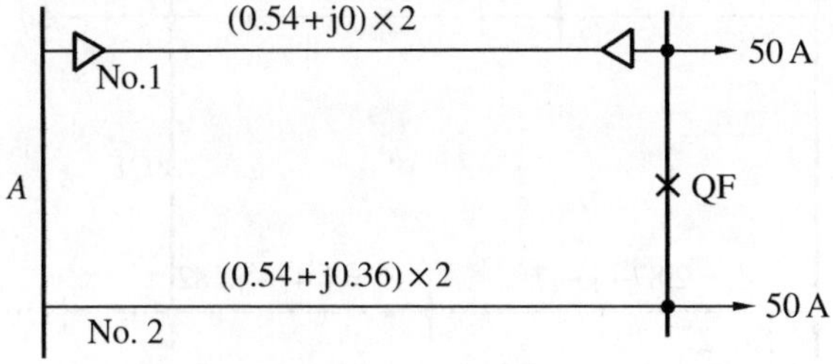

Figure 12.2 Operation of non-uniform power grids.

$$I_2 = (50+50) \times \frac{0.54\times 2}{0.54\times 2 + (0.54+j0.36)\times 2}$$

$$= 45 - j15 = 48/18.5°(A)$$

The power losses in the two lines are

$$\Delta P_1 = 3\times 57^2 \times 1.08\times 10^{-3} = 10.5\ (kW)$$
$$\Delta P_2 = 3\times 48^2 \times 1.08\times 10^{-3} = 7.45\ (kW)$$
$$\Delta P = \Delta P_1 + \Delta P_2 = 17.95\ (kW)$$

If breaker QF is open for open loop operation, the power losses in the lines are

$$\Delta P' = 2\times 3\times 50^2 \times 1.08\times 10^{-3} = 16.2\ (kW)$$

As the loss factor is $F = 0.7$, the line losses reduced as a result of a year's open loop operation are (17.95 – 16.2) × 0.7 × 8760 = 10720 (kW·h), but the reliability of electric supply is poor. According to the current distribution under closed loop operation, circulating current occurs in non-uniform grids under closed loop operation, leading to increased line losses.

12.1.4 Realization of Economic Power Distribution by Longitudinal and Transverse Voltage Regulating Transformer or Series Capacitor

To reduce the power losses and electric energy losses in the non-uniform loop net, forced circulating power can be superimposed in the naturally distributed power in the loop net, and the sum of the two types of power should be equal to the economic power distribution. To provide a forced circulating power in the loop net, there must be additional electromotive force. Because loads in the power grids change with time and the power distribution also changes with time, the circulating power should be adjustable. To produce an adjustable circulating power, there must be an adjustable additional electromotive force. Because the additional electromotive force has a longitudinal component whose phase is the same as that of the grid voltage and transverse component whose phase is 90° different from that of the grid voltage, the additional electromotive force should be able to change both the value and phase of the grid voltage. Such additional electromotive force is obtained by a series voltage regulating transformer connected to the loop net. Figure 12.3a shows a single-phase connection diagram of the transverse voltage regulating transformer; Figure 12.3b shows the

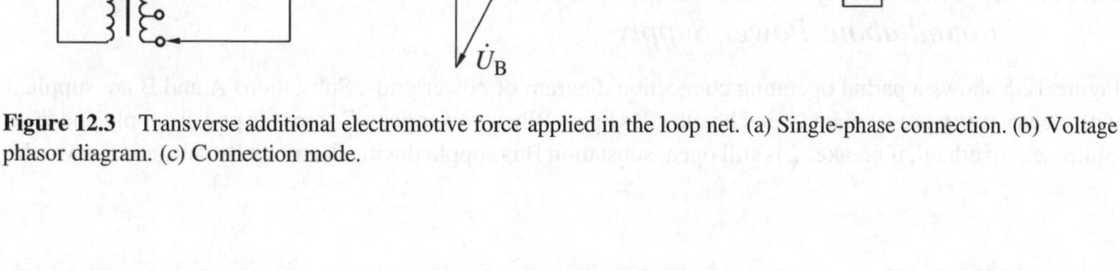

Figure 12.3 Transverse additional electromotive force applied in the loop net. (a) Single-phase connection. (b) Voltage phasor diagram. (c) Connection mode.

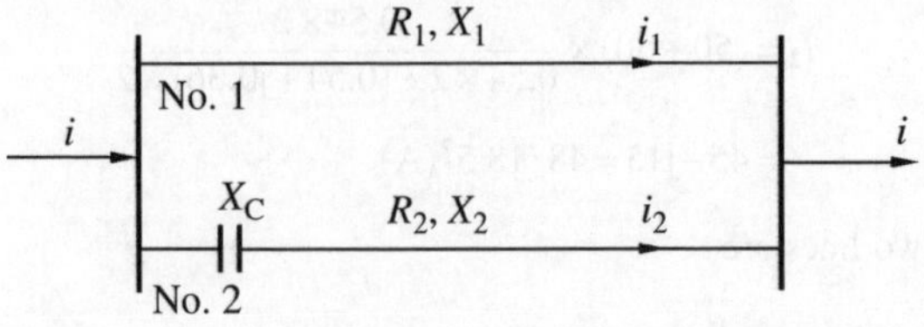

Figure 12.4 Series compensation in no-uniform loop net.

voltage phasor diagram; Figure 12.3c shows the connection mode of this voltage regulating transformer to the power grids. Only a transverse additional electromotive force can be added in the voltage regulating transformer shown in Figure 12.3. This longitudinal additional electromotive force can be obtained by changing the ratio of the original transformers in the power grids.

As longitudinal and transverse voltage regulating transformers carry a lot of investment costs, they are only installed in loop nets composed of lines with different voltage classes and through which considerable power flows. In a general non-uniform power grid, series capacitors can be used to compensate the partial impedance of lines to realize the economic power distribution.

Figure 12.4 shows the simplest loop net, composed of two lines with conductors of different cross-sections. In the two lines, the current is distributed in inverse proportion to the impedance, that is $\frac{\dot{I}_1}{\dot{I}_2} = \frac{Z_2}{Z_1} = \frac{R_2 + jX_2}{R_1 + jX_1}$, and $\dot{I}_1 + \dot{I}_2 = \dot{I}$.

Assume $\frac{X_2}{R_2} > \frac{X_1}{R_1}$. To satisfy the conditions of economic distribution, a capacitor can be connected in series to the line (No. 2 line) whose $\frac{X}{R}$ ratio is large to compensate its partial reactance, so that $\frac{X}{R}$ ratios in the two lines are equal, and the current distribution meets the conditions of minimum active power losses. The capacitive reactance X_C of the connected capacitor should satisfy the following formula:

$$\frac{X_2 - X_C}{R_2} = \frac{X_1}{R_1} \tag{12.2}$$

According to Formula (12.2), the compensated capacitive reactance is

$$X_C = X_2 - X_1 \frac{R_2}{R_1} \tag{12.3}$$

That is, the degree of compensation for No. 2 line is

$$K_C = \frac{X_C}{X_2} = 1 - \frac{X_1/R_1}{X_2/R_2} \tag{12.4}$$

12.1.5 *Prevention of Remote Supply by Nearby Power or Round about Power Supply*

Figure 12.5 shows a partial operating connection diagram of power grids. Substations A and B are supplied with electricity by power plant C, and breaker 2 is open. When power plant C stops the power supply due to an equipment overhaul, if breaker 2 is still open, substation B is supplied with electricity by substation A via the

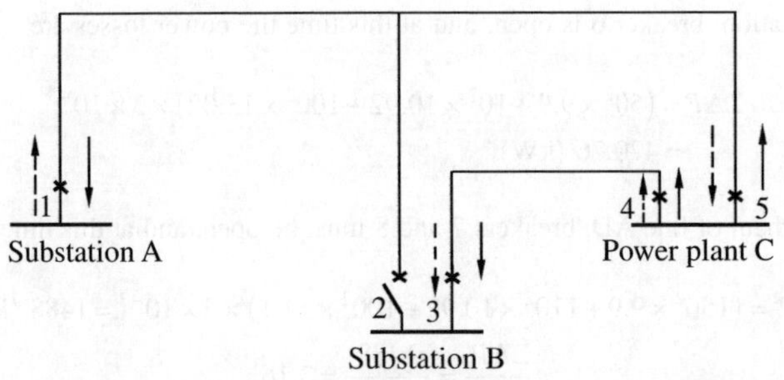

Figure 12.5 Partial operating connection diagram of power grids. The solid arrow refers to the load flow direction under normal operation. The dashed arrow refers to the load flow direction when power plant C is maintaining equipment.

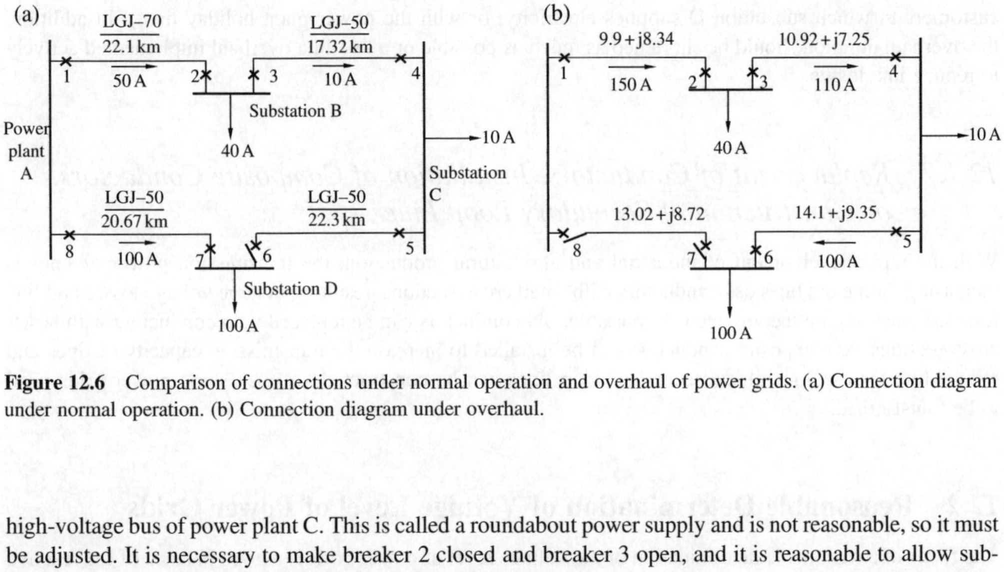

Figure 12.6 Comparison of connections under normal operation and overhaul of power grids. (a) Connection diagram under normal operation. (b) Connection diagram under overhaul.

high-voltage bus of power plant C. This is called a roundabout power supply and is not reasonable, so it must be adjusted. It is necessary to make breaker 2 closed and breaker 3 open, and it is reasonable to allow substation B to be switched to the tie line in order to receive a supply.

It must be pointed that, in 380/220 V low-voltage distribution grids, it is often necessary to adjust the range of power supply of distribution transformers to prevent the overload of such transformers. If care is ignored, a roundabout power supply may occur.

12.1.6 Reasonable Arrangement of Equipment Overhaul and Practice of Live-Line Overhaul

The connection mode of power grids under normal operation is generally safe, economical, and reasonable. In the case of equipment overhaul, the normal operation connection must not be changed. Any change in the connection mode will not only reduce the reliability of operation, but also increase line losses greatly. Figure 12.6a shows the normal operation connection of power grids, and line parameters and current are marked in the diagram.

Under normal operation, breaker 6 is open, and at this time the power losses are

$$\Sigma\Delta P = \left(50^2 \times 9.9 + 10^2 \times 10.92 + 100^2 \times 13.02\right) \times 3 \times 10^{-3}$$
$$= 470.76 \text{ (kW)}$$

In the case of an overhaul of line AD, breakers 7 and 8 must be open, and at this time the power losses are

$$\Sigma\Delta P' = \left(150^2 \times 9.9 + 110^2 \times 10.92 + 100^2 \times 14.1\right) \times 3 \times 10^{-3} = 1488 \text{ (kW)}$$
$$\frac{\Sigma\Delta P'}{\Sigma\Delta P} = \frac{1488}{470.76} = 3.16$$

According to calculations, under the same load conditions, the power losses under overhaul are over three times those under normal operation. Assume that the overhaul lasts 10 h, and the loss factor is 0.5. The increased electricity losses are $\Delta A = (1488 - 470.76) \times 0.5 \times 10 = 5086$ (kW·h). Therefore, the reasonable arrangement of equipment overhaul and an enhanced planning of overhaul are important loss reduction measures (the overhaul of line AD can be coordinated with the equipment overhaul of breakers 7 and 8 or of customers to which substation D supplies electricity, or with the power plant holiday time). In addition, the overhaul duration should be shortened as much as possible or a live-line overhaul implemented actively to reduce line losses.

12.1.7 Replacement of Conductors, Installation of Composite Conductors, or Construction of Secondary Loop Lines

With the rapid development of industrial and agricultural production, the transmission power of lines is increasing. Some old lines use conductors with small cross-sections, resulting in large voltage losses and line losses. If stepping up the voltage is impossible, old conductors can be replaced with conductors with larger cross-sections, or composite conductors can be installed to increase the transmission capacity of lines and reduce line losses. Sometimes, secondary loop lines can be constructed, or even some power grids need to be transformed.

12.2 Reasonable Determination of Voltage Level of Power Grids

The focus of a reasonable determination of the voltage level of power grids is a good balance of the reactive power, including the reasonable regulation of generator activity, increase in the voltage of generators, increase in the power factor of customers, and the use of reactive power compensation equipment and series capacitors. The next work is to adjust the tapping points of transformers. The influence of adjusting the tapping points of transformers or using transformers equipped with load voltage regulators on the reduction of line losses is mainly explained below.

Figure 12.7 shows a connection diagram of power grids and indicates the changes in all operating positions of transformers and tapping points. U_G is the generator voltage. Figure 12.7b shows the connection diagram after changing the tapping points.

Assume that the generator voltage and load remain unchanged before and after changing the tapping points. If the tapping point T1 is switched from 124 to 127 kV, then the voltage of 110 kV power grids is increased by 2.5%. If the tapping point T2 is switched from 115.5 to 112.8 kV, and the tapping point at the medium-voltage side is switched from 38.5 to 40.4 kV, then the voltage of 35 kV power grids is increased by 7.5%. The gross increase in voltage is 10%. The voltage at the 6 kV side of T2 is increased by (2.5% + 2.5% = 5%). To make the voltage at customers remain unchanged, the tapping point T3 should be switched from 33.3 to 36.7 kV, while the tapping point T4 should be switched from 6.0 to 6.3 kV.

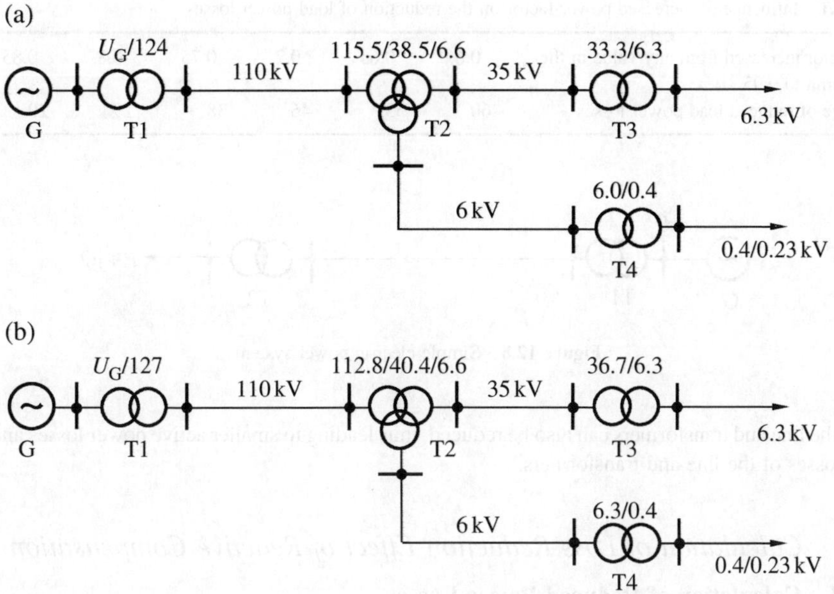

Figure 12.7 Connection diagram of power grids. (a) Before adjustment of tapping points. (b) After adjustment of tapping points.

Because the voltage levels of 6, 35, and 110 kV power grids are increased by 5, 10, and 2.5%, respectively, their load losses can be reduced. If the no-load losses of various voltages of power grids only account for 10% ($\theta = 0.11$) in total losses, then according to Formula (6.23), the total losses of the above three types of power grids can be approximately reduced by 7.5, 15, and 3.75%, respectively. This shows that the position of any tapping point in a transformer should be reasonably selected according to the requirement of voltage quality and considering the possibility of reducing line losses.

However, it must be pointed that, if the supply of reactive power in a system is tight, the method of adjusting the tapping points of transformers to increase the voltage of power grids may lead to a greater consumption of the reactive power of loads. Although at this time the charging power of 110 kV above lines is increased with the rise in voltage, the reactive power of the system cannot be balanced and the voltage cannot be maintained at the expected level of increase. Therefore, on the premise of a good balance of reactive power in power grids, the method of changing the tapping points of transformers can be used to increase the voltage level of power grids.

It must be also pointed that the voltage of rural power grids is generally higher during the non-drainage and irrigation season. The tapping points of 110 and 35 kV main transformers should be adjusted to reduce the voltage level of 10 kV agricultural distribution lines, thus reducing the no-load losses of the transformers.

12.3 Utilization of Reactive Power Compensation Equipment and Increase in Power Factor

Figure 12.8 shows a simple electric power system. According to Figure 12.8, on the condition that the active power P of a load remains unchanged, if the power factor of the load is increased, the reactive power Q required by the load can be reduced, and the reactive power output by a generator and the reactive power

Table 12.1 Influence of increased power factor on the reduction of load power losses.

Power factor increased from any value in the right column to 0.95	0.6	0.65	0.7	0.75	0.8	0.85	0.9
Percentage of reduced load power losses	60	53	46	38	29	20	10

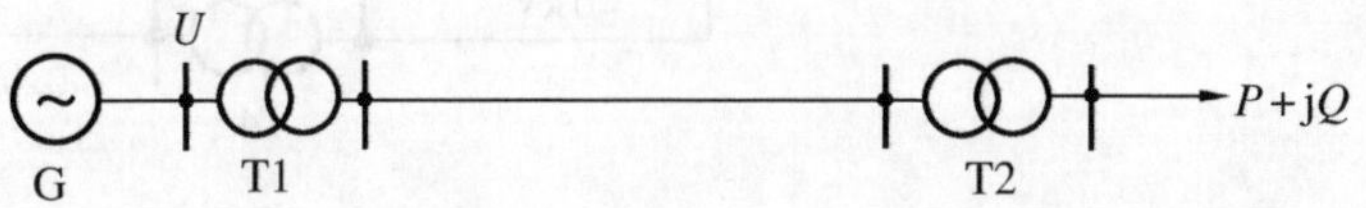

Figure 12.8 Simple electric power system.

passing the line and transformers can also be reduced, thus leading to smaller active power losses and electric energy losses of the line and transformers.

12.3.1 Calculation of Loss Reduction Effect of Reactive Compensation

12.3.1.1 Calculation of Reduced Power Losses

According to Formula (12.1), when the load power factor is increased from $\cos\varphi_1$ to $\cos\varphi_2$, the percentage of reduced active power losses can be expressed by the following simple relation:

$$\delta P\% = \left(1 - \frac{\cos^2\varphi_1}{\cos^2\varphi_2}\right) \times 100\% \tag{12.5}$$

Because no-load power losses are independent of the power factor, the influence of increased power factor on the reduction of load power losses is shown in Table 12.1.

The problem of achieving a degree of increase in the power factor which is economical and reasonable concerns the economic effect and the economical and reasonable arrangement of compensation equipment. The concept of equivalent reactive compensation power is usually introduced to analyze this problem.

In the simple system shown in Figure 12.8, assume that the maximum active power and maximum reactive power of the load occur at the same time. When no reactive compensation equipment is installed at the load, the maximum active power losses ΔP_{max} (kW) of the line and transformers in the system are

$$\Delta P_{max1} = \frac{P_{max}^2 + Q_{max}^2}{U^2} R \times 10^{-3}$$

Wherein P_{max}, Q_{max} – maximum active power (kW) and maximum reactive power (kvar) of the load;
R – system resistance included to voltage U (including gross resistance of transformers T1 and T2 and the line) (Ω).

When compensation equipment with capacity of Q_{com} is installed at the load, the maximum active power losses ΔP_{max2} (kW) of the line and transformers are

$$\Delta P_{max2} = \frac{P_{max}^2 + (Q_{max} - Q_{com})^2}{U^2} R \times 10^{-3}$$

Therefore, the reduced active power losses due to the installation of compensation equipment are

$$\Delta P_{\max 1} - \Delta P_{\max 2} = \frac{(2Q_{\max} - Q_{\text{com}})Q_{\text{com}}}{U^2} R \times 10^{-3} = C_{\text{bp}} Q_{\text{com}}$$
$$C_{\text{bp}} = \frac{2Q_{\max} - Q_{\text{com}}}{U^2} R \times 10^{-3} \tag{12.6}$$

Wherein C_{bp} – average of active power losses which can be reduced by reactive compensation equipment per unit capacity; this is called the equivalent reactive compensation power at the time of maximum load (kW/kvar).

According to Formula (12.6), the effect of installation of the first kilovar reactive compensation equipment is better than that of installation of subsequent kilovar reactive compensation equipment. In other words, the role of reactive compensation equipment in reducing the occupation of reactive load in power capacity is in decreasing order. Two conclusions can be drawn from Formula (12.6): (i) the longer the electrical distance between reactive compensation equipment and power source (the larger R is in the formula), the larger is the effect of the loss reduction of reactive compensation equipment; (ii) the larger the reactive load, the smaller is the effect of the loss reduction of reactive compensation equipment with the same capacity.

With the concept of equivalent reactive compensation power, the formula below can be directly used to calculate the effect of reduction of active power losses by compensation equipment.

$$\partial(\Delta P) = C_{\text{bp}} Q_{\text{com}} \tag{12.7}$$

12.3.1.2 Calculation of Reduced Electric Energy Losses

Reactive load varies during the measuring period. When the reactive compensation equipment is installed for the entire measuring period, the electric energy losses caused by reactive power after compensation can be calculated as per the following formula:

$$\Delta A_{2Q} = \frac{R}{U_{\text{av}}^2} \int_0^T (Q - Q_{\text{com}})^2 \mathrm{d}t \times 10^{-3}$$
$$= \frac{R}{U_{\text{av}}^2} \left[\int_0^T Q^2 \mathrm{d}t - \int_0^T 2QQ_{\text{com}} \mathrm{d}t + \int_0^T Q_{\text{com}}^2 \mathrm{d}t \right] \times 10^{-3}$$

Without the reactive compensation equipment, the electric energy losses caused by reactive power are

$$\Delta A_{1Q} = \frac{R}{U_{\text{av}}^2} \int_0^T Q^2 \mathrm{d}t \times 10^{-3}$$

If the electric energy losses of the reactive compensation equipment itself are not considered, the reduced electric energy losses due to the installation of reactive compensation equipment are

$$\partial(\Delta A) = \Delta A_{1Q} - \Delta Q_{2Q}$$
$$= \frac{R}{U_{\text{av}}^2} \left(\int_0^T 2QQ_{\text{com}} \mathrm{d}t - \int_0^T Q_{\text{com}}^2 \mathrm{d}t \right) \times 10^{-3}$$
$$= (2Q_{\max} f_Q - Q_{\text{com}}) \frac{R}{U_{\text{av}}^2} \times 10^{-3} \times Q_{\text{com}} T \tag{12.8}$$
$$= C_{\text{bA}} (Q_{\text{com}} T)$$

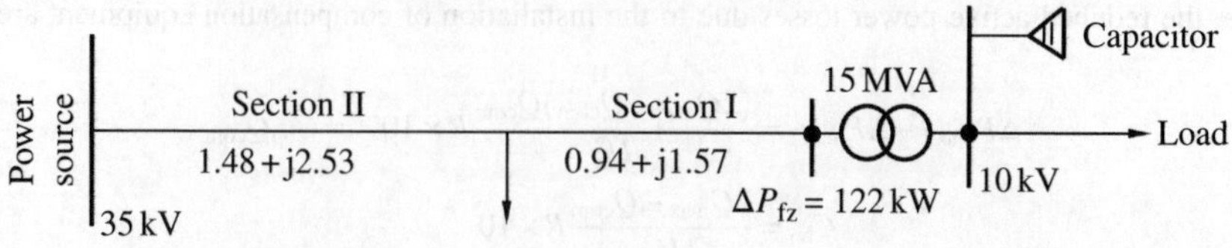

Figure 12.9 35 kV system and step-down substation.

$$C_{bA} = (2Q_{max}f_Q - Q_{com})\frac{R}{U_{av}^2} \times 10^{-3} \quad (12.9)$$

Wherein C_{bA} – equivalent reactive compensation electric energy (kW/kvar).

Due to $f_Q < 1$, $C_{bA} < C_{bp}$, that is the equivalent reactive compensation electric energy is smaller than the equivalent reactive compensation power. Then, the following relation exists

$$\partial(\Delta A) = C_{bA}(Q_{com}T) < C_{bp}(Q_{com}T)$$

This shows that in general situations, the equivalent reactive compensation electric energy should be used to calculate reduced electric energy losses; the use of the equivalent reactive compensation power to calculate reduced electric energy losses may result in larger results.

Example 12.3 Figure 12.9 shows that the primary 35 kV line in the power grids supplies electricity to a 35/10 kV substation.

According to measured data of loads, the maximum load in line I is 9.59 + j8.98 (MVA), with $f_{QI} = 0.90$, $\beta_{QI} = 0.65$. The maximum load in line II is 14.62 + j12.90 (MVA), with $f_{QII} = 0.85$, $\beta_{QII} = 0.60$. When a 1460 kvar capacitor is installed at the 10 kV side of the substation, the power factor of the main transformer is 0.73. *Try to calculate:*

1. *The equivalent reactive compensation power and equivalent reactive compensation electric energy at the 35 and 10 kV sides, and the annual electric energy losses saved by the capacitor.*
2. *The annual electric energy losses saved under the situation where a 3490 kvar capacitor is installed again at the 10 kV side of the substation and the power factor of the main transformer at the time of maximum load is increased to 0.9.*

Solution

1. Calculation of the equivalent reactive compensation power and the equivalent reactive compensation electric energy.

   Figure 12.9 indicates $\Delta P_L = 122$ kW for the main transformer, so the resistance of the main transformer (see Appendix B) as can be included to the 10 kV side is

$$R_T = \frac{\Delta P_L U_N^2 \times 10^3}{S_N^2} = \frac{122 \times 10^2 \times 10^3}{(15 \times 10^3)^2} = 0.054\,22(\Omega)$$

   According to Formula (12.6), the equivalent reactive compensation power at the high- and low-voltage sides of the substation is respectively

$$C_{bp.35}=\frac{(2Q_{maxII}-Q_{com})R_{II}}{U^2}\times10^{-3}+\frac{(2Q_{maxI}-Q_{com})R_{I}}{U^2}\times10^{-3}$$
$$=\frac{(2\times12\,900-1460)\times1.48}{35^2}\times10^{-3}+\frac{(2\times8980-1460)\times0.94}{35^2}\times10^{-3}$$
$$=0.029\,41+0.012\,66=0.042\,1\ (\text{kW/kvar})$$
$$C_{bp.10}=C_{com.35}+C_{com.b}=0.0421+\frac{(2\times8980-1460)\times0.054\,22}{10^2}\times10^{-3}$$
$$=0.042\,1+0.008\,94=0.051\,04\ (\text{kW/kvar})$$

Active power losses reduced by reactive compensation are

$$\partial(\Delta P)=C_{bp.10}Q_{com}=0.051\,04\times1460=74.52\ (\text{kW})$$

According to Formula (12.9), the equivalent reactive compensation electric energy at the low-voltage side of the substation is

$$C_{bA}=C_{bA.II}+C_{bA.I}+C_{bA.b}$$
$$=\left[(2Q_{zdII}f_{QII}-Q_{com})\frac{R_{II}}{U_{av1}^2}+(2Q_{maxI}f_{QI}-Q_{com})\left(\frac{RI}{U_{av1}^2}+\frac{R_b}{U_{av2}^2}\right)\right]\times10^{-3}$$
$$=\left[(2\times12\,900\times0.85-1460)\times\frac{1.48}{35^2}+(2\times8980\times0.90-1460)\left(\frac{0.94}{35^2}+\frac{0.054\,22}{10^2}\right)\right]\times10^{-3}$$
$$=(0.024\,73+0.011\,28+0.007\,97)$$
$$=0.043\,98(\text{kW/kvar})$$
$$\partial(\Delta A)_1=C_{bA}(Q_{com}T)=0.043\,98\times1460\times8760$$
$$=56.25\times10^4\ (\text{kW·h})$$

If C_{bp} is used to replace C_{bA}, then $\partial(\Delta A)'=0.051\,04\times1460\times8760=65.28\times10^4$ (kW·h), and the error is $\frac{65.28-56.25}{56.25}\times100\%=16.05\%$. This shows that such replacement will cause a big error, so the equivalent reactive compensation electric energy should be used to calculate electric energy losses reduced by the reactive compensation equipment.

2. Calculation of loss reduction effect of the installed compensation equipment.
   2.1 *First calculation method.* First, use relevant formulas in Section 5.1 of Chapter 5 to calculate the new parameters of the reactive load curves of all line sections after compensation, that is

$$p_I=\frac{Q_{com}}{\beta_{QI}Q_{maxI}}=\frac{1460}{0.65\times8980}=0.2501$$
$$p_{II}=\frac{1460}{0.60\times12\,900}=0.1886$$
$$f'_{QI}=\frac{f_{QI}-p_I\beta_{QI}}{1-p_I\beta_{QI}}$$
$$=\frac{0.90-0.250\,1\times0.65}{1-0.250\,1\times0.65}=0.8806$$
$$f'_{QII}=\frac{0.85-0.1886\times0.60}{1-0.1886\times0.60}=0.8309$$
$$Q'_{maxI}=8980-1460=7520\ (\text{kvar})$$
$$Q'_{maxII}=12\,900-1460=11\,440\ (\text{kvar})$$

Then, calculate the equivalent reactive compensation electric energy C'_{bA}, that is

$$C'_{bA} = \left[(2\times 11\,440\times 0.8309-3490)\times\frac{1.48}{35^2}+(2\times 7520\times 0.8806-3490)\times\left(\frac{0.94}{35^2}+\frac{0.054\,22}{10^2}\right)\right]\times 10^{-3}$$
$$=0.018\,75+0.007\,485+0.005\,289$$
$$=0.031\,524\,(\text{kW/kvar})$$

The reduced electric energy losses after the installation of reactive compensation equipment are

$$\partial(\Delta A)_2 = C'_{bA}(Q_{com2}T) = 0.031\,524\times 2490\times 8760$$
$$=96.38\times 10^4\ (\text{kW·h})$$

2.2 *Second calculation method.* Combine the two compensations and calculate the total saved electric energy losses minus the saved electric energy losses at the first compensation, that is

$$C_{bA\Sigma} = \left[(2\times 12\,900\times 0.85-4950)\times\frac{1.48}{35^2}+(2\times 8980\times 0.90-4950)\times\left(\frac{0.94}{35^2}+\frac{0.054\,22}{10^2}\right)\right]\times 10^{-3}$$
$$=0.020\,51+0.008\,61+0.006\,08$$
$$=0.0352\,(\text{kW/kvar})$$
$$\partial(\Delta A)_\Sigma = C_{bA\Sigma}(Q_{com\Sigma}T) = 0.0352\times 4950\times 8760$$
$$=152.63\times 10^4\,(\text{kW·h})$$
$$\partial(\Delta A)_2 = \partial(\Delta A)_\Sigma - \partial(\Delta A)_1 = 152.63\times 10^4 - 56.25\times 10^4$$
$$=96.38\times 10^4\ (\text{kW·h})$$

The results of the two calculation methods are totally the same.

12.3.2 Optimal Configuration of Reactive Compensation Equipment in Power Grids

The reactive comprehensive optimization of large electric power systems or regional power grids means minimizing the electric energy losses of electric power systems or regional power grids by regulating the distribution of reactive power flow and meeting the constraint conditions of all state variables (load node voltage, generator reactive output) and control variables (reactive compensation capacity, generator end voltage, ratio of on-load regulating transformer). In recent years, thanks to the wide application of computers, computer programs for reactive optimal configuration have been developing fast and increasingly improving. These programs generally consider the reactive voltage static characteristics of load buses and do a series of calculations to give the positions of compensation points, compensation capacity, and optimal positions of the tapping points of transformers, all of which comply with the optimization objective. Some reactive optimal configuration programs involve many optimization objective functions, such as minimum losses, minimum total compensation capacity, minimum compensation costs, and maximum composite economic results. In the objective of maximum composite economic results, the investment and depreciation of compensation equipment and the time-sharing calculation of electricity prices are taken into account.

In distribution power grids, the problem of calculating for the optimal configuration of fixed capacitors has been basically resolved. According to the literature [26], these computer programs introduce voltage constraint conditions to avoid over-high voltage after compensation, and they also consider the influence of voltage changes along the line on load losses and transformer no-load losses. Their objective function includes the net saved present value as a result of the capacitor investment; and the net saved present value is composed of the saved present value from reduced electric energy losses, the saved present value from reduced peak power, and the comprehensive investment and operating cost (negative value) of the capacitor. Such computer programs are applicable to radial distribution grids with multiple sections, multiple branches, any conductor cross-section, and any load distribution; but they address neither the problem of coordination in reactive power optimization for power grids and distribution grids at 35 kV and above, nor the problem of the optimal configuration of adjustable capacitors. With the gradual implementation of power distribution automation, these problems have been of great concern and have become new topics for the optimal configuration of reactive compensation equipment in power grids.

12.3.3 Exploitation of Reactive Potential and Reduction of Reactive Consumption

These measures include:

1. The reactive compensation equipment of customers should be put into use as much as possible.
2. Industrial enterprises should use motors whose power capacities match those of the mechanical equipment to be driven, so as to avoid the power factor from being reduced due to under-loading of the motors.
3. Asynchronous motors that are continuously running with constant rotating speed and large capacity should be replaced with synchronous motors which should be made under over-excitation operation.

12.4 Economical Operation of Transformers

12.4.1 Economical Operation of Two-Winding Transformers of the Same Model

To improve the reliability of electric supply and to meet the needs of load development, two transformers (occasionally more than two) operating in parallel with the same capacity are usually installed in a newly built substation. In the case of a failure or overhaul of one transformer, the other transformer (or transformers) can sustain the power supply. In the case of a light load, if the number of transformers in parallel operation is the same, the resistance losses in the windings are very small, but the iron core losses account for a large proportion in the total losses. On the condition that overload does not occur in partial transformers, the other transformers can be shut down to reduce the transformer total losses. In this way, the iron core losses of transformers will be greatly reduced, while the resistance losses in the transformer windings will be only slightly increased. If the reduced iron core losses exceed the increased winding losses, it is economically reasonable to shut one (or several) transformers.

If n two-winding transformers of the same model are in parallel operation, when the total loads are S (MVA), the total active power losses are

$$\begin{aligned}\Delta P_{\Sigma} &= n\Delta P_0 + n\Delta P_{\mathrm{LN}}\left(\frac{S}{nS_{\mathrm{N}}}\right)^2 \\ &= n\Delta P_0 + \frac{1}{n}\Delta P_{\mathrm{LN}}\left(\frac{S}{S_{\mathrm{N}}}\right)^2\end{aligned} \tag{12.10}$$

Wherein ΔP_0 – no-load losses of one transformer (kW);
ΔP_{LN} – rated load losses of one transformer (kW);
S – total load of n transformers (kVA);
S_{N} – rated capacity of one transformer (kVA).

After one transformer is shut down, the total power losses are

$$\Delta P'_{\Sigma} = (n-1)\Delta P_0 + \frac{1}{n-1}\Delta P_{LN}\left(\frac{S}{S_N}\right)^2$$

If the total power losses of n transformers in parallel operation are equal to those of $n-1$ transformers in parallel operation, the total loads at this time are called critical loads S_{cr1}, that is

$$n\Delta P_0 + \frac{1}{n}\Delta P_{LN}\left(\frac{S_{cr1}}{S_N}\right)^2 = (n-1)\Delta P_0 + \frac{1}{n-1}\Delta P_{LN}\left(\frac{S_{cr1}}{S_N}\right)^2$$

So critical loads are

$$S_{cr1} = S_N\sqrt{n(n-1)\frac{\Delta P_0}{\Delta P_{LN}}} \tag{12.11}$$

If the active power losses caused by the reactive power losses (no-load reactive power losses ΔQ_k and reactive power losses in reactance of on-load transformers) of transformers are considered, then Formula (12.11) can be rewritten to

$$S_{cr1} = S_N\sqrt{n(n-1)\frac{\Delta P_0 + C_{bp}\Delta Q_0}{\Delta P_{LN} + C_{bp}\Delta Q_{LN}}} \tag{12.12}$$

Wherein C_{bp} – equivalent coefficient used to convert reactive power losses to active power losses, that is equivalent reactive compensation power at the joint of transformer (kW/kvar);
ΔQ_0 – no-load reactive power losses of one transformer, approximately equal to the product of the no-load current percentage and the rated transformer capacity (kvar);
ΔQ_{LN} – no-load reactive power losses at short-circuit test of one transformer, approximately equal to the product of short-circuit voltage percentage and the rated transformer capacity (kvar).

When the total loads are reduced to below the critical load, it is economical to shut one transformer. When the total loads exceed the critical load, it is economical to have n transformers in parallel operation. Figure 12.10

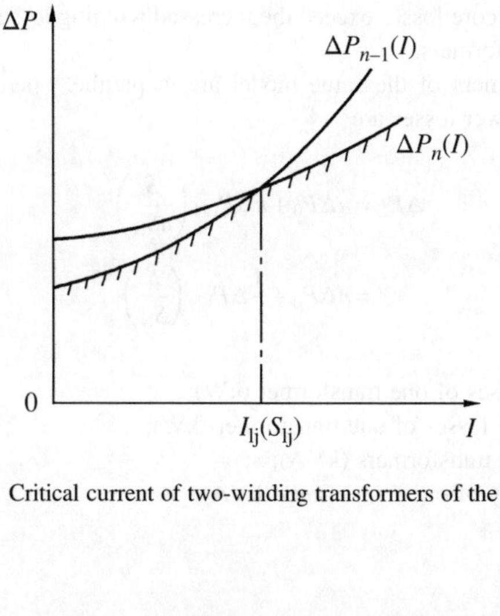

Figure 12.10 Critical current of two-winding transformers of the same model.

shows how to use critical load to decide the startup/shutdown of transformers, wherein $\Delta P_n(I)$ is the total loss curve of n transformers in operation, and ΔP_{n-1} is the total loss curve of $n-1$ transformers in operation.

12.4.2 Economical Operation of Two-Winding Transformers of Different Models

For a capacity expansion resulting from an increase in electric loads, the first step is usually to replace one transformer, so the situation where the models and parameters of two transformers in a substation are different is frequently encountered. Therefore, domestic scholars once explored the problem of economical operation of two-winding transformers of different models [28]. In the analysis below, one transformer with larger no-load losses is regarded as No. 1 transformer.

12.4.2.1 Economical Load Distribution of Transformers

As is known, when the loads of two transformers are distributed in inverse proportion to their resistance, the total losses are minimal. According to Appendix B, the resistance of a two-winding transformer is $R=\dfrac{\Delta P_{\mathrm{LN}}U_{\mathrm{N}}^2\times 10^3}{S_{\mathrm{N}}^2}$, so the economical load distribution coefficient of the transformer is

$$\left.\begin{aligned}
a_1 &= \frac{S_{\mathrm{cr1}}}{S_\Sigma} = \frac{R_2}{R_1+R_2} \\
&= \frac{\Delta P_{\mathrm{LN2}}\left(\dfrac{S_{\mathrm{N1}}}{S_{\mathrm{N2}}}\right)^2}{\Delta P_{\mathrm{LN1}}+\left(\dfrac{S_{\mathrm{N1}}}{S_{\mathrm{N2}}}\right)^2\Delta P_{\mathrm{LN2}}} = \frac{\Delta P'_{\mathrm{LN2}}}{\Delta P_{\mathrm{LN1}}+\Delta P'_{\mathrm{LN2}}} \\
a_2 &= \frac{S_{\mathrm{cr2}}}{S_\Sigma} = \frac{\Delta P_{\mathrm{LN1}}}{\Delta P_{\mathrm{LN1}}+\Delta P_{\mathrm{LN2}}} \\
\Delta P'_{\mathrm{LN2}} &= \left(\frac{S_{\mathrm{N1}}}{S_{\mathrm{N2}}}\right)^2\Delta P_{\mathrm{LN2}}
\end{aligned}\right\} \tag{12.13}$$

Wherein ΔP_{LN1} – rated load losses of No. 1 transformer (kW);
$\Delta P'_{\mathrm{LN2}}$ – rated load losses of No. 2 transformer after inclusion (kW).

If the transformers are in parallel operation, the actual load distribution is conducted in inverse proportion to the short-circuit voltage $\Delta u_{\mathrm{k}}\%$, that is

$$\left.\begin{aligned}
a'_1 &= \frac{\Delta u'_{\mathrm{k2}}}{\Delta u_{\mathrm{k1}}+\Delta u'_{\mathrm{k2}}} \\
a'_2 &= \frac{\Delta u'_{\mathrm{k1}}}{\Delta u_{\mathrm{k1}}+\Delta u'_{\mathrm{k2}}}
\end{aligned}\right\} \tag{12.14}$$

$$\Delta u'_{\mathrm{k2}} = \left(\frac{S_{\mathrm{N1}}}{S_{\mathrm{N2}}}\right)\Delta u_{\mathrm{k2}}$$

Wherein Δu_{k1} – short-circuit voltage percentage of No. 1 transformer;
$\Delta u'_{\mathrm{k2}}$ – short-circuit voltage percentage of No. 2 transformer after inclusion.

If a'_1 and a_1 are close, parallel operation is more economical. If the difference between the two is large, given the economical reduction of energy losses, the two transformers should be in separate operation, with loads being distributed based on the economical load distribution coefficients a_1 and a_2; and technical measures for improving the reliability of the electric supply should be taken.

12.4.2.2 Calculation of Critical Current Used to Determine Start-up/Shut-down of Transformers

$\Delta P_1(I)$ and $\Delta P'_2(I)$ are the respectively loss curves of No. 1 and No. 2 transformers after inclusion, and $\Delta P_\Sigma(I)$ is the total loss curve of these two transformers in operation. If I_1 is the point of intersection between the $\Delta P_1(I)$ and $\Delta P'_2(I)$ curves, I_{11} being the point of intersection between $\Delta P_\Sigma(I)$ and $\Delta P'_2(I)$ curves, and I_{12} being the point of intersection between $\Delta P_\Sigma(I)$ and $\Delta P_1(I)$ curves, then the following three situations can be discussed.

1. The load losses of No. 1 transformer are larger, that is $\Delta P_{LN1} > \Delta P'_{LN2}$. As shown in Figure 12.11, in this case, $\Delta P_1(I)$ and $\Delta P'_2(I)$ have no point of intersection, and there is only one kind of critical current I_{11}.
   Obviously, when $I > I_{11}$, two transformers should be allowed to operate; when $I < I_{11}$, No. 1 transformer with larger losses should be shut. The critical current I_{11} can be calculated as per the following formula:

$$I_{11} = I_{1N}\sqrt{\Delta P_{01}\frac{\Sigma\Delta P_{LN}}{\left(\Delta P'_{LN2}\right)^2}} \tag{12.15}$$

   Wherein I_{1N} – rated current of No. 1 transformer (A);
   $\Sigma\Delta P_{LN}$ – sum of rated load losses of transformers after inclusion (kW).

2. The load losses of No. 1 transformer are smaller, that is $\Delta P_{LN1} < \Delta P'_{LN2}$ and $I_1 > I_{11}$, $I_1 > I_{12}$. As shown in Figure 12.12, when $I > I_{11}$, two transformers should be allowed to operate; when $I < I_{11}$, No. 1 transformer with larger no-load losses should be shut. I_{11} can also be calculated as per Formula (12.15).
3. The load losses of No. 1 transformer are smaller, that is $I_1 < I_{11}$, $I_1 < I_{12}$. As shown in Figure 12.13, two kinds of current I_1 and I_{12} exist in this case. When $I > I_{12}$, two transformers should be allowed to operate; when $I_1 < I < I_{12}$, No. 2 transformer with larger load losses should be shut; when $I < I_1$, No. 1 transformer with larger no-load losses should be shut. To avoid frequent operations, when the difference between I_1 and

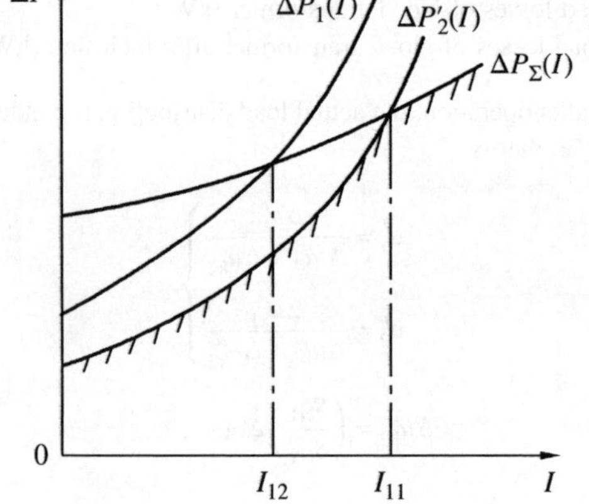

Figure 12.11 Schematic diagram under $\Delta P_{LN1} > \Delta P'_{LN2}$ situation.

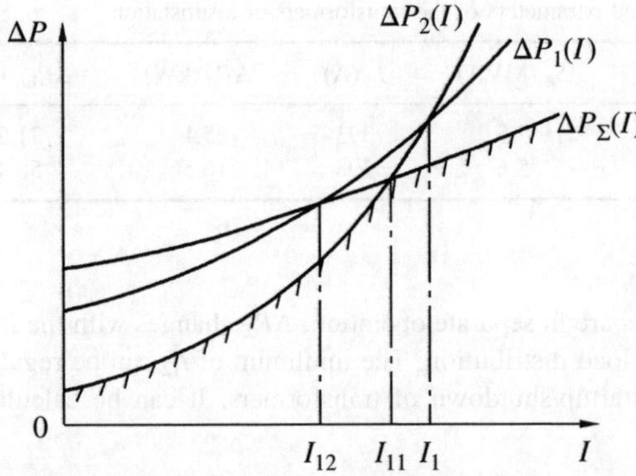

Figure 12.12 Schematic diagram under $I_1 > I_{11}$ and $I_1 > I_{12}$ situations.

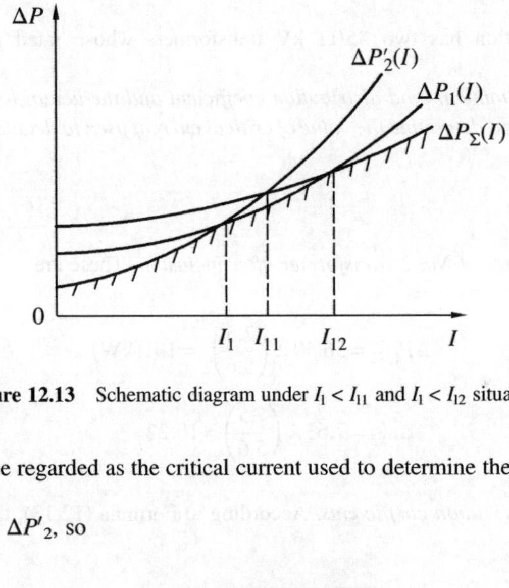

Figure 12.13 Schematic diagram under $I_1 < I_{11}$ and $I_1 < I_{12}$ situations.

I_{12} is not large, I_1 can be regarded as the critical current used to determine the startup/shutdown of No. 1 transformer.

When $I = I_1$, $\Delta P_1 = \Delta P'_2$, so

$$I_1 = I_{1N}\sqrt{\frac{\Delta P_{01} - \Delta P_{02}}{\Delta P'_{LN2} - \Delta P_{LN1}}} \tag{12.16}$$

When $I = I_{11}$, $\Delta P'_2 = \Delta P_\Sigma$; when $I = I_{12}$, $\Delta P_1 = \Delta P_\Sigma$, so

$$I_{11} = I_{1N}\sqrt{\frac{\Delta P_{01}}{a'_1\left(2\Delta P'_{LN2} - a'_1\Sigma\Delta P_{LN}\right)}} \tag{12.17}$$

$$I_{12} = I_{1N}\sqrt{\frac{\Delta P_{02}}{a'_2\left(2\Delta P'_{LN1} - a'_2\Sigma\Delta P_{LN}\right)}} \tag{12.18}$$

Wherein load distribution coefficients a'_1 and a'_2 when the two transformers are in parallel operation are calculated as per Formula (12.14).

Table 12.2 Rated parameters of the transformers of a substation.

Transformer No.	S_n (MVA)	I_n (A)	ΔP_0 (kW)	ΔP_{Ln} (kW)	Δu_k
No. 1	7.5	393	25.1	71.25	7.28
No. 2	5.6	308	16.56	56.4	7.63

If the two transformers are in separate operation, ΔP_Σ changes with the load distribution, so I_{11} and I_{12} also change with the load distribution. The minimum of I_{12} can be regarded as the critical current used to determine the startup/shutdown of transformers. It can be calculated as per the following formula:

$$I_{12.\min} = I_{1N}\sqrt{\Delta P_{02}\frac{\Sigma \Delta P_{LN}}{\Delta P_{LN1}^2}} \qquad (12.19)$$

Example 12.4 A substation has two 35/11 kV transformers whose rated parameters are shown in Table 12.2.

Try to compare the economical load distribution coefficient and the actual load distribution coefficient under parallel operation, and determine the value of critical current used to decide the start-up/shut-down of transformers.

Solution

1. *Calculation of parameters of No. 2 transformer after inclusion.* These are

$$\Delta P'_{LN2} = 56.40 \times \left(\frac{7.5}{5.6}\right)^2 = 101(\text{kW})$$

$$\Delta u'_{k2} = 7.63 \times \left(\frac{7.5}{5.6}\right) = 10.22$$

2. *Calculation of load distribution coefficients.* According to Formula (12.13), the economical load distribution coefficients are

$$a_1 = \frac{101}{101 + 71.25} = 0.586$$

$$a_2 = 1 - 0.586 = 0.414$$

If the transformers are in parallel operation, the actual load distribution coefficients can be calculated as per Formula (12.14), that is

$$a'_1 = \frac{10.22}{10.22 + 7.28} = 0.584$$

$$a'_2 = 1 - 0.584 = 0.416$$

The results show that the actual load distribution coefficients are close to the economical load distribution coefficients, so this parallel operation is economically reasonable.

3. *Calculation of critical current.* According to Formulas (12.16) to (12.18),

$$I_1 = 393 \times \sqrt{\frac{25.1 - 16.56}{101 - 71.25}} = 210.5(\text{A})$$

$$I_{l1} = 393 \times \sqrt{\frac{25.1}{0.584 \times [2 \times 101 - 0.584 \times (101 + 71.25)]}} = 256(\text{A})$$

$$I_{l2} = 393\sqrt{\frac{16.56}{0.416 \times (2 \times 71.25 - 0.416 \times 71.25)}} = 294.6(\text{A})$$

The calculation results show that the relationship among the three kinds of current belongs to the type shown in Figure 12.13.
4. *Startup/shutdown conditions of transformers.* The normal mode is parallel operation of two transformers. When the total loads are $I > 294.6$ A, both transformers should be operating. When $210.5 < I < 294.6$ (A), No. 1 transformer should be operating and No. 2 transformer should be shut down. When $I < 210.5$ A, No. 2 transformer should be operating and No. 1 transformer should be shut down.

12.4.3 *Economical Operation of Three-Winding Transformers of Different Models*

Because the economical operation of three-winding transformers involves many factors, the problem is complicated. Here, calculation formulas of load distribution coefficient and critical current are introduced, and one example is given to explain the calculation process [28].

12.4.3.1 Determination of Economical Load Distribution and Normal Operation Mode

In this example, subscripts *h*, *m*, and *l* are used to represent high-voltage, medium-voltage, and low-voltage windings. The transformer with a larger ΔP_0 is regarded as No. 1 transformer, and the load losses and short-circuit voltage parameters of No. 2 transformer are included with the No. 1 transformer. If the ratio between the low-voltage side loads and total loads is *b*, and a_{h1} and a_{l1} are respectively the economical load distribution coefficients of the high- and low-voltage windings of No. 1 transformer, then the above three independent coefficients can be used to calculate the other four economical load distribution coefficients. that is

$$a_{\text{m1}} = a_{\text{h1}} - a_{\text{l1}}, \quad a_{\text{h2}} = 1 - a_{\text{h1}}$$

$$a_{\text{m2}} = (1 - b) - a_{\text{m1}} = 1 - b - a_{\text{h1}} + a_{\text{l1}}, \quad a_{\text{l2}} = b - a_{\text{l1}}$$

1. Calculation of economical load distribution coefficients. On the condition that the sum of power losses at the three windings of two transformers is the minimum, with the ratio *b* as the parameter, the formulas of the economical load distribution coefficients are

$$a_{\text{h1}} = \frac{1}{N_P}(L_P \Delta P_{\text{h}\cdot\text{m2}} - M_P \Delta P_{\text{m2}}) + \frac{b}{N_P}(M_P \Delta P_{\text{m}\cdot\text{l2}} - L_P \Delta P_{\text{m2}}) \tag{12.20}$$

$$a_{\text{l1}} = \frac{1}{N_P}(M_P \Delta P_{\text{h}\cdot\text{m2}} - K_P \Delta P_{\text{m2}}) + \frac{b}{N_P}(K_P \Delta P_{\text{m}\cdot\text{l2}} - M_P \Delta P_{\text{m2}}) \tag{12.21}$$

$$K_P = \Delta P_{\text{h-m1}} + \Delta P_{\text{h-m2}}$$

$$L_P = \Delta P_{\text{m-l1}} + \Delta P_{\text{m-l2}}$$

$$M_P = \Delta P_{\text{m1}} + \Delta P_{\text{m2}}$$

$$N_P = K_P L_P - M_P^2$$

Wherein K_p, L_p, M_p, N_P – calculation parameters related to load power losses.

2. Calculation of actual load distribution coefficients under parallel operation of medium-voltage and low-voltage windings. According to the actual loads and short-circuit voltage parameters at three sides of the transformers, the loop voltage equations can be listed, and the calculation formulas of the actual load distribution coefficients a'_{h1} and a'_{l1} are

$$a'_{\text{h1}} = \frac{1}{N_{\text{u}}}(L_{\text{u}}\Delta u_{\text{h-m2}} - M_{\text{u}}\Delta u_{\text{m2}}) + \frac{b}{N_{\text{u}}}(M_{\text{u}}\Delta u_{\text{m-l2}} - L_{\text{u}}\Delta u_{\text{m2}}) \tag{12.22}$$

$$a'_{\text{l1}} = \frac{1}{N_{\text{u}}}(M_{\text{u}}\Delta u_{\text{h-m2}} - K_{\text{u}}\Delta u_{\text{m2}}) + \frac{b}{N_{\text{u}}}(K_{\text{u}}\Delta u_{\text{m-l2}} - M_{\text{u}}\Delta u_{\text{m2}}) \tag{12.23}$$

$$K_{\text{u}} = \Delta u_{\text{h-m1}} + \Delta u_{\text{h-m2}}$$

$$L_{\text{u}} = \Delta u_{\text{m-l1}} + \Delta u_{\text{m-l2}}$$

$$M_{\text{u}} = \Delta u_{\text{m1}} + \Delta u_{\text{m2}}$$

$$N_{\text{u}} = K_{\text{u}} L_{\text{u}} - M_{\text{u}}^2$$

If the two transformers are both step-down short-circuit voltage combinations, due to $\Delta u_{\text{m1}} \approx 0$, $\Delta u_{\text{m2}} \approx 0$, $\Delta M_{\text{u}} \approx 0$, Formulas (12.22) and (12.23) can be simplified to

$$a'_{\text{h1}} = \frac{\Delta u_{\text{h-m2}}}{K_{\text{u}}} = \frac{\Delta u_{\text{h-m2}}}{\Delta u_{\text{h-m1}} + \Delta u_{\text{h-m2}}} \tag{12.24}$$

$$a'_{\text{l1}} = b\frac{\Delta u_{\text{m-l2}}}{L_{\text{u}}} = b\frac{\Delta u_{\text{m-l2}}}{\Delta u_{\text{m-l1}} + \Delta u_{\text{m-l2}}} \tag{12.25}$$

3. High- and medium-voltage windings are in parallel operation, but low-voltage winding cannot be in parallel operation due to limitations of system conditions. The load distribution at the high- and medium-voltage sides is determined by loop voltage equations, and the economical load distribution coefficient at the low-voltage side is calculated on the condition that power losses are minimal. If the two transformers are both step-down short-circuit voltage combinations, a'_{h1} can also be calculated as per Formula (12.24), and a_{l1} can be calculated as per the following formula:

$$a_{\text{l1}} = b\frac{\Delta P_{\text{m-l2}}}{L_P} + \frac{a'_{\text{h1}} M_P - \Delta P_{\text{m2}}}{L_P} \tag{12.26}$$

4. Conversion of load distribution coefficients of three windings. The above load distribution coefficients are relevant to the total loads. If respective load distribution coefficients need to be calculated for medium- and low-voltage windings, conversion is required, that is

$$\bar{a}_{\text{m1}} = \frac{a_{\text{m1}}}{1-b}, \quad \bar{a}_{\text{l1}} = \frac{a_{\text{l1}}}{b}$$

$$\bar{a}_{\text{m2}} = 1 - \bar{a}_{\text{m1}}, \quad \bar{a}_{\text{l2}} = 1 - \bar{a}_{\text{l1}}$$

12.4.3.2 Determination of Critical Current Used to Decide the Startup/Shutdown of Transformers

1. Calculation of critical current I_l corresponding to the point of intersection of loss curves of two transformers. When $(\Delta P_{m\cdot l2} - \Delta P_{m\cdot l1}) < 0$, the calculation formula of I_l should be selected according to the low-voltage load ratio $b_j = \frac{\Delta P_{m2} - \Delta P_{m1}}{\Delta P_{m\cdot l2} - \Delta P_{m\cdot l1}}$ at the extreme point, that is

$$b_j < 0 \qquad I_{lmin} = I_{1N}\sqrt{\frac{\Delta P_{01} - \Delta P_{02}}{\Delta P_{h\cdot m2} - \Delta P_{h\cdot m1}}} \tag{12.27}$$

$$0 < b_j < 1 \qquad I_{lmin} = I_{1N}\sqrt{\frac{\Delta P_{01} - \Delta P_{02}}{(\Delta P_{h\cdot m} - \Delta P_{h\cdot m1}) - (\Delta P_{m2} - \Delta P_{m1})^2/(\Delta P_{m\cdot l2} - \Delta P_{m\cdot l1})}} \tag{12.28}$$

$$b_j > 1 \qquad I_{lmin} = I_{1N}\sqrt{\frac{\Delta P_{01} - \Delta P_{02}}{\Delta P_{h.l2} - \Delta P_{h.l1}}} \tag{12.29}$$

When $(\Delta P_{m\cdot l2} - \Delta P_{m\cdot l1}) > 0$, $b_j > 0.5$, I_{lmin} is calculated as per Formula (12.27); when $b_j \leq 0.5$, I_{lmin} is calculated as per Formula (12.29).

2. Calculation of current I_{l1} and I_{l2} corresponding to the point of intersection between the total operating loss curve of two transformers and the loss curve of one transformer. If the medium- and low-voltage windings of two transformers are in separate operation, when $b = 0$ (i.e. high- and medium-voltage windings are operating),

$$\left.\begin{aligned} I_{l1}^{(0)} &= I_{1N}\sqrt{\frac{\Delta P_{01}(\Delta P_{h\cdot m1} + \Delta P_{h\cdot m2})}{\Delta P_{h\cdot m2}^2}} \\ I_{l2}^{(0)} &= I_{1N}\sqrt{\frac{\Delta P_{02}(\Delta P_{h\cdot m1} + \Delta P_{h\cdot m2})}{\Delta P_{h\cdot m1}^2}} \end{aligned}\right\} \tag{12.30}$$

When $b = 1$ (i.e. high- and low-voltage windings are operating),

$$\left.\begin{aligned} I_{l1}^{(1)} &= I_{1N}\sqrt{\frac{\Delta P_{01}(\Delta P_{h\cdot l1} + \Delta P_{h\cdot l2})}{\Delta P_{h\cdot l2}^2}} \\ I_{l2}^{(1)} &= I_{1N}\sqrt{\frac{\Delta P_{02}(\Delta P_{h\cdot l1} + \Delta P_{h\cdot l2})}{\Delta P_{h\cdot l1}^2}} \end{aligned}\right\} \tag{12.31}$$

According to the results of the above two formulas, allow the smaller value between $I_{l1\cdot min}^{(0)}$ and $I_{l1\cdot min}^{(1)}$ to be $I_{l1\cdot min}$; allow the smaller value between $I_{l1\cdot min}^{(0)}$ and $I_{l1\cdot min}^{(1)}$ to be $I_{l2\cdot min}$.

Formulas (12.30) to (12.31) are also applicable to simplified calculations under several situations where medium- and low-voltage windings are in parallel operation; where high- and medium-voltage windings are in parallel operation, while a low-voltage winding is in separate operation; or where the impedance

Table 12.3 Rated parameters of the transformers of a substation.

S_N(MVA)	$I_{N\cdot g}$ (A)	ΔP_0 (kW)	ΔP_{LN} (kW)				Δu_k			
			h – m	m – l	h – l	m	h – m	m – l	h – l	m
60(100/100/100)	315	96.7	288	245	318.5	107	9.82	6.03	16.85	–0.5
31.5(100/100/100)	165	54.1	187.7	156.6	197.1	73.6	9.81	5.9	16.2	–0.25

combinations of two transformers are not of the same type. Errors caused by such simplifications generally do not exceed 10%, so such simplifications are acceptable to the calculation of economical operation of transformers.

3. Determination of start-up/shut-down of transformers. After $I_{l.min}$, $I_{l1.min}$ and $I_{l2.min}$ are obtained, the following determination can be made (see Figures 12.12 and 12.13):

   If $I_{l\cdot min} > I_{l1\cdot min}$, $I_{l2\cdot min}$, $I_{l1\cdot min}$ is the critical current used to determine the startup/shutdown of No. 1 transformer with larger ΔP_0.

   If $I_{l\cdot min} < I_{l1\cdot min}$, $I_{l2\cdot min}$, $I_{l2\cdot min}$ is the critical current used to determine the startup/shutdown of No. 2 transformer with larger ΔP_{LN}; $I_{l\cdot min}$ is the critical current used to determine the startup/shutdown of No. 1 transformer with larger ΔP_0.

Example 12.5 A substation has two 110/38.5/11 kV three-winding transformers whose rated parameters are shown in Table 12.3.

Due to limitations in the short-circuit capacity, the two transformers must be in separate operation. Assume that the ratio between low-voltage loads and total loads is $b = 0.3$. *Try to calculate economical load distribution coefficients and critical current, and determine the start-up/shut-down conditions of transformers.*

Solution

1. *Assume that the 60 MVA transformer is No. 1 transformer.* If the short-circuit parameters of No. 2 transformer are included, then

$$\Delta P_{h\cdot m2} = 681\,\text{kW}, \quad \Delta P_{m\cdot l2} = 568\,\text{kW}$$
$$\Delta P_{h\cdot l2} = 715\,\text{kW}, \quad \Delta P_{m2} = 276\,\text{kW}$$
$$\Delta u_{m2} = 18.7, \quad \Delta u_{m\cdot l2} = 11.2$$
$$\Delta u_{h\cdot l2} = 30.9, \quad \Delta u_{m2} = -0.48$$

2. *Calculation of economical load distribution coefficients.* According to Table 12.3 and inclusion results of No. 2 transformer, $K_P = 969$ kW, $L_P = 813$ kW, $M_P = 374$ kW, $N_P = 647\,921$ (kW)^2. Substitute these into Formulas (12.20) and (12.21) to obtain

$$a_{h1} = \frac{1}{647\,921} \times (813 \times 681 - 374 \times 267) + \frac{b}{647\,921} \times (374 \times 568 - 813 \times 267) = 0.7 - 0.007b$$

$$a_{l1} = \frac{1}{647\,921} \times (374 \times 681 - 969 \times 267) + \frac{b}{647\,921} \times (969 \times 568 - 374 \times 267) = -0.006 + 0.695b$$

When $b = 0.30$, $a_{h1} = 0.698$, $a_{l1} = 0.202$, $a_{m1} = 0.496$. So the load distribution coefficients at each side are obtained through conversion

$$\bar{a}_{m1} = \frac{0.496}{1-0.30} = 0.709$$

$$\bar{a}_{m2} = 1-0.709 = 0.291$$

$$\bar{a}_{l1} = \frac{0.202}{0.30} = 0.673$$

$$\bar{a}_{l2} = 1-0.673 = 0.327$$

3. *Calculation of critical current.* As $(\Delta P_{m\cdot 12} - \Delta P_{m\cdot 11}) > 0$, and $b_j = \frac{\Delta P_{m2} - \Delta P_{m1}}{\Delta P_{h\cdot 12} - \Delta P_{h\cdot 11}} = \frac{267-107}{568-245} = 0.495$, according to Formula (12.29),

$$I_{lmin} = I_{lN}\sqrt{\frac{\Delta P_{01} - \Delta P_{02}}{\Delta P_{h.12} - \Delta P_{h.11}}}$$

$$= 315 \times \sqrt{\frac{96.7-54.1}{715-318.5}} = 103.3\,(\text{A})$$

According to Formulas (12.30) and (12.31),

$$I_{l1}^{(0)} = 315 \times \sqrt{\frac{96.7 \times (288+681)}{681^2}} = 141.6\,(\text{A})$$

$$I_{l1}^{(1)} = 315 \times \sqrt{\frac{96.7 \times (318.5+715)}{715^2}} = 139.3\,(\text{A})$$

$$I_{l2}^{(0)} = 315 \times \sqrt{\frac{54.1 \times (288+681)}{288^2}} = 250.4\,(\text{A})$$

$$I_{l2}^{(1)} = 315 \times \sqrt{\frac{54.1 \times (318.5+715)}{318.5^2}} = 233.9\,(\text{A})$$

Take the smaller value between the two, obtaining $I_{l1\cdot min}$ = 139.3 A, $I_{l2\cdot min}$ = 233.9 A.

4. *Conclusion of economical operation.* The normal mode is separate operation of two transformers, wherein loads at the medium-voltage side are distributed by 0.71:0.29 and loads at the low-voltage side are distributed by 0.67:0.33. When the total load current of 110 kV is $I > 234$ A, two transformers should be operating. When 103 A $< I <$ 234 A, No. 1 transformer should be operating. When $I <$ 103 A, No. 2 transformer should be operating (see Figure 12.13).

12.5 Adjustment and Balancing of Loads

12.5.1 Adjustment of Load Curves

If the load curves of an electric power system fluctuate significantly, power generation equipment and power supply equipment with large capacities are needed, and line losses will be increased. In contrast, the adjustment of load curves can increase the load factor and reduce the maximum load, thus reducing line losses.

If the equivalent resistance of a common distribution line is R, then the start end maximum current and loss factor can be used to calculate electric energy losses ΔA (kW·h) within the measuring period as per Formula (1.39), that is

$$\Delta A = 3I_{max}^2 FRT \times 10^{-3}$$

Because the loss factor of the common distribution line can be calculated as per Formula (2.16), that is $F = 0.2f + 0.8f_2f^2$. As $f = I_{av}/I_{min}$, and $I_{av} = A/(\sqrt{3}U_{av}\cos\varphi_{av}T)$, so $I_{max} = A/(\sqrt{3}U_{av}\cos\varphi_{av}Tf)$. Substitute the formulas for F and I_{max} into ΔA to obtain

$$\Delta A = \frac{RA^2}{U_{av}^2\cos^2\varphi_{av}T}\left(\frac{0.2}{f} + 0.8\right) \times 10^{-3} \tag{12.32}$$

According to Formula (12.32), when the start end electric supply and operating parameters of the line remain unchanged, the electric energy losses are reduced with the increase in the load factor.

If f is increased from 0.8 to 0.88, $\delta f = \frac{0.08}{0.8} \times 100\% = 10\%$. According to Formula (12.32), ΔA will be reduced by 2.16%, that is $\delta(\Delta A)\% \approx -0.20\delta f$.

The adjustment of load curves can improve the load factor, reduce the electric energy losses, and reduce the line loss rate. The detailed calculation of the reduction of the line loss rate with the increase in the load factor can follow Formula (13.2).

12.5.2 Balancing the Loads of Lines or Transformers, and Adjusting the Power Sources of Dual Power Customers

In the low-voltage line shown in Figure 12.14, points A and B are two break points. Originally, point A is open and point B is closed. The load current of two adjacent 320 kVA distribution transformers is very non-uniform, and the line losses are very big. To save the line losses, point A is closed and point B is open. The maximum load current is marked in Figure 12.14.

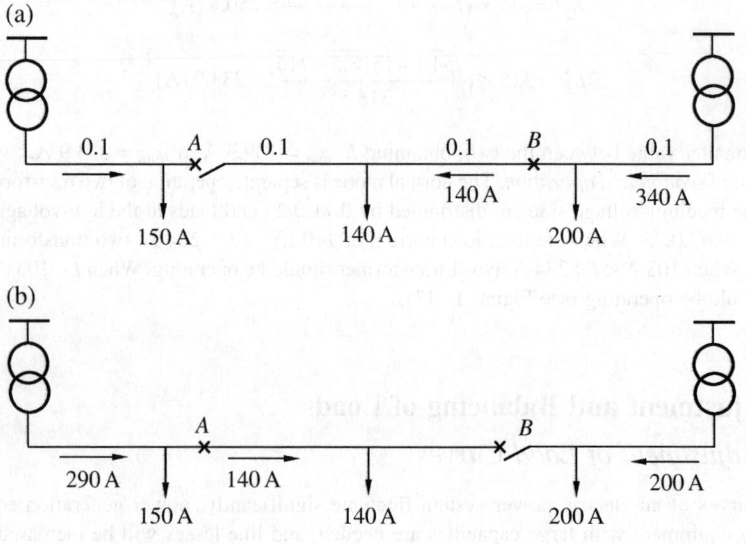

Figure 12.14 Balancing of loads of low-voltage lines. The figure on the line is line resistance (Ω); the figure under the line is current (A); the figure under the arrow is the load current. (a) Before change of break point. (b) After change of break point.

According to records of load current, the loss factor is calculated, that is $F = 0.47$. The rated load losses of a 320 kVA distribution transformer are 5.7 kW. After the change of break points, the reduced line losses are calculated as follows:

1. Iron core losses of the distribution transformer can be deemed as unchanged before and after the change of break points in the low-voltage line. The rated current of the distribution transformer is $I_N = 320/(\sqrt{3} \times 0.38 = 460\ (\text{A})$.
2. When point A is open and point B is closed before the change of break points, the load power losses of the transformer are

$$\Delta P_T = 5.7 \times \left[\left(\frac{150}{460}\right)^2 + \left(\frac{340}{460}\right)\right]^2 = 3.72\ (\text{kW})$$

The power losses of the line are

$$\Delta P_1 = 3 \times \left(150^2 \times 0.1 + 340^2 \times 0.1 + 140^2 \times 0.1\right) \times 10^{-3}$$
$$= 47.31\ (\text{kW})$$

The total power losses are

$$\Delta P_\Sigma = \Delta P_T + \Delta P_1 = 3.72 + 47.31 = 51.03\ (\text{kW})$$

3. When point A is closed and point B is open after the change of break points, the power losses of the transformer are

$$\Delta P_T' = 5.7 \times \left[\left(\frac{290}{460}\right)^2 + \left(\frac{200}{460}\right)^2\right] = 3.34\ (\text{kW})$$

The power losses of the line are

$$\Delta P_1' = 3 \times \left[290^2 \times 0.1 + 140^2 \times 0.1 + 200^2 \times 0.1\right] \times 10^{-3}$$
$$= 43.11\ (\text{kW})$$

The total power losses after the change of break points are

$$\Delta P_\Sigma' = \Delta P_T' + \Delta P_1' = 3.34 + 43.11 = 46.45\ (\text{kW})$$

4. After the change of break points, the monthly saved electricity line losses are

$$\Delta A_m = (51.03 - 46.45) \times 0.47 \times 720 = 1550\ (\text{kW·h})$$

Figure 12.15 shows a high-voltage distribution tie line between two substations, and there is an outdoor breaker QF in the line. QF is normally open. The industrial customer C close to breaker QF has two power sources which are respectively connected to both sides of breaker QF. It can be supplied with electricity by either substation A or substation B. Figure 12.15a and b shows two different operation modes.

The maximum current of each load along the line and the resistance of each line section are marked in Figure 12.15. According to measured records of load current, the loss factors F of outgoing lines of the two substations are both 0.6. To save line losses, the supplier of electricity to customer C is changed from

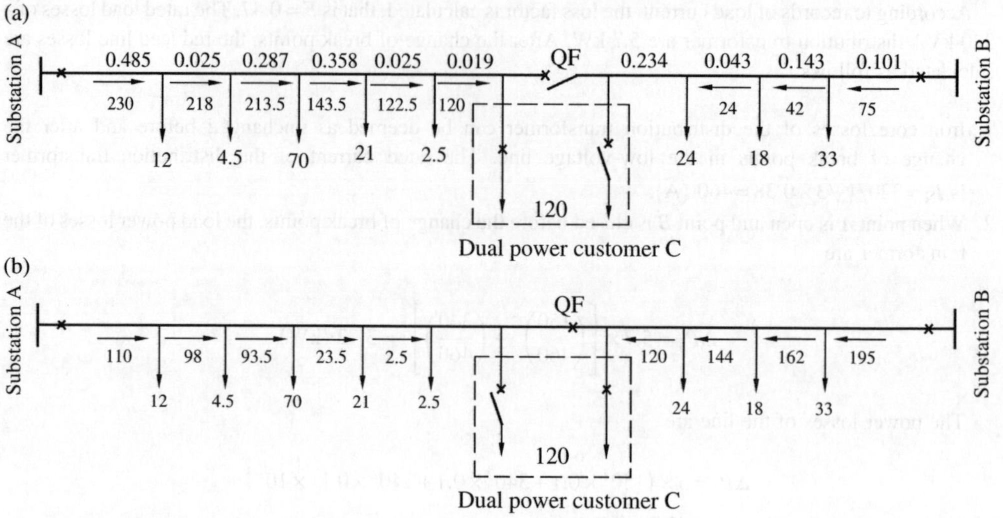

Figure 12.15 Change of power supply for dual power customers. The figure on the line is line section resistance (Ω); the figure under the line is current (A); the figure under the arrow is the load current (A). (a) Before change of power supply. (b) After change of power supply.

substation A to substation B, so the non-uniform loads of two outgoing lines have been greatly improved. The calculation shows that the monthly saved line losses are 37 200 kW·h.

12.5.3 Balancing Three-Phase Loads

During the operation of low-voltage distribution grids, it is often necessary to measure the three-phase load current and neutral current at leading-out terminals of distribution transformers and at some main lines, and to balance the three-phase load current. Unbalanced three-phase load current will affect the voltage quality of low-voltage grids and increase line losses.

To control the degree of unbalanced line loads and to take the necessary measures can reduce the losses resulting from unbalanced three-phase loads. The degree of imbalance (the ratio between neutral current and average three-phase current) in the outlet current of distribution transformers is generally required not to exceed 10%, and the degree of imbalance in the current of low-voltage main lines and start ends of major branch lines should not exceed 20%.

12.6 Strengthen Power Grid Maintenance

During the actual operation of power grids, leakage losses may occur due to the defective insulation of electrically charged equipment. Such losses can be reduced by strengthening the power grid maintenance. Measures in this regard include:

1. Clean the insulators and insulating sleeves of lines, transformers and breakers regularly.
2. Pay frequent attention to cutting off any tree branches touching line conductors.
3. Pay attention to removing bird nests and other foreign matter during patrols or overhauls.

4. Pay attention to the quality of conductor joints during the overhaul and constriction of lines. The resistance value of joints should meet the technical regulations to reduce losses due to larger joint resistance. For example, a line has three conductor joints. The resistance of each joint is 0.01 Ω, and 200 A current usually flows through the joints. The electricity losses in one year will reach $200^2 \times 0.01 \times 3 \times 8760 \times 10^{-3} =$ 10 410 (kW·h). Power grids have many conductor joints. If attention is omitted, the total electricity losses may be quite big. Poor joint connections can also affect the safe operation of power grids, so the connection of conductors must follow the technical requirements.
5. Gradually adopt energy saving hardware in line maintenance and transformation.

12.7 Strengthen Power Consumption Management and Measuring Management

The loss reduction measures in the previous sections can really reduce the electricity line losses of power grids. In other words, on the condition that the quantity of generated energy is certain, more electricity can be supplied to customers. This section describes the loss reduction measures relating to power consumption management and measuring management. Some measures can really save electric energy and reduce line losses, while others can ensure accurate statistics for electricity line losses and line loss rate and reduce the losses of enterprises. Relevant measures include:

1. To improve the accuracy of electric supply calculations, regularly examine the watt-hour meters of power plants and substations, and check the balance between the primary input power and output power of power plants and substations once a month, so as to identify abnormal measurement of watt-hour meters and take the necessary measures to eliminate them.
2. Strengthen the management of measuring equipment, including regularly examining the standard instrumentation of customers' watt-hour meters, regularly examining the current transformers, voltage transformers, and watt-hour meters of various customers, and strengthening the management and effect of watt-hour meter reading systems installed in recent years.
3. Strengthen the installation and replacement of customers' watt-hour meters. Implement the requirement that "power supply enterprises should reduce losses and wastes in power transmission and supply" as stipulated in Reference [77]. Determine measuring points reasonably and ensure the correct measurement of watt-hour meters.
4. Strengthen power consumption plans, adjust loads in time, and reduce line losses.
5. Strengthen meter reading and accounting work to improve the accuracy of statistics for the power sales quantity of power grids. For all large customers whose total power consumption quantity accounts for over 80% of total power sales quantity, it is generally necessary to do meter reading simultaneously at midnight on the last day of each month. The power consumption quantity of other general customers is read on a fixed date month by month, and the sum of the read electric quantity of the second half of the current month and the read electric quantity of the first half of the next month is used as the power sales quantity for calculating the statistical line loss rate of the current month.
6. Carry out a general survey of businesses and eliminate any power usage without a watt-hour meter and any illegal power usage.

To sum up, the reduction of line losses in power grids covers many fields and is a technical management work requiring the effort of all staff. Therefore, loss reduction can yield twice the result with half the effort through strengthened leadership and mobilizing the enthusiasm of all staff.

12.8 Application of New Designs, New Materials, and New Technologies

With progress in science and technology, endless technical measures for reducing line losses in the electric power system are emerging. This section only introduces the application of new designs, new materials, and new technologies since the last century.

12.8.1 New Design for Loss Reduction in Ground Wires of High-voltage Transmission Line

Transmission lines of 110 kV and above have double ground wires to prevent lightning strike. The conventional design is selecting steel stranded wires and connecting the power to the ground at each foundation. A short-circuit loop current results from the electromotive force which is generated when the ground wire between two towers is induced by the conductor load current, and this results in larger electric energy losses of the overhead ground wire. According to relevant calculations, the ground wire losses of a 220 kV single-loop line are about (5–10) × 10^4 kWh/year for each 100 km, 60 × 10^4 kWh/year for each 100 km for the 330 kV line, and 500 × 10^4 kWh/year for each 100 km for the 500 kV line. The new design selects two ground wires fully insulated across the line, installs a discharge gap, and takes ground wire transposition measures. The induced electromotive force of the ground wire is reduced by 90%, and therefore, ground wire losses are greatly reduced [57].

Since the middle of the 1980s, optical fiber composite overhead ground wires (OPGW) have been largely used in China. Due to the limitations of OPGW heat resistance, they cannot be used for the design of long-distance ground wire insulation. According to a relevant study, an effective new design program for use within a range of 1–2 km of a substation's incoming and outgoing ends is a ground wire which has an aluminum-coated steel good conductor with a large section and towers which are connected to the ground at each foundation. In the case of a line fault, most fault current flows to the ground via the good conductor, to prevent any damage to OPGW with its small thermal capacity. To reduce electric energy losses caused by the ground loop current in OPGW constructed along the full line, an option to conduct an insulated construction of OPGW and transposition has been realized using another ordinary steel strand ground wire outside the insulated construction of the good conductor. On the condition that the number of optical fiber connectors meets attenuation constraints along the full line, the small increase in investment and the difficulty in the transposition construction of OPGW for this new design are acceptable, given the long-term energy-saving benefits of the overhead ground wire [58].

The reduced electricity losses resulting from the new design of overhead ground wire in the transmission line can be estimated by the ground wire electricity losses per year per 100 km of lines under various voltage classes, line length, and percentage of construction and transposition of ground wire insulation. Before the line is put into operation it is also possible to calculate in detail (according to parameter test data) the average ground resistance, annual electric supply, annual average load factor, and specific conditions of the new design.

12.8.2 Application of Energy-saving Hardware and Energy-saving Conductors

12.8.2.1 Application of Energy-saving Hardware

Due to simple manufacturing techniques and low cost, for a long time most electric power hardware has been made from ferrous materials. Bolt type strain clamps and suspension clamps are largely used for directly fixing conductors in electric power lines. As eddy-current and hysteresis losses of conductor current are produced in these clamps, the electric power losses of electric power lines are increased, and the conductor temperature rises in the clamps.

To reduce the electric energy losses of this hardware, the relevant scientific research department in China in the 1970s developed new clamps with aluminum alloy clamps as the main body in combination with iron connecting bolts. According to tests, compared with the old NLD-4 strain clamp, the energy consumption of the new NNL-4 strain clamp is reduced by 60–65% under different current flow conditions, and the conductor temperature in the clamp is reduced by 53 °C under 400 A flow. Compared with the old XGU-5 suspension clamp, the energy consumption of the new CGL-5 suspension clamp is reduced by 73–77% under different current flow conditions, and the conductor temperature in the clamp is reduced by 40 °C under 400 A flow. If the annual maximum load utilization time is 4000 h, one energy-saving strain clamp can reduce 99 kWh of electricity per year. If the transmission price is 0.30 Yuan/kWh, it takes only 3.7 years to recover the investment in replacing the fittings [59].

12.8.2.2 Application of Energy-saving Conductors

Since the 1980s, by introducing and absorbing a variety of manufacturing technologies for energy-saving conductors, China has been able to implement the large-scale production of various types of energy-saving conductors and to apply them widely in transmission line transformation projects under various voltage classes. The most important three types of energy-saving conductors are steel-cored, high-conductivity, hard-drawn aluminum stranded conductors, aluminum alloy-cored aluminum stranded conductors, and moderate intensity full-aluminum alloy stranded conductors.

According to relevant calculations, compared with ordinary steel-cored aluminum stranded conductors with the same cross-section, the electric energy losses of the above three types of energy-saving conductors may be reduced by 3.08, 2.45, and 2.39%, respectively, due to different conductor structures and materials, different influences on line sag and tower load, and different amounts of investment and payback periods of line transformation. Among these, the payback period of aluminum alloy-cored aluminum stranded conductors is the shortest [60].

12.8.3 Application of Harmonic Control Technology and High-Temperature Superconducting Technology

12.8.3.1 Harmonic Control and Loss Reduction

Harmonics results from non-linear elements in power grids and electric equipment and is regarded as environmental pollution in power grids due to various damage harmonics may cause. According to theoretical analysis and actual measurement, when the harmonic current input to power grids by non-linear loads is lower than the national standard, increased electric energy losses caused by harmonic power are extremely small; power grid enterprises should pay more attention to the influence of harmonics on electric energy metering.

According to a relevant study, the fundamental wave watt-hour meter installed low-pass filter between current and voltage loops and AD conversion only measures fundamental wave electric energy. Subtracting the fundamental wave electric energy from the electric energy measured by the broadband full-electronic watt-hour meter (if the result is positive) shows that loads are linear and absorb harmonic electric energy, so they are victims of harmonics. If the result is negative, it shows that loads are non-linear and input harmonic electric energy into power grids, so they are harmonic sources [61]. For a 0.38 kV low-voltage system, due to the wide connection of non-linear loads, such as desktop computers, microwave ovens, induction cookers, and light switch illumination, if the broadband full-electronic watt-hour meter is used, the measured electric energy will be reduced. For example, the measured electric energy consumed by a desktop computer will reduced by up to 4.53%. If a city has one million desktop computers and the unit price of power sales is 0.50 Yuan/kWh, then the annual revenue will be reduced by RMB 3×10^6 due to the undercounted electric energy [62]. Moreover, the third harmonics resulting from non-linear loads in the low-voltage system greatly increases electric energy losses in the neutral line. The voltage drop in the neutral line affects the voltage quality and reduces the power supply reliability. The neutral line may be burnt out due to overheating, leading to a serious imbalance of three-phase voltage and even wide damage to electric equipment.

An effective harmonic control measure in a low-voltage system is to install an active filter in a suitable position. The active filter can track all orders of harmonic compensation, improve the voltage quality and power supply reliability, and reduce harmonic electric energy losses. To control harmonics in 6–10 kV systems, it is necessary to first measure harmonic load flows under different operating conditions and to take comprehensive control measures through economic and technical comparisons. The main connection of the higher level voltage system can be changed or dynamic reactive compensation equipment can be installed

Table 12.4 Parameters and refrigerating system conditions of high temperature superconducting cable.

Cable type	Three-phase split-phase AC electric cable	Operating temperature difference of terminal current head	70–330 K
Cable length	33.5 m (excluding terminal)	Operating altitude	1900 m
Rated voltage	35 kV	Cooling mode	Supercooling liquid nitrogen cycle
Rated current/ power	2000 A/121.24 MVA	System refrigerating capacity	2250 W/75 K
Conductor AC loss	0.75 W/kA·m	Inlet temperature of liquid nitrogen	71–74 K
Conductor heat loss	1.5 W	Outlet temperature of liquid nitrogen	76–79 K

if necessary. For harmonic sources like an electrified railway traction station and an electric-arc furnace, even if dynamic reactive compensation measures are taken, special attention should still be paid to measuring their electric energies. Using the fundamental wave watt-hour meter or the new electronic watt-hour meter which can measure both fundamental wave electric energy and harmonic electric energy can avoid any rise in the line loss rate of the power supply system due to a measured error of electric energy. Therefore, the correct configuration of watt-hour meters must be regarded as a technical measure for harmonic control and loss reduction.

12.8.3.2 Application of Superconducting Cables

The physical phenomenon that the resistance of conductors suddenly drops to zero around a temperature of absolute zero is called "superconduction". Thanks to various current effects, superconducting technology has been developed and applied in many fields. In the 1990s, Chinese and foreign scientists found the oxide superconductor yttrium barium copper oxide (YBCO) above 90 K, which makes the large-scale application of superconducting materials possible in a liquid nitrogen environment which is easily realized. On 19 April 2004, following the United States and Denmark, Chinese scientific and technical personnel put the world's third superconducting cable into operation at the Puji substation in Kunming, Yunnan. Its overall parameters and refrigerating system conditions are shown in Table 12.4 [63].

Compared with American and Danish superconducting cables connected to power grids, the rated voltage, current, and capacity of China's superconducting cable represented the highest level in the world at that time. This cable was also the first in the world to adopt a GM refrigerating machine as a cold source for superconducting cables connected to power grids and characterized by a convenient adjustment of refrigerating capacity and high reliability.

For the AC transmission cable used in the electric power system, the resistance of a superconducting band in the conducting layer can be regarded as zero during normal operation without joule heat losses, but losses occur during the operation of the superconducting cable. For example: AC losses occur due to skin effect and neighboring interference effect during AC transmission; dielectric losses occur in electric insulation after being charged with electricity; heat losses occur due to the resistance at the point of weld between the superconducting strip and the current lead; heat conduction and joule heat losses occur in the current lead of the cable terminal; heat leakage losses inevitably occur in the thermal insulation of the cable. Power losses of the cryogenic cooling device account for a large proportion. According to comparisons, the total heat losses of a superconducting cable with large capacity account for about 50% of the total losses of a normal cable with the same capacity [64].

Although superconducting cables can reduce a high proportion of losses, to promote the application of them in municipal grids with a large transmission power, it is necessary to develop new superconducting materials, to improve the superconducting temperature, and to reduce manufacturing costs. All of these require further efforts from scientific and technical personnel.

13

Line Loss Prediction and Loss Reduction Plan for Power Grids

Based on the analysis of change rules of no-load losses and load losses, Chapters 6 and 7 explain the theoretical calculation of line losses in loss calculation units, multi-branch distribution grids, and high-voltage power grids, and analyze the indicator of line loss rate from the aspect of: (i) electric supply structure, (ii) composition of power sales quantity, and (iii) optimal allocation of increased electric supply. Chapter 12 discusses technical measures for reducing line losses from seven aspects. As a result, the prediction of electricity line losses and the indicator of line loss rate can be analyzed in an in-depth manner, and the formulation and implementation of a loss reduction plan can be discussed.

13.1 Univariate Prediction of Electricity Line Losses and Line Loss Rate

With the load loss coefficient C as the core, this section analyzes the influence of load factor, power factor, and voltage level of regional grids on the line loss rate, and introduces univariate prediction methods for electricity line losses and line loss rate, with the loss constituent ratio θ and the correction coefficient of electricity line loss increase rate as objectives.

13.1.1 Basis for Predicting the Indicator of Line Loss Rate

Chapter 6 gives Formula (6.14) for calculating the load loss coefficient, that is

$$C=\frac{R}{U_{\text{av}}^2 f_P T}\left[1.278a^2 f_P+0.722a\left(1+\beta_P-\frac{\beta_P}{af_P}\right)\right]\times 10^{-3}$$

According to the above formula, the regional load factor, average power factor, and voltage level influence the line loss rate.

13.1.1.1 Influence of Regional Load Factor on Line Loss Rate

According to Formula (6.14), on the condition that other parameters remain unchanged, there is a functional relationship between C and f_P, so

$$\frac{dC}{df_P} = \frac{R}{U_{av}^2 T}\left\{0.722a\left[\frac{2\beta_P}{af_P} - \frac{(1+\beta_P)}{f_P^2}\right]\right\} \times 10^{-3}$$

thus

$$dC = \frac{-R}{U_{av}^2 T f_P^2}\left[0.722a\left(1+\beta_P - \frac{2\beta_P}{af_P}\right)\right] \times 10^{-3} df_P \tag{13.1}$$

According to Formula (13.1), the increase in the load factor can reduce the load loss coefficient. According to Formula (6.2),

$$\Delta A\% = \frac{B}{A} + CA$$

Then

$$\frac{d(\Delta A\%)}{dC}A, \quad d(\Delta A\%) = AdC$$

$$\Delta A\% = \Delta A_0\% + \Delta A_L\% = \left(\frac{1}{\theta}+1\right)CA$$

So

$$\frac{d(\Delta A\%)}{\Delta A\%} = \frac{AdC}{\left(\frac{1}{\theta}+1\right)CA} = \frac{-\frac{R}{U_{av}^2 T f_P^2}\left[0.722a\left(1+\beta_P - \frac{2\beta_P}{af_P}\right)\right] \times 10^{-3} df_P}{\left(\frac{1}{\theta}+1\right)\frac{R}{U_{av}^2 T f_P}\left[1.278a^2 f_P + 0.722a\left(1+\beta_P - \frac{\beta_P}{af_P}\right)\right] \times 10^{-3}}$$

After collection,

$$\delta(\Delta A\%) = -\frac{0.722a\left(1+\beta_P - \frac{2\beta_P}{af_P}\right)}{\left(\frac{1}{\theta}+1\right)\left[1.278a^2 f_P + 0.722a\left(1+\beta_P - \frac{\beta_P}{af_P}\right)\right]}\delta f_P \tag{13.2}$$

Formula (13.2) indicates that the rate of relative change in the line loss rate of regional power grids is related to not only the rate of relative change in the load factor of regional power grids, but also the average power factor, load factor, and loss constituent ratio.

Allowing $\cos\varphi_{av} = 0.80$, $a = 0.875$, if $f_P = 0.90$, $\beta_P = 0.60$, $\theta = 0.30$, then $\delta(\Delta A\%) = -0.008\delta f_P$. Allowing $\cos\varphi_{av} = 0.75$, $a = 0.9410$, if $f_P = 0.70$, $\beta_P = 0.40$, $\theta = 0.50$, then $\delta(\Delta A\%) = -0.032\delta f_P$. The calculation shows that, for urban power grids, the bigger f_P is, the smaller θ is, so the ratio between $\delta(\Delta A\%)$ and δf_P is very small; yet for rural power grids, the ratio is larger.

13.1.1.2 Influence of Average Power Factor on Line Loss Rate

According to Formula (6.14), on the condition that other parameters remain unchanged, there is a functional relationship between C and a, so

$$\frac{\mathrm{d}C}{\mathrm{d}a}=\frac{R\times 10^{-3}}{U_{\mathrm{av}}^2 f_P T}[2.556af_P+0.722(1+\beta_P)]$$

As

$$\delta(\Delta A\%)=\frac{A\mathrm{d}C}{\left(\frac{1}{\theta}+1\right)CA}$$
$$=\frac{2.556a^2 f_P+0.722a\,(1+\beta_P)}{\left(\frac{1}{\theta}+1\right)\left[1.278a^2 f_P+0.722a\left(1+\beta_P-\frac{\beta_P}{af_P}\right)\right]}\frac{\mathrm{d}a}{a}$$

$$a=\frac{1}{2}(1+\tan\varphi_{\mathrm{av}})$$

$$\frac{\mathrm{d}a}{a}=-\frac{1}{2\ \cos\varphi_{\mathrm{av}}\sqrt{1-\cos^2\varphi_{\mathrm{av}}}}\frac{\mathrm{d}(\cos\varphi_{\mathrm{av}})}{\cos\varphi_{\mathrm{av}}}$$

Substitute this formula to obtain

$$\delta(\Delta A\%)=-\frac{[2.556a^2 f_P+0.722a(1+\beta_P)]}{2\left(\frac{1}{\theta}+1\right)\left[1.278a^2 f_P+0.722a\left(1+\beta_P-\frac{\beta_P}{af_P}\right)\right]\cos\varphi_{\mathrm{av}}\sqrt{1-\cos^2\varphi_{\mathrm{av}}}}\delta(\cos\varphi_{\mathrm{av}}) \quad (13.3)$$

Formula (13.3) indicates that the rate of relative change in the line loss rate of regional power grids is related to not only the rate of relative change in the average power factor of regional power grids, but also the load factor, average power factor, and loss constituent ratio. If $\cos\varphi_{\mathrm{av}}=0.80$, $f_P=0.90$, $\beta_P=0.60$, $\theta=0.30$, then $\delta(\Delta A\%)=-0.472\,6\delta(\cos\varphi_{\mathrm{av}})$; if $\cos\varphi_{\mathrm{av}}=0.75$, $f_P=0.70$, $\beta_P=0.40$, $\theta=0.50$, then $\delta(\Delta A\%)=-0.640\delta(\cos\varphi_{\mathrm{av}})$. The calculation indicates that no matter for urban power grids or rural power grids, the ratio between $\delta(\Delta A\%)$ and $\delta(\cos\varphi_{\mathrm{av}})$ may be equal to or greater than 0.50, showing that the increase in the average power factor of regional power grids can significantly reduce line losses. Formula (13.3) provides conditions for estimating the effect of reactive compensation for the whole grid and calculating the recovery period of the investment in compensation equipment.

13.1.1.3 Influence of Average Voltage Level on the Line Loss Rate

According to Formula (6.14), the load losses of regional power grids are in inverse proportion to the square of the average voltage, but no-load losses generally increase with a voltage increase. The influence of any change in the voltage level on the total line loss rate depends on not only the component ratio of loads with different voltage characteristics, but also the ratio between no-load losses and load losses, that is the loss constituent ratio θ.

13.1.2 *Univariate Prediction of Electricity Line Losses*

13.1.2.1 Prediction of Change in Electricity Line Losses

Section 7.4 of Chapter 7 has clarified that the quantity of change in electricity line losses is approximately equal to the product of the electric energy loss structure coefficient, the original line loss rate, and the quantity of change in the electric supply, that is

$$\Delta A_2 - \Delta A_1 \approx H\Delta A\%(A_2 - A_1) = \left(\frac{2}{\theta+1}\right)\Delta A\%(A_2 - A_1) \tag{13.4}$$

For a short-term prediction, the loss constituent ratio θ can be deemed as approximately unchanged, and Formula (13.4) is the formula for predicting the electricity line losses, wherein the electric supply is the only variable. For a single distribution line, Formula (13.4) is simple and useful for a short-term prediction. From the perspective of assessment, if the changes in electric supply for the current and the next few months are known, Formula (13.4) provides a simple and effective standard for judging whether the change in electricity line losses is reasonable.

13.1.2.2 Prediction of Rate of Increase in Electricity Line Losses

Section 7.4 has explained that the rate of increase in electricity line losses is in linear function of the rate of increase in the electric supply, that is

$$(\Delta A_2 - \Delta A_1)/\Delta A_1 \approx \left(\frac{2}{\theta+1}\right)(A_2 - A_1)/A \tag{13.5}$$

When the no-load losses are larger than the load losses (i.e. $\theta > 1.0$) the proportionality coefficient (loss structure coefficient H) in Formula (13.5) is smaller than 1.0, and the rate of increase in the electricity line losses is smaller than the rate of increase in the electric supply, which is often encountered in agricultural grids. When the no-load losses are smaller than the load losses, $H > 1.0$, and the rate of increase in the electricity line losses is larger than the rate of increase in the electric supply, which is often encountered in urban distribution grids. For example, the range of θ for Shanghai urban and suburban areas in 1966 was roughly 0.31–0.45, and the approximate H range was 1.53–1.38. This shows that in urban grids the rate of increase in the electricity line losses is 38–53% higher than the rate of increase in the electric supply. Formula (13.5) is very simple and useful for judging whether the change in the increase in electricity line losses is normal.

13.1.3 *Univariate Prediction of Line Loss Rate*

13.1.3.1 Line Loss Rate Binomial

Section 6.1 of Chapter 6 defines the loss constituent ratio $\theta = B/CA^2$, and the line loss binomial $\Delta A = B + CA^2$ can be used to derive the following two formulas:

$$\Delta A = B + CA^2 = B\left(1 + \frac{1}{\theta}\right) = CA^2(1+\theta) \tag{13.6}$$

$$\Delta A\% = \frac{B}{A}\left(1 + \frac{1}{\theta}\right) = CA(1+\theta) \tag{13.7}$$

Formula (13.7) is called the line loss rate binomial.

13.1.3.2 Electric Quantity Growth Coefficient and Line Loss Rate Correction Coefficient

The ratio between the power sales quantity of the assessed year and the power sales quantity of the previous year is X, known as the electric quantity growth coefficient. When the line loss rate is small, the ratio of power sales quantity is approximately equal to the ratio of the power sales quantity of the adjacent two years, that is $X \approx A_2/A$.

If the acceptable ratio between the line loss rate of the assessed year and the line loss rate of the previous year is defined as the line loss rate correction coefficient E, then

$$\Delta A\% = \frac{B}{A_2}CA_2 = \frac{B}{XA_1} + CXA_1 = \frac{\theta CA_1^2}{XA_1} + CXA_1 = CA_1\left(\frac{\theta}{X} + X\right)$$

$$\Delta A_1\% = \frac{B}{A_1} + CA_1 = CA_1(\theta + 1)$$

$$E = \Delta A_2\%/\Delta A_1\% = \left(\frac{\theta}{X} + X\right)/(\theta + 1) = \left(\frac{\theta}{\theta + 1}\right)\frac{1}{X} + \frac{X}{\theta + 1} \tag{13.8}$$

If the loss constituent ratio θ of the previous year is calculated, and the power sales quantity growth coefficient of the next year is predicted as X, then the acceptable predicted line loss rate can be obtained through E

$$\Delta A_2\% = E\Delta A_1\% = \left[\frac{\theta}{(\theta + 1)X}\frac{1}{} + \frac{X}{(\theta + 1)}\right]\Delta A_1\% \tag{13.9}$$

In the 1990s, the electric power system once put forward the following formula for calculating E according to the power sales quantity growth coefficient X

$$E = 0.3\frac{1}{X} + 0.7X \tag{13.10}$$

According to Formula (13.10), the ratio between no-load losses and load losses is 3:7, so $\theta_{av} = 3/7 = 0.43$. For most urban distribution lines at that time, $\theta_{av} = 0.43$ is appropriate to a large extent; however, for agricultural lines where no-load losses account for a larger proportion and lines in urban central areas where load losses account for a larger proportion, the difference between actual θ and θ_{av} is big. For an accurate univariate prediction of the line loss rate, Formula (13.9) should apply.

For a prefectural power supply network consisting of several voltage systems, as these voltage systems have different electric supply growth coefficients and different loss constituent ratios, the univariate prediction of the line loss rate should be conducted for different voltage systems, and the line loss rate of the whole network can then finally be predicted.

13.2 Multivariable Prediction of Electricity Line Losses and Line Loss Rate

Formulas (13.4) and (13.9) can be used to calculate the change in line losses caused by the change in a single factor so as to realize the univariate prediction of the indicator of line loss rate. Over a long calculation period, in addition to the electric supply, the load factor, the average power factor, and the no-load losses change, so the methods for calculating changes in electricity line losses and line loss rate caused by changes in multiple factors should also be considered.

13.2.1 Multivariable Prediction of Electricity Line Losses

Section 7.4 of Chapter 7 has explained that the loss structure function can be used to calculate the total derivative of electricity losses, that is Formula (7.15). Now it is rewritten to

$$d(\Delta A)=dB+\left(\frac{\delta C/\delta A}{\theta+1}+\frac{2}{\theta+1}\right)\Delta A\%dA$$

When the quantity of changes in the three variables B, C, A is much smaller than the original values, the total derivative can be approximately replaced by increment, so the above formula can be rewritten to

$$\partial(\Delta A)=\partial B+\left(\frac{\delta C/\delta A}{\theta+1}+\frac{2}{\theta+1}\right)\Delta A\%\partial A \tag{13.11}$$

Wherein $\partial(\Delta A)$ – quantity of change in electricity line losses, that is $\partial(\Delta A) = \Delta A_2 - \Delta A_1$;
∂B – quantity of change in no-load electricity line losses, that is $\partial B = B_2 - B_1$;
δC – quantity of relative change in load loss coefficient, that is $\delta C=\frac{C_2-C_1}{C_1}$;
δA – quantity of relative change in electric supply, that is $\delta A=\frac{A_2-A_1}{A_1}$;
∂A – quantity of change in electric supply, that is $\partial A = A_2 - A_1$.

Formula (13.11) is the formula for predicting electricity line losses considering multiple factors such as the change in no-load losses, the operating parameters (load factor, minimum load rate, average power factor, average voltage), and the original loss conditions (loss constituent ratio θ, line loss rate $\Delta A\%$). Because Formula (13.11) is applicable to various voltage systems, the prediction of electricity line losses can be conducted by voltage class; as coefficients C of different voltage systems can be included (Section 6.3 of Chapter 6), the prediction of electricity line losses can be completed for the power grids of the entire region at the same time.

13.2.2 Multivariable Prediction of Line Loss Rate

According to Formula (13.7), $\Delta A\%=\frac{B}{A}+CA$, and the line loss rate is related to the three variables A, B, and C. Take the total derivative of $\Delta A\%$ to obtain

$$d(\Delta A\%)=\frac{1}{A}dB+\left(C-\frac{B}{A^2}\right)dA+AdC$$

Due to $\Delta A\% = CA(1 + \theta)$, the above formula is divided by it at both sides to obtain

$$d(\Delta A\%)/\Delta A\%=\frac{dB}{CA^2(1+\theta)}+\frac{(CA^2-B)dA}{CA^2(1+\theta)A}+\frac{AdC}{CA(1+\theta)}$$

so

$$\delta\,(\Delta A\%)=\frac{\delta B}{1+\frac{1}{\theta}}+\left(\frac{1-\theta}{1+\theta}\right)\delta A+\frac{\delta C}{1+\theta} \tag{13.12}$$

$$\delta(\Delta A\%) = \frac{\Delta A_2\% - \Delta A_1\%}{\Delta A_1\%}$$

$$\delta B = (B_2 - B_1)/B_1$$

Wherein $\delta(\Delta A\%)$ – rate of change in the line loss rate;
δB – rate of change in the no-load losses.

Formula (13.12) indicates that the rate of change in the line loss rate is composed of three components, all of which reflect the influence of four factors, namely no-load losses, electric supply, load loss coefficient, and original loss composition. In general situations $\theta < 1.0$, so the second term in Formula (13.12) has a positive value, that is the increase in the electric supply certainly leads to the increase in the line loss rate. To make the rate of change in the line loss rate be negative value, δB and δC must be negative values, that is the no-load losses and load loss coefficient must be reduced.

Formula (13.12) can directly predict the relative change in the line loss rate, so it is called the multivariable prediction formula of line loss rate.

Example 13.1 The electric supply, no-load losses, load losses, annual minimum load rate β_P, annual average load factor f_P, and annual average power factor $\cos\varphi_{av}$ of a region in 1984 are given in Table 13.1.

The rates of increase δA in electric supply in 1985 and 1986 are predicted to be 5.37 and 5.62%, respectively. The rate of increase in no-load losses is $\delta B = 0.8\delta A$. The increased values of β_P, f_P, and $\cos\varphi_{av}$ year by year are shown in Table 13.1. *Try to predict the increases in electricity line losses and the changes in line loss rates in 1985 and 1986.*

Solution

1. *Calculation of the quantity of relative change in the load loss coefficient* δC. As

$$C_{84} = \frac{\Delta A_L}{A_{84}^2} = \frac{2.497\times10^8}{(81.57\times10^8)^2} = 3.753\times10^{-12}[1/(\text{kW·h})]$$

According to Formula (8.14),

$$\frac{R}{U_{av}^2} = \frac{C_{84} f_P T}{\{1.278a^2 f_P + 0.722a[1+\beta_P - \beta_P/(af_P)]\}\times10^{-3}}$$

$$= \frac{3.753\times10^{-12}\times0.90\times8760}{\left\{1.278\times(0.875)^2\times0.90+0.722\times0.875[1+0.60-0.6/(0.875\times0.90)]\right\}\times10^{-3}}$$

$$= 2.098\times10^{-3}$$

According to the 1985 operating parameters and $\frac{R}{U_{av}^2}$ listed in Table 13.1, C_{85} is calculated

$$C_{85} = 2.098\times10^{-5}\times\frac{1}{0.92\times8760}\times\left[1.278\times0.836^2\times0.92 + 0.722\times0.836\times\left(1+0.62-\frac{0.62}{0.836\times0.92}\right)\right]\times10^{-3}$$

$$= 3.418\times10^{-12}[1/(\text{kW·h})]$$

$$\delta C = \frac{3.418-3.753}{3.753}\times100\% = -8.926\%$$

$$\frac{\delta C}{\delta A} = \frac{-8.926}{5.37} = -1.662$$

Table 13.1 Calculation table for an increase in line losses and a change in line loss rate within a region.

Year	Electric supply A ($\times10^8$ kW·h)	No-load losses B ($\times10^8$ kW·h)	Load losses CA^2 ($\times10^8$ kW·h)	Total losses ΔA ($\times10^8$ kW·h)	Line loss rate $\Delta A\%$ (%)	Loss constituent ratio θ	Annual minimum load rate β_P	Annual average load factor f_P	Annual average power factor $\cos\varphi_{av}$	Increase in electric supply ∂A ($\times10^8$ kW·h)	Increase in loss coefficient δC (%)	Rate of increase in electric supply δA (%)	Increase in no-load losses ∂B ($\times10^8$ kW·h)	Increase in electricity losses $\partial(\Delta A)$ ($\times10^8$ kW·h)
1984	81.57	0.563	2.497	3.06	3.75	0.2255	0.60	0.90	0.80					
1985	85.95	0.5872	2.5423	3.1295	3.641	0.231	0.62	0.92	0.83	4.83	–8.926	5.37	0.02419	0.0695
1986	90.78	0.6136	2.7496	3.3632	3.705	0.2232	0.62	0.935	0.84	4.38	–3.087	5.62	0.0264	0.2337

Note: When $\cos\varphi_{av} = 0.80$, $a = 0.875$; when $\cos\varphi_{av} = 0.83$, $a = 0.836$; when $\cos\varphi_{av} = 0.84$, $a = 0.823$.

2. *Calculation of the increase in no-load losses.* This is

$$\partial B = \delta B \times B = (0.8 \times 0.053\,7) \times 0.563 \times 10^8$$
$$= 0.024\,2 \times 10^8 \text{ (kW·h)}$$

3. *Calculation of the increase in total line losses.* According to Formula (13.11),

$$\partial(\Delta A) = 0.024\,2 \times 10^8 + \left(\frac{-1.662}{0.225\,5 + 1} + \frac{2}{0.225\,5 + 1}\right)$$
$$\times 0.037\,5 \times \left(0.053\,7 \times 81.57 \times 10^8\right)$$
$$= 0.024\,2 \times 10^8 + 0.045\,3 \times 10^8$$
$$= 0.069\,50 \times 10^8 \text{ (kW·h)}$$

The electricity line losses in 1985 are

$$\Delta A = (3.06 + 0.069\,50) \times 10^8 = 3.129\,5 \times 10^8 \text{ (kW·h)}$$

The line loss rate in 1985 is

$$\Delta A\% = \frac{3.129\,5 \times 10^8}{1.053\,7 \times 81.57 \times 10^8} = 3.641\%$$

4. *Accounting.* The accounting result is

$$\Delta A_0\% = \frac{(0.024\,2 + 0.563) \times 10^8}{1.053\,7 \times 81.57 \times 10^8} = 0.683\%$$

$$\Delta A_L\% = CA = 3.418 \times 10^{-12} \times \left(1.053\,7 \times 81.57 \times 10^8\right)$$
$$= 2.94\%$$
$$\Delta A\% = \Delta A_0\% + \Delta A_L\% = 0.683\% + 2.94\% = 3.623\%$$

The relative error is

$$\delta(\Delta A\%) = \frac{3.641 - 3.623}{3.623} \times 100\% = 0.5\%$$

This shows that the prediction formula of the line loss rate can satisfy the actual needs of line loss management.

5. *Calculation of loss constituent ratio.* This is

$$\theta = \frac{\Delta A_0}{\Delta A_L} = \frac{0.024\,2 + 0.563}{0.045\,3 + 2.497} = \frac{0.587\,2}{2.542\,3} = 0.231$$

According to the above calculations, the relevant data for 1986 can also be obtained

$$C_{86} = 3.312\,5 \times 10^{-12} [1/(\text{kW·h})], \quad \delta C = -0.030\,87$$
$$\frac{\delta C}{\delta A} = -0.549\,3, \quad \partial(\Delta A) = 0.233\,7 \times 10^8 \text{ (kW·h)}$$
$$\Delta A = 3.363\,2 \times 10^8 \text{ (kW·h)}, \quad \Delta A\% = 3.705\%, \quad \theta = 0.223\,2$$

The above data are listed in Table 13.1.

In this example, Formula (13.12) for multivariable prediction of the line loss rate can also be used to calculate the predicted line loss rate. The loss constituent ratio is calculated in this example, $\theta_1 = 0.2255$. $\delta A = 0.0537$ and $\delta B = 0.8 \times 0.0537 = 0.042\,96$ are given. The rate of relative change in the load loss coefficient is obtained according to the operating parameters in Table 13.1, that is $\delta C = -0.089\,26$. Substitute the four values into Formula (13.12) to obtain

$$\delta(\Delta A\%) = \frac{0.042\,96}{1+1/0.225\,5} + \frac{1-0.225\,5}{1+0.225\,5} \times 0.053\,7 - \frac{0.089\,26}{1+0.225\,5}$$
$$= 0.007\,905 + 0.033\,94 - 0.072\,83 = -0.030\,98$$

$\Delta A_1\% = 3.75\% = 0.037\,5$ is given in the example

$$\delta(\Delta A\%) = \frac{\Delta A_2\% - \Delta A_1\%}{\Delta A_1\%} = \frac{\Delta A_2\% - 0.037\,5}{0.037\,5} = -0.030\,98$$

$\Delta A_2\% = 3.634\%$ is obtained

Use to the method for predicting electricity line losses to obtain $\Delta A'_2\% = 3.641\%$ (see Table 13.1). So the error in the calculation results (i.e. line loss rate of 1985) of the two methods is

$$\delta(\Delta A\%) = \frac{3.634 - 3.641}{3.641} \times 100\% = -0.19\%$$

The error in the calculation results of the two methods is very small, indicating that both the methods can be used for predicting line losses and are selected based on the source of relevant data of actual line losses. The formula of multivariable prediction of line loss rate is more convenient for this example, because the calculation workload with Formula (13.12) is small as δA and δB are known.

In the calculation process of Example 13.1, R/U^2_{av} is assumed constant, which is acceptable to situations where power grids experience neither step-up transformation nor major line change construction. In the case of any line change construction or major change in the grid structure for a voltage system experiences, R will change, but the above methods can also be used to calculate the change in load losses caused by changes in the operating parameters and the increase in electric supply. In addition, reduced losses by each major loss reduction technical measure should be separately estimated to obtain the total changed line losses of the regional power grids. In this way, the predicted line loss rate is likely to be more close to the actual line loss rate.

13.2.3 Rolling Prediction Method

To make the approximate Formula (13.11) have a certain accuracy during the process of calculating the increase in line losses and predicting the indicator of line loss rate, the principle of step by step rolling should be followed. Generally speaking, there are two rolling prediction methods: (i) the electric supply is increased by small steps and the annual increase is much smaller than the base number of the previous year, which is the case in Example 13.1, (ii) small-step rolling within the prediction period, such as predicting the line loss rate in October based on the accumulated data at the end of September, predicting the line loss rate in November based on the predicted data at the end of October, and predicting the data of several months at the beginning of

Table 13.2 Table of prediction and calculation of line loss rate.

Period	Electric supply A ($\times 10^4$ kW·h)	No-load losses ΔA_0 ($\times 10^4$ kW·h)	Load losses ΔA_L ($\times 10^4$ kW·h)	Loss constituent ratio θ	Line loss rate $\Delta A\%$ (%)	Load factor f_P	Minimum load rate β_P	Average power factor $\cos\varphi_{av}$	Calculation coefficient $a=\frac{1+\tan\phi_{av}}{2}$
Jan. to Oct.	16 554.2	98.3	848.29	0.1088	5.68	0.823	0.396	0.842	0.88
Nov.	1324.3	9.109	55.35	0.165	4.87	0.773	0.396	0.842	0.82
Jan. to Nov.	17 878.5	101.409	903.64	0.1122	5.68	0.809	0.396	0.842	0.82

a month based on the data of the previous year. During the second rolling prediction, the difference between the prediction period and the accumulation period should be taken into account for calculating the load loss coefficient.

Example 13.2 For the regional power grids in Example 6.1, the accumulated load rates, minimum load rates,and average power factors of the total electric supply for January to October are shown in Table 13.2.

It is predicted that the electric supply in November is reduced by about 20% less than the average electric supply in the previous 10 months and the regional load rate reduced by 0.05, while the minimum load rate, the average power factor, and the no-load power losses are not changed. *Try to predict the line loss rate accumulated to the end of November.*

Solution

1. According to Example 6.1, the regional load loss coefficient is

$$C_{(1-10)} = 3.096\times 10^{-10}[1/(\text{kW·h})]$$

According to Formula (6.14), R/U^2_{av} is

$$\frac{R}{U^2_{av}} = (3.096\times 10^{-10}\times 0.823\times 7296)/\{\{1.278\times(0.82)^2\times 0.823+0.722\times 0.82$$
$$\times[1+0.396-0.396/(0.82\times 0.823)]\}\times 10^{-3}\}$$
$$=1.567\times 10^{-3}$$

2. According to the operating parameters and R/U^2_{av} in November, $C_{(11)}$ is

$$C_{(11)} = 1.567\times 10^{-3}\times\frac{1}{0.773\times 720}\times\Big[1.278\times(0.82)^2\times 0.773$$
$$+0.722\times 0.82\times\left(1+0.396-\frac{0.396}{0.82\times 0.773}\right)\Big]\times 10^{-3}$$
$$=3.156\times 10^{-9}[1/(\text{kW·h})]$$

Note that the measuring period T in November is 720 h, while T in January to October is 7296 h.

3. Calculation of electricity losses in November. No-load losses can be calculated according to the proportion of measuring period, that is

$$\Delta A_{0(11)} = 92.3 \times 10^4 \times \frac{720}{7296} = 9.109 \times 10^4 \text{ (kW·h)}$$

$$\Delta A_{L(11)} = C_{(11)} A_{(11)}^2 = 3.156 \times 10^{-9} \times \left(1324.3 \times 10^4\right)^2$$

$$= 55.35 \times 10^4 \text{ (kW·h)}$$

Wherein, the electric supply in November is

$$A_{(11)} = (1-0.2) \times \frac{16\,554.2 \times 10^4}{10} = 1324.3 \times 10^4 \text{ (kW·h)}$$

The loss constituent ratio is

$$\theta = \frac{9.109 \times 10^4}{55.35 \times 10^4} = 0.165$$

The monthly line loss rate is

$$\Delta A_{(11)}\% = \frac{(9.109 + 55.35) \times 10^4}{1324.3 \times 10^4} \times 100\% = 4.87\%$$

4. Calculation of line loss rate accumulated to the end of November. This is

$$\Delta A\% = \frac{\Delta A_\Sigma}{A_\Sigma} \times 100\%$$

Thus

$$= \frac{1005.05 \times 10^4}{(16554.2 + 1324.3) \times 10^4} \times 100\% = 5.62\%$$

In this example, Formula (13.11) for the line loss rate can also be used for the calculation, as follows:

1. Calculation of the load factor for January to November

$$P_{\max(1-10)} = \frac{16\,554.2 \times 10^4 / 7296}{0.823} = 2.756\,9 \times 10^4 \text{ (kW)}$$

$$P_{\max(11)} = \frac{1324.3 \times 10^4 / 720}{0.773} = 2.379\,4 \times 10^4 \text{ (kW)}$$

$$f_{P(1-11)} = \frac{(16\,554.2 + 1324.3) \times 10^4 / (7296 + 720)}{2.756\,9 \times 10^4}$$

$$= 0.809$$

2. Calculation of the load loss coefficient for January to November

$$C_{(1-11)} = 1.567\times10^{-3}\times\frac{1}{0.809\times8016}\times\left[1.278\times(0.82)^2\times0.809+0.722\times0.82\times\left(1+0.396-\frac{0.396}{0.82\times0.809}\right)\right]\times10^{-3}$$

$$= 2.823\,0\times10^{-10}[1/(\text{kW}\cdot\text{h})]$$

3. Calculation of the rates of relative change in load loss coefficient and electric supply. They are

$$\delta C = \frac{C_{(1-11)}-C_{(1-10)}}{C_{(1-10)}} = \frac{2.823\,0-3.096}{3.096}\times100\% = -8.818\%$$

$$\delta A = \frac{1324.3\times10^4}{16\,554.2\times10^4}\times100\% = 8\%$$

$$\frac{\delta C}{\delta A} = \frac{-8.818}{8} = -1.102$$

4. Calculation of the increase in line losses and the line loss rate. They are

$$\partial B = \Delta A_{0(11)} = 9.109\times10^4\ (\text{kW}\cdot\text{h})$$

$$\partial A = A_{(11)} = 1324.3\times10^4\ (\text{kW}\cdot\text{h})$$

According to Table 13.2, $\theta_{(1-10)} = 0.1088$, $\Delta A_{(1-10)} = 5.68\%$. So according to Formula (13.11),

$$\partial(\Delta A) = 9.109\times10^4+\left(\frac{-1.102}{0.108\,8+1}+\frac{2}{0.108\,8+1}\right)\times0.056\,8\times1324.3\times10^4$$

$$= 9.109\times10^4+60.92\times10^4$$

$$= 70.03\times10^4\ (\text{kW}\cdot\text{h})$$

$$\Delta A\% = \frac{\Delta A_{(1-10)}+\partial(\Delta A)}{A_{(1-10)}+A_{(11)}}\times100\%$$

$$= \frac{(92.3+848.29)\times10^4+70.03\times10^4}{16\,554.2\times10^4+1324.3\times10^4}\times100\%$$

The relative error of the above two prediction methods is

$$\delta(\Delta A\%) = \frac{5.65-5.62}{5.62}\times100\% = +0.5\%$$

The above results show that the new load losses and predicted line loss rate as obtained by calculating the load loss coefficient under the new state corresponding to the change in operating parameters, the extension of measuring period, or the change in the included resistance *R*, are consistent with those calculated by Formula (13.11). Therefore, in actual work, either method can be selected for prediction, depending on the conditions provided in a real problem.

The calculation example indicates that, when the measuring period rolls, the load loss coefficient varies; when the electric supply structure of power grids changes or conductors are replaced over a wide range, the resistance included for a voltage class also changes, thus leading to a change in the load loss coefficient. In other words, Formula (6.14) for calculating the load loss coefficient as derived by a combination of the equivalent load curve method and line loss binomial provides conditions for calculating the increased line losses due to changes in the electric supply structure parameter (R) and operating parameters (U_{av}, f_P, β_P, $\cos\varphi_{av}$) of power grids, so as to basically solve the difficult problem of predicting the indicator of line loss rate.

13.3 Main Content and Preparation Process of Loss Reduction Plan

As early as the late 1990s, the former State Planning Commission organized the Beijing–Tianjin–Tangshan Grid, the East China Grid, and the power supply departments of key cities experiencing urban grid transformation in Guangdong, twice holding meetings in Suzhou and Chengdu, and exchanging their processes and experiences concerning the preparation of a loss reduction pilot plan. Later, this developed into the large-scale implementation of transformation projects of urban and rural grids in 1998–2000. Therefore, from the perspective of the whole country, the preparation of a loss reduction plan has now promoted the implementation of transformation projects for urban and rural grids.

Under normal circumstances, after the process of increasing the electric supply is determined for the long-term planning of power grids, the loss reduction plan focuses on a comparison between the accrual value and the target value of the line loss rate, to evaluate the expected energy conservation, and provides basic conditions for the comprehensive arrangement of grid construction and transformation projects and the prediction of project benefits.

13.3.1 Content and Preparation Basis of the Loss Reduction Plan

13.3.1.1 Content Requirement for the Loss Reduction Plan

According to the *Circular on the Preparation of Energy Conservation and Loss Reduction Plan for Transmission and Transformation Grids* issued by the former State Planning Commission in January 1997, the energy conservation and loss reduction should mainly focus on the followings: optimizing the network and dispatching of power transmission, transformation and supply systems; simplifying voltage class and shortening low-voltage power supply distance; replacing transformers with high losses, using energy saving hardware in lines, and configuring reactive compensation equipment reasonably; strengthening power supply management, reducing unknown electricity losses, as well as eliminating power usage free of charge. Therefore, as to the loss reduction plan, on the one hand, it is necessary to make dispatching and operation plans to reduce main grid losses, carry out project construction to realize optimization of grid structure, and take technical measures such as replacing transformers with high losses and adding reactive compensation equipment to reduce no-load losses and load losses of medium-voltage and low-voltage grids; on the other hand, it is important to strengthen marketing management, eliminate power usage free of charge, and reduce management line losses significantly.

13.3.1.2 Preparation Basis and Basic Method of the Loss Reduction Plan

The loss reduction plan is an annual comprehensive arrangement from the current situation to the expected objective. Starting with the detailed analysis of current situations of electric energy losses in the whole network, the plan, based on the predicted increase in the electric supply, predicts the change in electric energy losses year by year from the three aspects of operating measures, engineering and technical transformation measures, and management measures, and compares them with target values; the basic method is "detailed in near term and rough in long term", rolling year by year until the end of the planning period.

13.3.1.3 Determination of the Target Value of Line Loss Rate

There are two methods for determining the target value of line loss rate in the loss reduction plan: one is the simple ratio reduction method, where the line loss rate should be specified to be a certain value at the end of the planning period, and it is uniformly reduced by a percentage per year during the whole planning period. The other is the sophisticated double parameter determination method. The first parameter is the correction coefficient E of the increase in line loss rate and is calculated by the rate of increase X in power sales quantity, as shown in Formula (13.8). The second parameter is the line loss rate decreasing coefficient K, $K \geq 0.007$ in general, and the relationship between the planned annual line loss rate $\Delta A_2\%$ and last year's line loss rate $\Delta A_1\%$ is determined by the following formula:

$$\Delta A_2\% = (1-K)E(\Delta A_1\%) \tag{13.13}$$

13.3.2 *Preparation of the Loss Reduction Plan*

13.3.2.1 Calculation of Accrual Value of Line Loss Rate

According to Formula (13.12), if neither no-load losses nor system operating parameters and grid structure changes, $\delta B = 0$ and $\delta C = 0$. At this time, $\delta(\Delta A\%) = \left(\frac{1-\theta}{1+\theta}\right)\delta A$, that is the rate of change in the line loss rate is only related to the rate of change in the electric supply. In this case, the calculated line loss rate is called accrual value of line loss rate, expressed by $\Delta A\% \mid_{na}$, which can be calculated as per the following formula

$$\Delta A\% \mid_{na} = \left[1 + \left(\frac{1-\theta}{1+\theta}\right)\delta A\right]\Delta A_1\% \tag{13.14}$$

With the increase in electric supply year by year, as the number of equipment rises and operating parameters and grid structure change, load losses are also increased, and its ratio loss constituent ratio θ is relatively stable. Therefore, the accrual value of line loss rate within the planning period can be calculated according to the rolling increase in the electric supply as per Formula (13.14).

13.3.2.2 Calculation of Expected Reduction of Electricity Losses and Calculation of Change in No-load Losses

1. *Comparison between accrual value and target value of line loss rate within the planning period.* The planned annual line loss rate determined by the loss reduction plan is smaller than the accrual value of line loss rate, and two factors (the difference between the two values, the electric supply) determine the expected reduction of electricity losses, that is

$$\Delta A_{ex} = \left(\Delta A\%|_{na} - \Delta A\%|_{aim}\right)A_i \tag{13.15}$$

Wherein $\Delta A\% \mid_{aim}$ – target value of line loss rate within the planning period, as determined by Formula (13.13);

A_i – predicted electric supply within the planning period.

2. *Calculation of the change in no-load losses.* In near-term planning, the different influences of the replacement of transformers with high losses and the increase in capacity of transformers on no-load losses can be calculated. In long-term planning, only the increase in no-load losses caused by the increase in transformer capacity is considered, and the common calculation method is that the capacities of newly added main

transformers and distribution transformers are estimated according to the electric supply corresponding to per 10 000 KVA capacity of existing main transformers or distribution transformers, and then the increase in no-load losses within the planning period is calculated according to the no-load power losses of per unit capacity of main transformers or distribution transformers.

13.3.2.3 Calculation of Engineering Reduced Electricity Losses and Selection of Engineering Projects

According to the multivariable prediction formula of line loss rate, when δB and δA are positive values, the year by year decrease in the line loss rate can only be realized by the year by year reduction of the load loss coefficient. There are two ways to reduce the load loss coefficient: (i) improve the operating parameters; (ii) carry out new construction or transformation to improve the grid structure. To determine the value of engineering loss reduction, the first step is to estimate the effect of operating loss reduction.

1. *Calculation of operating loss reduction component of load losses.* According to Formula (6.14), the load loss coefficient is related to several operating parameters (f_P, β_P, U_{av}). Based on the statistical information of line losses in the start year of planning, the value of operating reduction of electricity losses which may realized by the improvement in operating parameters can be calculated. Arrows are used below to indicate the calculation process of operating loss reduction component (expressed by the subscript I):

$$
\begin{gathered}
(\Delta A_1-\Delta A_{01})\rightarrow\Delta A_{\mathrm{L1}}\rightarrow C_1=\frac{\Delta A_{\mathrm{L1}}}{A_1^2}\\
C(f_{P2},\beta_{P2},U_{\mathrm{av2}},\cos\varphi_{\mathrm{av2}},R_1)-C(f_{P1},\beta_{P1},U_{\mathrm{av1}},\cos\varphi_{\mathrm{av1}},R_1)\rightarrow(\Delta C)_{\mathrm{I}};\\
(\delta C)_{\mathrm{I}}=(\Delta C)_{\mathrm{I}}/C_1;\ \ \delta A=(A_2-A_1)/A_1\\
\mathrm{d}(\Delta A_{\mathrm{L}})_{\mathrm{I}}=\frac{(\delta C)_{\mathrm{I}}/\delta A}{\theta+1}\Delta A_1\%(A_2-A_1)
\end{gathered}
\tag{13.16}
$$

Wherein ΔA_{L} – load electricity losses.

2. *Calculation of engineering loss reduction component of load losses.* The subscript II is used to indicate the engineering loss reduction component, and the following formula is derived:

$$
\mathrm{d}(\Delta A_{\mathrm{L}})_{\mathrm{II}}=(\Delta A_{\mathrm{ex}}+\mathrm{d}B)-\mathrm{d}(\Delta A_{\mathrm{L}})_{\mathrm{I}} \tag{13.17}
$$

Wherein ΔA_{ex} – expected reduction of electricity losses, as calculated by Formula (12.15);
$\mathrm{d}B$ – increase in no-load losses.

3. *Selection of loss reduction engineering projects.* One of the main contents of the theoretical calculation and analysis of annual line losses is to give suggestions of major loss reduction engineering projects, which provides the possibility for determining loss reduction engineering projects during loss reduction planning. The ranking principle of project selection is to give priority to a project whose comprehensive benefits per unit investment are the best, and the other principle is that any expansion or transformation project that does not increase land occupation for substations and line corridors is preferred. The reduced electricity losses included in the loss reduction engineering projects of the planning year can be estimated by theoretical calculation methods for units, multi-branch power grids, and high-voltage power grids introduced in this book. It is important to note that no-load losses and load losses should be calculated separately. The new installation or added installation of reactive compensation equipment can be included in this type of engineering loss reduction, and Formula (12.9) of equivalent reactive compensation electric energy can be used for calculation. If the compensation capacity is added for several substations, it can be included in this type of operating loss reduction, and the reduced load losses can be estimated according to the change in the average power factor in the load loss coefficient.

13.3.3 Implementation and Monitoring of the Loss Reduction Plan

13.3.3.1 Significance of Implementation and Monitoring of the Loss Reduction Plan

As early as the end of 1990, according to Reference [76], provincial power grid bureaus and prefectural power grid enterprises were to prepare and implement loss reduction plans, but this stipulation has still not been comprehensively executed. The management of electric energy losses varies under different power grid enterprise departments, and this complexity is still a challenge for comprehensive energy conservation and loss reduction management. The loss reduction plan is explicitly stipulated in relevant national regulations and rules. Against a background of concern on energy conservation nationwide, it is very important to take practical measures to prepare, implement. and monitor the loss reduction plan well.

13.3.3.2 Feasible Conditions and System Design for Implementation and Monitoring of the Loss Reduction Plan

Currently, according to relevant provisions, all power grid enterprises conduct a theoretical calculation of line losses for their entire network once a year and for county distribution grids or urban distribution grids two or three times a year, which provides extremely favorable conditions for the preparation, implementation, and monitoring of the loss reduction plan. Thanks to the rapid development of electric power system informatization, the effect of energy conservation and loss reduction has been gradually transformed from confirmation by calculation to confirmation by measuring information, indicating that the basic conditions for the implementation and monitoring of the loss reduction plan are ready.

If existing regulations or systems for energy conservation and loss reduction further stipulate that the loss reduction plan should be prepared on a regular interval (such as five years), this can be combined with the annual theoretical calculation of line losses for the entire network. Such system design does not add more workload, but can significantly improve the level of theoretical calculation of line losses and accelerate the pace of energy conservation and loss reduction.

In recent years, the State Grid Corporation of China conducted detailed management of line losses in East China and the China Southern Power Grid carried out detailed "four division" management (by voltage, by region, by line, by distribution area) as regulatory measures. These have achieved good benefits in loss reduction, providing experience for the implementation and monitoring of the loss reduction plan.

13.3.3.3 Indicators of Implementation and Monitoring of the Loss Reduction Plan

The implementation and monitoring of the loss reduction plan need several simple and effective indicators which can be compared between power grid enterprises of the same scale. According to experience in historical line loss management, the following indicators are provided for selection:

1. Line loss rates with loss for the entire network and various classes of voltage, excluding electric supply without loss.
2. Line loss rate for the main network, excluding power usage by substations.
3. Rate of theoretical calculation value consistency for line losses of the main network and 10 kV distribution grids with statistical values, which can be calculated as per the following formula:

   Rate of consistency for line loss rate = 100% – | [(theoretical calculation value of line loss rate – statistical value of line loss rate)/statistical value of line loss rate] × 100% |

4. Rate of realization of reduced electricity losses during the planning period, which can be calculated as per the following formula:

   Rate of realization of reduced electricity losses = (statistical reduction of electricity losses within planning period/planned reduction of electricity losses within planning period) × 100%

5. Deviation coefficient of the entire network. The concept of deviation coefficient is introduced on the condition that the line loss rate is minimal in Section 6.1 of Chapter 6, and calculated as per the following formula:

$$K_{\text{no}\cdot\text{op}} = 0.50\left(\frac{1}{\theta} - 1\right)$$

Deviation coefficients at the beginning and end of the planning period are calculated and compared. If the deviation coefficient is reduced, the line losses of the entire network approach the condition of minimum line loss rate, and the change trend of electric energy losses is good as a whole.

13.3.4 Introduction of an Example of the Loss Reduction Plan

The table of contents for a loss reduction feasibility report of a power grid enterprise (hereinafter referred to XX Grid) within the "ninth five-year plan" period is as follows:

1. Overview
   - 1.1 Purpose and method of preparing the loss reduction feasibility report within the "ninth five-year plan" period
   - 1.2 Loads and power supply areas of XX Grid
   - 1.3 Composition, production scale and technical indicators of XX Grid
   - 1.4 Electric energy metering system of XX Grid
2. Change and analysis of line loss rate within the "eighth five-year plan" period
   - 2.1 Increase in electric supply and change in line loss rate within the "eighth five-year plan" period
   - 2.2 Analysis of technical factors influencing line losses
   - 2.3 Analysis of management factors influencing line losses
3. Loss reduction objective for 1996–2000
   - 3.1 Prediction of electric supply
   - 3.2 Prediction of accrual increase in electricity line losses
   - 3.3 Reduced electricity losses and indicator requirements
4. Loss reduction measures and input–output analysis
   - 4.1 Improvement in measuring equipment
   - 4.2 Construction and transformation of power grids
   - 4.3 Transformation of 10 kV and below distribution grids
   - 4.4 Increase in system reactive compensation
   - 4.5 Strengthening of business management of power usage
   - 4.6 Input–output analysis
5. Possibility and suggestions for financing
   - 5.1 Idea of financing
   - 5.2 Suggestions

14

Analysis of the Influence of Power Grid Line Losses on Power Grid Enterprises

Like power consumption in the production process of general industrial products, electric energy losses are a kind of production costs. Similar to the situation where a few goods are inevitably damaged during circulation, electric energy losses should be reduced as much as possible, because such losses directly reduce merchantable goods and sales profits. Electric energy losses include no-load losses and load losses. No-load losses are independent of the quantity of the electric supply and seem to be included in the fixed costs, while load losses are a function of the square of the electric supply quantity (power sales quantity) and should be included in the variable costs. Due to these characteristics, the influence of electric energy losses on enterprises attracts more attention with the increasing transformation of the electric power industrial system. This chapter will discuss: (i) the influence of line losses on the profits of power grid enterprises, (ii) a model for evaluating and managing the internal link cost and link electricity price of power grid enterprises, (iii) the multisection electricity price composition model, and (iv) the coal–electricity price linkage, for the purpose of helping line loss workers learn the significance and role of managing electric energy losses from a broader perspective.

14.1 Influence of Line Losses on the Profits of Power Grid Enterprises

14.1.1 Calculating the Profits of Power Grid Enterprises

The weighted average of unit price of power purchase under various voltages is expressed by *a*. Various taxes irrelevant of the power sales quantity are included in enterprise fixed expenses. Fixed expenses, composed of maintenance expenses, overhaul expenses, staff wages, depreciation of fixed assets, and enterprise administration expenses, are expressed by *b*. The tax rate in proportion to the power sales quantity is expressed by *C*. The calculation model for the profits *W* of power grid enterprises is shown in the following formula:

$$W=(1-C)\chi y-\frac{a\chi}{1-z}-b \qquad (14.1)$$

Line Loss Analysis and Calculation of Electric Power Systems, First Edition. Anguan Wu and Baoshan Ni.

Wherein x – power sales quantity within the calculation period (kW·h);
y – average unit price of power sales within the calculation period [Yuan/(kW·h)];
z – enterprise comprehensive line loss rate within the calculation period; if the electric supply quantity is expressed by A, then

$$A = \chi/(1-z)$$

According to Formula (14.1), $(1-C)xy$ is the after-tax sales revenue, and $ax/(1-z)$ is the power purchase expense paid by the quantity of electric supply, that is variable cost. b is fixed costs.

14.1.2 Break-Even Point Power Sales Quantity

When $W = 0$, the break-even point (BEP) power sales quantity can be calculated by Formula (14.1), that is

$$(1-C)\chi y - a\chi/(1-z) = b, \quad \chi_{b1} = b/[(1-C)y - a/(1-z)] \tag{14.2}$$

According to Formula (14.2), a higher line loss rate, larger value z, and smaller denominator indicate a larger BEP power sales quantity, so this formula is useful for power grid enterprises running in the red.

14.1.3 Profit and Tax Amount per Unit Power Sales Quantity

The ratio between profit and sales revenue is the rate of profit, expressed by γ_0. Then Formula (14.1) is changed to

$$(\gamma_0 + C)\ \chi y = \chi y - a\chi/(1-z) - b$$

If $\gamma_0 + C = \gamma$ is the rate of profit and tax, the above formula is divided by x on both sides to obtain

$$\gamma y = y - a/(1-z) - b_0 \tag{14.3}$$

Wherein γy – profit and tax amount per unit power sales quantity [Yuan/(kW·h)];
b_0 – fixed expense per unit power sales quantity [Yuan/(kW·h)];
$a/(1-z)$ – average unit price of power purchase considering line loss factor [Yuan/(kW·h)].

According to Formula (14.3), the profit and tax amount per unit power sales quantity of power grid enterprises is the average unit price of power sales minus the average unit price of power purchase, considering electric energy losses minus the fixed expense per unit power sales quantity.

14.1.4 Analysis of Factors Affecting Profits

Take the total derivative of profits W in Formula (14.1) to obtain

$$\mathrm{d}W = [(1-C)y - a/(1-z)]\mathrm{d}x + (1-C)x\mathrm{d}y - \frac{ax}{(1-z)^2}\mathrm{d}z - \mathrm{d}b - xy\mathrm{d}c \tag{14.4}$$

According to Formula (14.4), the tax rate C is far small than 1.0, so $(1-C)y > a/(1-z)$, and dx, dy, and dW are of the same direction, indicating that the increase in power sales quantity and average unit price of power sales can lead to higher profits; there are negative signs before the dz, db, and dc components, and profits can be increased only when these three differential components are negative values, that is, the line loss rate, fixed expenses, and proportional tax rate are reduced.

According to the component of line loss rate in the total derivative of profits

$$dW_z = -\left[ax/(1-z)^2\right]dz \tag{14.5}$$

The increased profits due to a reduction of the line loss rate are related to the power sales quantity x, the average unit price of power purchase a, and the original value of line loss rate z. If the range of loss reduction is the same, a larger power sales quantity and higher average unit price of power purchase indicate a greater increase in profits.

14.2 Link Cost and Link Electricity Price

14.2.1 Significance of Division of Internal Links of Power Grid Enterprises

Within a power grid enterprise, other than Level 3 dispatching, the division of the three links between power transmission, transformation, and distribution and sales is clear in terms of operation and maintenance, division of personnel, and definition of production areas. Electricity losses occur in the above three links, and the sum of output energy and electricity losses should be balanced with the input energy. Each link has a link cost, and the sum of these link costs and internal power purchase expenses should be balanced with the "internal sales revenue" of output energy. In this way, the double control of costs and energy consumptions in the internal three links can be used to improve the overall benefits of a power grid enterprise.

14.2.2 Calculation Model of Link Electricity Price Under a Simple Electric Supply Structure

The simplest electric supply structure is shown in Figure 14.1. The input energy in the power transmission link is all from the thermal power plant, and the output energy in the power transmission link is the input energy in the power transformation link. Likewise, the output energy in the power transformation link is the input energy in the power distribution and sales link.

If the line loss rates in the three links are respectively z_2, z_3, z_4, then the following formulas can be derived according to the energy balance:

$$x_1 = \frac{x_2}{1-z_2}, x_2 = \frac{x_3}{1-z_3}, x_3 = \frac{x_4}{1-z_4} \tag{14.6}$$

As profits are 0 in the power transmission and transformation links, the sales revenue of internal links = link power purchase expense + link cost, that is

$$\left.\begin{aligned} a_2x_2 = a_1x_1 + B_2 \\ a_3x_3 = a_2x_2 + B_3 \end{aligned}\right\} \tag{14.7}$$

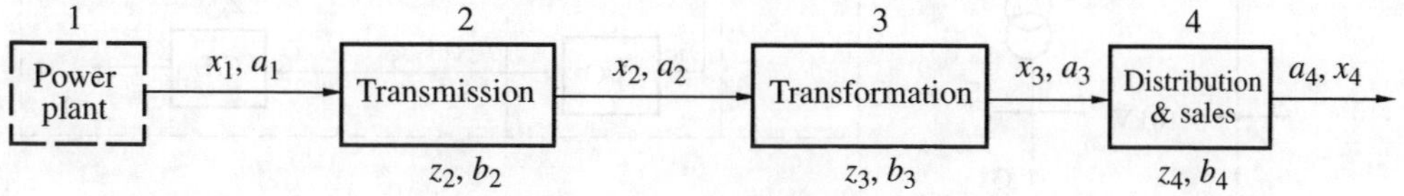

Figure 14.1 Simple electric supply structure.

Thus the formulas for calculating output prices in the power transmission and transformation links are

$$\left.\begin{aligned} a_2 &= \frac{a_1x_1 + B_2}{x_2} = \frac{a_1x_1 + B_2}{x_1(1-z_2)} = \frac{a_1 + b_2}{1-z_2} \\ a_3 &= \frac{a_2x_2 + B_3}{x_3} = \frac{a_2x_2 + B_3}{x_2(1-z_3)} = \frac{a_2 + b_3}{1-z_3} \end{aligned}\right\} \tag{14.8}$$

If the profits of the whole power grid enterprise are W_4 and are all included in the power distribution and sales link, then

$$a_4x_4 = a_3x_3 + B_4 + w_4 \tag{14.9}$$

As $x_4 = x_3(1 - z_4)$, substitute it into the above to obtain

$$a_4 = \frac{a_3 + b_4 + w_4}{1-z_4} \tag{14.10}$$

In Formulas (14.8) and (14.10), b_2, b_3, b_4 are the respective fixed expenses per unit input energy, and w_4 is the profit per unit electric supply.

According to Formulas (14.9) and (14.10), link output prices a_2, a_3, a_4 are determined by the link input price, the fixed expense per unit input energy, and the profit per unit electric supply. After the on-grid price a_1 of the power plant is determined, output prices in the power transmission, transformation, and distribution and sales links can be determined and assessed according to the allowable link fixed expenses and link line loss rates, while the expected value of the unit price of power sales can be determined and assessed according to the target profit of the whole enterprise and the line loss rates and fixed expenses in all links.

14.2.3 Calculation Model of Link Electricity Price Under a Complicated Electric Supply Structure

The actual electric supply structures of power grid enterprises are complicated. One situation is a complicated electric supply structure with repeated step-downs, as shown in Figure 14.2.

According to the electric connection diagram in Figure 14.2a, the link connection diagram in b can be obtained, wherein uppercase letters represent the buses of different links. Generally speaking, no matter

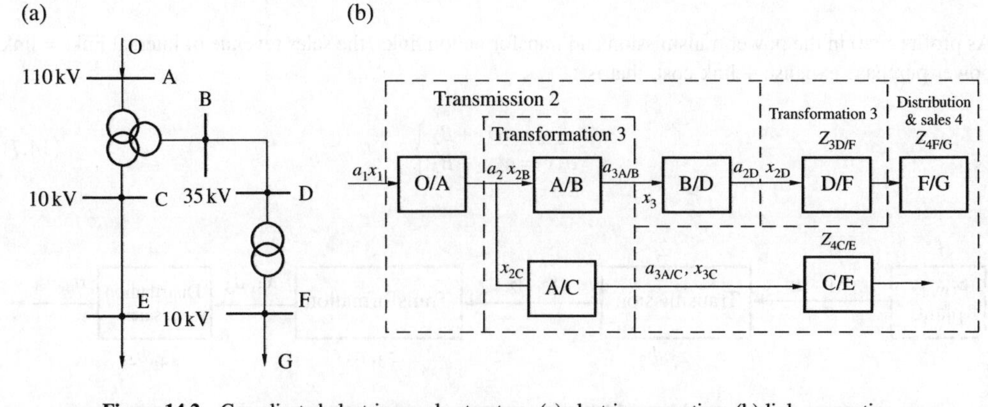

Figure 14.2 Complicated electric supply structure: (a) electric connection; (b) link connection.

how complicated the electric supply structure, the corresponding link connection diagram can be obtained. Subscripts 2, 3, 4 respectively represent power transmission, transformation, and distribution and sales links. According to Figure 14.2, the transmission link has two sub-links: O/A and B/D. According to Formula (14.8),

$$\begin{cases} a_{2A} = (a_1 + b_{2O/A})/(1 - z_{2O/A}) \\ a_{2D} = (a_{3A/B} + b_{2B/D})/(1 - z_{2O/A}) \end{cases} \tag{14.11}$$

The transformation link has three sub-links: A/B, A/C, and D/F. According to Formula (14.8),

$$\left.\begin{array}{l} a_{3B} = (a_2 + b_{3A/B})/(1 - z_{3A/B}) \\ a_{3C} = (a_2 + b_{3A/C})/(1 - z_{3A/C}) \\ a_{3F} = (a_{2D} + b_{3D/F})/(1 - z_{3D/F}) \end{array}\right\} \tag{14.12}$$

The distribution and sales link has two sub-links: F/G and C/E. According to Formula (14.8),

$$\begin{array}{l} a_{4G} = (a_{3D} + b_{4F/G} + w_4)/(1 - z_{4F/G}) \\ a_{4E} = (a_{3C} + b_{4C/E} + w_4')/(1 - z_{4C/E}) \end{array} \tag{14.13}$$

Wherein ω_4, ω'_4 – profits per unit input energy (electric supply) in the distribution and sales link.

14.2.4 Equivalent Merging of Parallel Electric Supply Structure

A parallel electric supply structure in which the start- and tail-end voltages are the same can be merged, as shown in Figure 14.3.

Because the energy flow remains unchanged before and after the merging, $x_{1eq} = x_{1I} + x_{1II}$, $x_{2eq} = x_{2I} + x_{2II}$; the capital flow also remains unchanged before and after the merging, so $a_{2eq}x_{2eq} = a_{2I}x_{2I} + a_{2II}x_{2II}$, thus

$$\left.\begin{array}{l} a_{2eq} = (a_{2I}x_{2I} + a_{2II}x_{2II})/(x_{2I} + x_{2II}) \\ a_{3eq} = (a_{3I}x_{3I} + a_{3II}x_{3II})/(x_{3I} + x_{3II}) \end{array}\right\} \tag{14.14}$$

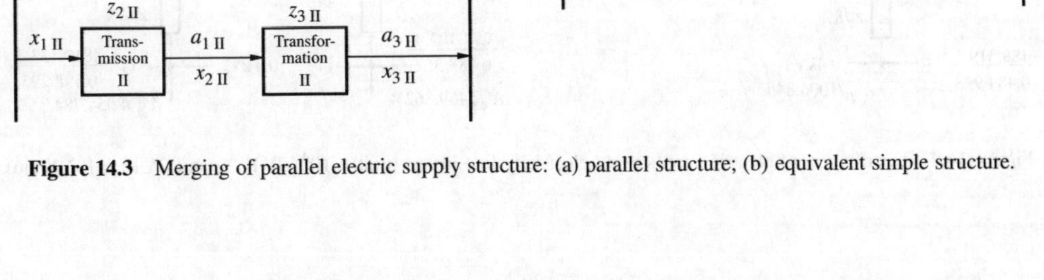

Figure 14.3 Merging of parallel electric supply structure: (a) parallel structure; (b) equivalent simple structure.

Electricity line losses in the same link should remain unchanged before and after the merging, that is

$$x_{1I}z_{2I} + x_{1II}z_{2II} = x_{1eq}z_{2eq}$$

$$x_{2I}z_{3I} + x_{2II}z_{3II} = x_{2eq}z_{3eq}$$

so

$$\left.\begin{aligned} z_{2eq} &= (z_{2I}x_{1I} + z_{2II}x_{1II})/(x_{1I} + x_{1II}) \\ z_{3eq} &= (z_{3I}x_{2I} + z_{3II}x_{2II})/(x_{2I} + x_{2II}) \end{aligned}\right\} \tag{14.15}$$

According to Formula (14.14) and Formula (14.15), when the parallel electric supply structure is merged, the equivalent values of output price and line loss rate are both calculated by the weighted average values of the output price and line loss rate of the parallel electric structure with the output energy and input energy as weights.

Example 14.1 Example 6.1 gives the calculation results of the distribution of electric supplies and electricity line losses of a county-level power grid enterprise for 10 months, as shown in Figure 14.4; the on-grid price is 0.148 Yuan/(kW·h); the fixed expenses per unit electric supply in the three links of power transmission, transformation, and distribution and sales are respectively 0.042, 0.038, and 0.117 Yuan/(kW·h); the profit per unit electric supply is 0.10 Yuan/(kW·h); the line loss rate in the distribution and sales link is 8%.

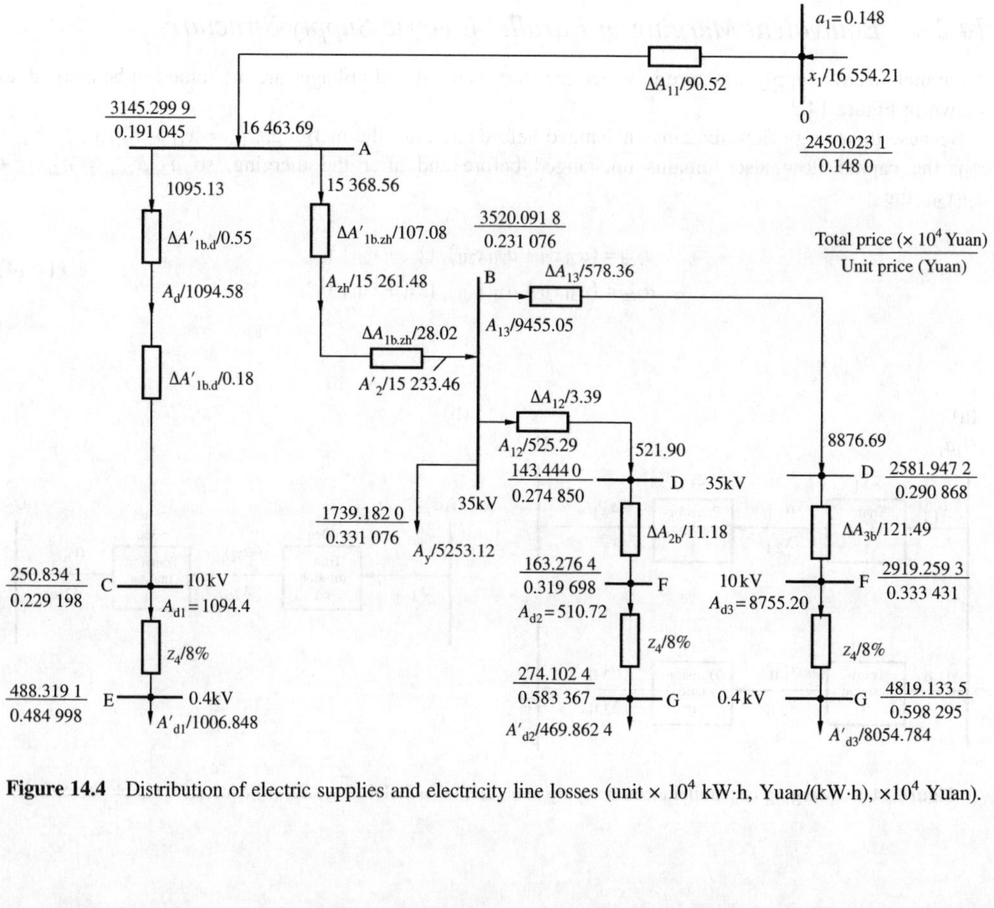

Figure 14.4 Distribution of electric supplies and electricity line losses (unit × 10^4 kW·h, Yuan/(kW·h), ×10^4 Yuan).

Try to calculate output prices in each link for 35, 10, and 0.4 kV systems; calculate link output prices for the equivalent three-link electric supply structure; calculate equivalent parameters in each link of the equivalent simplified electric supply structure; draw the diagram of the equivalent simplified electric supply structure.

Solution

1. *No-load losses and load losses of the high-voltage winding are decomposed to medium-voltage and low-voltage windings.* Losses of the high-voltage winding are allocated based on the square of electric supplies of medium- and low-voltage windings. Losses allocated to medium- and low-voltage windings are expressed by $\Delta A'_{1b\cdot 2h}$ and $\Delta A'_{1b\cdot d}$, so

$$\Delta A'_{1b.zh} = \frac{107.63\times 10^4 \times 15\,261.48^2}{15\,261.48^2 + 1094.58^2} = 107.08\times 10^4\ (\text{kW}\cdot\text{h})$$

$$\Delta A'_{1b.d} = (107.63 - 107.08)\times 10^4 = 0.55\times 10^4\ (\text{kW}\cdot\text{h})$$

2. *Calculation of capital flow from power source node to sales node.* According to Formula (14.8), $(a_1 + b_2)x_1 = x_1(1 - z_2)a_2$. The left side of the equation is the sum of unit price of power purchase at start end of power source and fixed expense per unit input energy in the transmission link, multiplied by the input energy, and the product can be regarded as the value of capital flow at the input end of the line. The right side of the equation is the product of output energy and output unit electricity price. The link output price a_2 at node A can be obtained by the value of capital flow at node A divided by the output energy at node A. Numbers on and under the horizontal line respectively refer to the value of the capital flow of each node and the value of the link output price, as shown in Figure 14.4.
3. *Checking calculation of capital flow.* First calculate the sales revenue, link cost, and power purchase expense, and then calculate the profit. Compare the profit with the scheduled profit. If they are consistent, the calculation of capital flow is correct. See Figure 14.4 and the following calculations can be obtained:

Gross sales revenue $= a_{Ay}A_y + a_{d1}A'_{d1} + a_{d2}A'_{d2} + a_{d3}A'_{d3}$
$= (1739.1820 + 488.319\,1 + 274.102\,4 + 4819.133\,5)\times 10^4$
$= 7320.737\times 10^4$ (Yuan)

Link cost = fixed expense per unit input energy × link input energy

So, distribution link cost $= 0.117\times(8755.20 + 510.72 + 1094.4)\times 10^4 = 1212.157\,4\times 10^4$ (Yuan)

Transformation link cost $= 0.038\times(8876.69 + 521.90 + 16\,463.69)\times 10^4 = 982.766\,6\times 10^4$ (Yuan)

Transmission link cost $= 0.042\times(9455.05 + 525.29 + 16\,554.21)\times 10^4 = 1114.451\,1\times 10^4$ (Yuan)

Total cost of three links $= 3309.3751\times 10^4$ (Yuan)

Power purchase expense $= a_1x_1 = 0.148\times 16\,554.21\times 10^4 = 2450.0231\times 10^4$ (Yuan)

Profit = gross sales revenue − total cost of three links − power purchase expense
$= (7320.7370 - 3309.3751 - 2450.0231)\times 10^4$
$= 1561.3388\times 10^4$ (Yuan)

Sales profit given in the example = sales profit per unit electric supply × sum of electric supplies in the sales link under various voltages
$= 0.10\times(1094.4 + 510.72 + 8755.2 + 5253.12)\times 10^4$
$= 1561.344\times 10^4$ (Yuan)

According to the above calculations, the sales profit calculated by link electricity prices is consistent with the sales profit given in the example.

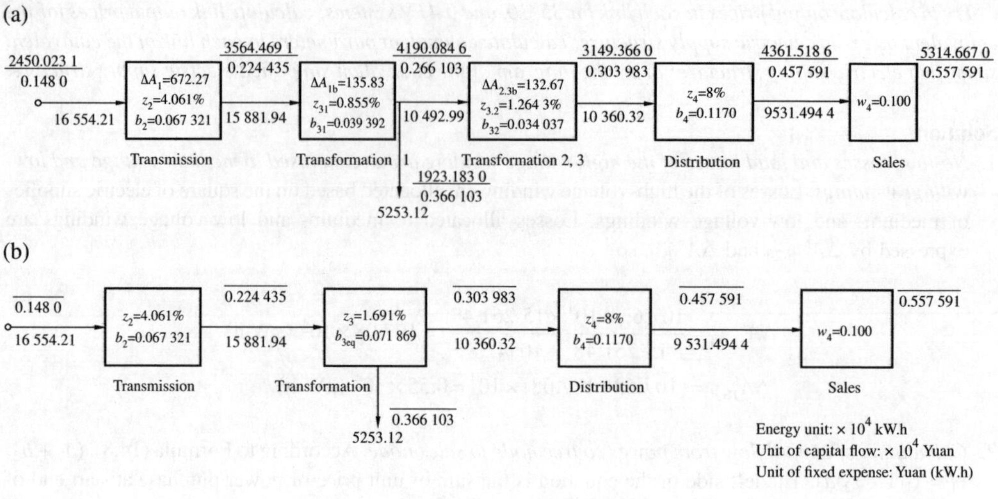

Figure 14.5 Simplification of complicated electric supply structure: (a) two transformation sub-links; (b) simplified electric supply structure.

4. *Relevant calculations of the simplified electric supply structure* (see Figure 14.5).

Equivalent line loss rate in the transmission link $z_2 = (\Delta A_{11} + \Delta A_{12} + \Delta A_{13})/x_1 = (90.52 + 3.39 + 578.36)/16\,554.21 = 672.27/16\,554.21 = 4.061\%$

Equivalent fixed expense per unit input energy in the transmission link b_2 = transmission link cost/x_1 = 1114.4511/16 554.21 = 0.067 321[Yuan/(kW·h)]

Equivalent line loss rate in the transformation sub−link 1 $z_{31} = \Delta A_{1b}/\left(x_1 - \sum \Delta A_1\right) = (107.08 + 28.02 + 0.55 + 0.18)/(16\,554.21 - 672.27) = 0.855\%$

Equivalent fixed expense per unit input energy in the transformation sub-link 1 $b_{31} = (0.038 \times 16\,463.69)/(16\,554.21 - 672.27) = 625.620\,22/15\,881.94 = 0.039\,39$ [Yuan/(kW·h)]

Equivalent line loss rate in transformation sub-links 2 and 3

$$z_{32} = (\Delta A_{2b} + \Delta A_{3b})/\left(x_{2eq} - \Delta A_{1b}^{zh} - A_y\right) = (11.18 + 121.49)/(15\,881.94 - 135.10 - 5253.12) = 132.67/10\,493.72 = 1.2643\%$$

Wherein $\Delta A_{1b}^{zh} = \Delta A'_{1b.zh} + \Delta A_{1b.zh} = (107.08 + 28.02) \times 10^4 = 135.10 \times 10^4$ (kW·h)

Fixed expense per unit input energy in transformation sub-links 2 and 3 b_{32} = 0.038(8876.69 + 521.90)/10 492.72 = 357.1464/104 92.72 = 0.034 037 [Yuan/(kW·h)]

According to the example, $z_4 = 8\%$, $b_4 + w_4 = 0.117 + 0.10 = 0.217$ [Yuan/(kW·h)]. Likewise, according to $x_1(a_1 + b_2) = x_2a_2$, mark the link input and output energy. Based on $a_2 = x_1(a_1 + b_2)/x_2$, calculate the link output prices and the values of capital flow at the link output node. See Figure 14.5a for the calculation results. In the figure, the number under the arrow is the energy flow; the number on the horizontal line is the value of node capital flow; the number under the horizontal line is the link output price.

The further simplification of two transformation sub-links should meet the requirement that the values of capital flow at the three nodes of input end, output end, and direct supply load end A_y in the transformation links should remain unchanged before and after the simplification, and only the equivalent fixed expense per unit energy in the transformation sub-links after simplification needs to be calculated. At this time, $(a_2 + b_{3eq})(x_2 - A_y) = a_3(1 - z_3)x_2$ is still satisfied. According to the numbers in Figure 14.5a,

$$(0.224435 + b_{3eq})(15\,881.94 - 5253.12) \times 10^4 = 0.303\,983 \times 10\,360.324 \times 10^4$$
$$= 3149.3660 \times 10^4$$
$$b_{3eq} = (3149.3660/10\,628\,82) - 0.224\,435 = 0.071\,869\ \text{Yuan/kW·h}$$

See Figure 14.5b for the final simplification results.

The calculations in this example show that, following the basic principle of balanced energy flow and capital flow among links, the equivalent simplification of the complicated electric supply structure can be completed, as well as the double control of energy consumptions and costs in major production links of power grid enterprises, thereby providing a mathematical model and theoretical basis for the informatization and lean management of enterprises.

14.3 Influence of Line Losses on the Composition of Multi-section Electricity Prices

As early as December 1998, Xie Songlin, former Chief Economist of the State Grid Corporation of China, put forward in the electricity price forum held in Changsha [78] that it was necessary to establish a new electricity pricing mechanism based on continuous rectification, to gradually improve the electricity price structure based on the adjustment of electricity price level, to straighten out the electricity price management relationship based on the deepening reform, and to promote electricity price legislation during the exploration and practice of the electricity price reform. In this speech, he was the first to propose: (i) establishoing the three-section electricity model; (ii) changing the formation mechanism of on-grid price, transmission-distribution price, and sales price, including changing the pricing mechanism of the on-grid price and executing ex-ante pricing according to the average cost of social power generation; (iii) changing the situation where the power grid development lags behind the power source development an improving the quality and safety of grid operation; (iv) implementing the separation between power plants and power grids, separately approving the transmission-distribution price of each grid according to the principles of reasonable compensation cost, fair benefits and legal calculation of taxes, and adopting one standard for one grid; (v) scientifically formulating the sales price according to the principle of fair burden, and reasonably establishing the electricity price structure by customer voltage class and load factor under the aggregate level approved by grid; (vi) establishing the electricity price balance system, balancing the balance of total revenue calculated by electricity sales and sales price minus payment to independent power plants and power grid enterprises through dedicated account, so as to maintain a relatively stable sales price and avoid frequent adjustment.

It has been several years since the separation between power plants and power grids was realized, and most provinces and regions have implemented the on-grid of partial energy bidding for many years. The coal–electricity price linkage has made progress after years of appeal. However, multi-section electricity prices have yet not been implemented, and the measurement and calculation of the transmission–distribution price can only be found in documents like yearbooks. Therefore, the electricity price reform still has a long way to go.

14.3.1 Type of Electricity Price and Comparison of Calculation Methods

14.3.1.1 Type of Electricity Price and Calculation Methods

According to relevant domestic materials, different types of electricity prices should be calculated by different methods. Electricity prices are divided into the following three types by the electricity price update cycle t and calculated by different methods.

1. $t = 0.5$–1.0 h, real-time price. This is calculated by the short-term marginal cost method, and is corrected according to the power generation balance of payment and network balance of payment before a real-time electricity price settlement.

2. t = 1a, peak–valley time of use price. Average electric energy costs are calculated for peak, waist, and valley periods by the stochastic production simulation method and included in the profit and tax adjustment coefficient to determine peak, waist, and valley time of use prices.
3. t > 1a, average electricity price for many years, also known as non-real-time price or basic price. The aggregate cost method considering the electric power development planning and the time value of money is recommended in China. Similar to this method, the system total production cost present value method uses the loss of load probability (LOLP) as an indicator to determine the commissioning schedule of expansion (or construction) of generator units, and it adopts the traditional present value mathematical method to analyze the system cost and calculate the electricity price. The other method is the long-term marginal cost method which can theoretically gain maximum social benefits, but its price result is much higher than that calculated by the aggregate cost method and needs to be adjusted manually.

14.3.1.2 Comparison and Discussion

1. Due to the needs of budding on-grid pilot projects for power plants, recent years have witnessed in-depth studies of the real-time price (including transit expenses) in China. But few studies have been seen regarding how to determine the long-term average electricity price. The level of electricity price expected in the marketization reform of the electric power industry is reduced to a certain extent, and more attention should be paid to the analysis and study of the composition of a long-term average electricity price.
2. With the planning period of an electric power system as the period for designing a long-term average electricity price, the commissioning time (considering LOLP requirements) of capacity expansion or construction of generator units is included as an investment time factor. This total production cost present value method is simple, intuitive, and can be easily understood and accepted by general electric power customers. In addition, mature foreign experience in this regard can be learnt.
3. Currently, the aggregate cost method has been used for calculating the cost and electricity price of electric power systems in China. This method uses two components to determine the cost, namely variable cost and fixed cost, directly corresponding to energy price and capacity price. The in-depth study of changes in these two components after the separation between power plants and power grids and the separation among transmission, distribution, and sales is of important significance for analyzing the degree of leeway for electricity prices and determining the long-term average electricity price. Likewise, as for definite production simulations, the composition of multi-section electricity prices can be analyzed below by the volume–cost–profit principle with a fiscal year as the period.

14.3.2 Analysis of Composition of Two-Section Electricity Prices Under the Single Electricity Purchaser Model

A possible operation model after the separation between power plants and power grids is that a power grid enterprise serves as the single electricity purchaser of the electric power system within a region. It is necessary to analyze the composition of the on-grid price and sales price under this model.

14.3.2.1 Calculating the On-Grid Price of a Power Generation Enterprise

For an independent power generation enterprise, the on-grid energy x_1, whole coal consumption in power supply c_0, fixed expense B_1, and rate of profit and tax from sales r_1 (excluding VAT) are given. If the on-grid price is a_1, use the volume–cost–profit analysis principle to obtain

$$r_1 x_1 a_1 = x_1 a_1 - x_1 c_0 p_0 - B_1 \tag{14.16}$$

Wherein c_0 – whole coal consumption in power supply (g/kWh), which is calculated according to total variable cost and included in the power consumption rate η, that is $c_0 = c'_0/(1-\eta)$; c'_0 is whole coal consumption in power generation;
p_0 – unit price of standard coal (Yuan/g).

Use Formula (14.16) to directly obtain the formula for the on-grid price of the power generation enterprise

$$a_1 = (c_0 p_0 + B_1/x_1)/(1 - r_1) \tag{14.17}$$

14.3.2.2 Calculation of Average Unit Price of Power Sales of Power Supply Enterprise

If the internal profit, power sales quantity, average unit price of power sales, internal unit price of power purchase, and line loss rate of the power supply enterprise are respectively w_g, x_m, y, a_g, B_g, and z_g, then the internal profit can be calculated as per the following formula

$$w_g = x_m y - x_m a_g/(1 - z_g) - B_g \tag{14.18}$$

According to Formula (14.18), the average unit price of power sales of the power supply enterprise is

$$y = [a_g/(1 - z_g) + B_g/x_m] + w_g/x_m \tag{14.19}$$

Considering the influence of line loss rate on the electric quantity, the formulas for electric supply and on-grid energy are

$$x_g = x_m/(1 - z_g); \quad x_1 = x_g/(1 - z_1) \tag{14.20}$$

Wherein z_1 – loss rate.

14.3.2.3 Analysis of the Paid Profit and Tax and the Composition of Unit Price of Power Sales of On-Grid Enterprises

If the paid profit and tax, rate of profit and tax, and fixed expense of an on-grid enterprise are respectively w_0, r_0, B_0, then $w_0 = r_0 x_m y$, and $w_0 = w_g$ B_0, so

$$r_0 x_m y = x_m y - a_g x_m/(1 - z_g) - B_g - B_0$$

And the average unit price of power sales is

$$y = [a_g/(1 - z_g) + (B_g + B_0)/x_m]/(1 - r_0) \tag{14.21}$$

If the condition $a_g = a_1/(1 - z_1)$ is satisfied when determining the internal power purchase price, then $a_g x_g = a_1 x_1$, indicating that the power purchase fee collected by the power grid enterprise from the power supply enterprise is equal to the expense paid to the power plant as per the on-grid price. In other words, the internal power purchase price and the on-grid price have a "seamless connection" without intermediate profit. On this condition, Formula (14.21) can be rewritten to

$$y = \frac{\dfrac{c_0 p_0}{(1 - r_1)(1 - z_g)(1 - z_1)} + \left[\dfrac{b_1}{(1 - r_1)(1 - z_g)(1 - z_1)} + b_g + b_0\right]}{1 - r_0} \tag{14.22}$$

Wherein $b_1 = B_1/x_1$, $b_g = B_g/x_m$, $b_0 = B_0/x_m$ are respectively the fixed expense per unit on-grid energy, the fixed expense of power supply per unit power sales quantity, and the fixed expense of power grid enterprise per unit power sales quantity.

According to Formula (14.22), the average unit price of power sales is composed of unit variable cost and unit fixed cost. The third factor that affects it is the rate of profit and tax from sales of the power grid enterprise. To control the level of unit price of power sales, it is necessary not only to reasonably determine the rates of profit and tax of the power plant and the power grid enterprise, but also to raise funds to improve the technical level of equipment of the power generation and supply enterprises, strengthen management, improve efficiency, reduce (i) coal consumption, (ii) power consumption rate, and (iii) loss rate of the power plant, and reduce the line loss rate of the power supply and the unit fixed expenses of the power plant, power supply enterprise, and power grid enterprise. Another factor that affects the level of electricity price is the unit price of standard coal, and as long as the increase in the electricity price matches the decrease in coal consumption, the level of electricity price can be controlled.

In addition, if there is no seamless connection between the internal power purchase price and the on-grid price, and the former is higher than the latter, then the difference between the two constitutes the main source of a balancing fund in the electricity price balance system.

14.3.3 Recursive Calculation of Multi-Section Electricity Prices

One of the main contents of the electricity price reform is "to form an electricity price system composed of generation price, transmission price, distribution price, and sales price", so it is necessary to theoretically analyze the relationship between energy and price in the four links. The four links have three types with different characteristics. The transmission and distribution links both have input and output energy as well as energy directly sold to customers (hereinafter referred to as direct supply energy). The generation link only has output energy, that is on-grid energy, and some power plants also have little direct supply energy which is subtracted from the generated energy, so only on-grid energy is considered. As a terminal link, the sales link has output energy that constitutes the power sale quantity, and losses in this link include low-voltage line losses and losses resulting from the sales process.

14.3.3.1 Balance of Link Energy

For convenient expression, the following symbols are used to express different items: energy: x; link electricity price: a; unit price of power sales: y; line loss rate: z; four links of generation, transmission, distribution, and sales: subscripts 1, 2, 3, 4; direct supply energy: on the upper right corner; ratio between direct supply energy and link output energy: α; ratio between direct supply sales price and link electricity price: β. The schematic diagram of link energy flow is shown in Figure 14.6.

If the line loss rate of direct supply energy in the transmission link is z'_2, then the energy balance in the transmission link is

$$x_2/(1-z_2) + a_2 x_2/(1-z'_2) = x_1$$

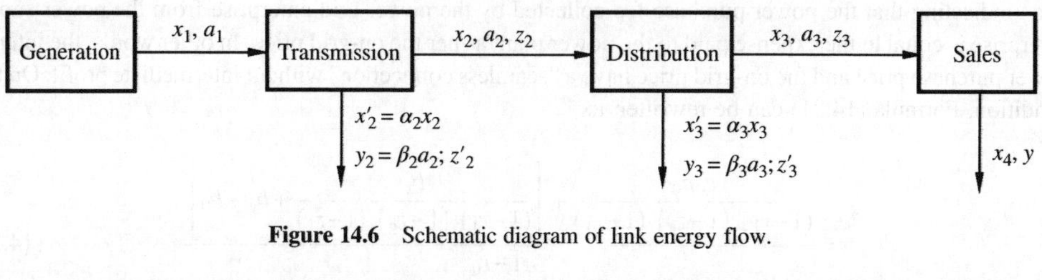

Figure 14.6 Schematic diagram of link energy flow.

The difference coefficient of the line loss rate is $k_2 = (1 - z_2)/(1 - z'_2)$. Substitute this into the above formula to obtain

$$x_1 = (1 + k_2 a_2)x_2/(1 - z_2) \tag{14.23}$$

Likewise

$$x_2 = (1 + k_3 a_3)x_3/(1 - z_3) \tag{14.24}$$

As for the sales link, $x_3 = x_4/(1 - z_4)$, that is the third term in Formula (14.6).

The above energy recursion formulas can be used to derive the formula of on-grid energy and power sales quantity

$$x_1 = (1 + k_2 a_2)\ (1 + k_3 a_3)x_4/[(1 - z_2)\ (1 - z_3)\ (1 - z_4)] \tag{14.25}$$

14.3.3.2 Calculation of Link Electricity Price

As for the generation link, the fixed cost per unit on-grid energy is $b_1(B_1/x_1)$, so Formula (14.17) can be rewritten to

$$a_1 = (c_0 p_0 + b_1)/(1 - r_1) \tag{14.26}$$

This is a standard "cost–profit" expression of the electricity price, that is

Electricity price = (unit variable cost + unit fixed cost)/(1 – rate of profit and tax)

Line losses caused by direct supply energy in the transmission link are smaller than those caused by output energy in this link. This is the immediate cause that the direct supply electricity price is lower than the link output price. For convenience, set

$$\beta_2 = (1 - z_2)/(1 - z'_2) = k_2 \tag{14.27}$$

Likewise, set $\beta_3 = k_3$.

The sales revenue in the transmission link is composed of the revenue from electricity price output to the distribution link and the revenue from the direct supply electricity price

$$a_2 x_2 + (k_2 a_2)\alpha_2 x_2 = (1 + k_2 \alpha_2) a_2 x_2$$

The value of profit and tax is the sales revenue minus the fixed expense of power purchase. Given Formula (14.23) obtained from the energy balance, the formula for calculating the transmission link output price can be obtained

$$a_2 = [a_1/(1 - z_2) + b_2(1 + \alpha_2)/(1 + k_2 \alpha_2)]/(1 - r_2) \tag{14.28}$$

Similar to the above, the formula for calculating the distribution link output price can be obtained

$$a_3 = [a_2/(1 - z_3) + b_3(1 + \alpha_3)/(1 + k_3 \alpha_3)]/(1 - r_3) \tag{14.29}$$

For the sales link, $a_4 = 0$, so the formula for calculating the sales link electricity price can be obtained

$$y = a_4 = [a_3/(1 - z_4) + b_4]/(1 - r_4) \tag{14.30}$$

Formulas (14.26) and (14.28) to (14.30) are recursion formulas of the link electricity price.

According to the above recursion formulas: (i) the on-grid price in the generation link is composed of the whole unit variable cost (which is mainly spent on fuel) and the unit fixed cost, and is related to the rate of profit and tax in the generation link; (ii) the difference between the output price and input price in the transmission and distribution links and the link line loss rates are related to the link unit fixed cost, link rate of profit and tax, and the proportion of direct supply energy; (iii) the proportion of direct supply energy influences the difference in link electricity prices, thus affecting the direct supply electricity price. In a word, after the electric power market is fully open, the structural design of an electricity price system is a complicated multivariable prediction and control problem.

14.3.4 Controlling the Aggregate Level of Electricity Price

The control of the average unit price of power sales in the terminal sales link is one of the core contents of an electricity price reform. To analyze the composition of the average unit price of power sales, it is necessary to analyze the connection between link electricity price and on-grid price.

14.3.4.1 Analysis of Composition of Link Electricity Price and Calculation of Direct Supply Electricity Price

Substitute Formula (14.26) for on-grid price into Formula (14.28) for transmission link electricity price to obtain

$$a_2 = (v_2 + f_2)/(1 - r_{eq2}) \tag{14.31}$$

$$v_2 = c_0 p_0/(1 - z_2)$$

$$f_2 = b_1/(1 - z_2) + b_2(1 + \alpha_2\ (1 - r_1)/(1 + k_2\alpha_2)$$

$$r_{eq2} = r_1 + r_2 - r_1 r_2$$

Wherein v_2 – unit variable cost accumulative to transmission link output price;
f_2 – unit fixed cost accumulative to transmission link output price;
r_{eq2} – equivalent rate of profit and tax accumulative to transmission link.

By means of recursion, the distribution output price and the average unit price of power sales y can be expressed by a "cost–profit" expression

$$a_3 = (v_3 + f_3)/(1 - r_{eq3}) \tag{14.32}$$

$$y = (v_4 + f_4)/(1 - r_{eq4}) \tag{14.33}$$

To facilitate comparison, v_i, f_i, and r_{eqi} are listed in Table 14.1.

Table 14.1 v_i, f_i, r_{eqi} values.

i	v_i	f_i	r_{eqi}
Generation 1	$c_0 p_0$	b_1	r_1
Transmission 2	$v_1/(1 - z_2)$	$f_1/(1 - z_2) + b_2(1 + a_2)(1 - r_1)/(1 + k_2 a_2)$	$r_1 + r_2 - r_1 r_2$
Distribution 3	$v_2/(1 - z_3)$	$f_2/(1 - z_3) + b_3(1 + a_3)(1 - r_{eq2})/(1 + k_3 a_3)$	$r_{eq2} + r_3 - r_{eq2} r_3$
Sales 4	$v_3/(1 - z_4)$	$f_3/(1 - z_4) + b_4(1 - r_{eq3})$	$r_{eq3} + r_4 - r_{eq3} r_4$

According to Table 14.1, the variable cost per unit energy in the link electricity price rises with the increase in the link line loss rate. The fixed cost per unit energy in a link is accumulated step by step, given the increase in line losses, and is related to the fixed expenses in this link and the fixed expenses in the last link. The rate of profit and tax accumulative to the output price of a link depends on not only the rate of profit and tax in this link, but also the rate of profit and tax in the last link.

If the link output price is given, then the direct supply electricity price can be calculated

$$y_2 = k_2 a_2 = \left(v_2' + f_2'\right)/\left(1 - r_{\text{eq2}}\right) \tag{14.34}$$

Wherein

$$v_2' = c_0 p_0/\left(1 - z_2'\right)$$
$$f_2' = b_1/\left(1 - z_2'\right) + k_2 b_2 (1 + \alpha_2)\ (1 - r_1)/(1 + k_2 \alpha_2)$$

Likewise

$$y_3 = k_3 a_3 = \left(v_3' + f_3'\right)/\left(1 - r_{\text{eq3}}\right) \tag{14.35}$$

Wherein

$$v_3' = c_0 p_0/\left[(1 - z_2)\left(1 - z_3'\right)\right],$$
$$f_3' = f_2/\left(1 - z_3'\right) + k_3 b_3 (1 + \alpha_3)\ \left(1 - r_{\text{eq2}}\right)/(1 + k_3 \alpha_3)$$

The result of the comparison between v'_3 and v'_2 and between f'_3 and f'_2 is the same as the change rule reflected in Table 14.1.

14.3.4.2 Control Way of Aggregate Level of Electricity Price

According to Formulas (14.33) to (14.35), the main ways to control the average unit price of power sales and the direct supply electricity price include:

1. To reduce the variable cost per unit power sales, it is optional to reduce coal consumption in the power supply of power plants, unit price of standard coal, and link line loss rate.
2. To reduce the fixed cost per unit power sales, it is only possible to reduce the link fixed expense.
3. To control the equivalent rate of profit and tax r_{eq4} of power sales, it is necessary to coordinate and determine the rates of profit and tax in all links from generation to sales. From the perspective of future necessary improvements in power grids, the rates of profit and tax in the transmission and distribution links should account for a larger proportion.
4. When other conditions are the same, the reduction of line loss rates z'_2 and z'_3 of direct supply energy can directly reduce the level of direct supply electricity price.

Example 14.2 Basic data and calculation requirements: the basic data of a municipal power supply enterprise are as follows: power sales quantity 7.543 74 billion kW·h, comprehensive line loss rate 4.19%, average unit price of power sales 0.2444 Yuan/(kW·h), provincial average cost of power sales 0.2187 Yuan/(kW·h), coal consumption in power supply 393 g/(kW·h), and unit price of standard coal 187.23 Yuan/t. The power sales quantities in transmission (110 kV and above), distribution (35 and 10 kV) and sales (0.4 kV) links are shown in Figure 14.7, and the line loss rates in transmission and sales links are shown in Figure 14.7.

Calculate:

1. If the total profit and tax remain unchanged, and the profits and taxes and costs in the four links are allocated by 0.50/0.3/0.15/0.05 and 0.65/0.2/0.12/0.03, design multi-section electricity prices and calculate the rates of profit and tax in the links.
2. If the average unit price of power sales is reduced by 5%, and the coal consumption in power supply, unit price of standard coal, and total profit and tax remain unchanged, try to calculate the multi-section electricity prices and the rates of profit and tax in the links.

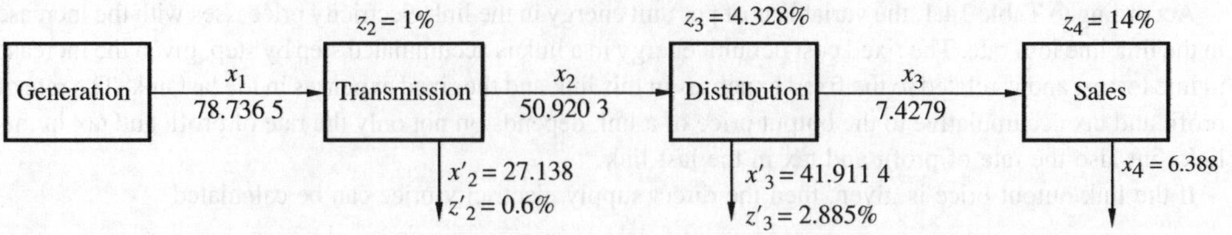

Figure 14.7 Distribution of power sales quantities in a city and line loss rates (underscored values are known; the unit of electric quantity is 100×10^6 kW·h).

Table 14.2 Intermediate data and results of calculation (part 1).

Link	Profit and tax (RMB 10 000)	Calculation coefficient of total revenue	$\gamma_1 a_i$ [Yuan/(kW·h)]	Fixed expense (RMB 10 000)	Calculation coefficient of fixed expense	Fixed expense per unit, b_i [Yuan/ kW·h)]	a_i [Yuan/ (kW·h)]
Generation 1	969.7059	1.0	0.0 123 115 784	49 302.6249	1.0	0.062 172 4	0.148 509 82
Transmission 2	581.2235	$1+k_2\alpha_2$ 1.530 805 4	0.007 461 568 8	32 996.3188	$1+\alpha_2$ 1.532 95	0.042 271 39	0.199 802 06
Distribution 3	290.1118	$1 + k_3\alpha_3$ 6.558 589 6.558 589	0.005 969 454 3	19 797.7913	$1 + \alpha_3$ 6.642 43	0.040 125 80	0.255 448 89
Sales 4	969.706	1.0	0.015 174 869	4949.478	1.0	0.077 480 4	0.389 688 86

Solution

1. Calculation process: determine z_2 and z'_3 (statistics shows that $z_3 = 1.5z'_3$) in the distribution link and calculate the output energy in the links according to the total electricity line losses; allocate profits and taxes and costs according to the given ratios; calculate electricity prices in the links; calculate taxes by total revenue minus total expenditure; do accounting of revenue and expenditure; and calculate the rates of profit and tax in the links.
2. The intermediate data and results of calculation are shown in Tables 14.2 and 14.3, and the calculation results when the electricity price is reduced by 5% are shown in Table 14.4.

14.3.5 Analysis and Discussion

1. The formula for calculating the average unit price of power sales under the single electricity purchaser model provides a quantitative basis for analyzing the degree of leeway of electricity price. The change rule for variable and fixed costs illustrated by the recursion formulas of multi-section electricity prices in this section can serve as a comparative reference for the calculation of transit expenses after the electric power market is open, and also provides an analysis model for determining the price ratio relationship between energy price and capacity price, or further studying the price ratio relationship in peak–valley periods and the price difference relationship during different seasons among customers under different voltages.
2. Example 14.2 shows that, if the total profit and tax remain unchanged and the aggregate level of electricity price is reduced by 5%, then the rate of profit and tax of the whole system will be increased to 5.31%, exceeding the range of electricity price reduction. If only one measure of reducing the fixed coast is taken to reduce the electricity price, the fixed expense of the whole system must be decreased by 8.61%, which is

Table 14.3 Intermediate data and results of calculation (part 2).

Link	Total revenue (RMB 10 000)	Direct supply revenue (RMB 10 000)	Output expenditure (RMB 10 000)	Fuel or electricity purchase expenditure (RMB 10 000)	Fixed expense (RMB 10 000)	Calculated profit and tax (RMB 10 000)	Rate of profit and tax (%)
Generation 1	116 931.4344		116 931.4344	57 935.4 111	49 302.6249	9693.3984	8.29
Transmission 2	155 743.8481	54 004.0397	101 739.8084	116 431.4344	32 996.3188	5816.0949	3.7345
Distribution 3	124 445.8689	105 471.3808	18 974.4881	101 739.8084	19 797.7913	2908.2692	2.337
Sales 4	248 993.3244	24 893.3244		18 974.4881	4949.4478	969.3885	3.894
Total	422 014.4758	184 368.7449	237 645.7309	295 581.142	107 046.1828	19 387.151	4.594

Table 14.4 Calculation results when the electricity price is reduced by 5%.

	Comparison of link electricity price (RMB)			Comparison of fixed expense (RMB 10 000)			Comparison of rate of profit and tax		
Link	a_i	a'_i	Declined (%)	B_i	B'_i	Declined (%)	$r_i\%$	$r'_i\%$	±%
Generation 1	0.148 509 82	0.140 899 6	5.12	49 302.624 9	4331.6322	12.15	8.29	8.738	5.4
Transmission 2	0.199 802 06	0.189 749 72	5.03	32 996.318 8	3115.6287	5.59	3.7345	3.93	5.23
Distribution 3	0.255 448 89	0.242 671 07	5.00	19 797.791 3	1869.5772	5.59	2.337	2.46	5.26
Sales 4	0.389 688 86	0.370 501 66	4.924	4949.447 8	4672.8943	5.59	3.894	4.096	5.19
Total	0.244 399 7	0.232 180	5.0	107 046.182 8	9782.7324	8.61	4.594	4.838	5.31

1.72 times the range of electricity price reduction. This indicates that it is very difficult to realize the objective of electricity price reduction through only one measure. Solutions such as reducing taxes imposed on electric power enterprises, accelerating the pace of technological progress of electric power enterprises, or continuously improving indicators like coal consumption in power supply, availability ratio of generator units, and line loss rate, can provide a stronger technical basis for reducing the level of electricity price and worth being discussed as they produce enlarged positive results. Example 14.2 also shows that the aggregate cost method with volume–cost–profit analysis as the core can serve as a major method for determining the basic electricity price.

14.4 Analysis of Coal–Electricity Price Linkage

This section interprets the existing policy of coal–electricity price linkage, explains relevant factors to the linkage between coal price and power plant on-grid price, analyzes several issues related to the successive linkage between coal price and multi-section electricity prices through formulas for the composition of multi-section electricity prices in Section 14.3, and gives a calculation example.

14.4.1 Interpretation of the Existing Policy of Coal–Electricity Price Linkage

14.4.1.1 Connotation and Calculation Method of the Policy of Coal–Electricity Price Linkage

The coal–electricity price linkage mechanism is established according to Reference [79]. The core content is that, taking six months as the cycle for a coal–electricity price linkage, if the average coal price within a cycle is higher by 5% or above than that within the last cycle, then the electricity price should be adjusted accordingly.

As of the end of 2008, China has implemented a total of "two and a half" coal–electricity price linkages. May 2005 saw the first linkage when the national sales price was increased by 0.0252 Yuan/(kW·h) on average. July 2006 marked the second linkage when the national sales price was increased by 0.0245 Yuan/(kW·h) on average. In August 2008, the power plant on-grid price was only adjusted with an increase of 0.02 Yuan/(kW·h) on average, while the sales price was not adjusted, thus being known as a "half" linkage.

Domestic experts in the field of energy generally believe that the coal–electricity price linkage is only a pricing mechanism and serves as an interim measure to reduce the coal–electricity price contradiction. The real solution for the contradiction is complete marketization of the electricity price. When the electric power system reform has not achieved enough conditions, the coal–electricity price linkage acts as an adjustable compromise measure and plays an important role in ensuring the normal production of coal and electricity in a stable macroeconomic operation.

Since the completion of the reform of separation between power plants and power grids, power plants do not directly face electric power end users but power grid enterprises instead, so the on-grid price of power plants changes with the coal price and the sales price changes with the on-grid price. As a result, the coal–electricity price linkage includes two levels of contents and each has a different calculation method.

1. Calculation method for the linkage between on-grid price and coal price

$$\text{On-grid price adjustment standard} = \text{coal price change} \times \text{conversion coefficient} \tag{14.36}$$

Wherein, the conversion coefficient depends on factors such like standard coal consumption in power supply, calorific value and price adjustment digestion ratio and is calculated as follows

$$\begin{aligned}\text{Conversion coefficient} &= (1 - \text{digestion ratio}) \times \text{standard coal consumption in power supply} \\ &\times (7000/\text{natural coal calorific value}) \times (1 + 17\%)/(1 + 13\%)\end{aligned} \tag{14.37}$$

Wherein standard coal consumption in power supply – average standard coal consumption in the power supply of domestic thermal power plants in the last cycle [g/(kW·h)]; digestion ratio – proportion of reduction of generation cost by power generation enterprise to the rise in price; it is 30% in the first coal–electricity price linkage; 7000 (kilocalories) – standard coal calorific value; natural coal calorific value – average of actual calorific value generated by steam coal in power plant; (7000/natural coal calorific value) represents the conversion of steam coal to standard coal; (1 + 17%)/(1 + 13%) – tax rate where VAT in the electric power industry is 17% and 13% in the coal industry.

2. Linkage between sales price and on-grid price

$$\text{Sales price adjustment standard} = \text{on-grid price adjustment standard} \times \text{proportionality coefficient} \tag{14.38}$$

Wherein proportionality coefficient = 1/(1 – transmission and distribution loss rate); it is an increase coefficient in the sales price mark-up, given actual electric energy losses in transmission and distribution.

14.4.1.2 Relevance Between Coal Cost and Rise in Coal Price

According to analyses by relevant professionals, the occurrence conditions of coal resources in China are relatively poor, and large quantities of resources need to be input in coal mining.

1. The geologic structure of a coal seam is complicated and coal resources are generally buried deep; key state-owned coal mines have only 23% with simple structure, 41% with moderately complicated structure, and 36% with extremely complicated structure; predicted coal resources with a vertical burial depth of below 100 0m only account for about 50%; currently, the average mining depth of large and medium-sized coal mines is approximately 600 m, while the mining depth of most coal mines in Australia is about 250 m.
2. The geologic structure of coal fields in China was formed by the collision and jamming of several plates, through several strong crustal movements and geological transformations, and there are many types of mining disasters which are widely distributed.
3. There are few coal resources which are suitable for opencast mining; the quantity of opencast mining only accounted for about 5% in 2007, while the proportion of opencast mining in the eastern region of the United States is always about 60%, and in Australia this proportion reaches around 76%.
4. The future increase in coal production in China mainly relies on Shanxi, Shaanxi, Inner Mongolia, and Xinjiang, and these provinces lack water resources and have fragile ecological environments, so the government will take more administrative and economic measures to protect the local environment.
5. The production and demand of coal in China are characterized by reverse distribution and relatively high transportation costs.

Currently, China has made insufficient safety investments in coal mines, and the safety situation is still very grim. The death rate per million tonnes of coal mined in China was reduced to 1.184 in 2008, which was 47 times that of the United States, while Australia has achieved a zero death rate since 2004. The average years of education for coal mine workers is 9.2, lower than the average 11.7 years nationwide. In 2007, the overall labor productivity of key state-owned coal mines was 5.1 t/worker, accounting for only one-ninth of the per capita productivity in the United States over the same period.

The occurrence conditions of coal resources and the current situations of the coal mining industry show that there is an imminent cause for the gradual increase in coal cost, which has been specifically studied by a relevant domestic professional institution [80]. According to this study report, according to the coal price and production in 2007, external losses caused by coal amounted to about RMB 1790.3 billion, equivalent to 7.3% of the national GDP. The internalization of coal external costs required additional costs of RMB 821.8 billion, resulting in an increase in social wealth by about RMB 968.5 billion, and the coal price rose by about 23.1%, leading to a decrease in GDP by about 0.07%. If this internalization of external costs was realized in five years, the disposable income of urban and rural residents would be respectively reduced by 0.02% and 0.013%, which is small. According to the study, the internalization of external costs has advantages outweighing disadvantages, and has an extremely weak negative influence if implemented appropriately. Overall, the conclusion of this economic study can be interpreted as: the internalization of coal external costs within five years and the rise in price by about 23.1% are acceptable to both the overall society and economy and to urban and rural residents.

14.4.1.3 Requirement of Electricity Price Reform For Electric Power Enterprises

According to the explicit stipulation in Reference [81], the long-term objective of electricity price reform is that power generation and sales prices are formed through market competition, and that transmission and distribution prices are set by the government, known as "decontrol in both ends, and control in the middle". The short-term objective is to establish an on-grid price mechanism appropriate to moderate competition in the generation link, to initially set up a transmission and distribution price mechanism, and to realize the linkage between sales price and on-grid price. This shows that the government has already planned that the short-term objective of the electricity price reform is that the on-grid price fluctuates with the coal price and the sales prices changes with the on-grid price.

According to the calculation method of coal–electricity price linkage, such linkage requires that electric power enterprises should digest partial price rise factors, which puts forwards explicit requirements for electric power enterprises to deepen reform, reduce costs and improve efficiency. During 1985–1996 when power plants were not separated from power grids, the annual increase in labor efficiency of the electric power industry was 5.1%, and the increase speed was the same as that of the agricultural labor productivity over the same period [82], only accounting for 60% of the nationwide average annual growth rate of GDP per capita over the same period. As a technology-intensive electric power industry, such work efficiency was not satisfactory. Since the reform of separation of power plants and power grids, especially in recent years, power generation enterprises have seized the favorable opportunity when the contradiction between power supply and demand has been basically solved, installing large generator units and eliminating a large number of small generator units with high coal consumption, so that the nationwide average coal consumption in generation has been greatly reduced year by year, providing a technical basis for power generation enterprises to digest the rise in coal price. Driven by the economic policy of stimulating domestic demand, power grid enterprises have expanded the scale of grid construction and provided a new technical basis for reducing power supply costs. With the improvement in the intensification level of power grid enterprises, the reduction of costs has reached a new level. According to statistics, the overall labor productivity of the State Grid Corporation of China was increased by 22.1% in 2007 over 2006, setting a new record.

To predict the trend and requirement of the electricity price reform in the future, it is necessary to learn the experience of developed countries in the control of electricity price. There are two methods for controlling the sales price overseas: one is the "fair rate of return method", whose core is "allowable cost + allowable profit"; the other is the "upper limit method", also known as "price-cap regulation", which is a new electricity price regulation idea. For example, the formula for "price-cap regulation" of the British distribution price is price adjustment ratio (%) = $RPI - x$, wherein RPI is retail price index, representing the inflation rate; x is the rate of increase specified by the government in the production efficiency of electric power enterprises, subject to regular adjustment depending on scientific and technological progress and other factors. This shows that the price reform of developed countries clearly specifies the requirement for the cost factor as a component of price in monopolized industries. In other words, the costs are reduced with technological progress, including not only equipment costs, but also labor costs, reflecting the level of labor productivity.

The separation between main business and ancillary businesses and the direct supply to large customers are detailed projects in the electric power system reform, which are under progress or pilot implementation. The implementation process must be difficult, but the reduction of costs for electric power enterprises is a long-term requirement according to the macroeconomic requirements and the reality of coal–electricity price linkage and it needs a new awareness and effort from the leaders and employees of electric power enterprises.

Overall, two basic judgments can be made from the interpretation of the coal–electricity price linkage: first, the coal price must be gradually increased with the internalization of external costs; second, to reduce the overall influence of sales price on the national economy, the nation and society must put forward expectations and requirements of cost reduction on electric power enterprises that are downstream in the energy industry chain. Electric power enterprises are challenged with the following problems: What are the factors concerning the space for adjusting the on-grid price? How should power generation enterprises make efforts? What are the factors related to the space for changing the sales price? Is there any space for national policy? All these problems need to be considered and explored by technical and economic personnel in electric power enterprises.

14.4.2 Analysis of the Linkage between On-Grid Price and Coal Price

14.4.2.1 Relationship Between On-Grid Price and Unit Price of Standard Coal

Section 13.3 of Chapter derived the formula for calculating the on-grid price of the power generation enterprise when analyzing two-section electricity prices under the single electricity purchaser model

$$a_1 = [C_0 p_0 + B_1/x_1]/(1 - r_1)$$

If $b_1 = B_1/x_1$ is called the fixed cost per unit on-grid energy, then

$$a_1 = (C_0 p_0 + b_1)/(1 - r_1) \tag{14.39}$$

$$C_0 = C_0'/(1 - \eta)$$

Wherein C_0 – whole coal consumption in power supply [kg/(kW·h)];
C'_0 – whole coal consumption in generation calculated by total variable cost;
η – power consumption rate;
p_0 – unit price of standard coal (Yuan/kg);
r_1 – rate of profit and tax of the power generation enterprise.

Formula (14.39) shows that the on-grid price of the power generation price depends on four factors, namely whole coal consumption in power supply C_0, unit price of standard coal p_0, fixed cost per unit on-grid energy b_1, and rate of profit and tax r_1.

14.4.2.2 Analysis of Rate of Relative Change in On-Grid Price

Use Formula (14.39) to calculate the total increment (total derivative) in the on-grid price

$$da_1 = (p_0 dC_0 + C_0 dp_0 + db_1)/(1 - r_1) + (C_0 p_0 + b_1) dr_1/(1 - r_1)^2$$

This is divided by Formula (14.39) on both sides to obtain

$$da_1/a_1 = (p_0 dC_0 + C_0 dp_0 + db_1)/(C_0 p_0 + b_1) + dr_1/(1 - r_1)$$

If $b_1/C_0 p_0 = \theta$ is called the cost constituent ratio of the power generation enterprise, then the above formula can be simplified to

$$da_1/a_1 = dC_0/[C_0(1+\theta)] + dp_0/[p_0(1+\theta)] + db_1 \Big/ \left[b_1 \left(1 + \frac{1}{\theta} \right) \right] + dr_1/(1 - r_1)$$

If the rates of relative change da_1/a_1, dC_0/C_0, dp_0/p_0, db_1/b_1, dr_1/r_1 are respectively expressed by δa_1, δC_0, δp_0, δb_1, δr_1, the formula below can be finally obtained [55]

$$\delta a_1 = \delta C_0/(1+\theta) + \delta p_0/(1+\theta) + \delta b_1 \Big/ \left(1 + \frac{1}{\theta} \right) + \delta r_1 \Big/ \left(\frac{1}{r_1} - 1 \right) \quad (14.40)$$

In other words, the rate of relative change in the on-grid price depends on five factors, namely whole coal consumption in power supply, unit price of standard coal, fixed cost per unit, rate of relative change in the rate of profit and tax, and cost constituent ratio.

14.4.2.3 Analysis of Factors Affecting the Rate of Change in On-Grid Price

1. *Influence of change in the cost constituent ratio.* According to Formula (14.40), a smaller θ indicates a larger influence of the same rate of change δp_0 in the unit price of standard coal on δa_1. With improvements in the management and efficiency of the power generation enterprise, θ is reduced, and the effect where the increase in the unit price of standard coal will lead to rise in the on-grid price will be increasingly stronger, so the coal–electricity price linkage must be implemented to control the level of electricity price effectively.
2. *Influence of change in the fixed cost per unit.* As b_1 and θ are reduced, $1/(1 + 1/\theta)$ will be significantly decreased, and the effect where the reduction of the fixed cost per unit will result in lower on-grid price will be increasing weaker.
3. *Influence of the rate of profit and tax.* When the rate of profit and tax is reduced, the on-grid price will also be decreased, but the proportionality coefficient between the two $k = 1/(1/r_1 - 1)$ is small as r_1 is small. This shows that the policy of reduction of the rate of profit and tax has no obvious effect on controlling the on-grid price.

14.4.3 Linkage Between Sales Price and On-Grid Price

The changed on-grid price can be calculated as per the following formula

$$a'_1 = a_1 + \Delta a_1 = a_1 + \delta a_1 a_1 = a_1(1+\delta a_1)$$

Use Formula (14.28) to calculate the output price a'_2 in the transmission link

$$a'_2 = \left[a'_1/(1-z_2) + b_2(1+\alpha_2)/(1+k_2\alpha_2)\right]/(1-r_2)$$

Use Formula (14.29) to calculate the output price a'_3 in the distribution link

$$a'_3 = \left[a'_2/(1-z_3) + b_3(1+\alpha_3)/(1+k_3\alpha_3)\right]/(1-r_3)$$

Use Formula (14.30) to calculate the sales price a'_4

$$y' = a'_4 = \left[a'_3/(1-z_4) + b_4\right]/(1-r_4)$$

In conclusion, according to calculation formulas for the composition of multi-section electricity prices, transmission and distribution prices and sales prices can be respectively calculated, thus completing the linkage from the on-grid price to the sales price.

Because the sales price is generally applicable to provincial and regional levels, to realize relevant calculations from the change in on-grid price to the change in sales price, it is necessary to collect a proportion of the direct supply energy from transmission and distribution links, and line loss rates and fixed costs per unit energy in transmission and distribution links within the pricing territory, as well as parameters specified by the state such as the rates of profit and tax in transmission, distribution, and sales links.

Example 14.3 From 2000 to 2008, the coal consumption in generation for the region specified in Example 14.2 was reduced from 393 g/(kW·h) to 364 g/(kW·h); the unit price of standard coal was increased from 187.23 Yuan/t to 350 Yuan/t; the fixed cost per unit on-grid power was reduced by 10% from 62.174 Yuan/(MW·h) to 55.9551 Yuan/(MW·h); the rate of profit and tax of the power generation enterprise was reduced from 8.29% to 5%; the rate of profit and tax in transmission, distribution and sales links adopted the return on investment in transmission and distribution, that is 8.39% [83]; considering urban construction tax and educational tax, the rate of profit and tax was 9.84%; the electric supply structure and line loss rates in all links remained unchanged.

1. *Try to calculate the changed on-grid price a'_1 of the power generation enterprise.*
2. *If the fixed cost per unit power in transmission, distribution and sales links is reduced by 5%, try to calculate the rate of change in the sales price, and compare it with the rate of change in the on-grid price.*

Solution

1. *Calculation of on-grid price.* According to the changed coal consumption in power supply, unit price of standard coal, fixed cost per unit energy, and rate of profit and tax, use Formula (14.17) to directly calculate the changed on-grid price a'_1, that is

$$\begin{aligned} a'_1 &= (C'_0 p'_0 + b'_1)/(1-r'_1) \\ &= (0.364 \times 0.350 + 0.055\ 955\ 1)/(1-0.05) \\ &= 0.193\ 005\ 368[\text{Yuan}/(\text{kW·h})] \end{aligned}$$

If Formula (14.40) for the rate of relative change in the on-grid price is used for calculation, then $\delta a_1 = 34.5\%$, $a'_1 = 0.199\,745\,7$ Yuan/(kW·h). The error of the two calculation methods is 3.49%, indicating that the calculation result of Formula (14.40) is within the allowable range of error.

2. *Calculation of sales price.* According to Example 14.2, $1 + k_2\alpha_2 = 1.530\,805\,4$, $1 + \alpha_2 = 1.532\,95$, $1 + k_3\alpha_3 = 6.558\,589$, $1 + \alpha_3 = 6.642\,43$, $z_2 = 0.01$, $z_3 = 0.043\,28$, $z_4 = 0.14$. Formulas (14.28) to (14.30) can be used to calculate output prices and sales prices in transmission and distribution links

$$a'_2 = [0.193\,005\,4/(1-0.01) + 0.95 \times 0.042\,271\,4 \times 1.532\,95/0.530\,805\,4]/(1-0.098\,4)$$
$$= 0.260\,835\,2[\text{Yuan}/(\text{kW·h})]$$
$$a'_3 = [0.260\,835\,2/(1-0.043\,28) + 0.95 \times 0.040\,125\,8 \times 6.642\,43/6.558\,589]/(1-0.098\,4)$$
$$= 0.311\,241\,6[\text{Yuan}/(\text{kW·h})]$$
$$y' = a'_4 = [0.311\,241\,6/(1-0.14) + 0.95 \times 0.077\,480\,4]/(1-0.098\,4)$$
$$= 0.483\,047\,1[\text{Yuan}/(\text{kW·h})]$$

Rate of change in the on-grid price δa_1 = (0.193 005 368 – 0.148 509 8)/0.148 509 8 = 29.96%
Rate of change in the sales price δa_4 = (0.483 047 1 – 0.389 688 9)/0.389 688 9 = 23.95%

Example 14.3 shows that, when the electric supply structure is not changed greatly, changes in the fixed cost per unit energy and the rate of profit and tax in transmission, distribution, and sales links will affect the terminal sales price to a certain extent, and can digest part of the influence caused by the rise of coal price. The rate of change in the sales price may be smaller than the rate of change in the on-grid price. In this example, the rate of change in the sales price is 80% of the rate of change in the on-grid price, indicating that the coal–electricity price linkage plays a very obvious role in affecting the influence of change in the on-grid price on the sales price.

14.5 Analysis of Electricity Price Factor in Post Project Evaluation

It is traditionally believed that, according to the theory of project cycle and the procedures of project construction, the post project evaluation should be conducted as the project benefit and influence gradually appear after the project is put into operation and receives completion acceptance. Based on the project initiation decision and the technical and economic requirements for design, it is necessary to analyze results and problems during the project implementation, to evaluate the effect, efficiency, role, and influence of the project, to judge the realization degree of the project objective, to summarize experience and lessons, and to provide suggestions for guiding projects to be constructed, adjusting projects under construction, and improving constructed projects.

In response to a request from the State-owned Assets Supervision and Administration Commission, the State Grid Corporation of China issued a set of *Rules* in October 2005 for post project evaluation [84]. Article 13 of the *Rules* stipulates that any project should be evaluated for its technical level, operation, and operating management, and tha, to evaluate the economic benefits of the project, it is necessary to "conduct financial analysis of actual financial data of the project, calculate the cost–profit ratio, return on assets, debt asset ratio, interest coverage ratio and debt service coverage ratio, and evaluate the project for its profitability and debt paying ability."

For the post project evaluation of a transmission and transformation project, it is necessary to predict the electric power and energy conditions of the project within its term of validity and to calculate the effect and benefit within its term of validity according to the actual power supply loads and energies one year after the project is put into operation. Therefore, it is important to calculate the link input and output prices under the voltage class of the project according to the electric supply structure of the electric power system in the region where the project is located, and to analyze the year by year benefits of the project, so as to complete the calculation of the various benefit indicators necessary for the post project evaluation.

Due to the needs of raising funds for power grid construction and supporting the development of green energy, China's economic policy has stipulated several provisions concerning the mark-up of the price of electricity at the sales end. The role of the transmission, distribution, and sales links serving as a carrier for the mark-up should be recognized by currency expression. This requires that, to carry out the post project evaluation of a specific 220 or 110 kV project, it is important to calculate the link benefits of a voltage class project during the positive process of obtaining the electricity price by starting with the on-grid price of the power generation enterprise and considering the link costs and reasonable profits and taxes. It is also necessary to consider the reverse process of sales end mark-up which is not dependent on the on-grid price and to express mark-up benefits from links where the project is located. Finally, benefits in the positive and reverse processes should be added up and analyzed. This reverse expression is a reverse derivation from result to reason, and is a reverse form of the discussion of multi-section electricity prices. It solves the difficulty in expressing this reverse form, so that virtual (calculated) mark-up revenue included in the links can be used to comprehensively evaluate the benefits of the transmission and transformation project and to resolve the core problem challenging the post project evaluation.

14.5.1 Reverse Calculation of Mark-Up in Link Output End

For a link in a power system, the input price is a_0 and the output price is a_1. Similar to Formula (14.30),

$$a_1 = [a_0/(1-z_0) + b_0]/(1-r_0)$$

then

$$a_0 = [a_1(1-r_0) - b_0](1-z_0)$$

If a_1 is the electricity price with mark-up in the output end, then a_0 is the electricity price with mark-up as reverse calculated in the input end. Calculate the derivative of the above formula to obtain the incremental relationship

$$\Delta a_0 = \Delta a_1(1-r_0)\ (1-z_0) \tag{14.41}$$

According to Formula (14.41), the mark-up Δa_1 in the output end can be used to reverse calculate the mark-up Δa_0 in the input end.

14.5.2 Calculation of Mark-Up Allocation Coefficient of Simplified Electric Supply Network

A regional electric supply network can be simplified to a 220/110/10/0.38 three-level step-down form. A'_1, A'_2, A'_3 are annual direct power sales quantities of buses under the first three types of voltages, and A_4 is the annual low-voltage power sales quantity, as shown in Figure 14.8.

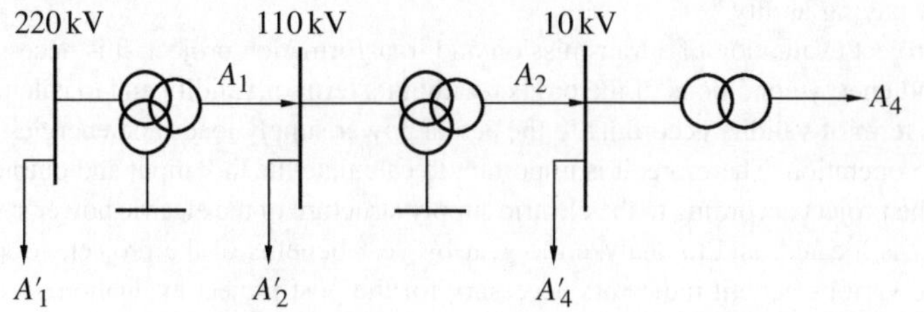

Figure 14.8 Simplified electric supply network.

14.5.2.1 Reverse Derivation of Mark-up Calculation Formula From 0.38 kV End to 10 kV Bus

Assume that the markup per unit power sales quantity is d [Yuan/(kW·h)]. The analysis is conducted through step-down from 10 kV to both ends of 0.38 kV. According to Formula (14.41), $\Delta a_4 = d(1 - r_4)(1 - z_4)$. The mark-up revenue of 10 kV direct power sales quantity is dA'_4, so the gross mark-up revenue C_{10} of total 10 kV input energy is

$$C_{10} = dA'_4 + \Delta a_4[A_4/(1-z_4)] = d[A'_4 + A_4(1-r_4)]$$

For the total 10 kV input energy, its mark-up revenue per unit energy can be calculated as per the following formula

$$\Delta a_{40} = C_{10}/[A'_4 + A_4/(1-z_4)] = K_4 d$$

Mark-up allocation coefficient $$K_4 = [A'_4 + A_4(1 - r_4)]/[A'_4 + A_4/(1 - z_4)] \quad (14.42)$$

Obviously, $K_4 < 1.0$, indicating due to link line losses and profits and taxes, the mark-up in the link input end is lower than that in the output end.

14.5.2.2 Reverse Derivation of Mark-up Calculation Formula From 10 kV Input End to 110 kV Bus

For the step-down from 110 kV to both ends of 10 kV, $\Delta a_2 = \Delta a_{40}(1 - r_2)(1 - z_2)$, and the gross mark-up revenue C_{110} of total 110 kV input energy is

$$C_{110} = dA'_2 + \Delta a_{40} A_2(1-r_2) = d[A'_2 + K_4 A_2(1-r_2)]$$

For the total 110 kV input energy, its mark-up revenue per unit energy can be calculated as per the following formula

$$\Delta a_{20} = C_{110}/[A'_2 + A_2/(1-z_2)] = K_2 d$$

Mark-up allocation coefficient $$K_2 = [A'_2 + K_4 A_2(1 - r_2)]/[A'_2 + A_2/(1 - z_2)] \quad (14.43)$$

14.5.2.3 Reverse Derivation of Mark-up Calculation Formula From 110 kV Input End to 220 kV Bus

Similar to the above, $\Delta a_{10} = C_{220}/[A'_1 + A_1/(1 - z_1)] = K_1 d$

Mark-up allocation coefficient $$K_1 = [A'_1 + K_2 A_1(1 - r_1)]/[A'_1 + A_1/(1 - z_1)] \quad (14.44)$$

Obviously, $K_1 < K_2 < K_4$. Due to the influence of line losses in several links and profits and taxes, the mark-up per unit energy is smaller at the bus side under higher voltage.

14.5.3 *Mark-up Calculation of Complicated Electric Supply Structure*

Some urban grids have two distribution systems of 35 and 10 kV and thus have complicated electric supply structures, as shown in Figure 14.9.

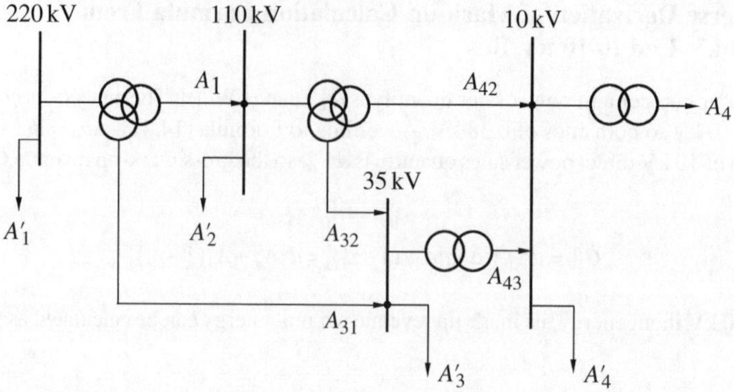

Figure 14.9 Complicated electric supply network.

The annual direct power sales quantities of buses under the four voltages are respectively A'_1, A'_2, A'_3, A'_4. The annual power sales quantity at the 0.38 kV side is A_4. The 220 and 110 kV buses have two output ends whose annual output energy is respectively A_1, A_{31} and A_{32}, A_{42}. The influence of two types of voltage output on the same input end should be considered simultaneously for the calculation of the power sales mark-up allocation coefficient.

14.5.3.1 Calculation of Power Sales Mark-up Allocation Coefficients at Input Ends of 10 and 35 kV Buses

Through similar analysis processes, the calculation formulas for power sales mark-up allocation coefficients at the input ends of 10 and 35 kV buses are obtained

$$K_4 = \left[A'_4 + A_4(1-r_4)\right] / \left[A'_4 + A_4/(1-z_4)\right] \tag{14.45}$$

$$K_3 = \left[A'_3 + K_4 A_{43}(1-r_3)\right] / \left[A'_3 + A_{43}/(1-z_3)\right] \tag{14.46}$$

14.5.3.2 Calculation of Power Sales Mark-up Allocation Coefficients at Input Ends of 110 and 220 kV Buses

Because the 10 and 35 kV output ends are recovered to a 110 kV bus, considering the direct power sales quantity, the power sales mark-up revenue of this bus is the sum of three parts: $C_{110} = dA'_2 + K_4 dA_{42}(1 - r_2) + K_3 dA_{32}(1 - r_3)$. The input energy of the 110 kV bus is also composed of three parts: A'_2, $A_{42}/(1 - z_{42})$ and $A_{32}/(1 - z_{32})$. Therefore, the mark-up revenue per unit input energy of the 110 kV bus can be calculated as per the following formula

$$\Delta a_{20} = C_{110} / \left[A'_2 + A_{42}/(1-z_{42}) + A_{32}/(1-z_{32})\right] = K_2 d$$

Wherein, the mark-up allocation coefficient K_2 can be calculated as per the following formula

$$K_2 = \left[A'_2 + K_4 A_{42}(1-r_2) + K_3 A_{32}(1-r_3)\right] / \left[A'_2 + A_{42}/(1-z_{42}) + A_{32}/(1-z_{32})\right] \tag{14.47}$$

The mark-up allocation coefficient K_1 for the input energy of the 220 kV bus can be calculated as per the following formula

$$K_1 = [A'_1 + K_2A_1(1-r_2) + K_3A_{31}(1-r_3)] / [A'_1 + A_1/(1-z_{21}) + A_{31}/(1-z_{31})] \quad (14.48)$$

14.5.4 *Calculation of Annual Power Sales Mark-up Revenue of a Single Transmission and Transformation Project*

For a single 220 kV transmission and transformation project, if its annual output energy is A_{220}, then its annual power sales mark-up revenue is

$$W_{J220} = K_1 dA_{220} \quad (14.49)$$

For a single 110 kV transmission and transformation project, the above formula is changed to

$$W_{J110} = K_2 dA_{110} \quad (14.50)$$

Example 14.4 For a 110 kV system, the on-grid energy of the power plant is $A_1 = 1.49 \times 10^8$ kW·h, and the 110 and 10 kV direct supply energies A'_2 and A'_4 are 0.1×10^8 and 0.2×10^8 kW·h; the 110 kV link line loss rate is $z_2 = 2\%$, and the 10 kV link line loss rate is $z_4 = 14\%$; the power sales quantity is $A_4 = 1 \times 10^8$ kW·h. The rate of profit and tax in transmission and distribution links is $r_2 = r_4 = 10\%$, and the mark-up on the power sales side and the direct supply side is $d = 0.04$ Yuan/(kW·h). *Try to calculate the mark-up value in 110 kV links and draw the diagram of mark-up capital flow of the whole system.*

Solution

1. Calculate the link input energy according to given link line loss rates

$$A'_2 = \frac{A_4}{1-z_4} = 1 \times 10^8/(1-0.14) = 1.162\ 8 \times 10^8 (\text{kW·h})$$
$$A_2 = A'_2 + A'_4 = (1.162\ 8 + 0.2) \times 10^8 = 1.362\ 8 \times 10^8 (\text{kW·h})$$
$$A'_1 = A_2/(1-z_2) = 1.3628 \times 10^8/(1-0.02) = 1.390\ 6 \times 10^8 (\text{kW·h})$$
$$A_1 = A'_1 + A'_2 = (1.390\ 6 + 0.10) \times 10^8 = 1.490\ 6 \times 10^8 (\text{kW·h})$$

   The system energy flow diagram is shown in Figure 14.10a.

2. According to Formula (14.42),

$$\begin{aligned} K_4 &= [A'_4 + A_4(1-r_4)] / [A'_4 + A_4/(1-z_4)] \\ &= (0.2 + 1 \times 0.9) \times 10^8 / (0.2 + 1/0.86) \times 10^8 \\ &= 1.1/1.362\ 8 = 0.8072 \end{aligned}$$
$$\Delta a_{40} = K_4 d = 0.807\ 2 \times 0.04 = 0.032\ 29 [\text{Yuan}/(\text{kW·h})]$$

   According to Formula (14.43),

$$\begin{aligned} K_2 &= [A'_2 + K_4 \times A_2(1-r_2)] / [A'_2 + A_2/(1-z_2)] \\ &= (0.1 + 0.807\ 2 \times 1.362\ 79 \times 0.9)/(0.1 + 1.362\ 79/0.98) \\ &= 0.7313 \end{aligned}$$
$$\Delta a_{20} = K_2 d = 0.7313 \times 0.04 = 0.029252 [\text{Yuan}/(\text{kW·h})]$$

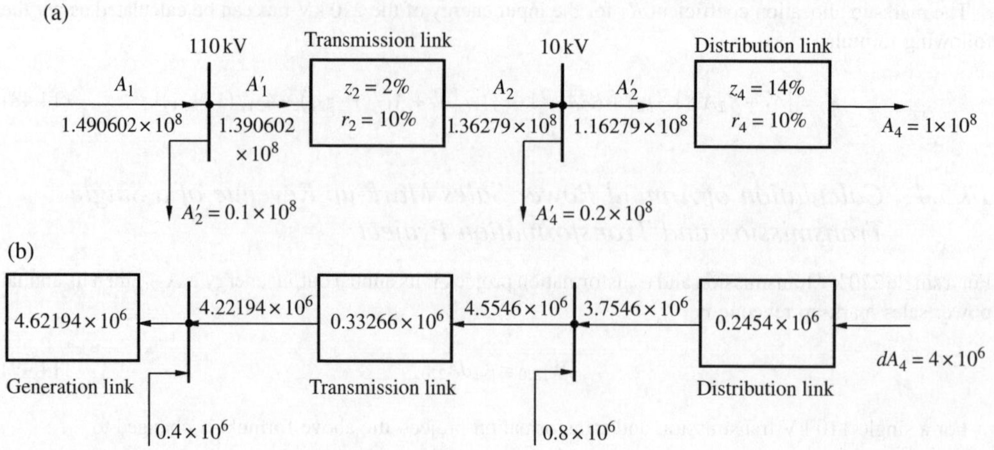

Figure 14.10 Diagram for Example 14.4: (a) system energy flow diagram; (b) mark-up capital flow diagram.

3. Mark-up difference in the distribution link $= dA_4 - K_4 dA'_2 = 0.04 \times 1 \times 10^8 - 0.032\,29 \times 1.162\,79 \times 10^8 = 0.2454 \times 10^6$ (Yuan)

   Mark-up difference in the transmission link $= K_4 dA_2 - K_2 dA'_1 = 0.032\,29 \times 1.362\,79 \times 10^8 - 0.029\,252 \times 1.390\,602 \times 10^8 = 0.332\,66 \times 10^6$ (Yuan)
4. Check calculation

   Sum of difference in the two links $= (4.621\,94 + 0.332\,66 + 0.245\,4) \times 10^6 = 5.20 \times 10^6$ (Yuan)
   This is consistent with the sum of the mark-up in the three power sales ends.

15

Management and Utilization of Line Loss Mass Information for an Electric Power System

Chapters 1–6 of this book analyze the basic theory and change rule of line losses, and Chapters 7–13 explain the analysis and calculation of line losses and the selection of loss reduction measures, accompanied by many calculation methods and examples. This shows that the calculation of line losses needs a lot of electric power system operating data and statistical line loss data. In addition, the line loss-oriented analysis and calculation of energy, electricity price, and transmission–distribution–sales production links lead to many intermediate and final data. Due to the wide relevance between line losses and production and operation, the information on line losses is "massive".

Managers of power grid enterprises should develop the general management information system to a higher application level so as not to be overwhelmed by this mass of information. This chapter starts with the functional analysis of two existing line loss management information systems, and analyzes the enterprise conceptual model, the business conceptual model, and multi-dimensional data model which may be provided in the basic theory of line losses according to the management concept, value creation, and support process put forward in Reference [85]. This chapter also gives a preliminary exploration of the utilization of data warehouse, data mining, and other information-based technological means used for the management of line loss mass information, in the hope that the management and in-depth utilization of line loss information can again draw the attention of managers of power grid enterprises and play a bigger role in improving the performance of enterprises.

15.1 Evaluation and Functions of Two Management Information Systems under the Guidelines

15.1.1 Functional Design Requirements for Two Types of Software

Compared with the trial Guidelines issued in 1988 [84], Reference [86] issued in February 2002 mainly added the application of electronic computers to calculate line losses and attached an appendix which addresses the functional design requirements for two types of software of the theoretical calculation of electric energy losses and the management of electric energy losses.

Line Loss Analysis and Calculation of Electric Power Systems, First Edition. Anguan Wu and Baoshan Ni.

As for the software of theoretical calculation of electric energy losses, the updated Guidelines stipulate three functional requirements for loss statistics: (i) calculate and analyze in the form of a block diagram: (ii) adopt the load flow calculation method and the simplified equivalent resistance method which are respectively applicable to the main network and the distribution network, (iii) list details in the three levels of input interface, core calculation, and output interface.

As for the software for the management of electric energy losses, the Guidelines determine a four-level software structure: level 1 includes county power supply subsidiaries within prefectural power supply branches, regional dispatching, and large electricity consumption customers; level 2 includes prefectural power supply branches; level 3 includes power plants under provincial dispatching and substation dispatching; and level 4 includes regional and provincial power grid enterprises. The Guidelines also give seven functional modules, such as system code input, calculation target registration, calculation relationship setting, statistical calculation, line loss analysis, and output in the form of block diagram. Such functional design requirements for the two types of software for theoretical calculation and management of line losses indicate the direction in which software developers and regional and provincial electric power bureaus develop management information systems for line losses, and significantly promote such development.

15.1.2 Functions of Line Loss Calculation and Management Information Systems Developed by Provincial Power Grid Enterprises

As a whole entity assessing line loss indicators, a provincial power grid enterprise should have its own independent management system, so line loss calculation and management information systems developed by the enterprise in cooperation with some college or university are the best examples which can be studied to learn the actual level of such types of software.

Guangdong Power Grid Corporation and the School of Electric Power, SCUT, jointly developed line loss theoretical calculation software in July 2001. Based on this, Guangdong Power Grid Corporation, Guangzhou Power Supply Bureau, and the School of Electric Power, SCUT, jointly developed the "line loss theoretical calculation and management systems for power grids in Guangdong" [47] in 2004. Thanks to continuous improvement during the application process, the current operating software has the following characteristics:

1. Practical selection and expansion of calculation models and calculation methods of various loss calculation units or combinations. The line losses of each single loss calculation unit and distribution network are calculated by the rms current, which is obtained according to actual active and reactive electric quantity from meter reading. The line losses of a medium-voltage distribution network are calculated by the equivalent resistance method based on the capacity of the distribution transformer or the load flow calculation method. The line losses of 0.40 kV low-voltage power grids are calculated by the typical distribution area loss conversion method or the voltage loss method.
2. It provides supplementary solutions to the line loss calculation of distribution networks with several small thermal power plants and small hydroelectric power plants, and the loss calculation of backup transmission of main transformers, dual power distribution network, and the receiving and transmission line. The line losses of a distribution network with a small power source are calculated by the equivalent resistance method with the average apparent power calculated from active and reactive electric quantity calculations within the calculation period as an equivalent capacity.
3. It puts forward a rational structure of basic data and operating data of units, and the graphic unit data structure is included for a distribution network.
4. With the AutoCAD secondary development technique, it realizes the bilateral transformation between a single line diagram of the distribution network and an Access database for the first time, and it correlates Excel with Auto CAD and Access, thus greatly reducing the workload of entry of distribution network data during the generation of connection diagram and improving the accuracy of data entry.
5. It has a series of new functions, such as a browsing and automatic debugging function before calculation, a function for merging multiple original data files into one data file (which facilitates the decomposition and composition of the line loss calculation), an excellent result analysis function, including automatic

generation of theoretical calculation analysis report, and sorting of line loss calculation results of units by attribute, a formation of three unified reporting modes, that is reporting, re-reporting, and virtual reporting, and a function for generating and printing reports under multiple modes.

15.1.3 Integrated Management System for Theoretical Calculation of Line Losses Developed by Regional Power Grid Enterprise

The "integrated management system of the Beijing–Tianjin–Tangshan grid line loss theoretical calculation" [48] developed by North China Electric Power Research Institute has already realized the line loss theoretical calculation and integrated management covering the Beijing–Tianjin–Tangshan power grids, Hebei southern power grids, Shanxi power grids, Shandong power grids, and Inner Mongolia western power grids in 2000. This system has been operating stably so far.

15.1.3.1 Relationships Between Three Functional Modules

This system is composed of three modules: (i) power grid graphics, (ii) line loss theoretical calculation, and (iii) statistical analysis and summary. The relationships between the three functional modules and with the database are shown in Figure 15.1.

The relationships shown in Figure 15.1 include: (1) entry of power grid data by power grid graphics system; (2) line loss calculation by power grid topological relation; (3) acquisition of other calculation data from database; (4) storage of intermediate data and result data of line loss calculation in database; (6) display of result data of line loss calculation in power grid graphics; (6) query, statistics, analysis and summary of data.

15.1.3.2 System Characteristics

1. *Good graphical user interface, with convenient and intuitive graphical editing module.* The system provides abundant dialogue boxes. During data entry, it provides the data change curve display interface

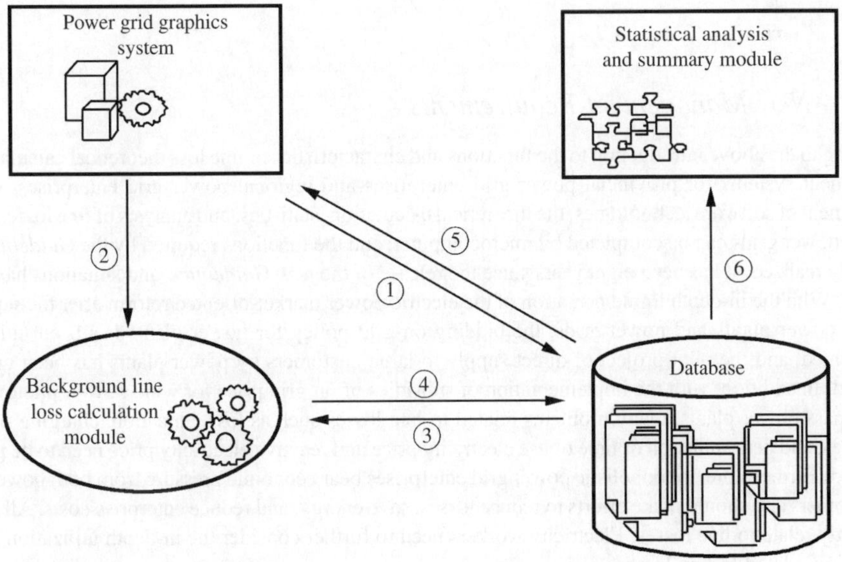

Figure 15.1 Relationships between three functional modules and with the database.

for users to correct errors. The system uses the graphical editing tool to generate primary system and provide intuitive topological relation, featuring high editing efficiency. Input of unit parameters is subject to visual management, ensuring the consistency and accuracy of input parameters.

2. *Fast and accurate load flow and line loss calculation function.* The load flow and line loss theoretical calculation module is compiled in a standard C++ program and extensively uses class, template, custom data type, operator overloading, and other important features of the C++ program, so that the program has good readability and inheritance, and the code execution speed is greatly improved. The system uses the sparse technique and node numbering optimization technique to save a lot of storage space and reduce the system demand for memory. Calculation without zero is adopted in iterative calculation, significantly improving the speed of load flow calculation. The system can automatically select the load flow iterative algorithm or execute combined iteration according to the topology information and parameter structure of the network, so that the load flow convergence and calculation speed can get optimal processing.

   By means of text output, this system can analyze the topological structure of power grids, check attribute parameters and operating data of various units, and provide alarms in case of error. The system can also print and output intermediate variables in the iterative process of load flow calculation, such as node voltage amplitude and degree of balance between active and reactive power, for reference by debugging personnel. The error correction function and prompt message of the system provide favorable conditions for improving load flow convergence.

   According to the actual measurement of large power grids with nearly 5000 nodes in Central China, North China, and Northeast China by this system, it takes only 30 s to complete the load flow and line loss calculations. According to a comparison with calculation results from the IEEE (Institute of Electrical and Electronic Engineers) 9-node and 300-node power grids and Beijing–Tianjin–Tangshan power grids, under the same network topology and unit parameters, the difference between the maximum load flow and line loss calculation results obtained by this system and those obtained by the internationally recognized BPA load flow calculation procedure does not exceed 0.5%.

3. *Convenient and comprehensive management of line loss data and reports.* The system adopts the design mode of data acquisition, data transmission, data auditing, and data centralized storage, and is able to store the calculation and analysis data of different power grid enterprises under different representative days together through a data loading process, automatically generating various analysis reports by a flexible condition-setting function, so that researchers can conduct statistical analyses and print them on demand.

15.1.4 New Management Requirements

According to the above introduction to the functions and characteristics of line loss theoretical calculation and management systems for provincial power grid enterprises and regional power grid enterprises, with the development of software technologies, the theoretical calculation, statistics, and analysis of line losses in most regional power grids can be completed by microcomputer, and the functions required by the *Guidelines* have been fully realized. It has been eight years since the release of the new *Guidelines*, and situations have changed a lot. With the in-depth implementation of the electric power market-oriented reform after the separation between power plants and power grids, the bidding on-grid policy for power plants has been universally implemented, and the pilot project of direct supply to large customers by power plants has been gradually promoted. In addition, with the implementation of subsidies of on-grid price for wind power, photovoltaics, and biomass power plants, many problems related to line losses such as loss allocation, charging of power transit fees, and determination of time of use electricity price and reactive electricity price need to be properly addressed. Furthermore, monopolistic power grid enterprises bear economic pressure from both power plants and customers and should make efforts to reduce losses, save energy, and reduce enterprise costs. All of these are directly relate to line losses. Electricity workers need to further consider the in-depth utilization of mass information related to line losses.

15.2 Value Creation and Support Processes for Power Grid Enterprises

15.2.1 Information-Oriented Development of Large Enterprises and Application of Enterprise Resource Planning

During the process of management modernization, many large enterprises in China have experienced long processes of management system certification and information-based management. When learning from global large enterprises, most of them resorted to the application of the Enterprise Resource Planning (ERP) system. There was no exception for the State Grid Corporation of China, which is the most important power grid enterprise in China. The information-oriented development plan SG186 (constructing an integrated enterprise information integration platform, establishing eight business application systems, and creating and improving six guarantee systems) implemented from 2006 is fully transformed from the trial implementation in regional and provincial power grid enterprises and prefectural branches to the application of an ERP system. Years of effort have witnessed significant results.

As one of the most advanced enterprise management models in the world, ERP not only reflects the world's most advanced enterprise management concept, but also provides the best solution to enterprise information integration. ERP unifies the enterprise's logistics, capital flow, and information flow for management; and it makes the most of the enterprise's existing resources, thus maximizing the enterprise's economic benefits.

15.2.1.1 Three Core Management Concepts of ERP

1. The ERP system reflects the concept of management of all supply chain resources. In the era of a knowledge-based economy, an enterprise must gather relevant parties in the operation process (such as suppliers, manufacturing plants, distribution network, and customers) into a close supply chain, so as to effectively arrange the production, supply, and marketing activities of the enterprise, and quickly and efficiently meet the needs of production and operation by utilizing all market resources in the entire society. ERP realizes the management of the enterprise's entire supply chain, and adapt to the enterprise's need of market competition in the era of knowledge-driven economy.
2. An ERP system reflects the idea of lean production, simultaneous engineering and agile manufacturing. Lean production is a new enterprise operation strategy system. There are no longer simple business transaction relationships between the enterprise and its sales agents, customers, and suppliers. Instead, they maintain benefit-sharing partnerships. The core concept of lean production is to retain and reduce costs while ensuring profits. When the enterprise is faced with specific market and product demands, its basic partners may not be able to satisfy the new product development and production needs. In this case, the enterprise will organize specific suppliers and sales channels to constitute a short-term or one-time supply chain and form a "virtual factory". The core concept of "agile manufacturing" is to organize production by simultaneous engineering, launch new products to the market in the shortest time, and always keep high quality, diversity, and flexibility of products.
3. The ERP system reflects the idea of planning ahead and in-process control. The planning system of an ERP system mainly includes production planning, material requirements planning, capacity planning, procurement plan, sales execution plan, profit plan, finance budget, and human resource planning, and these planning functions and value control functions have been fully integrated into the whole supply chain system. The ERP system defines accounting subjects and accounting methods related to transaction processing, so as to automatically generate accounting entries at the same time of transaction processing, thus ensuring synchronous recording and data consistency of capital flow and logistics.

15.2.1.2 Core Content of ERP

The core content of ERP includes the following four modules:

1. Finance management module, including financial accounting and financial management.

2. Production control management module, including master production schedule, material requirements planning, capacity requirements planning, shop floor control, and manufacturing standards.
3. Logistics management module, including distribution management, inventory control, and purchasing management.
4. Human resource management module, including decision support of human resource planning, recruitment management, wage accounting, and travel business accounting.

15.2.1.3 Profound Connotation of ERP

With the improvement of ERP systems and the development of IT technology, ERP has been given more profound connotation. It focuses on the management of supply chain and has added functions such as distribution management, human resource management, transportation management, quality management, equipment management, and decision support. It supports collectivized, cross-regional, and cross-border operations. The main purpose of ERP is to fully allocate and balance enterprise resources and help the enterprise fully play against fierce competition, thus gaining better economic benefits.

15.2.2 Value Creation and Support Processes of Power Grid Enterprises

One of the internationally recognized ways of enterprise management modernization is to establish and implement a quality management system. In the 1990s, China started to implement quality management system certification, which significantly promotes the quality management level of enterprises and public institutions in China and provides access conditions for Chinese products and building industry into the international market. At the end of the twentieth century, China established the latest version of the standard GB/T 19001-2000 *Quality Management System*, equivalent to the international standard, driving the construction of quality management systems to a new and higher stage. Now, the latest version is GB/T 19001-2008 (ISO 9001-2008), equivalent to the 2008 version from ISO.

In 2004, China formulated the voluntary standard GB/T 19580-2004 *Criteria for Performance Excellence* on the basis of the experience in quality management we learnt from the United States, Japan, and European countries. From the perspective of informatization-driven management modernization, GB/T 19580-2004 attaches importance to the enterprise strategic planning which clearly illustrates the vision and objective centering on customer, market, and finance to staff. The identification, design, implementation, and improvement of the value creation process and support process are realized through the development, utilization, and optimal configuration of six kinds of resources, namely human resource, financial resource, infrastructure, information, technology, and related party relationship. New operating results can be achieved in customer and market, finance, resources, process effectiveness, organization governance, and social responsibility through continuous efforts for the pursuit of excellence. Compared with the business system planning (BSP) method for information system construction, the strategic data planning method, and the information engineering method, GB/T 19580-2004 can become the most specific guideline to the planning and construction of DSS.

In terms of the enterprise, the foundation of generating performance is the identification, design, implementation, and improvement of the value creation and support processes. GB/T 19580-2004 clarifies connotations of the value creation process and the support process.

The value creation process is the process of creating benefits for the customers of an organization and for the operation of the organization. It is the most important process for organization operation, and most employees are involved in this process to generate the products and services of the organization and bring actual operating results to the shareholders and other major stakeholders of the organization.

The support process is the process of supporting the daily operation, production, and service delivery of the organization, and includes finance and statistics, equipment management, legal service, human resource service, public relations, and other administrative services. This process does not add value or create value for customers directly, but guarantees and supports the implementation of the value creation process. The organization needs to identify all support processes and to determine key support processes if necessary.

The production and operation processes of a power grid enterprise can be summarized as: (i) the transmission, transformation, distribution, and sales of electric energy purchased from regional power grids or regional power plants by maintaining transmission and transformation equipment in good order; (ii) the timing measurement of electricity consumption of various customers and the collection of fees; (iii) the final generation of enterprise profit, deducting power purchase expense, equipment depreciation, maintenance, staff wages and welfare, and other expenses. The core management processes of the enterprise include power purchase management process, electricity sales, and service implementation process, production equipment operation and maintenance process, financial management process as currency reflection of the production and operation process, human resource management process as support to the production and operation process, and the power grid development and construction management process. The enterprise cultural construction can be considered in the human resource management process; the original logistics management process can be omitted through logistics socialization. Therefore, the three processes of power purchase management, electricity sales and service implementation, and production equipment operation and maintenance can be identified as the value creation processes; the three processes of power grid construction, human resource management, and financial management can be identified as support processes; the power grid enterprise model represented by matrix of the above six core business processes is shown in Figure 15.2, where solid lines refer to correlations between value creation processes and with support processes, and dotted lines refer to the role of support processes in supporting and guaranteeing value creation processes [49].

15.3 Composition of Model Driven Decision Support System

15.3.1 Structure and Functions of Decision Support System

The decision support system (DSS) emerged in the middle of the 1970s. It is generally believed that DSS combines and utilizes the powerful information processing capability and flexible judgment capability of computer and supports decision makers in solving semi-structured and unstructured decision problems in an interactive manner [50]. In the 1980s, a model base and method base were added in DSS, constituting the triangular distribution structure composed of database, model base, and method base, as shown in Figure 15.3.

15.3.1.1 Dialogue Management Subsystem

The dialogue management subsystem serves as the interface between the user and computer in DSS and plays an important role in transmitting and transforming commands and data between the operator, model base, method base, and database. Its core is the man–machine interface. The system user wants to benefit from the system output but learns little about the system, so the man–machine dialogue subsystem serves the DSS window. Whether it is strong or weak directly represents the level of the system.

15.3.1.2 Database Management System

The database management system stores, manages, provides, and maintains data used for decision support and serves as the foundation for supporting the model base subsystem and the method base subsystem; it consists of database, data extraction module, data dictionary, database management system, and data query module. The database storing specific business system data is the source database. The difference between the source database and DSS database is that data stored in the former is for a single purpose and is utilized in a lower level, while data stored the latter basically can be used for decisions and is information in a true sense.

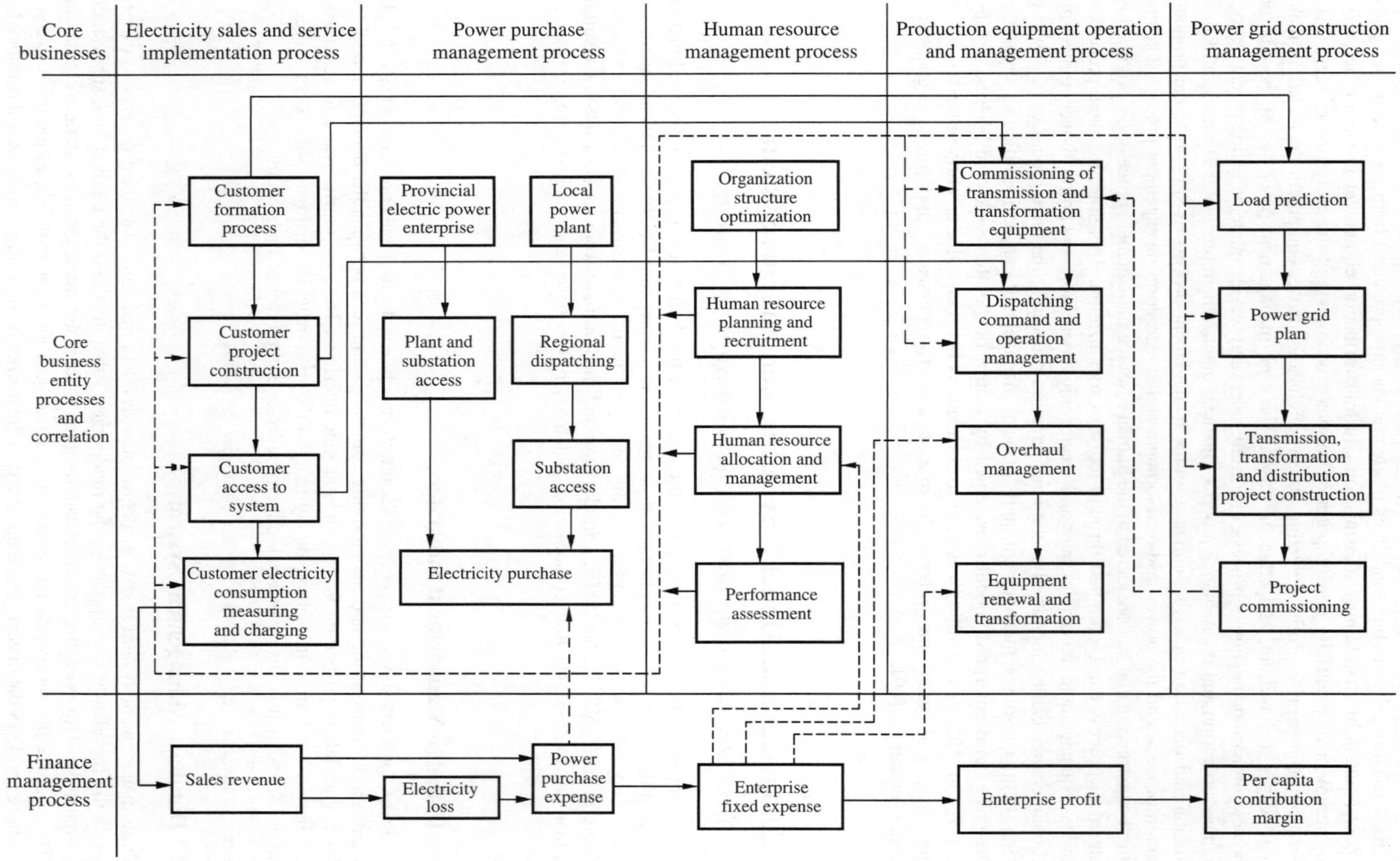

Figure 15.2 Power grid enterprise model (matrix of core business processes).

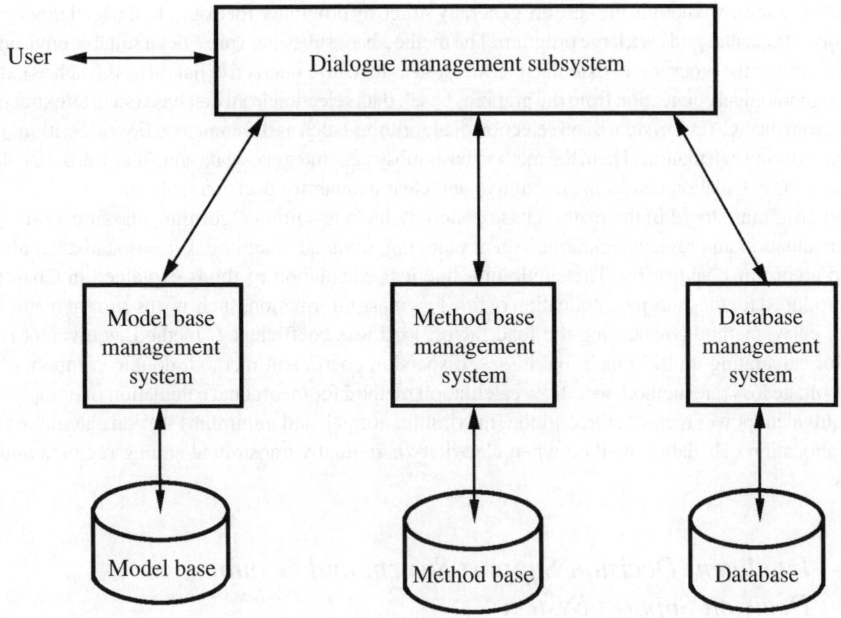

Figure 15.3 Triangular structure of DSS.

The function of the data extraction module is to extract data from the database that can be used for decision support, and the extraction process features the selection, enrichment, and transformation of data in the source database.

15.3.1.3 Model Base Management System

The model base management system is a computer software system used to build and manage models. It is the most complicated and most difficult to be realized in DSS and is composed of a model base and a model base management system. The model base stores basic model modules, cell models, and their relationships which can be shared by various decision problems or be used for specific problems. In terms of economy, decision support models can be classified into the following types by cell model structure:

1. Prediction model, such as electric quantity prediction model, and load (power) prediction model.
2. Comprehensive balance model, such as electric quantity balance model of various classes of voltage systems, and electric quantity balance model of line losses of loss calculation units.
3. Structure optimization model, such as power sales structure optimization model, and time of use power purchase and sales optimization model.
4. Economic control model, such as cost control model necessary for lean production.

Basic model elements are stored in the model base mainly by four means (subprogram, statement, data, and logical relationship), where the logical way is mainly used for the intelligent decision support system introduced below.

15.3.1.4 Method Base Management System

The method base management system stores, manages, calls, and maintains generic algorithms, standard functions, and other methods to be used in DSS components; it is composed of a method base and a method base management system. Methods in the base are generally stored by programs; for example, the load flow calculation method for complicated grids is a large program. The method base subsystem provides a suitable environment for DSS by describing the programs of external interfaces, so as to realize interactive data access (such as extracting a power grid topological connection from the graphics base), data selection from database (such as extracting operating data from the SCADA system), and selection of algorithms (such as different load flow algorithms) from the method base during calculation. Then, the method base subsystem integrates data and algorithms for calculation, and displays the calculation results in an intuitive and clear manner for decision makers.

Method programs stored in the method base generally include sorting algorithm, classification algorithm, program evaluation and review technique, linear planning, dynamic planning, various statistical algorithms, combined algorithms, and so on. The single-unit line loss calculation methods explained in Chapters 1–13 can be introduced for the in-depth application of line loss mass information, such as rms current method, equivalent load curve method considering the load factor, load loss coefficient C method, equivalent resistance method for calculating multi-branch line losses, dispersion coefficient method, double component balance method, voltage loss rate method, load flow calculation method for theoretical calculation of complicated grid losses, equivalent power method, three-mode (maximum, normal, and minimum) section calculation method, and loss allocation calculation method when electricity is mutually transmitted among regional and provincial grids.

15.3.2 Intelligent Decision Support System and Group Decision Support System

15.3.2.1 Intelligent Decision Support System

The intelligent decision support system (IDSS) is formed based on the traditional DSS in combination with the expert system (ES). The ES is a knowledge system used to explain and solve problems through a computer by utilizing expert knowledge and knowledge reasoning.

With a knowledge base as the core, IDSS introduces a heuristic method and other artificial intelligence calculation methods on the basis of models of numerical calculation and allows partial or more qualitative analysis tasks originally undertaken by operators to be done by computer. The reasoning mechanism of knowledge can obtain new knowledge and the accumulation of knowledge helps with the constant enhancing of system capabilities. The man–machine dialogue subsystem of IDSS adopts natural language processing technology to form an intelligent man–computer interaction. Such an interactive interface enables users to raise decision problems in natural language. The natural language processing function converts them into problem descriptions that can be understood by a computer, which then delivers solutions. With natural language processing technology, decision makers do not have to rely on computer professionals, but can use IDSS directly.

In a typical IDSS structure, knowledge base and inference engine are added on the basis of the traditional three-base DSS, and a natural language processing system is added in the man–machine dialogue system. In addition, a problem-processing system is added. In this way, they constitute a "four-base" system structure.

15.3.2.2 Group Decision Support System

With the development of economic regionalization and globalization, a group decision support system (GDSS) has emerged against the backdrop where several decision makers seek for the optimal solution to a problem through asynchronous and allopatric cooperation and consultation within a period. Based on DSS and by means of computer networks and communication technology, GDSS enables several decision makers to collaboratively seek semi-structured or unstructured solutions to a decision problem by complying with common standards and specifications.

Compared with traditional DSS, GDSS must be established in an LAN or WAN; and a rules base, communication base, shared common database, model base, and method base are then added to GDSS. The State

Grid Corporation of China which has established a WAN; and regional power grids composed of several provincial power grids have conditions for establishing GDSS. There are objective requirements for establishing GDSS for the purpose of safe and economical operation of grids over a large range. Therefore, it is possible and necessary to establish GDSS within a power grid enterprise.

15.3.3 Conceptual Model of Power Grid Enterprise

The conceptual model of a power grid enterprise should be able to reflect the main production and operation characteristics of the enterprise and be able to evaluate the operating results of the enterprise. It can be a group of mathematical models reflecting production and operation characteristics, or a comprehensive mathematical model reflecting enterprise characteristics by total quantity or unit quantity. As the core model of DSS, the enterprise conceptual model plays an important role in the top-down structural design of a decision support system.

15.3.3.1 Conceptual Model of Power Grid Enterprise with Total Power Sales Quantity as Core

Assume that x is total power sales quantity under various voltages; y is average unit price of power sales under various voltages (called average sales price for short); z is comprehensive line loss rate; a is average unit price of power purchase calculated according to different unit prices of power purchase and electric quantities under different voltages and different time of use periods; b is enterprise fixed expense (including depreciation, maintenance fee, personnel wages and welfare, taxes not related to power sales quantity, and other fees); c is tax rate of taxes imposed on power sales; W is enterprise profit. The conceptual model of the power grid enterprise can be obtained as shown in Formula (15.1)

$$W=(1-c)xy-\frac{ax}{1-z}-b$$

In other words, the pre-tax profit (profit before payment of income tax) of the enterprise is equal to the sales revenue after proportional tax minus the variable cost (power purchase expense) minus the fixed expense.

15.3.3.2 Conceptual Model of Power Grid Enterprise Per Unit Power Sales Quantity

Formula (15.1) is divided by the power sales quantity x on both side to derive the formula of profit per unit power sales quantity

$$w_0=W/x=(1-c)y-\frac{a}{1-z}-b_0 \qquad (15.1)$$

Wherein b_0 – fixed expense per unit power sales quantity [Yuan/(kW·h)];

$\frac{a}{1-z}$ – average unit price of power purchase considering line loss factor, known as "unit price of power purchase with loss" [Yuan/(kW·h)]for short.

In other words, the profit per unit power sales quantity is equal to the average price of power sales after proportional tax minus the unit price of power purchase with loss minus the fixed expense per unit power sales quantity.

15.3.3.3 Basic Analyses which can be Realized by the Conceptual Model of Power Grid Enterprise

According to Section 13.1 of Chapter 13, the conceptual model of power grid enterprise can be used to conduct the following basic analyses: (i) break-even point analysis, as shown in Formula (14.2); (ii) analysis of profit and tax amount per unit power sales quantity, as shown in Formula (14.3); (iii) analysis of factors affecting profits, as shown in Formula (14.4); (iv) analysis of strong correlation of line losses to profits, as shown in Formula (14.5), which indicates that the effect of increase in profits by reducing the line loss rate depends on three factors of power sales quantity, average unit price of power purchase and original line loss rate, revealing the special significance of loss reduction for profit increase.

15.3.4 Business Conceptual Model of Power Grid Enterprise

For all core business processes, the power grid enterprise can create business conceptual models reflecting the basic features of these processes, and the models may be simple or complex. In benchmarking work conducted in recent years, the State Grid Corporation of China listed many indicators which provide basic conditions for the combination and establishment of various business conceptual models.

15.3.4.1 Business Conceptual Model of Value Creation Process – Decomposition of Per Capita Contribution Margin

As one of the core indicators put forward by the State Grid Corporation of China in benchmarking, the per capita contribution margin is an integrated indicator used to measure the enterprise profitability and overall quality of staff, so it is suitable to use it as a core content of the business conceptual model in the value creation process. Because this indicator is a ratio, the continued multiplication of a ratio can be used to decompose and expand this indicator. In other words,

$$\text{Per capita contribution margin} = \text{gross margin/headcount} = \frac{\text{gross margin}}{\text{gross sales}} \times \frac{\text{gross sales}}{\text{total assets}} \times \frac{\text{total assets}}{\text{headcount}}$$
$$= \text{pre-tax profit ratio of sales} \times \text{sales per unit asset} \times \text{per capita asset}$$

The formula expressed by symbols is

$$w_{\text{rj}} = \frac{W_{\text{sh}\cdot\text{q}}}{yx} S_{\text{z}\cdot\text{sh}} z_{\text{rj}} \tag{15.2}$$

Wherein $W_{\text{sh}\cdot\text{q}}$, $S_{\text{z}\cdot\text{sh}}$, z_{rj} – pre-tax total profit, sales revenue per unit asset, per capita asset value;
y – average unit price of power sales [Yuan/(kW·h)];
x – power sales quantity (kW·h).

Substitute the pre-tax profit calculation Formula (15.1) into Formula (15.2) to obtain

$$w_{\text{rj}} = [(1-c) - a'/y - b_0/y] S_{\text{z}\cdot\text{sh}} z_{\text{rj}} \tag{15.3}$$
$$a' = a/(1-z)$$

Wherein c – proportional tax rate, only including urban construction tax and additional tax of education;
a' – unit price of power purchase considering line losses;
z – line loss rate.

In Formula (15.3), inside the square brackets is a ratio; $S_{\text{z}\cdot\text{sh}}$ is also a ratio; z_{rj} is an actual value.

Formula (15.3) is the decomposition result of the per capita contribution margin indicator and is the business conceptual model in the value creation process of the power grid enterprise. According to

Formula (15.3), in addition to increasing the average unit price of power sales y, and reducing the unit price of power purchase with loss a' and the fixed expense per unit power sales, increasing the sales per unit asset, reducing the headcount and improving the per capita asset value are effective ways to improve the per capita contribution margin.

15.3.4.2 Support to Value Creation Process

1. *Power grid construction management process.* The actual performance of the invested annual construction fund is mainly analyzed in terms of the following indicators

$$\text{Rate of performance } K_{\text{jsh}} \text{ of construction fund in current year} = \text{settled fund of operated projects in current year/used construction fund in current year} \tag{15.4}$$

Unit investment benefit W_{jsh} from grid construction = added power supply benefit from operated equipment in current year/average revenue in payback period of operated projects in current year (15.5)

Numerator in Formula (15.5) = added electric supply from newly operated equipment in current year × power supply reliability rate × (1 – repeated electric supply rate) × (1 – z_2) × average link price difference in production system; z_2 is average line loss rate of operated equipment in current year.

$$\text{Power grid construction process efficiency } W_{\text{jsh.Gsh}} = K_{\text{jsh}} W_{\text{jsh}} \tag{15.6}$$

2. *Human resource management process.* The contribution of human resource increase to enterprise benefit increase is mainly analyzed as follows

$$\text{Human resource increase coefficient } K_{\text{RL}} = \text{personnel equivalent density at end of current year/personnel equivalent density at end of last year} \tag{15.7}$$

In Formula (15.7), personnel equivalent density refers to the ratio between the sum of education background, academic degree, professional title, and skill level of each permanent employee which are converted by relevant coefficients and the number of permanent employees in the enterprise. Wherein: conversion coefficients of education background and academic degree: doctoral candidate (including doctoral degree) 1.5, master degree candidate (including master degree) 1.2, undergraduate (including bachelor's degree) 1.0, junior college 0.80, technical secondary school, technical school and vocational high school 0.60, high school or below 0.40; conversion coefficient of professional title: senior 1.2, intermediate 1.0, junior 0.60, no title 0.40; conversion coefficient of skill level: senior technician 1.30, technician 1.0, senior worker 0.80, intermediate worker 0.60, junior worker or below 0.40.

$$\begin{aligned}\text{Rate of human cost contribution} &= K_{\text{RL}} K_2 \text{ rate of profit creation per unit wage}\\ &= K_{\text{RL}} K_2 \times \text{cost} - \text{profit ratio} \times \text{cost} - \text{wage ratio}\\ &= K_{\text{RL}} K_2 (\text{enterprise total profit/total cost})\\ &\quad \times (\text{total cost/total wages and welfares})\end{aligned} \tag{15.8}$$

Wherein K_2 – corporate culture appreciation coefficient, evaluated by the management team year by year.

15.3.4.3 Modification and Integration of Model Group

The business conceptual model mentioned in this book is just a preliminary exploration. The coordination and consistency of calculation results of various business conceptual models need to be checked by macro-indicators such as enterprise total profit, and to be gradually realized through various coefficient adjustments. This is what is done during operation and commissioning of DSS and is an inevitable process to form the enterprise decision support knowledge base.

15.4 Utilization of Line Loss Mass Information

15.4.1 Basic Concept of Data Warehouse

15.4.1.1 Definition and Characteristics of Data Warehouse

The data warehouse (DW) is a decision subject-oriented, integrated, time-varying, non-volatile, and read-major data set, and generally can be divided into Web data warehouse, parallel data warehouse, multi-dimensional data warehouse, and packed data warehouse.

Decision subject orientation means that DW, centering on some subjects, eliminates useless data to decision, and provides a concise overview specific to the subjects. Integrating data from one or more disparate sources creates a DW with consistent naming convention, coding structure, and attribute quality. Time-varying means data storage provides historical information, and DW implicitly or explicitly includes time elements. Non-volatile means DW always physically separates stored data and needs no transaction processing, recovery, and concurrency control due to such separation. Read-major means DW data are mainly queried for decision-making and generally do not have to be updated in a timely manner. Such data can be refreshed regularly or on demand. Generally DW only needs two types of data processing, namely data initial loading and data access.

15.4.1.2 Conceptual Structure and Hierarchical Structure of Data Warehouse

The conceptual structure of DW includes data source, data staging area, data warehouse databases, data market (or mart), knowledge mining base, and a variety of management tools and applications.

The overall hierarchical structure of DW is generally divided into DW basic function layer, DW management layer, and DW environment support layer.

The DW basic function layer includes extracting data from data source, transforming extracted data, and loading transformed data to DW, known as the extract/transformation/load (ELT) process; establishing data markets (marts) according to the needs of users; completing DW functions such as complex query, decision analysis, and knowledge mining.

The DW management layer includes data management and metadata management and is mainly responsible for extracting, transforming, loading, updating, and refreshing data in DW. Metadata refers to critical data generated during the creation of DW regarding data source definition, object definition, and transformation rule, so it is "data about data".

The DW environment support layer includes data transmission and the data warehouse base.

15.4.1.3 Data Organization and Basic Data Model of DW

The data organization of DW is different from a database and is usually implemented in a hierarchical way, generally including four levels of early detailed data, current detailed data, mildly integrated data, and highly integrated data. Highly integrated data is completely refined and is a kind of quasi decision data. The organization structure of the four levels of data is uniformly completed by metadata.

The basic data model of DW includes a star-like model (fact-centered), a snowflake model (multi-layer diffusion), and a multi-dimensional model (cube).

15.4.1.4 Design Steps of DW

DW design generally has the following nine steps: collect and analyze business requirements; build the data model and physical design of DW; define the data source; choose the DW technology and platform; extract data from the operational database, transform data and load data to the DW; choose access and reporting tools; choose the database connection software; choose data analysis and data display software; update the DW.

15.4.2 Basic Concepts of Data Mining and Online Analysis

15.4.2.1 Concept and Process of Data Mining

Data mining (DM) means extracting information and knowledge that are unknown but have potential value from a large quantity of incomplete, noisy, fuzzy, and random practical application data.

The process of DM is as follows: as for the identified business object, clearly define the business problem and recognize the purpose of DM; prepare data, including data selection, data preprocessing, and data transformation; mine the transformed data with an applicable algorithm; analyze, interpret, and evaluate the result (visualization technology is usually used to display the result of analysis); integrate the knowledge resulted from analysis into the organization structure of the business information system for knowledge assimilation.

15.4.2.2 Relationship and Difference Between DW and DM

1. *Relationship.* DW provides better and wider data sources for DM; DW provides a new support platform for DM; DW provides convenient conditions for better usage of DM. DM provides better decision support for DW; DM puts forwards higher requirements for the data organization of DW; DM provides wider technical support for DW.
2. *Difference.* DW is a kind of data storage and data organization technology and provides data sources; DM is a kind of data analysis technology and conducts in-depth analysis of data in DW.

15.4.2.3 Concept and Characteristics of Online Analysis Processing

Online analysis processing (OLAP) is a kind of software technology and enables analysts, managers or operators to rapidly, consistently and interactively access the information that is transformed from original data, can be really understood by users and truly reflect enterprise characteristics from multiple perspectives, thus getting a deep understanding of the data.

OLAP has the following characteristics: rapid – OLAP can respond to the users' most analytical requests in 5 s; analyzable – OLAP can process any application-related logical analysis and statistical analysis; multi-dimensional – OLAP must provide a multi-dimensional view and analysis of data, including full support to hierarchy dimension and multiple hierarchy dimension; informative – OLAP must be able to obtain information in time and manage high-capacity information regardless of the quantity of data and the place where data is stored.

15.4.2.4 Technical Core of OLAP – Basic Concept of "Dimension"

1. *Concept of dimension.* Dimension is a specific perspective from which people observe data and is a kind of attribute for considering or analyzing problems; "dimension" can be defined as a set of attributes, such as time dimension, geographical dimension, product dimension, and price dimension.
2. *Levels and members of dimension.* The specific perspective from which people observe data is a dimension. Dimension can also be described from various aspects with different detailed degrees, known as levels of dimension. For example, day, month, quarter, and year are four levels of the time dimension.

   A member of a dimension is one value of this dimension and means a description of the position of a data item in this dimension. For example, some year, some month, or some day is the description of a position in the time dimension and is one value of the time dimension.

3. *Multi-dimensional array and data unit.* Multi-dimensional array refers to a combination of dimensions and variables. A multi-dimensional array can be expressed by dimension 1, dimension 2, … dimension *n*; variables, such as time, region, product, sales volume. Data unit, also called unit cell, is value of a multi-dimensional array, such as February 2006, Taiyuan, Printer, RMB 100 000.

15.4.3 Application of Data Warehouse Technology in Electric Power Dispatching and Marketing Systems

The electric power system is one of the industries with high application level of informatization, and electric power dispatching and marketing systems are the biggest two departments with the largest amount of information in the power grid enterprise. A good knowledge of management and the application of mass information in these two departments is of essential reference for the in-depth application of line loss mass information and the construction of an enterprise data warehouse.

15.4.3.1 Structure and Function of Electric Power Dispatching Data Platform

1. *Structure of internal and external data platforms.* According to the guiding thought and general framework of integration of dispatching system data and the requirements of security protection architecture of electric power secondary system of China National Power Dispatching and Communication Center, the State Grid Automation Research Institute put forward a reference architecture for a dispatching data platform [51] which is generally divided into internal platform and external platform, located in the production control area and the management information area, respectively. Transversely speaking, internal and external platforms execute forward and reverse isolations through dedicated physical isolation device; longitudinally speaking, the internal platform realizes longitudinal data exchange through SPDNet, and the external platform realizes longitudinal data exchange through SPTNet. To ensure overall integrity, the composition of modules is basically equivalent in internal and external platforms, both including an underlying data model, a unified dispatching system, a data acquisition system, a storage system, a data processing system, a data output system, a data transmission system, a data display system, and a data service. The reference architecture of a dispatching data platform is shown in Figure 15.4.
2. *Security protection based on isolation device.* The information network of dispatching department is usually divided into security areas I, II, III, and IV. In general, power grid monitoring is in security area I; electric power trading is in security area II; dispatching management information system (DMIS) is in security area III; and application systems are in security area IV. The data platform should be located in security area III. Because security isolation device is installed between security area II and security area III, and TCP/IP direct connection is set, the data exchange platform crossing the "isolation device" should meet the following three requirements: (i) real-time data transmission is realized from security area II to security area III, mainly used for transmitting SCADA and EMS real-time data and abnormal accident alarm, (ii) database synchronization is allowed, and the platform provides data transmission functions such as timing transmission, batch synchronization, write trigger, modify trigger, and instant call, and (iii) during file transmission, after the user designates the file directory to be synchronized in internal and external servers, the platform should automatically maintain synchronization of contents under directories of the two servers and maintain the consistency between the external and internal files within the specified time.
3. *Technologies and functions of the data platform.*
   a. Data acquisition is the input part of the data platform, and common data acquisition modes include receiving data files, such as txt, Excel, XML (extensible markup language), E language files (E language is an independently developed Chinese programming language that is suitable for domestic situations); extracting historical databases, including Oracle relational database, DB2, SQL, Server, Sybase, and so on; receiving real-time messages via specific protocol, including EMS real-time data

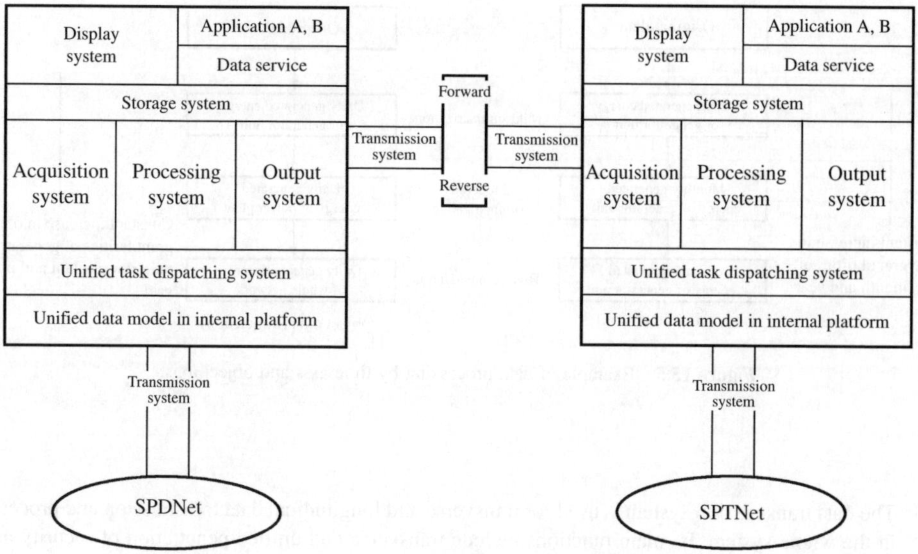

Figure 15.4 Reference architecture of dispatching data platform.

(104 protocol), quasi real-time energy data (102 protocol). To be general-purpose, the data acquisition system is usually realized by forward mapping technology, that is, specific mapping is used to satisfy requirements of diversified data formats. Data processed by the data acquisition system enters the background storage system according to unified regulations and standards.

b. The storage system serves as a concentrated place to store all data in the data platform; data types include grid model, equipment parameters, power generation data (unified dispatching, unified distribution, peak regulation, local power plant, self-generation power plant, etc.), power utilization data, power reception data, load data, system frequency data, power generation plan, circuit parameters, maintenance data, large customer load, flow gate energy, meteorological data, and so on. The key to the storage system is the data model design. The data platform generally provides a definition of the data model and management and maintenance tools (metadata management) and has instrumentalization-based model modification and expansion capabilities which are very necessary for the flexibility and expansion capability of the whole data platform. The data platform should hold a lot of quasi real-time data (1, 5, 15 min sampling data), so the data model must give consideration to the storage efficiency of the mass data and the ease of use of program processing. A preferable mode is a hybrid storage mode which realizes the balance between efficiency and user friendliness. In other words, the integrated design between the data model of the platform and the common information model (CIM) in IEC 61970 can realize an automatic synchronization of power grid models in EMS, internal platform, external platform, and DMIS, providing these systems with consistent data and allowing convenient data maintenance.

c. Data output. The output system outputs data in the storage system in the required external form through reverse mapping technology. XML and other international standardized technologies can be adopted, or the simple and efficient E language specification recommended by the China National Power Dispatching and Communication Center can also be followed. The output system can unify the external caliber of dispatch data, transmit it to the unified data center in a centralized manner, and distribute it to information nodes on demand, thus achieving better integration effect and meeting the need of enterprise decision support.

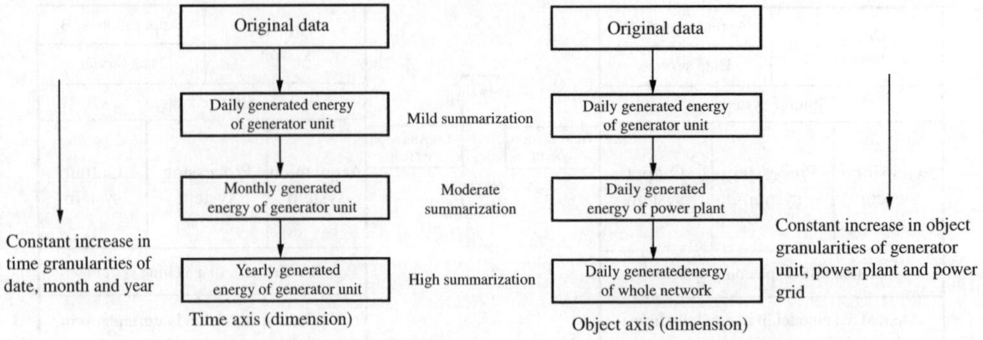

Figure 15.5 Example of data processing by time axis and object axis.

d. The data transmission system is used for transverse and longitudinal data transmission and processing in the whole system. Its main functions include transverse and unified penetration of security areas, longitudinal and unified spanning of higher and lower levels, unified authentication, encryption and decryption necessary in transmission, and guaranteeing of unified quality of service (QOS). The data transmission system supports the transmission of not only conventional files, but also data messages and other data transmission modes with high real-time requirements.

e. Data processing means the secondary processing of original data in the storage system, such as sorting, sum, average, extreme value, year on year comparison, and other operations, and processed data can be reprocessed, thus forming a processing cycle. Data processing allows the formation of a higher level of comprehensive subject data model in the storage system, and provides sufficient data resource reserves for subsequent statistical analysis, report generation, data mining (DM), and other advanced applications. Data processing is generally realized by rule-driven technology and supports user-defined rules. The principal line of data processing is attribute axis, time axis, and object axis, and various data processing algorithms basically can be decomposed to any combination of the three basic processing algorithms. An example of data processing by time axis and object axis is shown in Figure 15.5.

With regard to complex data processing requirements (such as relevant formulas of rate of increase), flexible scripting techniques can be adopted for definition and description, and the definition results are automatically executed by a dedicated script engine.

f. Data exchange and sharing. The above data acquisition, transmission, and output technologies can be collectively referred to as data exchange technology, and the formed platform can be called a data exchange platform. Combining the data exchange platform with the data storage technology based on the unified data model can form a higher level of data sharing platform, which not only meets the demands of data exchange, but also further provides a perfect data integration function, thus providing support for a wider range of data sharing.

g. Data service. The platform offers standard API interfaces for peripheral online applications and provides data service for various applications. In addition to the standard service, the platform allows interface expansion based on the Component Interface Specification (CIS) of IEC 61970 and provides non-standard cross-platform service encapsulation.

h. Data display. Based on metadata from the data model, data can be displayed from multi-angles and multi-levels through a variety of data representation and interaction means, such as form, bar chart, curve, and report. Many types of reports are generated through the data platform, and common reports include load report, electric quantity report, voltage report, loss report, load and electric quantity analysis report, planning report, power generation and utilization balance report, evaluation report, and various brief reports and daily reports.

Advanced display systems can also provide functions like data verification, data backfill, and data log, and can even provide a custom display which can be realized automatically without administrator intervention, that is a custom desktop display. This enables desktop users to get rid of administrator predefined displays and satisfies their temporary and special data requirements. It is of great importance to in-depth utilization from data market (or mart) to enterprise data.

4. *Normalization and standardization of data integration.* One objective of integrating the dispatch data is to formulate a set of dispatching specifications and standards that comply with application requirements through the construction of data platform. The core is to unify the naming specification and standard of objects (devices) and data, providing a basis for automatic data acquisition, automatic model synchronization, and automatic data correlation. The China National Power Dispatching and Communication Center established an instructive specification for naming power grid objects, giving standard names for objects in many power grids above provincial level, and providing a normative definition of names, semantic meanings, and dimensions of common data in the reports of a dispatching production system. All of these provisions should be followed in the development of the data platform. Before national industrial standards are established, regional and provincial power grid enterprises may define their own regulations and standards based on the principle of "compatible with existing standards, making standards for the future" to meet the actual needs of data integration.

15.4.3.2 Dimensional Modeling of Marketing Department

1. *Fact properties and related dimensions of electric power marketing subjects.* According to the analysis of business data and business processes of the marketing department of the power grid enterprise, relevant decision subjects can be preliminarily determined, and then factual properties and related dimensions of these subjects can be listed [52], as shown in Table 15.1.
2. *Data markets connected by common dimensions.* The dimensional model allows the formation of a bus structure for data markets (or marts) based on consistent dimensions and facts, thereby closely relating

Table 15.1 Fact properties and related dimensions of electric power marketing subjects.

Decision subject	Fact properties	Dimensions
Analysis of electric quantity	Time of use active energy, total active energy, total reactive energy	Time, customer, type of power use, voltage class
Analysis of electricity fee	Basic electricity fee, energy electricity fee, reward and punishment electricity fee, peak–valley increase, overdue fee	Time, customer, payment, preferential, voltage class
Analysis of customers	Number of customers, capacity of power use, customer credit	Time, customer, type of power use, voltage class
Analysis of load	Peak–valley load, fiducial point load, load factor	Time, customer, industry, type of power use, voltage class, weather, date type
Analysis of measurement	Quantity of measuring equipment, precision type, pre-calibration qualified rate, verification rate	Time, customer, industry, type of power use, voltage class
Analysis of business expansion	Quantity applied, quantity to be installed, capacity to be installed	Installation location, meter category, manufacturer, use state Time, customer, industry, type of power use, voltage class
Analysis of repair statistics	Rate of request for repair, workload of request for repair	Time, customer, request for repair, settling result
Analysis of compliant against power supply service	Complaint rate, complaint quantity, settling rate	Time, customer, complaint type, settling situation, satisfaction degree

different data markets through dimensions and realizing the integration of data in subjects. As for two decision analysis subjects of electric quantity analysis and electricity fee analysis, data markets of the two subjects are connected through three dimensions of customer, voltage class, and time, as shown in Figure 15.6.

In fact, the dimensional model of one subject can be used to establish a unit data market. When a new data market is established and expanded gradually to the whole power grid enterprise, the enterprise data warehouse can be established.

15.4.4 In-Depth Utilization of Line Loss Mass Information – Integration of Data in Dispatching and Marketing Systems

15.4.4.1 Analysis of DSS in Position for Power Grid Enterprise

According to analyses of the value creation process and enterprise model for a power grid enterprise, two parameters that are most closely related to the enterprise operating profit are sales energy and average sales price (also known as average price of power sales) in the marketing system. Therefore, the setting of DSS in the external data platform of the marketing system is an appropriate locating plan, so as to protect information on customers and electricity fees.

During April 2005 to March 2006, the State Grid Corporation of China conducted three representative day load measurements and theoretical line loss calculations [53] for its electric power systems according to maximum load mode, normal load mode, and minimum load mode. According to statistics and analysis, the electricity losses of 110(66) kV grids account for 17% of total losses, and the electricity losses of 220 kV and 500 (330) kV grids account for 23.2% and 8.2% in total losses, respectively. This shows that the line losses of loss calculation units managed by the dispatching system account for about 48.4% in total losses, while the line losses of loss calculation units managed by the marketing system, that is total losses of 0.38 kV low-voltage grids and 10–35 kV distribution networks, account for about 51.6% in total losses. If the proportion of electricity line losses and the work of loss reduction and energy saving are only considered for the analysis, the dispatching system and the marketing system have equivalent responsibilities within the power grid enterprise. When considering the locating of DSS in the power grid enterprise, it is reasonable and necessary to further consider the data integration of the dispatching system and the marketing system, provided one accepts that main network losses and distribution line losses are almost equal.

15.4.4.2 Planning Idea of "Two-Base" Structure for Line Loss Information of Main and Distribution Networks

The DSS of a power grid enterprise is established in the terminal link of business operation and can directly connect the overall parts of the enterprise. Just considering the in-depth utilization of line loss information, however, two data warehouses can be established in the main and distribution networks. This "two-base" structure has not only a separate work division but also a mutual connection, facilitating the in-depth utilization of information in each warehouse. According to the requirements of DSS, analysis subjects that connect "two bases" can be determined, so that one data market with analysis subjects that belong to dispatching system is connected with the other data market with analysis subjects that belong to the marketing system, thus forming a larger data market with subjects of analyzing enterprise operating results, and promoting the comprehensive construction of an enterprise data warehouse.

15.4.4.3 Analysis Subjects of Main Network Loss Data Warehouse and Distribution Network Line Loss Data Warehouse

1. *Analysis subjects of main network losses.* Taking a regional power grid enterprise as an example, the fact properties and related dimensions of analysis subjects of main network losses are shown in Table 15.2.

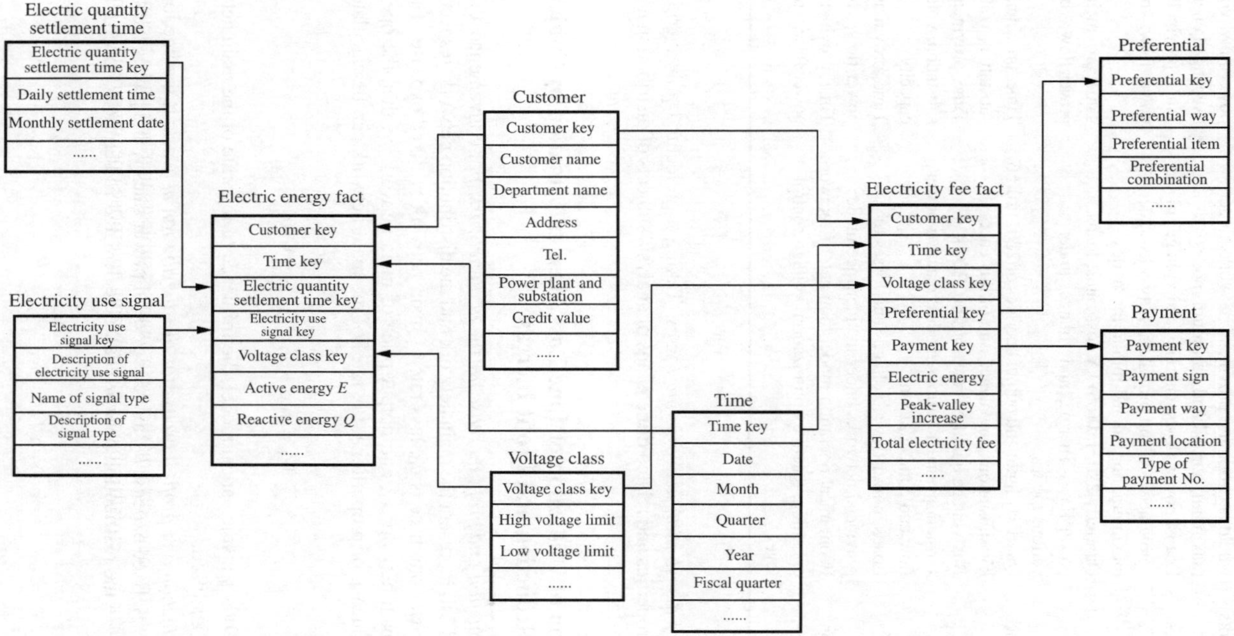

Figure 15.6 Two data markets (or marts) connected by common dimensions.

Table 15.2 Fact properties and related dimensions of analysis subjects of main network losses.

Decision analysis subject	Fact properties	Dimensions
Analysis of loss allocation among provincial networks	Loads, power flows, energies and voltages of the whole network and provincial networks under maximum, normal and minimum modes	Time, provincial network, load, power flow, voltage, energy, exchanged energy
Analysis of 500 kV losses	Loads, power flows, energies and voltages of the whole network and provincial networks under maximum, normal and minimum modes	Time, provincial network, load, power flow, voltage, energy
Analysis of loss rate by voltage	Quarterly 220, 110(66) kV loss rates; loads, power flows, energies and voltages under various modes	Time, provincial network, load, power flow, voltage, energy
Analysis of energy balance	Quarterly input and output energies of 220, 110(66) kV substations; unbalance rate of bus energy	Time, provincial network, substation, voltage, energy
Analysis of voltage and reactive power	Bus voltage qualified rates of 500, 220 kV substations under various modes; reactive power flow and compensation	Time, provincial network, substation, voltage, compensation capacity
Analysis of losses under maintenance mode	Loads, power flows, voltages and losses of provincial networks under maintenance mode	Time, provincial network, load, power flow, voltage, loss
Analysis of loss reduction and energy saving in the main network	Economical operation mode, loop net, off-network operation, reactive optimization, voltage control level	Time, provincial network, load, power flow, voltage

2. *Analysis subjects of distribution network line losses.* Taking a prefectural power grid enterprise as an example, the fact properties and related dimensions of analysis subjects of distribution network line losses are shown in Table 15.3.

15.4.4.4 Expansion of Utilization of Line Loss Mass Information by Subjects of Loss Reduction and Profit Increase

1. *Analysis of loss reduction and profit increase for the main network.* For convenience, this paragraph uses the subscripts 1 and 2 to represent the main network and the distribution network, respectively. If the average unit price of power purchase is a_0, loss rate of main network z_1, fixed expense of main network b_1, output energy of main network x_1, and output price of main network a_1, then the operating profit (not considering the influence of proportional tax rate) of the main network can be calculated as per the following formula

$$W_1 = a_1 x_1 - a_0 x_1/(1 - z_1) - b_1 \tag{15.9}$$

   With Formula (15.9) as a mathematical model, the influence of loss rate of the main network on its operating profit can be analyzed.
2. *Analysis of loss reduction and profit increase for the distribution network.* If the line loss rate of the distribution network is z_2, fixed expense b_2, average price of power sales y, and power sales quantity x_2, then the operating profit of the distribution network can be calculated as per the following formula

$$W_2 = y x_2 - \frac{a_1 x_2}{1 - z_2} - b_2 \tag{15.10}$$

3. *Analysis of loss reduction and profit increase for the whole network by DSS.*

$$W = W_1 + W_2 \tag{15.11}$$

Table 15.3 Fact properties and related dimensions of analysis subjects of distribution network line losses.

Decision analysis subject	Fact properties	Dimensions
Analysis of line losses in various power supply areas	Quarterly electric supplies, power sales quantities and statistical line loss rates in 220, 110 kV power supply areas	Time, power supply area, energy, line loss rate
Analysis of line loss rate by voltage	Quarterly electric supplies, power sales quantities and statistical line loss rates in 35, 10, 0.38 kV power supply areas	Time, power supply area, energy, line loss rate
Analysis of electric supply structure	Quarterly secondary step-down electric supply ratio and electricity line losses of three-winding main transformer	Time, main transformer, electric supply, line loss rate
Analysis of voltage level	Monthly distribution export voltage, electric supply, power sales quantity, dispersion compensation	Time, line, electric supply, voltage, compensation capacity
Analysis of loss constituent ratio	Quarterly 35, 10 kV no-load electricity losses, load electricity losses and θ values	Time, line, electric supply, line loss rate
Analysis of comprehensive line loss rate	Quarterly electric supply with loss, electric supply without loss, line loss rate with loss, comprehensive line loss rate	Time, area, electric supply, line loss rate
Analysis of loss reduction and energy saving in the distribution network	Quarterly reduced electricity losses, average price of power purchase, average price of power sales, profit increase	Time, area, electric supply, planned line loss rate and actual line loss rate

Due to

$$W = yx_2 - a_0x_1/(1-z_1) - (b_1 + b_2) \quad (15.12)$$

$$W = yx_2 - a_0x_2/(1-z) - (b_1 + b_2) \quad (15.13)$$

Formula (15.12) calculates the electric supply of the main network according to its output energy and loss rate, thereby calculating the power purchase expense. Formula (15.13) calculates the power purchase expense of the whole network according to the final power sales quantity of the distribution network and the comprehensive line loss rate z of the whole network. The results of Formulas (15.12) and (15.13) should be consistent. Obviously, neither formula considers the actual influence of direct supply energy provided by the main network to customers.

Decision support analysis shows a change in the proportions of W_1 and W_2. This change can be compared with the change in z_1 and z_2, so from the level of enterprise one finds the degree of influence of line losses in the main network and the distribution network on the increase in enterprise profit, thus adjusting the direction of loss reduction and energy saving. This analysis of the influence on enterprise macro-benefit is obviously more thorough than a comparison of electricity losses, and this is the significance of decision support.

15.4.4.5 Expansion of Utilization of Line Loss Mass Information by Data Mining

1. *Calculation of reduced payment of electricity fees due to unqualified energy.* As early as October 1996, Reference [87] stipulated that, if a customer has explicit requirements for voltage and frequency quality in a power supply contract, 20% of electricity fees of the energy purchased by the customer within any period when the voltage quality or frequency is unqualified should be deducted for compensation. Actually there

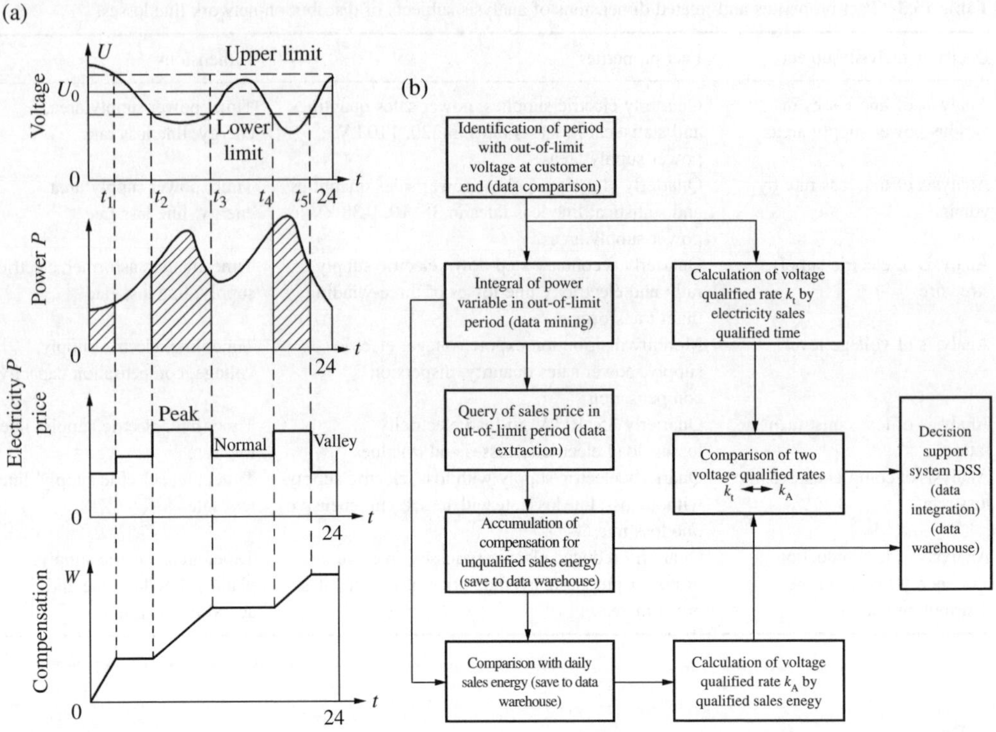

Figure 15.7 Process of integration of voltage quality and load data: (a) integration relationship dimension; (b) integration process and result.

are few such compensation cases, but this provision is reasonable from the perspective of power grid enterprises' social responsibilities. To measure unqualified energy and to calculate compensation, it is necessary to integrate the voltage, frequency, load, and electricity price of the dispatching system and the marketing system. See Figure 15.7 for the integration process.

If the ratio between qualified sales energy and total sales energy is k_A, given the characteristics of load curve of a customer sensitive to voltage quality and the provision of time of use electricity price, the average price of power sales p_{av} of the customer can be calculated, so a compensated electricity fee when the voltage quality exceeds the upper or lower limit can be calculated as per the following formula

$$\overline{W}_{psh} = 0.20 p_{av}(1 - k_A)A \tag{15.14}$$

2. *Calculation of loss reduction and energy saving benefits.* According to the analysis in Section 6.4 of Chapter 6, a higher voltage level can reduce the total electric energy losses of most distribution lines whose proportion of no-load losses is small.

According to Formula (6.1), $\Delta A = CA^2 + B = CA^2(1 + \theta)$. If it is assumed that the loss constituent ratio θ remains unchanged within a short time, then the rate of change in total losses with the load loss coefficient is $d(\Delta A)/dC = A^2(1 + \theta)$; according to Formula (6.14), the partial derivative of load loss coefficient C to voltage is $\partial C/\partial U_{av} = -(2C/U_{av})$, and substitute it into the just mentioned formula to obtain

$$d(\Delta A) \approx dCA^2(1+\theta) = \left(\frac{\partial C}{\partial U_{av}}\right) dU_{av} A^2(1+\theta) = -\left(\frac{2C}{U_{av}}\right) A^2(1+\theta) dU_{av}$$

Since $C = \Delta A/[A^2(1 + \theta)]$, substitute it into the above to obtain

$$\mathrm{d}(\Delta A) \approx -(2\Delta A/\mathrm{d}U_{\mathrm{av}})/U_{\mathrm{av}}$$

that is

$$\frac{\mathrm{d}(\Delta A)}{\Delta A} \approx -2\frac{\mathrm{d}U_{\mathrm{av}}}{U_{\mathrm{av}}} \tag{15.15}$$

According to Formula (15.15), when θ remains unchanged within a short time, the rate of relative change in electric energy losses is two times the rate of change in voltage level, and their change directions are opposite.

If the average price of power purchase is a_{av}, then the loss reduction and energy saving benefits of distribution lines after the voltage level is improved can be calculated as per the following formula

$$\overline{W}_{\mathrm{jsh}} = a_{\mathrm{av}}\mathrm{d}(\Delta A) \approx 2a_{\mathrm{av}}\Delta A \delta U_{\mathrm{av}} \tag{15.16}$$

3. *Analysis of benefits from information integration and voltage control for several voltage-sensitive customers.* If several voltage-sensitive customers are in different distribution lines of a 10 kV bus in a step-down substation and signed power supply quality contracts, it is necessary to integrate voltage qualified rates $k_{\mathrm{A}i}$ at all customer ends, average prices of power sales $p_{\mathrm{av}.i}$ of all customers during peak, normal, and valley periods, and monthly power sales quantities to the marketing system and to calculate the reduced payment $\sum_{i=1}^{n} W_{\mathrm{p.sh}.i}$ of electricity fees due to the poor quality of the outgoing voltage in the 10 kV bus.

   Total losses $\sum_{j=1}^{m} \Delta A_j$ can be obtained according to statistics of line losses in all 10 kV outgoing lines of this bus in last month, and the reduced electricity line losses can be calculated after the voltage level is improved according to the average voltage level in the previous month. Loss reduction benefits are calculated as per Formula (15.16).

   Under an acceptable voltage level, the change in total quality compensations to voltage-sensitive customers for the next month is collected and compared with the loss reduction benefits due to the increase in voltage level, so the new understanding of benefit optimization by voltage control can be obtained. As a result, the in-depth utilization of line loss mass information is necessary and possible, and this is helpful for the establishment and improvement of the enterprise operation decision support knowledge base. Enterprise lean operation is endless, and like various kinds of enterprise information, line loss mass information is expected to be utilized in an in-depth manner with renewed efforts from the managers. During the progress of information-driven modernization, line loss managers and researchers still shoulder heavy responsibilities.

Appendix A

Calculation Curve of Corona Loss Power ΔP_{cor}

Assume that E_M is conductor surface maximum electric field intensity, δ is relative air density, r is calculation radius of conductor, and n is number of conductors per phase. Then, $\frac{\Delta P_{cor}}{n}$ is corona loss power of each conductor per phase and per kilometer under all kinds of weather conditions.

With E_M/δ (kV/cm) as x-coordinate and $\frac{\Delta P_{cor}}{n}$ [kW/(km·phase)] as y-coordinate, different δr values correspond to different calculation curves of corona loss power. See Figures A.1 to A.4 for calculation curves of corona loss power under good days, rainy days, foggy days, and icy and snowy days.

Line Loss Analysis and Calculation of Electric Power Systems, First Edition. Anguan Wu and Baoshan Ni.

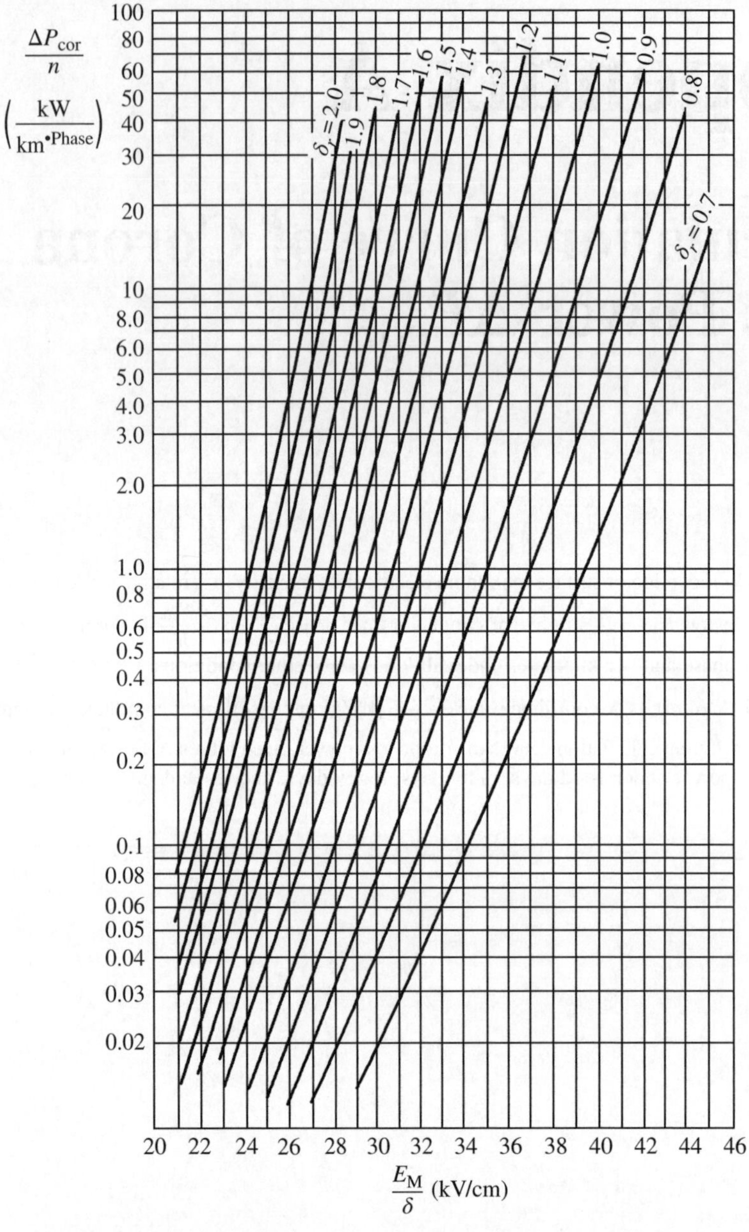

Figure A.1 Calculation curve of corona loss power (good days).

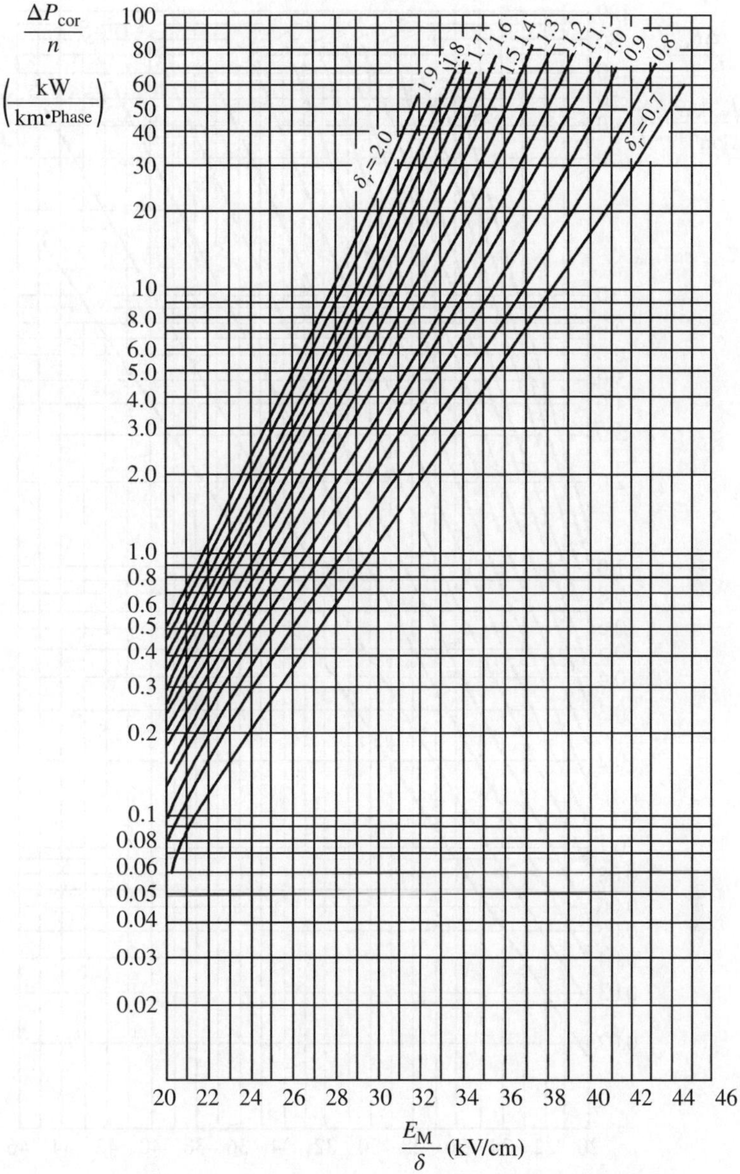

Figure A.2 Calculation curve of corona loss power (rainy days).

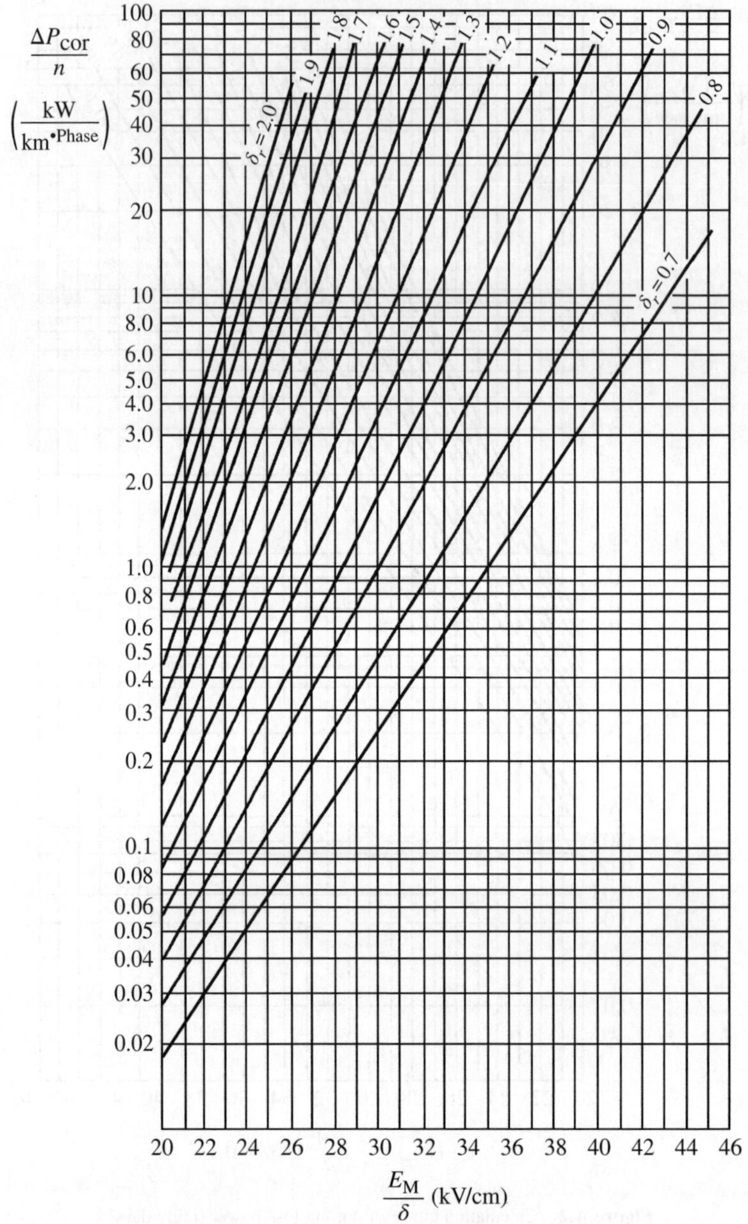

Figure A.3 Calculation curve of corona loss power (foggy days).

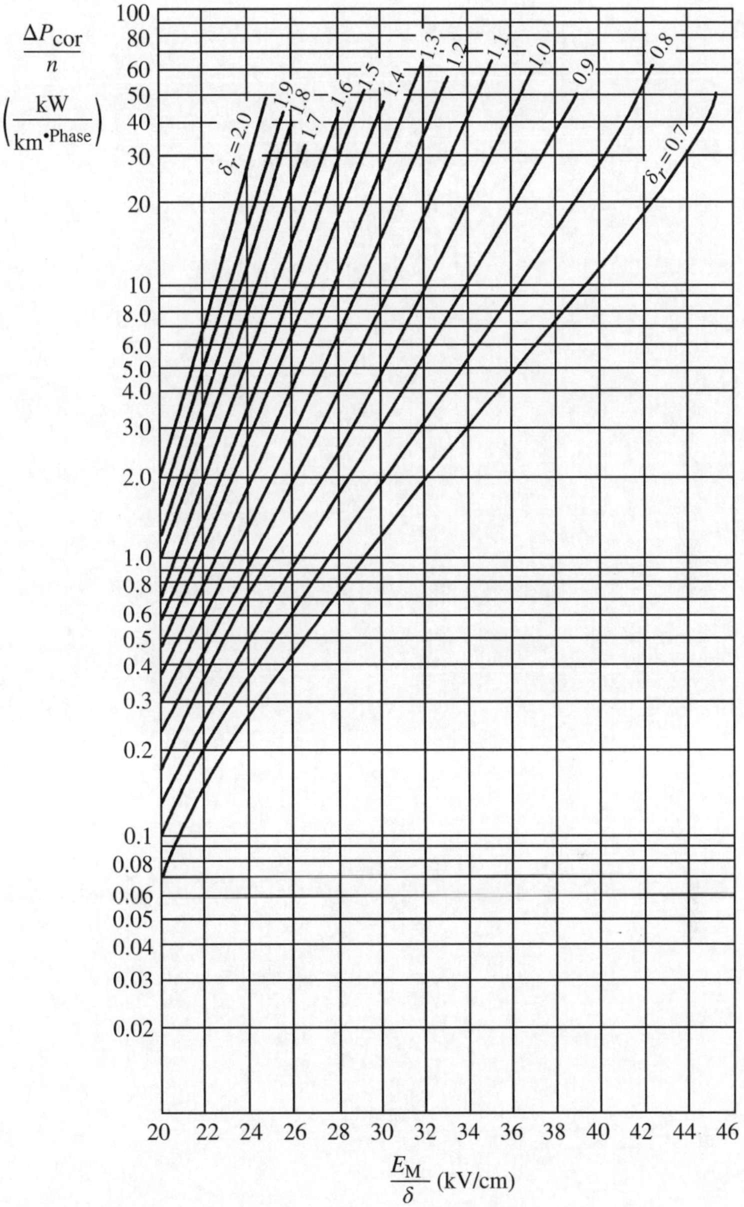

Figure A.4 Calculation curve of corona loss power (icy and snowy days).

Appendix B

Calculation of Electrical Parameters of Power Grid Units

B.1 Parameters of Overhead Lines

B.1.1 Parameters of Overhead Transmission Lines

The equivalent circuit of an overhead transmission line is shown in Figure B.1.

The parameters of the line shown in Figure B.1 are calculated as follows:

The resistance R is

$$R=\frac{r_0}{n}l=\frac{\rho}{nS}l \tag{B.1}$$

Wherein r_0 – conductor resistance per unit length (Ω/km);
L – line length (km);
n – the number of bundled conductors; n = 1 means single conductor;
ρ – calculation electrical resistivity of conductor (Ω·mm²/km); ρ = 31.5 for aluminum conductors and ρ = 18.8 for copper conductors;
S – nominal sectional area of conductor (mm²).

Reactance X and susceptance B are respectively

$$X=x_0l=\left(0.144\ 5\lg\frac{D_{\text{av}}}{r_{\text{eq}}}+\frac{0.0157}{n}\right)l \tag{B.2}$$

$$B=b_0l=\left(\frac{7.58}{\lg\dfrac{D_{\text{av}}}{r_{\text{eq}}}}\times 10^{-6}\right)l \tag{B.3}$$

Line Loss Analysis and Calculation of Electric Power Systems, First Edition. Anguan Wu and Baoshan Ni.

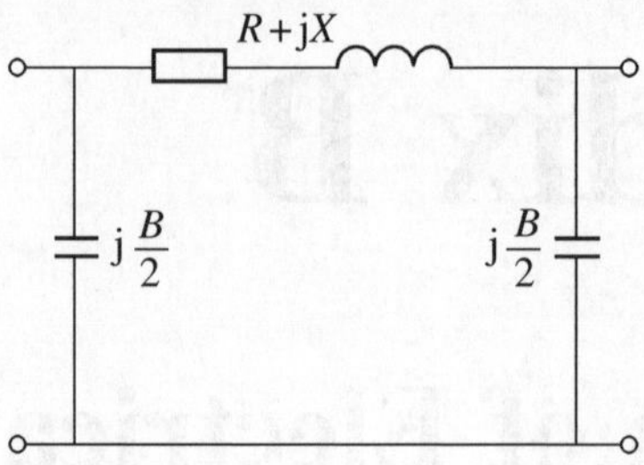

Figure B.1 Equivalent circuit of overhead transmission line.

$$D_{av} = \sqrt[3]{D_{ab}D_{bc}D_{ca}}$$

$$r_{eq} = \sqrt[n]{r a_{av}^{n-1}}$$

Wherein x_0, b_0 – reactance (Ω/km) and susceptance (1/(Ω·km) per phase and per kilometer of line;
D_{av} – average geometrical spacing (cm);
D_{ab}, D_{bc}, D_{ca} – distances among three-phase conductors (cm);
r_{eq} – equivalent radius of conductor per phase (cm);
r – calculation radius of conductor (cm);
a_{av} – geometric mean of split spacing (cm; depending on the number of bundled conductors n; see Table 8.2 for details).

If the length of the line is more than 300 km but does not exceed 1000 km, the R, X, B values calculated by the above formulas should be corrected before serving as parameters of the circuit. The correction coefficients of resistance, reactance, and susceptance are

$$K_r = 1 - \frac{1}{3} x_0 b_0 l^2 \tag{B.4}$$

$$K_x = 1 - \frac{l^2}{6}\left(x_0 b_0 - r_0^2 \frac{b_0}{x_0} \right) \tag{B.5}$$

$$K_b = 0.5 \times \frac{3 + K_r}{1 + K_r} \tag{B.6}$$

The corrected parameters are

$$R' = K_r R \tag{B.7}$$

$$X' = K_x X \tag{B.8}$$

$$B' = K_b B \tag{B.9}$$

The calculation of conductance G is omitted. If the length of the line is smaller than 80 km and the voltage is 35 kV and below, parameters G and B can both be omitted.

B.1.2 Parameters of Steel Conductor Overhead Lines

The resistance r_0 changes with current, which can be referred to in the relevant manual. The reactance is $x_0 = x'_0 + x''_0$, wherein x'_0 is the external reactance of the steel conductor, that is

$$x'_0 = 0.144\,5 \lg \frac{D_{av}}{r} \tag{B.10}$$

x''_0 is the internal reactance of steel conductor and depends on the quantity of current, which can be referred to in the relevant manual.

B.1.3 Parameters of Two-Wire One-Ground Overhead Lines

Parameters of the conductor phase of the two-wire one-ground overhead line are similar to those of the three-phase system line, that is

$$\begin{aligned} r_0 &= \frac{\rho}{S} \\ x_0 &= 0.144\,5 \lg \frac{D}{r} + 0.015\,7 \end{aligned} \tag{B.11}$$

Wherein r_0 – conductor resistance per unit length (Ω/km);
x_0 – conductor reactance per unit length (Ω/km);
ρ – resistance coefficient of conductor (Ω·mm²/km);
S – sectional area of conductor (mm²);
D – spacing between conductors (cm);
r – calculation radius of conductor (cm).

The resistance r_d and reactance x_d of the grounding phase per kilometer are respectively

$$r_d = \frac{1}{3}(r_0 + 2r_{di}) \tag{B.12}$$

$$x_d = \frac{1}{3}(x_0 + 2x_{di}) \tag{B.13}$$

Wherein r_{di} – ground resistance per unit length (Ω/km), generally $r_{di} \approx 0.05$ Ω/km;
x_{di} – ground reactance per unit length (Ω/km), generally $x_{di} = 0.43$ Ω/km for ordinary soil.

B.2 Parameters of Transformer

B.2.1 Parameters of Two-Winding Transformer

The equivalent circuit of a two-winding transformer is shown in Figure B.2.

In the equivalent circuit of a two-winding transformer, three-winding transformer, or auto-transformer, there are no-load power losses ΔP_0 and ΔQ_0 of transformer in the power supply side. The resistance R (Ω) and reactance X (Ω) in the equivalent circuit of the two-winding transformer are respectively

$$R = \frac{\Delta P_{LN} U_N^2 \times 10^3}{S_N^2} \tag{B.14}$$

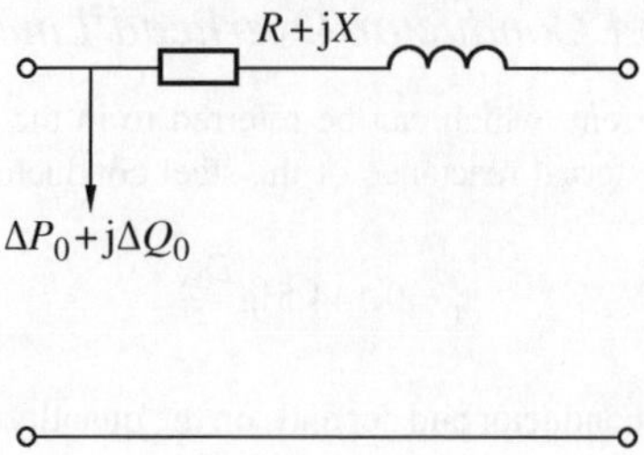

Figure B.2 Equivalent circuit of two-winding transformer.

$$X = \frac{\Delta u_x U_N^2 \times 10}{S_N} \tag{B.15}$$

$$\Delta u_x = \sqrt{(\Delta u_x)^2 - (\Delta u_r)^2} \tag{B.16}$$

$$\Delta u_r = \frac{\Delta P_{LN}}{S_N} \times 100 \tag{B.17}$$

Wherein ΔP_{LN} – rated load losses (short-circuit losses) of transformer (kW);
U_N – rated voltage of transformer (kV);
S_N – rated capacity of transformer (kVA);
Δu_x – percentage of voltage drop in transformer reactance to rated voltage;
Δu_k – percentage of short-circuit voltage (impedance voltage) of transformer[1].

As for a transformer whose capacity is above 1000 kVA, its reactance X is much larger than its resistance, so $\Delta u_x \approx \Delta u_k$, and

$$X = \frac{\Delta u_k U_N^2 \times 10}{S_N}$$

No-load reactive losses ΔQ_0 (kvar) in the equivalent circuit are

$$\Delta Q_0 = \frac{i_0 S_N}{100} \tag{B.18}$$

Wherein i_0 – percentage of no-load current of transformer.

B.2.2 Parameters of Three-Winding Transformer

The equivalent circuit of a three-winding transformer is shown in Figure B.3.

To calculate the parameters of a three-winding transformer, it is important to know the capacity ratio of windings on three sides of the transformer and the capacity under which the load losses ΔP_{LN} and the percentage of short-circuit voltage Δu_k are determined.

[1] Short-circuit voltage Δu_k is usually expressed by percentage of rated voltage, that is $\Delta u_k = \left(\frac{\Delta u_{dl}}{U_n}\right) \times 100\%$.

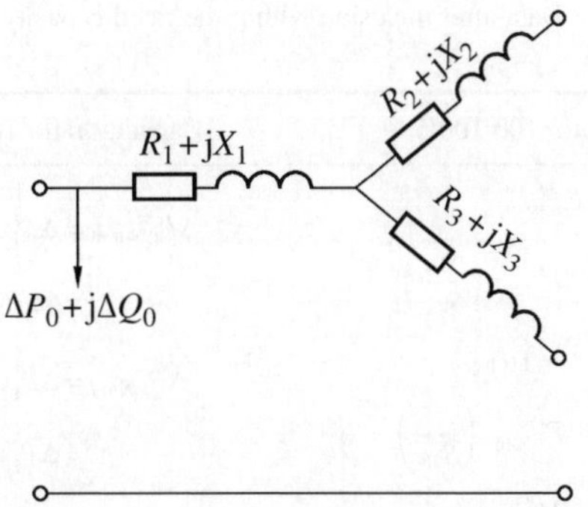

Figure B.3 Equivalent circuit of three-winding transformer.

1. When the capacity ratio of the three windings of the transformer is 100/100/100, and the load losses in the windings are $\Delta P_{LN\cdot h\cdot m}$, $\Delta P_{LN\cdot h\cdot l}$, $\Delta P_{LN\cdot m\cdot l}$, the load losses of each winding are respectively

$$\Delta P_{LN\cdot h}=\frac{1}{2}(\Delta P_{LN\cdot h\cdot m}+\Delta P_{LN\cdot h\cdot l}-\Delta P_{LN\cdot m\cdot l}) \tag{B.19}$$

$$\Delta P_{LN\cdot m}=\frac{1}{2}(\Delta P_{LN\cdot h\cdot m}+\Delta P_{LN\cdot m\cdot l}-\Delta P_{LN\cdot h\cdot l}) \tag{B.20}$$

$$\Delta P_{LN\cdot l}=\frac{1}{2}(\Delta P_{LN\cdot h\cdot l}+\Delta P_{LN\cdot m\cdot l}-\Delta P_{LN\cdot h\cdot m}) \tag{B.21}$$

The equivalent resistance of each winding that is included into the high-voltage side rated voltage U_{1N} is respectively

$$R_1=\frac{\Delta P_{LN\cdot h}U_{1N}^2\times 10^3}{S_N^2} \tag{B.22}$$

$$R_2=\frac{\Delta P_{LN\cdot m}U_{1N}^2\times 10^3}{S_N^2} \tag{B.23}$$

$$R_3=\frac{\Delta P_{LN\cdot l}U_{1N}^2\times 10^3}{S_N^2} \tag{B.24}$$

Wherein subscripts h, m, l – high-, medium-, and low-voltage sides of the three-winding transformer;
S_N – rated capacity of the transformer.

If short-circuit voltages among the windings are respectively $\Delta u_{k\cdot h\cdot m}$, $\Delta u_{k\cdot h\cdot l}$, $\Delta u_{k\cdot m\cdot l}$, then the percentage of short-circuit voltage drop of each winding is respectively

$$\Delta u_{k\cdot h}=\frac{1}{2}(\Delta u_{k\cdot h\cdot m}+\Delta u_{k\cdot h\cdot l}-\Delta u_{k\cdot m\cdot l}) \tag{B.25}$$

$$\Delta u_{k\cdot m}=\frac{1}{2}(\Delta u_{k\cdot h\cdot m}+\Delta u_{k\cdot m\cdot l}-\Delta u_{k\cdot h\cdot l}) \tag{B.26}$$

$$\Delta u_{k\cdot l}=\frac{1}{2}(\Delta u_{k\cdot m\cdot l}+\Delta u_{k\cdot h\cdot l}-\Delta u_{k\cdot h\cdot m}) \tag{B.27}$$

Table B.1 Data after inclusion within the rated capacity of the transformer.

Capacity ratio 100/100/50	Capacity ratio 100/50/100
$\Delta P_{\text{LN·h·m}} = \Delta P'_{\text{LN·h·m}}$	$\Delta P_{\text{LN·h·m}} = \Delta P'_{\text{LN·hm}}\left(\frac{S_{1N}}{S_{2N}}\right)^2 = 4\Delta P'_{\text{LN·h·m}}$
$\Delta P_{\text{LN·h·l}} = \Delta P'_{\text{LN·h·l}}\left(\frac{S_{1N}}{S_{3N}}\right)^2 = 4\Delta P'_{\text{LN·h·l}}$	$\Delta P_{\text{LN·h·l}} = \Delta P'_{\text{LN·h·l}}\left(\frac{S_{1N}}{S_{2N}}\right)^2 = \Delta P'_{\text{LN·h·l}}$
$\Delta P_{\text{LN·m·l}} = \Delta P'_{\text{LN·m·l}}\left(\frac{S_{2N}}{S_{3N}}\right)^2 = 4\Delta P'_{\text{LN·m·l}}$	$\Delta P_{\text{LN·m·l}} = \Delta P'_{\text{LN·m·l}}\left(\frac{S_{3N}}{S_{2N}}\right)^2 = 4\Delta P'_{\text{LN·m·l}}$
$\Delta u_{\text{k·h·m}} = \Delta u'_{\text{k·h·m}}$	$\Delta u_{\text{k·h·m}} = \Delta u'_{\text{k·h·m}}\left(\frac{S_{1N}}{S_{2N}}\right) = 2\Delta u'_{\text{k·h·m}}$
$\Delta u_{\text{k·h·l}} = \Delta u'_{\text{k·h·l}}\left(\frac{S_{1N}}{S_{3N}}\right) = 2\Delta u'_{\text{k·h·l}}$	$\Delta u_{\text{k·h·l}} = \Delta u'_{\text{k·h·l}}$
$\Delta u_{\text{k·m·l}} = \Delta u'_{\text{k·m·l}}\left(\frac{S_{2N}}{S_{3N}}\right) = 2\Delta u'_{\text{k·m·l}}$	$\Delta u_{\text{k·m·l}} = \Delta u'_{\text{k·m·l}}\left(\frac{S_{3N}}{S_{2N}}\right) = 2\Delta u'_{\text{k·m·l}}$

Then, the equivalent reactance of each winding that is included into the high-voltage side rated voltage U_{1N} is respectively

$$X_{\text{h}} = \frac{\Delta u_{\text{k·h}} U_{1N}^2 \times 10}{S_{\text{N}}} \tag{B.28}$$

$$X_{\text{m}} = \frac{\Delta u_{\text{k·m}} U_{1N}^2 \times 10}{S_{\text{N}}} \tag{B.29}$$

$$X_{\text{l}} = \frac{\Delta u_{\text{k·l}} U_{1N}^2 \times 10}{S_{\text{N}}} \tag{B.30}$$

2. When the capacity ratio is 100/100/50 (or 100/50/100), and the load losses and percentages of the short-circuit voltage in the windings are obtained from the test, which is conducted based on the smaller winding capacity, they must be included in the rated capacity data of the transformer before being used to calculate the equivalent resistance and reactance according to the above formulas.

   If the load losses before the inclusion are $\Delta P'_{\text{LN·h·m}}$, $\Delta P'_{\text{LN·h·l}}$, $\Delta P'_{\text{LN·m·l}}$, and percentages of short-circuit voltage drop are $\Delta u'_{\text{k·h·m}}$, $\Delta u'_{\text{k·h·l}}$, $\Delta u'_{\text{k·m·l}}$, the data after inclusion to the rated capacity of the transformer are shown in Table B.1.

   With the calculated $\Delta P_{\text{LN·h·m}}$, $\Delta P_{\text{LN·h·l}}$, $\Delta P_{\text{LN·m·l}}$ and $\Delta u_{\text{k·h·m}}$, $\Delta u_{\text{k·h·l}}$, $\Delta u_{\text{k·m·l}}$, the equivalent resistance and equivalent reactance can be calculated by the formulas for the capacity ratio of 100/100/100.
3. Auto-transformer. The parameter calculation and equivalent circuit of auto-transformers are the same as those of general two-winding or three-winding transformers.

Appendix C

Derivation of Loss Factor Formula by Subsection Integration Method

Chinese electric power scholar Liu Yingkuan derived a loss factor formula by the subsection integration method when he worked with the Xining Administration of Power Supply in 1981, which was an outstanding high-level achievement. Due to the narrow transmission of the original materials [14] few people knew about them, so this appendix quotes such materials for sharing with the readers.

It is well-known that if a load curve is given, its β and f values can be uniquely determined. However, if the β and f values are given, it is impossible to uniquely determine a load curve. Instead, a family of load curves can be determined.

As shown in Figure C.1, the double-step load curve *BCDE* represents the critical state under which a maximum change in the family of load curves occurs, while the horizontal linear load curve *AF* represents the critical state under which a minimum change in the family of load curves occurs.

The actual family of load curves is within the range of rectangles *ABCN'* and *N'DEF* formed by the two critical states. The key to identify the two rectangles is to find the projection positions of the step points in the horizontal axis. Allow $\overline{ON} = n$, and then $\overline{NK} = 1 - n = m$. Because the area of the double-step load curve should be equal to the area under the horizontal linear load curve, and the y-coordinate $\overline{OA}$ is equal to the average load f, so

$$\beta n + m \times 1.0 = f$$

Substitute $m = (1 - n)$ into the above formula to obtain

$$\left.\begin{aligned} n &= \frac{1-f}{1-\beta} \\ m &= \frac{f-\beta}{1-\beta} \end{aligned}\right\} \tag{C.1}$$

After the position of N is determined, to derive the loss factor formula comes down to the above two rectangles. Do average calculation of the integrated square of the family of load curves.

Line Loss Analysis and Calculation of Electric Power Systems, First Edition. Anguan Wu and Baoshan Ni.

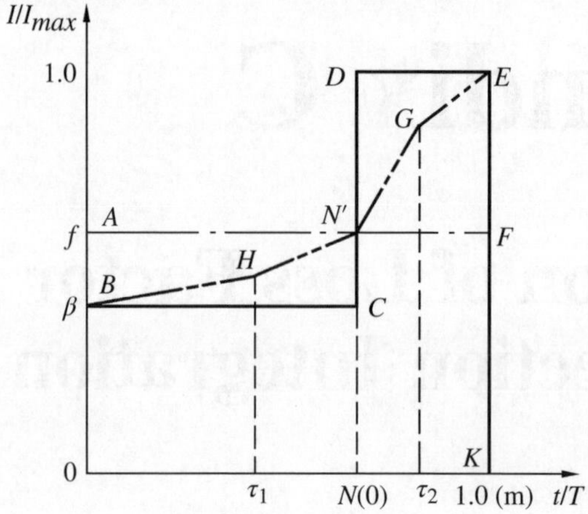

Figure C.1 Range of the family of load curves with four sections of broken lines.

Assume that a moving point $H(\tau_1, y_1)$ is in the rectangle $ABCN'$. Accordingly, there is a moving point $G(\tau_2, y_2)$ in the rectangle $N'FED$. Here, the origin of coordinates for τ_2 is N for convenient analysis. To ensure that the area under the broken lines is equal to the load factor f, the position of G must be subject to the position of H. For example, if H is located in B, G should be located in E accordingly, so a definite relationship occurs between G and H in the coordinates, that is

$$\left.\begin{aligned} y_2 &= f + \frac{n}{m}(f - y_1) \\ \tau_2 &= m\left(1 - \frac{\tau_1}{n}\right) \end{aligned}\right\} \tag{C.2}$$

According to Formula (C.2), functional expressions of the four sections of broken lines can be listed

$$\left.\begin{aligned} &\overline{BH} \text{ section} \quad I_1(t) = \beta + (y_1 - \beta)\frac{t}{\tau_1} \quad (0 \le t \le \tau_1) \\ &\overline{HN'} \text{ section} \quad I_2(t) = y_1 + \frac{f - y_1}{n - \tau_1}(t - \tau_1) \quad (\tau_1 \le t \le n) \\ &\overline{N'G} \text{ section} \quad I_3(t) = f + (y_2 - f)\frac{t}{\tau_2} \quad (0 \le t \le \tau_2) \\ &\overline{GE} \text{ section} \quad I_4(t) = y_2 + \frac{1 - y_2}{m - \tau_2}(t - \tau_2) \quad (\tau_2 \le t \le m) \end{aligned}\right\} \tag{C.3}$$

In the per unit value coordinate system, loss factor $F = \int_0^1 I^2(e)\mathrm{d}t$. To get an average of the integrated square of the family of load curves, do an integral calculation of the square of the linear function in each section by the time variable t. Then, do an average calculation of the integral interval by x-coordinate parameters τ_1, τ_2. Finally, do an average calculation of the integral interval by y-coordinate parameters y_1, y_2. See Table C.1 for the three integral processes.

Table C.1 Table of results of three integral calculations on four sections of linear functions.

Linear section	Integral of square of load curve by t $\int I_1^2(t)\mathrm{d}t$	Average of integral interval by τ $\frac{1}{(a-b)}\int_b^a \varphi(\tau)\mathrm{d}\tau$	Average of integral interval by y $\frac{1}{(c-d)}\int_d^c \psi(y)\mathrm{d}y$
$\overline{BH}$	$\int_0^{\tau_1} I_1^2(t)\mathrm{d}t=\varphi_1(\tau_1)$ $=\frac{1}{3}\tau_1(\beta^2+\beta y_1+y_1^2)$	$\frac{1}{n}\int_0^n \varphi_1(\tau_1)\mathrm{d}\tau_1=\psi_1(y_1)$ $=\frac{n}{6}(\beta^2+\beta y_1+y_1^2)$	$\frac{1}{(f-\beta)}\int_\beta^f \psi_1(y_1)\mathrm{d}y_1$ $=\frac{n}{36}(11\beta^2+5\beta f+2f^2)$
$\overline{HN'}$	$\int_{\tau_1}^{n} I_2^2(t)\mathrm{d}t=\varphi_2(\tau_1)$ $=\frac{1}{3}(n-\tau_1)(f^2+fy_1+y_1^2)$	$\frac{1}{n}\int_0^n \varphi_2(\tau_1)\mathrm{d}\tau_1=\psi_2(y_2)$ $=\frac{n}{6}(f^2+fy_1+y_1^2)$	$\frac{1}{(f-\beta)}\int_\beta^f \psi_2(y_1)\mathrm{d}y_1$ $=\frac{n}{36}(11f^2+5\beta f+2\beta^2)$
$\overline{N'G}$	$\int_0^{\tau_2} I_3^2(t)\mathrm{d}t=\varphi_3(\tau_2)$ $=\frac{1}{3}\tau_2(y_2^2+y_2f+f^2)$	$\frac{1}{m}\int_0^m \varphi_3(\tau_2)\mathrm{d}\tau_2=\psi_3(y_2)$ $=\frac{m}{6}(y_2^2+y_2f+f^2)$	$\frac{1}{(1-f)}\int_f^1 \psi_3(y_2)\mathrm{d}y_2$ $=\frac{m}{36}(11f^2+5f+2)$
$\overline{GE}$	$\int_{\tau_2}^{m} I_4^2(t)\mathrm{d}t=\varphi_4(\tau_2)$ $=\frac{1}{3}(m-\tau_2)(1+y_2+y_2^2)$	$\frac{1}{m}\int_0^m \varphi_4(\tau_2)\mathrm{d}\tau_2=\psi_4(y_2)$ $=\frac{m}{6}(1+y_2+y_2^2)$	$\frac{1}{(1-f)}\int_f^1 \psi_4(y_2)\mathrm{d}y_2$ $=\frac{m}{36}(2f^2+5f+11)$

Add up the four formulas in the four columns of Table C.1 and substitute Formula (C.1) to obtain

$$F=\int_0^1 I^2(t)\mathrm{d}t=\frac{23}{36}f^2+\frac{13}{36}(f+f\beta-\beta)$$
$$=0.639f^2+0.361(f+f\beta-\beta) \quad (C.4)$$

Appendix D

Actual Measurement Analysis of No-load Power Losses and Relationship between No-load Current and Voltage of Distribution Transformers

D.1 Actual Measurement Analysis of $\Delta P_0(U)$ of General Transformers

Test data are quoted from Reference [23], and the test samples are two 50 kVA general transformers (hot-rolled silicon-steel sheet). See Table D.1 for actual measurement data and relevant calculated values (analysis I).

If $\Delta P_{0*} = aU^{b}_{*}$, then $\lg\Delta P_{0*} = \lg a + b\lg U_{*}$. Use the correlational analysis method to obtain $\lg a = 0.035848$. So $a = 1.086$, $b = 4.041$, correlation coefficient $r = 0.9514$; the approximate formula is

$$\Delta P_{0*} \approx 1.08U_{*}^{4.0} \quad (U_{*} \neq 1.0)$$

D.2 Actual Measurement Analysis of $\Delta P_0(U)$ of Low Loss Transformers

Test data are quoted from Reference [23], and the test sample is one 100 kVA transformer. See Table D.2 for actual measurement data and relevant calculated values (analysis II).

If $\Delta P_{0*} = aU^{b}_{*}$, then $\lg\Delta P_{0*} = \lg a + b\lg U_{*}$. Use the correlational analysis method to obtain $a = 1.0118 \approx 1.012$, $b = 1.836 \approx 1.84$, correlation coefficient $r = 0.932$. So the approximate formula is

$$\Delta P_{0*} \approx 1.012U_{*}^{1.84} \quad (U_{*} \neq 1.0)$$

Line Loss Analysis and Calculation of Electric Power Systems, First Edition. Anguan Wu and Baoshan Ni.

Table D.1 Actual measurement data and relevant calculated values (analysis I).

Voltage		No-load losses		Voltage		No-load losses	
Actual measurement U(V)	$\lg U_*$	Actual measurement ΔP_0(W)	$\lg \Delta P_{0*}$	Actual measurement U(V)	$\lg U_*$	Actual measurement ΔP_0(W)	$\lg \Delta P_{0*}$
9013	–0.0462	408	–0.1437	9000	–0.045 76	248	–0.076 84
9525	–0.0223	432	–0.1189	9488	–0.022 83	265	–0.048 05
10 025	0	568	0	10 000	0	296	0
10 550	0.0222	704	0.0932	10 500	0.021 19	392	0.121 99
10 988	0.0398	952	0.2243	11 000	0.041 39	528	0.2513

Table D.2 Actual measurement data and relevant calculated values (analysis II).

Voltage		No-load losses		Voltage		No-load losses	
Actual measurement U(V)	$\lg U_*$	Actual measurement ΔP_0(W)	$\lg \Delta P_{0*}$	Actual measurement U(V)	$\lg U_*$	Actual measurement ΔP_0(W)	$\lg \Delta P_{0*}$
9193	–0.0244	280	–0.03	10 008	0.012 46	300	0
9455	–0.012 23	290	–0.0147	10 282	0.024 20	340	0.054 36
9725	0	300	0	10 463	0.031 77	360	0.079 18

Table D.3 Actual measurement data and relevant calculated values (analysis III).

U(V)	9013	9525	10 025	10 55	10 988	9000	9488	10 0	10 5	11 0
U_*	0.8991	0.9501	1.0	1.0524	1.097	0.9	0.9488	1.0	1.05	1.1
I_0(A)	0.088	0.113	0.145	0.188	0.245	0.082	0.102	0.132	0.163	0.22
I_{0*}	0.6069	0.7793	1.0	1.2966	1.6897	0.6212	0.7727	1.0	1.2348	1.6667

D.3 Actual Measurement Analysis of $I_0(U)$ of General Transformers

Test data are quoted from Reference [23], and the test samples are two 50 kVA general transformers (hot-rolled silicon-steel sheet). See Table D.3 for actual measurement data and relevant calculated values (analysis III).

If $I_{0*} = aU^{b}{}_*$, then $\lg I_{0*} = \lg a + b\lg U_*$. Use the correlational analysis method to obtain $a = 1.016$, $b = 4.973 \approx 5.0$, correlation coefficient $r = 0.997\,2$. So the approximate formula is

$$I_{0*} \approx 1.00U_*^{5.0}$$

References

1. Power Generation Teaching and Research Group of Zhejiang University, Nanjing Administration of Power Supply (1980) *Theoretical Calculation of Line Loss and Technical Measures of Loss Reduction*. China Electric Power Press, Beijing. This book was written by Ni Baoshan of Zhejiang University during 1974–1977.
2. Institute of Electrical Engineers of Japan (1984) *Electrical Engineering Manual, Volume 2 – Chapter 11: Wires and Cables* (translated by Lei Z.). China Machine Press, Beijing.
3. Mystaphae, R., Idoshina N.V., Kydiev D.A. (1982) [in Russian]. *Elektr. Stan.* **4**: 58–60.
4. Pytein K.A. (1960) [in Russian]. *Acad. Riga Nauk. Natr. USSR* **1960**: 117–127.
5. Wang X. (1963) *Discussion of calculation methods of electric energy losses of electric power system*. In: *Report of the Annual Meeting of the Chinese Society for Electrical Engineering*. Chinese Society for Electrical Engineering, Beijing.
6. Xiao Y. (1982) *Analysis of form coefficient of load curve*. In: Science and Technology Literature of Jilin Polytechnic College, Jilin.
7. Jausen B. (1926) Ùber die Querschnittsberechnung in Hochspannungs Fernleitungen. *E. T. Z.* **1926**, S 820.
8. Buller F.H., Woorow C.A. (1928) Load factor–equivalent hour values compared. Means of computing loss on line when maximum load and load factor are known. *Electr. W.*, **92**(2), 59–60.
9. Tröger R. (1920) Groβkraftubertragung. *E. T. Z.* **1920**, H46 905–H47 927.
10. Wolf M. (1931) Die Anwendung von symbolischen Belastung – skurven für Elektrizitutswerke. *E. T. Z.* **1931**, H41 1267–H41 1270; Wolf M. (1958) Arbeitsverlustfaktor Verluststunde und Verwandte Begriffe – Geschichte und Theorie. *Elektrizitäswirtschaft* **1958**, 11.
11. Kezevish, V.V. (1948) [in Russian]. Elektr. Stan. **9**: 29.
12. Raymond A. (1980) Loss factor evaluation. *Trans. and Distrib.* **3**, 70.
13. Pabla A.S. (1981) *Electric Power Distribution Systems*. Tata McGraw-Hill, New Delhi.
14. Liu Y. (1981) Calculation of theoretical losses with energy and simplified calculation method. *Information Office of Qinghai Electric Power Center Laboratory. Qinghai Electric Power* **2**: 1–12.
15. Barth J.B. (1965) [in Swedish]. *Elektroteknisk Tidsskrit* **78**: 441–453.
16. Glasynov A.A. (1947) [in Russian]. *Elektrishestvo* **1947**: 2.
17. Cweik J. (1962) Using the values of active and reactive electricity meters to calculate energy loss of electricity network. *Energetyka* **16**, 151–154, 173–177, 249–251.
18. Huang Y. (1981) Calculation of line (conductor) loss rate by current density under maximum load. *Rural Electrification* **4**: 30–37.
19. Planning Department of Dandong Electric Power Bureau. (1977) Calculation of distribution network losses. *Dandong Electric Power Communication* **1977**, 2.

20. Hu Q. (1992) Discussion of conductor selection by corona effect for 330 kV line in high attitude area. *Electric Power Technology* **12**: 52–57.
21. Zheng Z., Wang K. (1983) *High Voltage Cable Line*. Water Resources and Electric Power Press, Beijing.
22. Zheng S. (1987) *Chapter 10: Transformer, Mutual Inductor, Voltage Regulator, Reactor and Capacitor*. In: *Standard Handbook for Electrical Engineers*. China Machine Press, Beijing.
23. Wang W. (1981) Operating voltage and loss reduction and electricity saving. *Rural Electrification* **4**: 38–42.
24. Xu D. (1983) *Control of Power Supply Voltage and Energy Conservation*. Tianjin Society for Electric Power, Tianjin.
25. Xing S., Zhang S. (1988) Discussion of functional relationship between grid losses and load changes in Tianjin. *Distribution and Utilization* **6**: 28–30.
26. Wang L., Yang X., Xu G., *et al.* (1987) Optimal compensation of shut capacitor in distribution network – optimization of compensation capacity, generator unit quantity and position. *Electric Power Technology* **11**: 26–30.
27. Grainge J.J., Lee S.H. (1979) Optimum size and location of shunt capacitors for reduction of losses on distribution feeders. *IEEE Transmission and Distribution Committee of the IEEE PES Winter Meeting*, F80, 246–249.
28. Xu D. (1982) Further discussion and improvement of calculation of economical operation of power transformer. Electric power science and technology information network in Shanxi, Hebei, Inner Mongolia, Tianjin and Ningxia. *Electric Power Technology Communication* **1982**, 4.
29. Yang X. (1985) *Theoretical Calculation and Analysis of Line Losses of Power Grids*. Water Resources and Electric Power Press, Beijing.
30. State Economic and Trade Commission of the PRC (1999) DL/T 686-1999 *Guidelines for Calculation of Electric Energy Losses of Power Grids*. State Economic and Trade Commission of the PRC, Beijing.
31. (1990) *Provisions on the Management of Electric Energy Loss in Power Grids*. National Energy Administration of the PRC, Beijing.
32. Shen H. (1982) *Probability Theory Handout*. Higher Education Press, Beijing.
33. Transformation Equipment Energy Saving Special Committee, Japan (1987) *Energy Saving Measures and Energy Efficiency of Transformation Equipment*. Shenyang Transformer Research Institute, Shenyang.
34. Tang Y. (1990) *Electric Energy Metering Manual*. Henan Science and Technology Press, Zhengzhou.
35. Baylor J.S., Gustafson M.W. (1989) Power system loss calculations are updated losses are determined on the basis of load factor. *Transmission and Distribution* **1989**: 60; translated by Luo B. (1991) *1991 Collection of Electric Power Translations of the Information Office of Gansu Electric Power Test Research Institute*.
36. Hua L. (1965) *Explanation and Supplement of Overall Planning Method*. China Industry Press, Beijing.
37. Fu B. (1993) Monopolar operating corona losses of DC power transmission lines in Gezhou Dam–Nanqiao line. *Power System Technology* **3**: 16–21.
38. Chen Y., Chang Q. (1999) Study of GPS-based real-time loss calculation of electric power system. *Power System Technology* **23**(12): 24–27.
39. Ni L. (2000) Method and practice of theoretical calculation of losses by state estimation software. *Hangzhou Electric Power* **1**: 31–32.
40. Xu H., Hou J., Shi L., *et al* (2003) Theoretical calculation method of line losses of power grids based on state estimation data. *Power System Technology* **27**(3): 59–62.
41. Li C., Ding X., Liu X., *et al* (2005) Analysis method of comprehensive line losses of power grids based on real-time system data and its application. *Electric Power Automation Equipment* **25**(3): 47–50.
42. Guo Q., Sun H., Zhang B., *et al* (2007) *Study of Coordinated Secondary Voltage Control. Funding Projects of the Natural Science Foundation of China* (50595414, 50323002), Tsinghua University, Tsinghua.
43. Liu Z., Cheng H., Liu X., *et al* (2003) Method of loss allocation in electric power market of foreign typical areas. *Automation of Electric Power Systems* **27**(4): 29–31.
44. Xu H., Liu B., Yu C., *et al* (2007) Practical research of loss allocation of power grids in Central China. *Central China Electric Power* **20**(1): 6–8.
45. Hu C., Han Z. (2003) New method of loss allocation based on Shapley value. *Automation of Electric Power Systems* **27**(7): 32–35.
46. Zhou X., Du S. (2005) Loss allocation method based on kernel theory under bilateral trading electric power market. *Proceedings of the CSEE* **25**(1): 60–65.
47. Deng C., Chen H., Liu M., *et al* (2006) Development of line loss theoretical calculation and management systems for power grids in Guangdong. *Southern Power System Technology Research* **2**(3): 45–49.
48. Fan H. (2008) Integrated management system of Beijing–Tianjin–Tangshan grid line loss theoretical calculation. *Power Technology of North China* **2**: 34–38.
49. Wu J. (2008) Analysis of "information island" phenomenon in power supply enterprise and its solution. *Shanxi Electric Power* **2008** Suppl. 1.

50. Zuo M., Kuang K. (2001) *Information System Development and Management Tutorial.* Tsinghua University Press, Beijing.
51. Lin F., Hu M., Jiang Y., *et al* (2007) Electric power dispatching data platform architecture and related technologies. *Automation of Electric Power Systems* **31**(1): 61–64.
52. Ye B., Zeng W., Xiao Z. (2005) Application of data warehouse in electric power system. *Automation of Electric Power Systems* **17**(3): 10–14.
53. Yu W., Xiaong Y., Zhou X., *et al* (2006) Analysis of technical losses of power grids and loss reduction measures. *Power System Technology* **30**(18): 54–57.
54. Wu A. (2011) Three-mode division of loads under multiplying probability assumption and its application. *Distribution and Utilization* **28**(2): 1–4.
55. Wu A. (2003) The make-up of electricity price with multiple sections and its application. *Power Demand Side Management* **2003**(3): 18–20.
56. Yang Y. (2011) *Distributed Energy System.* Chemical Industry Press, Beijing.
57. Wu B., Zhang X., Fang Y. (1997) Study of insulated ground wire of ultrahigh-voltage line. *Electric Power* **30**(3): 11–12, 52.
58. Zhou X. (2001) Exploration of relationship between OPGW losses and transmission line losses. *Telecommunications for Electric Power System* **6**: 14–16.
59. Ma W., Ao M., Cui M., *et al* (2007) Test and analysis of energy-saving effect of energy-saving fittings. *Electric Power Construction* **28**(7): 81–83.
60. Zhang R., Zhao X., Li M., *et al* (2012) Promotion of use of new energy-saving conductors in transmission line. *Electric Power Construction* **6**: 89–92.
61. Zhang J., Luo S. (2002) Exploration of electric energy metering mode under influence of harmonic. *Guangdong Electric Power* **4**: 55–57.
62. Wang S. (2006) *Influence of Harmonic on Electric Energy Metering.* Master's Thesis, Shandong University, Shandong.
63. Ding H., S. N., Chen D., *et al* (2006) 30 M–35 kV/2000 high temperature superconducting cable refrigeration system. *Low Temperature Electronic Technology* **33**(4): 60–63.
64. Lin L., Xiao L. (2007) Current situation and development of high temperature superconducting transmission cables. *Electrical Equipment* **8**: 1–4.
65. Jiang R. (2009) *Cross-linked Polyethylene Power Cable Lines, 2nd edn.* China Electric Power Press, Beijing.
66. British Standards Institute (1994) GB 50217-1994, *Code for the design of cables of electric engineering.* British Standards Institute, London; and British Standards Institute (2007) GB 50217-2007, *Code for the design of cables of electric engineering.* British Standards Institute, London.
67. British Standards Institute, London (1999) GB/T 17886.1-1999 *Shunt power capacitors of the non-self-healing type for A.C. systems having a rated voltage up to and including 1 kV – Part 1: General – Performance, testing and rating – Safety requirements – Guide for installation and operation.* British Standards Institute, London.
68. British Standards Institute (2010) GB/T 11024.1-2010 *Shunt capacitors for A.C. power systems having a rated voltage above 1000 V.* British Standards Institute, London.
69. IEEE (2007) *Manual of Electrical Design of Power Engineering: Electrical Primary Part.* IEEE, New York.
70. IEEE (2006) *1547 Standard for Interconnecting Distributed Resources with Electric Power Systems.* IEEE, New York.
71. BDEW (2007) *Generating Plants Connected to the Medium-voltage Network.* BDEW, Heidelberg.
72. State Grid Corporation of China (2010) Q/GDW480-2010 *Technical Rule for Distributed Resources Connected to Power Grid.* State Grid Corporation of China, Beijing.
73. State Electricity Regulatory Commission (2009) *Circular on Relevant Matters to Improvement in the Pilot Direct Trading between Power Consumers and Power Generation Enterprises.* State Electricity Regulatory Commission, Beijing.
74. State Electricity Regulatory Commission (2004) the *Entrusted Transmission Service Contract Concerning Power Purchase and Sales in Direct Trading between Large Customers and Power Generation Enterprises (Template).* State Electricity Regulatory Commission and State Administration for Industry and Commerce, Beijing.
75. State Grid Corporation of China (2004) *Provisions of State Grid Corporation of China on the Management of Electric Energy Loss in Power Grids.* State Grid Corporation of China, Beijing.
76. National Energy Administration (1990) *Amendment to the Provisions on the Management of Electric Energy Loss in Power Grids.* National Energy Administration, Beijing.
77. State Electricity Regulatory Commission (2010) *Measures of Electric Power Supply.* State Electricity Regulatory Commission, Beijing.

78. Xie S. (1998) The Chief Economist of State Grid Corporation of China, puts forward a proposal in the electricity price forum held in Changsha, *China Electric Power News* **1999**: 9.
79. State Council (2004) *Opinions on the Establishment of Coal–electricity Price Linkage Mechanism*. State Council,Beijing.
80. Mao Y. (2009) *Report of Coal Cost, Price Formation and Internalization of External Costs*. State Electricity Regulatory Commission, Beijing.
81. General Office of the State Council (2003) *Plan of Electricity Price Reform*. General Office of the State Council, Beijing.
82. Academy of Macroeconomic Research (1997) *Study of Price Reform of Monopoly Industries*. Academy of Macroeconomic Research, State Development Planning Commission, Beijing.
83. Zhao H. (2005) Determination of return rate of investment under tariff regulation of power transmission and distribution, *Power System Technology* **21**: 61–66.
84. State Grid Corporation of China (2005) *Rules for Implementation of Post Project Evaluation of Investment in Fixed Assets of State Grid Corporation of China (Trial)*. State Grid Corporation of China, Beijing.
85. British Standards Institute (2004) GB 19580-2004 *Criteria for Performance Excellence*. British Standards Institute, London.
86. State Economic and Trade Commission of the PRC (2002) DL/T 686-1999 *Guidelines for Calculation of Electric Energy Losses of Power Grids*, State Economic and Trade Commission of the PRC, Beijing.
87. State Ministry of Electric Power Industry (1996) *Power Supply Business Regulation No. 89, Articles 96 and 97*, State Ministry of Electric Power Industry, Beijing.
88. Energy Conservation Office (1990) *Provisions on the Management of Electric Energy Loss in Power Grids No. 1149*, Energy Conservation Office, National Energy Administration, Beijing.

Index

Line Loss Analysis and Calculation of Electric Power Systems, First Edition. Anguan Wu and Baoshan Ni.